PLEASE STAMP DATE DUE, BOTH BELOW AND ON CARD

DATE DUE	DATE DUE	DATE DUE	DATE DUE
JUN 1 3 .2008			

GL-15

Topics in Applied Physics
Volume 103

Available **online** at
<u>Springer Link.com</u>

Topics in Applied Physics is part of the SpringerLink service. For all customers with standing orders for Topics in Applied Physics we offer the full text in electronic form via SpringerLink free of charge. Please contact your librarian who can receive a password for free access to the full articles by registration at:

springerlink.com → Orders

If you do not have a standing order you can nevertheless browse through the table of contents of the volumes and the abstracts of each article at:

springerlink.com → Browse Publications

Topics in Applied Physics

Topics in Applied Physics is a well-established series of review books, each of which presents a comprehensive survey of a selected topic within the broad area of applied physics. Edited and written by leading research scientists in the field concerned, each volume contains review contributions covering the various aspects of the topic. Together these provide an overview of the state of the art in the respective field, extending from an introduction to the subject right up to the frontiers of contemporary research.

Topics in Applied Physics is addressed to all scientists at universities and in industry who wish to obtain an overview and to keep abreast of advances in applied physics. The series also provides easy but comprehensive access to the fields for newcomers starting research.

Contributions are specially commissioned. The Managing Editors are open to any suggestions for topics coming from the community of applied physicists no matter what the field and encourage prospective editors to approach them with ideas.

Managing Editors

Dr. Claus E. Ascheron

Springer-Verlag GmbH
Tiergartenstr. 17
69121 Heidelberg
Germany
Email: claus.ascheron@springer.com

Dr. Hans J. Koelsch

Springer-Verlag New York, LLC
233, Spring Street
New York, NY 10013
USA
Email: hans.koelsch@springer.com

Assistant Editor

Adelheid H. Duhm

Springer-Verlag GmbH
Tiergartenstr. 17
69121 Heidelberg
Germany
Email: adelheid.duhm@springer.com

Katrin Kneipp   Martin Moskovits
Harald Kneipp   (Eds.)

Surface-Enhanced Raman Scattering

Physics and Applications

With 221 Figures, 3 in color

Springer

Katrin Kneipp
Wellman Center for Photomedicine
Harvard-M.I.T. Division of
Health Sciences & Technology
Harvard University, Medical School
Boston, MA 02114, USA
kneipp@usa.net

Harald Kneipp
Wellman Center for Photomedicine
Harvard University, Medical School
Boston, MA 02114, USA
kneipp@usa.net

Martin Moskovits
Department of Chemistry and Biochemistry
University of California Santa Barbara
Santa Barbara, CA 93105, USA
mmoskovits@ltsc.ucsb.edu

Library of Congress Control Number: 2006928060

Physics and Astronomy Classification Scheme (PACS):
33.20.Fb, 42.62.Fi, 73.20.Mf, 87.64.-t

ISSN print edition: 0303-4216
ISSN electronic edition: 1437-0859
ISBN-10 3-540-33566-8 Springer Berlin Heidelberg New York
ISBN-13 978-3-540-33566-5 Springer Berlin Heidelberg New York

Springer is a part of Springer Science+Business Media

springer.com

© Springer-Verlag Berlin Heidelberg 2006
Printed in Germany

Typesetting: DA-TeX · Gerd Blumenstein · www.da-tex.de
Production: LE-TeX Jelonek, Schmidt & Vöckler GbR, Leipzig
Cover design: *design & production* GmbH, Heidelberg

Printed on acid-free paper   57/3100/YL   5 4 3 2 1 0

Preface

Recently, almost 30 years after the discovery of surface-enhanced Raman scattering (SERS) interest in SERS has exploded for two major reasons. The first is the realization that under favorable circumstances, Raman enhancements as large as 14 orders of magnitude can be achieved, i.e., a level of enhancement capable of single-molecule detection. The other is the recent interest in creating an ultrasensitive sensing platform based on SERS with molecular identification capabilities, especially as a sensor for biological molecules. Although a rough physical understanding of the SERS mechanism, which is capable of producing such giant enhancements exists, a detailed theory of SERS is often said to be lacking, by virtue of the fact that most SERS-active systems are random, and often disordered, nanosystems whose properties vary from experiment to experiment making quantitative comparison between theory and observation difficult. However, although much remains to be discovered both theoretically and experimentally, we understand a great deal about the origin of SERS, certainly for the knowledgeable experimentalist to design workable SERS systems and successful SERS experiments and otherwise make useful predictions. The notion that the basic physics behind SERS is either absent or wrong, is therefore a bewildering shibboleth that often creates doubt in the ability to design SERS-based applications, rationally. The sudden surge of new workers in the field compounds the effect of this misconception. Clearly the time is right for a book that summarizes the current major trends and the range of current thinking about SERS, pointing out what we understand, where our understanding is deficient and what opportunities exist for harnessing SERS towards basic nano- and surface science as well as for application in a variety of fields. This is our goal in assembling this book. It is also our hope that with the level of understanding of the physics of SERS one can begin to design SERS systems rationally, predictably and reproducibly.

The book is topical for yet another reason. Although now independent of SERS, the fields of near-field optics and plasmonics owe their existence to the attention drawn to the plasmon both as a polariton and as a localized excitation following its implication in the mechanism of SERS. Likewise, the near-field nature of the optical fields implicated in SERS was recognized in the early 1980s, soon after the discovery of SERS. This, too, we believe, turned people's attention to near-field optics more generally, resulting in the robust area of research it is today. Finally, it is interesting to note that SERS is one of the few, truly nanoscale effects. That is the SERS intensity decreases significantly as one constructs structures that are either much larger than ~100 nm, or much smaller than ~10 nm.

The book is divided into two parts. The first section begins with the basic physics behind SERS, providing both an overview and a summary of recent developments in the field. The electromagnetic theory of SERS is introduced by Martin Moskovits. George Schatz, Matthew Young and Richard Van Duyne continue the discussion of the electromagnetic-enhancement mechanism drawing upon both computational and experimental studies. Mark Stockman considers the electromagnetic enhancement factor by small, properly designed nanoclusters as well as by nanostructures with fractal symmetry. George Schatz and Shengli Zou discuss the extremely high field-enhancement factors possible when plasmonic effects are coupled with photonic resonances. Mikael Käll and Hongxing Xu present a rather complete treatment of SERS by silver-particle aggregates using generalized Mie theory, while Stefano Corni and Jacopo Tomasi apply continuum models to the study of SERS from nanoparticles and nanoparticle aggregates. Finally, Zhong-Qun Tian, Zhi-Lin Yang, Bin Ren, and De-Yin Wu discuss the observation of SERS from transition metals under ultraviolet excitation. Andreas Otto and Masayuki Futamata summarize the other large class of theories for SERS that are often referred to as the chemical mechanism as well as more generalized electronic mechanisms. Many of the chapters dealing with the electromagnetic mechanism in SERS also touch upon the broader issues of plasmonics in optical and spectroscopic contexts where the enhanced local optical fields are important independently of SERS.

Several chapters in the first section concentrate on modern experimental aspects of SERS as well as more general consequences of field enhancement. Harald and Katrin Kneipp describe the novel effects that are observable in the extremely high local optical fields produced in silver and gold nanostuctures, such as vibrational pumping and surface-enhanced hyper-Raman scattering. Enhancement of resonance Raman scattering and fluorescence by strongly coupled metallic nanostructures is the subject of the chapter by Ricardo Aroca, Paul Goulet, and Nicholas Pieczonka. Raman scattering can also be strongly enhanced in the vicinity of the sharp tips of scanning probe microscopes. Bruno Pettinger describes the development of, and recent results from tip-enhanced Raman spectroscopy (TERS). Tip-enhanced Raman spectroscopy as well as tip-enhanced CARS experiments are also discussed by Satoshi Kawata, Yasushi Inoue, and Verma Prabhat. Katrin Kneipp, Harald Kneipp and Henrik Bohr introduce single-molecule SERS spectroscopy, describing single-molecule Raman of nonresonant target molecules. Finally, Anna Rita Bizzarri and Salvatore Cannistraro consider the statistics of temporal fluctuations observed in single-molecule SERS experiments.

The second part of the book concentrates on some of the exciting applications of SERS. Single-molecule surface-enhanced resonance Raman spectroscopy and its application to green fluorescence protein is the topic of the chapter by Johan Hofkens and Satoshi Habuchi. Peter Hildebrandt and Daniel Murgida describe "surface-enhanced vibrational spectroelectrochemistry", that is the redox and redox-coupled processes of heme proteins at an electrode surface as visualized by SERS. Janina Kneipp describes a new SERS nanosensor as an ultra-sensitive chemical probe in living cells. Adaptive plasmonic nanostructures are introduced and used for biomolecular sensing by Vladimir M. Shalaev and Vladimir P. Drachev. Another promising sensing application of SERS – glucose sensing – is described by Richard van Duyne, Chanda Ranjit Yonzon, Olga Lyandres, Nilam C. Shah, and Jon A. Dieringer. The potential for quantitative surface-enhanced resonance Raman spectroscopy as an analytical tool is presented by W. Ewen Smith, Karen Faulds, and Duncan Graham in their chapter, while the use of SERS for rapid analysis in microbiological systems is described by Royston Goodacre, Roger Jarvis, and Sarah Clarke. Tuan Vo-Dinh, Fei Yan, and Musundi B. Wabuyele discuss surface-enhanced Raman scattering as a biomedical diagnostics and molecular imaging tool. In a similar vein, Marc Porter, Hye-Young Park, Jeremy D. Driskell, Karen M. Kwarta, Robert J. Lipert, Christian Schoen, John D. Neill, and Julia F. Ridpath describe the application of SERS for carrying out ultrasensitive immunoassays using immunogold labels. Finally, Stuart Farquharson, Frank E. Inscore, and Steve Christesen report the use of SERS for detecting chemical agents and their hydrolysis products in aqueous environments.

We cordially thank all our colleagues who have contributed to this book for the time and effort they invested in writing such excellent chapters. Additionally, we are grateful to Henrik Bohr and Salim Abdali who suggested that we edit a book such as this one following a SERS workshop at the Quantum Protein Center, Technical University of Denmark. We hope this book stimulates thought, leading to a deeper understanding of the physics behind SERS and its multiple uses both in promoting a deeper understanding of basic science as well as in applications in analysis and sensing, in nano sciences, surface chemistry, as a biological probe as well as in daughter disciplines such as plasmonics and near-field optics.

Boston, MA and Santa Barbara, CA, January 2006   *Katrin Kneipp,*
Martin Moskovits,
Harald Kneipp

Contents

SERS From Transition Metals and Excited by Ultraviolet Light

Electronic Mechanisms of SERS

Single-Molecule Surface-Enhanced Resonance Raman Spectroscopy of the Enhanced Green Fluorescent Protein EGFP

Surface-Enhanced Vibrational Spectroelectrochemistry: Electric-Field Effects on Redox and Redox-Coupled Processes of Heme Proteins

Nanosensors Based on SERS
for Applications in Living Cells

Biomolecule Sensing with Adaptive Plasmonic
Nanostructures

Glucose Sensing with Surface-Enhanced Raman Spectroscopy

Quantitative Surface-Enhanced Resonance Raman Spectroscopy for Analysis

Rapid Analysis of Microbiological Systems Using SERS

Surface-Enhanced Raman Scattering for Biomedical Diagnostics and Molecular Imaging

Ultrasensitive Immunoassays Based on Surface-Enhanced Raman Scattering by Immunogold Labels

Hye-Young Park, Jeremy D. Driskell, Karen M. Kwarta,
Robert J. Lipert, Marc D. Porter, Christian Schoen, John D. Neill,

Detecting Chemical Agents and Their Hydrolysis Products in Water

Surface-Enhanced Raman Spectroscopy: a Brief Perspective

Martin Moskovits

Department of Chemistry and the California Nanosystems Institute,
University of California, Santa Barbara, CA 93105
mmoskovits@ltsc.ucsb.edu

1 Introduction

Surface-enhanced Raman spectroscopy (SERS) was first observed by *Fleischman* et al. in 1974 [1], and discovered by *Jeanmarie* and *Van Duyne* [2] and *Albrecht* and *Creighton* [3] in 1977. I make the distinction between observation and discovery because, when first observed for pyridine adsorbed on an electrochemically roughened silver electrode [1], the unusual intensity of the Raman signals was attributed to the increased surface area of the rough substrate. It was Van Duyne and Creighton who pointed out that the intensification of the effective Raman cross section was far in excess of the increased number of molecules interrogated as a result of the surface's roughness factor. In 1978 I had the good fortune of proposing that the huge increase in Raman cross section was a result of the excitation of surface plasmons [4]. That insight immediately led to a number of predictions such as the expectation that SERS should be observable in metal colloids and the hierarchy of intensification that should be observed, all else being equal, with silver and the alkali metals providing the most intense SERS signals followed by gold and copper in that order, then other good conductors such as aluminum, indium and platinum, and finally the transition metals and the other more poorly conducting metals. Of course one had to factor into all of this such important parameters as the excitation wavelength, the polarization of the exciting and scattered radiation with respect to symmetry axes of the nanostructures illuminated, the precise structural features of the SERS-active system and so on. Within a few years of the discovery of SERS, essentially all of these predictions were shown to be true. The very close qualitative agreement between the SERS enhancement and the intensity and quality factor of the surface plasmon is a baseline feature of essentially everything that has been learned about SERS over the last 30 years, a fact that needs to be borne in mind as new theories of SERS are proposed.

Work in SERS reached a plateau approximately 10 years ago and became invigorated once again by the reports by *Kneipp* and coworkers [5,6,7,8,9,10] and *Nie* and coworkers [11, 12, 13, 14, 15, 16] who simultaneously and independently reported that intense enough SERS emissions could be recorded under favorable circumstances to detect single molecules, that together with the quest for high-sensitivity molecular- and especially biomolecular-sensing

K. Kneipp, M. Moskovits, H. Kneipp (Eds.): Surface-Enhanced Raman Scattering – Physics and Applications, Topics Appl. Phys. **103**, 1–18 (2006)
© Springer-Verlag Berlin Heidelberg 2006

platforms, has returned SERS as a research field to the front burner so that there are arguably now more people working in SERS than ever before. An indirect barometer of its recent popularity can be gauged from the fact that the review article I wrote in 1985 [17] has had more citations in 2005 than in any previous year. Moreover, two active fields can arguably be said to have evolved largely out of SERS: near-field optics [18] and plasmonics [19]. And all of this has benefited from the increased availability of high-performance computing that has allowed large-scale and high-level computations to be carried out on large nanoparticle aggregates, rough-surface models and other nanostructured systems pertinent to SERS.

2 The Electromagnetic Theory of SERS

The simplest model that translates the excitation of surface plasmons into a SERS mechanism is the so-called electromagnetic model simultaneously enunciated in 1980 by *Gersten* [20, 21], *Gersten* and *Nitzan* [22, 23] and *McCall* et al. [24, 25] and expanded upon by *Kerker* et al. [26, 27, 28, 29, 30] who consider the electromagnetic fields surrounding a small illuminated metal particle. Although a survey of the SERS literature leads to the conclusion that SERS from a single, isolated metal nanoparticle has likely never been credibly reported, a preliminary analysis of the fields surrounding a small isolated metal nanoparticle is useful in defining some of the basic criteria one needs to fulfill in order to see intense SERS.

A small, isolated, illuminated metal sphere will sustain oscillating surface plasmon multipoles of various order induced by the time-varying electric-field vector of the light. The surface plasmons are collective oscillations of the conduction electrons against the background of ionic metal cores [17]. In addition, light can induce a host of other excitations in the metal particle including interband transitions. For a particle much smaller than the wavelength of the exciting light, all but the dipolar plasmon can be ignored. Systems with free or almost free electrons will sustain such excitations; and, the freer the electrons the sharper and the more intense the dipolar plasmon resonance will be. When the exciting laser light is resonant with the dipolar plasmon the metal particle will radiate light characteristic of dipolar radiation [31]. This radiation is a coherent process with the exciting field and is characterized by a spatial distribution of field magnitudes (that reaches steady state a few femtoseconds after the light is turned on) in which the light intensity from certain portions of space surrounding the particle is depleted, while the intensity at certain portions near the metal particle is enhanced. Although this has been known for a very long time, the recent images of this process produced by *Käll* and coworkers [32, 33] illustrate this process vividly.

Let us call the field enhancement averaged over the surface of the particle g. The average magnitude of the field radiated by the metal particle E_s will be: $E_s = gE_0$, where E_0 is the magnitude of the incident field. One should keep in mind that E_s is the average local *near field* at the particle surface. The average molecule adsorbed at the surface of the metal particle will therefore be excited by a field whose magnitude is E_s, and the Raman-scattered light produced by the molecule will have a field strength $E_R \propto \alpha_R E_s \propto \alpha_R g E_0$, where α_R is the appropriate combination of components of the Raman tensor. (The arguments in this section are meant to give a clear pictorial idea of the SERS enhancement process. To carry out the problem correctly one must carry out the tensor product properly, taking into account the vectorial nature of the fields involved that includes both their wave and polarization vectors.)

The Raman-scattered fields can be further enhanced by the metal particle in exactly the same manner as the incident field was. That is, the metal particle can scatter light at the *Raman-shifted* wavelength enhanced by a factor g'. (The prime is used to indicate the fact that the field enhancement at the Raman-shifted wavelength will, in general, differ from its value at the incident wavelength.) The amplitude of the SERS-scattered field will therefore be given by $E_{SERS} \propto \alpha_R gg' E_0$, and the average SERS intensity will be proportional to the square modulus of E_{SERS}. That is, $I_{SERS} \propto |\alpha_R|^2 |gg'|^2 I_0$, where I_{SERS} and I_0 are the "intensities" of the SERS-scattered and incident fields, respectively. For low-frequency bands when $g \cong g'$ the SERS intensity will be enhanced by a factor proportional to the fourth power of the enhancement of the local incident near field, i.e., $|E_L|^4 = |g|^4$. (For higher-frequency Raman modes the SERS intensity will be a more complicated function of the plasmon resonant properties of the metal particle according to the precise wavelengths at which the incident and Raman-scattered light fall. Likewise, it has been shown by *Stockman* and others [34, 35, 36, 37] that for fractal aggregates the correct sum over the cluster leads to a $|E_L|^3$ rather than the fourth power of the local field.) It is helpful to define the "SERS enhancement" G as the ratio of the Raman-scattered intensity in the presence of the metal particle to its value in the absence of the metal particle $G = \left|\frac{\alpha_R}{\alpha_{Ro}}\right| |gg'|^2$ where α_{Ro} is the Raman polarizability of the isolated molecule.

Three important points must be noted. 1. The major contribution to SERS is scattering *by the metal particle* rather than by the molecule whose Raman spectrum is, however, reflected in the SERS spectrum of the light scattered by the metal. 2. Although the SERS intensity (for low-frequency Raman modes) varies as the fourth power of the local field, the effect is a linear optical effect, which depends on the first power of I_0. (Of course, nonlinear optical phenomena are also enhanced [38]. But these will not be discussed here.) However, the fourth-power dependence on g is key to the inordinate enhancements SERS provides. For silver at 400 nm, for example, g is only ~ 30, yet that implies a Raman enhancement $G \sim 8 \times 10^5$ assuming the Raman polarizability to be unchanged from that of the isolated molecule. (In fact the field enhancement by an isolated particles is likely one or more orders of magnitude lower than that on account of physical aspects that are often omitted in carrying out this calculation, such as the fact that retarded

fields must be used and nonlocal dielectric function values for the metal [17].)
3. Although one speaks loosely of α_R as the Raman polarizability "of the molecule", in fact it is the Raman polarizability of the scatterer that includes the molecule but, when the molecule is adsorbed on the metal particle's surface, will include contributions from the metal and may, as a result, be greatly altered both in its magnitude, symmetry and resonant properties from the Raman polarizability of the isolated molecule. This will be particularly important in systems where metal-to-molecule or molecule-to-metal charge transfer occurs, altering dramatically the resonances of the system thereby contributing to so-called chemical enhancement. Experience shows that such resonances often occur. 4. SERS excitation is a near-field phenomenon. The near field, especially near a metal surface, will have spatial components that decay more rapidly with distance than the spatial variation in the far field (where the spatial "structure" in the field is of the order of the wavelength). Hence, one expects to see (and indeed does see) effects such as relaxation of dipole-selection rules [39, 40, 41], which causes normally forbidden vibrational modes to occur in the SERS spectrum and dipole-forbidden fluorescences to be observed [42, 43, 44]. This point needs to be taken into account also in the description of the field response of the particles themselves, especially when (as will be discussed below) one deals with electromagnetic and other physical phenomena such as quantum-mechanical tunneling taking place in interparticle gaps and interstices that are much smaller than the wavelength, in which the spatial variation of the electromagnetic field could be very large over a rather small distance.

The foregoing contains all of the seminal characteristics of SERS. It indicates that essentially all systems possessing free carriers can show SERS; hence, the observation of SERS from "unusual" systems such as Si or transition metals is not unexpected (if one excites with light of the appropriate wavelength). The intensity of such emissions will depend, to zeroth order, on the magnitude of g, which will be much larger that unity for metals such as silver, gold and the alkalis, larger, but not much larger than unity for other good conductors (Al, In, Pt) and only a little larger than unity for most other metals (in all cases we assume optimal choice of excitation wavelength). Improvements in the throughput of Raman spectrometers and in multiplex spectroscopic detection, have made SERS measurements from most metals, indeed, all Raman measurements, enhanced or otherwise, far more accessible nowadays.

SERS is one of the few phenomena that can truly be described as nanoscience. This is because for it to occur, the metal particles or metal features responsible for its operation must be small with respect to the wavelength of the exciting light. This normally means that the SERS-active systems must ideally possess structure in the 5 nm to 100 nm range. Likewise, the dimensions of the active structure cannot be much smaller than some lower bound, which is normally larger than the average molecule. The upper dimensional bound of the SERS-active system is determined by wavelength.

As features of the order of the wavelength or larger are used, the optical fields no longer excite dipolar plasmons almost exclusively; instead progressively higher-order multipoles are excited. Unlike the dipole, these modes are nonradiative, hence they are not efficient in exciting Raman (or other dipole-driven excitations). Accordingly, the SERS efficiency drops until, for large enough particles, so much of the exciting radiation is locked up in higher-order plasmon multipoles that SERS is all but extinguished.

At the other end of the dimensional scale as the nanostructure responsible for SERS becomes too small, the effective conductivity of the metal nanoparticles diminishes as a result of electronic scattering processes at the particle's surface [45, 46, 47]. As a result, the quality factor of the dipolar plasmon resonance is vitiated and the reradiated field strength reduced. When the metal particle becomes small enough, the pseudo bulk description implicit in the definition of the surface plasmon no longer applies. Instead, one needs to treat the metal particle as a fully quantum object whose electronic properties show so-called quantum-size effects. Reducing the size of the metal particle even further so that the particle is composed of only a few metal atoms, one passes into a regime in which a molecular description expresses the particles' properties best.

Some of these points are illustrated using the following rudimentary model. The polarizability of a small metal sphere with dielectric function $\varepsilon(\lambda)$ and radius R, surrounded by vacuum is given by:

$$\alpha = R^3 \frac{\varepsilon - 1}{\varepsilon + 2}. \tag{1}$$

Combining this expression with the expression for the dielectric function of a Drude metal slightly modified for interband transitions we obtain:

$$\varepsilon = \varepsilon_b + 1 - \frac{\omega_p^2}{\omega^2 + i\omega\gamma}, \tag{2}$$

in which ε_b is the (generally wavelength-dependent) contribution of interband transitions to the dielectric function, ω_p is the metal's plasmon resonance whose square is proportional to the electron density in the metal and γ is the electronic-scattering rate that is inversely proportional to the electronic mean-free-path and therefore also inversely proportional to the metal's DC conductivity. Substituting (2) into (1) yields the expression

$$\alpha = \frac{R^3(\varepsilon_b \omega^2 - \omega_p^2) + i\omega\gamma\varepsilon_b}{[(\varepsilon_b + 3)\omega^2 - \omega_p^2] + i\omega\gamma(\varepsilon_b + 3)}. \tag{3}$$

The real and imaginary parts of the expression for α given in (3) have a pole when the frequency ω is equal to $\omega_R = \frac{\omega_p}{\sqrt{\varepsilon_b + 3}}$. The width of that resonance is given by $\gamma(\varepsilon_b + 3)$.

Hence, when γ is large either because of the inherent poor conductivity of the metal or due to the fact that the metal nanofeatures are so small

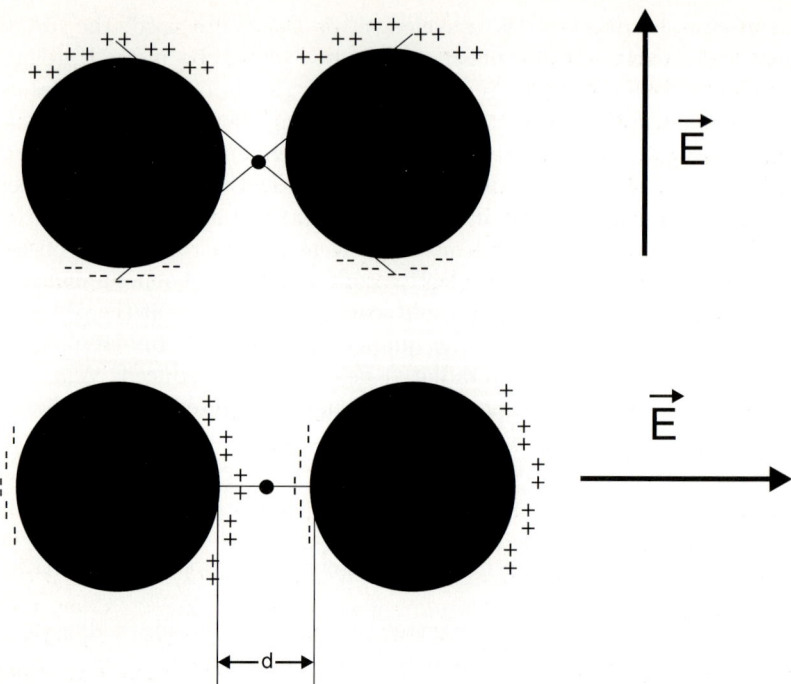

Fig. 1. Simple graphical illustration of the reason that light polarized with the **E**-vector along the interparticle axis can result in huge enhancements in the gap between the two nanoparticles while the orthogonal polarization cannot. For light polarized along the interparticle axis the proximity of the charges (induced by the optical fields) to the molecule can be made arbitrarily small and hence the field sensed by the molecule commensurately large as the nanoparticles are brought closer together. That capability is not available for light polarized orthogonally to the interparticle axis

that electronic scattering at the particle's surfaces become the dominant electron-scattering process, the quality of the resonance is reduced and with it the SERS enhancement. Likewise, for metals whose dielectric properties are greatly modified by interband transitions in the wavelength range under consideration, i.e., for which the value of the function ε_b is large, the width of the resonance is increased and the SERS-enhancement decreases. This explains why, all things being equal, the SERS enhancement of silver exceeds that of gold, which, in turn, exceeds that of copper. The participation of interband transitions in the dielectric function of those metals in the visible range of the spectrum increases in that order. Most transition metals are poor SERS enhancing systems because, for them, the two effects combine to reduce their SERS enhancement ability, i.e., their conductivity is low (γ is large) and the interband contribution to the dielectric function is great (ε_b is large).

The interband contribution to the dielectric function also affects the location of the dipolar plasmon resonance. The quantity $\omega_p \sim 9$ eV for the coinage metals, but the dipolar surface resonance ω_R occurs in, or near the visible for Cu, Ag and Au as a result of the contribution of ε_b in $\omega_R = \frac{\omega_p}{\sqrt{\varepsilon_b + 3}}$.

Summarizing, for a given metal system the SERS intensity will depend, to first order, on the size of the nanostructure responsible for its enhancement. It will be optimal when this size is small with respect to the wavelength of the exciting light so long as that size is not much smaller than the electronic mean free path of the conduction electrons. For the coinage metals this optimal range will span ~ 10 nm to 100 nm.

3 Assemblies of Interacting Nanostructures: The Ubiquitous SERS-Active Systems

The foregoing considered the situation for a single metal particle. Almost all effective SERS-active systems, however, consist of assemblies of interacting particles. Examples are nanoparticle aggregates, rough metals surfaces and island films. More recently, the quest has been for closely spaced, well-engineered systems of interacting metal nanostructures (either particles or cavities) that fulfill the double goals of providing high field enhancement and doing so with highly reproducible and controllable SERS platforms [48,49,50]. For assemblies of isolated particles the interaction is through-space electromagnetic coupling that will be dominated by dipolar coupling, but will include multipolar coupling for closely spaced particles. A firm understanding that the SERS intensity can be greatly enhanced when two or more nanoparticles are brought closely together was already in place in the early 1980s though the work of *Metiu* and coworkers [51,52,53] and of *Nitzan* and coworkers [54], who showed that at an appropriate wavelength and polarization and for a small enough interparticle gap, Raman enhancements $\sim 10^{10}$ could be obtained for molecules localized in a small volume within the interparticle interstice. Others have added seminally to our understanding of these interparticle-coupling effects [55, 56], in some cases including a treatment in which the role of the adsorbed molecules is considered as more than a bystander in the enhanced fields [57, 58, 59]. Those calculations carried out in the 1980's were of a high level, taking account of the effect of multipoles out to order 60 or more. This type of calculation was repeated recently by *Käll* and coworkers [32,33] and others whose evocative figures show that when two nanoparticles are brought close together the optical field strength in the interstitial "hot spot" can reach $\sim 10^{11}$ or more, provided the two nanoparticles are brought within ~ 1 nm or less, light of the appropriate wavelength is used and the exciting electric field vector is polarized along the interparticle axis. This giant enhancement, which exceeds that at isolated metal particles by ~ 6 orders of magnitude, falls off rapidly as the interparticle gap increases.

For light polarized across the interparticle axis, the enhancement is almost negligibly different from its value at a single, isolated particle.

These sorts of calculations were recently performed for core-shell particles [60, 61, 62, 63, 64, 65] composed of a dielectric inner core surrounded by a thin continuous metal shell, which show that the further localization achievable in this geometry can make the SERS enhancement within the interparticle hot spot greater still by some 3 orders of magnitude potentially resulting in enhancements of 10^{13} or more. These predictions should be looked upon with some skepticism since a number of physical requirements such as those listed previously have been ignored. I also note that the (local) dielectric function for silver or gold is seldom corrected for the reduction in electronic mean free path, an effect that will be rather pronounced in thin metal shells. In constructing experimental versions of these core-shell particles one also needs to take care that the assumption of a continuous and smooth metal shell (as assumed in the calculation) is in fact realized. Discontinuous and rough metal features may result in rather intense SERS activity, but not of a kind that is consistent with the analogous calculation. Moreover, since such experiments are carried out in the name of replacing the previously studied, random SERS-active systems with well-designed and properly engineered SERS platforms, spurious roughness on these more regular structures simply replaces one type of random system with another. This is a rather general caveat. A number of "engineered" or lithographically produced SERS active systems that are described in terms of their *designed* properties probably show SERS activity that is an amalgam of those "designed" properties plus the effect of spurious roughness, where, at times, the latter is the dominant component.

A number of seminal aspects of the effect of interparticle coupling can be understood in terms of some very simple physics. Referring to Fig. 1, a molecule (indicated by a small circle in the figure) located in the interstice between two metal nanospheres is flanked by two sets of (time-varying) conjugate charges arising from the polarization of the individual nanoparticles. (The model also describes the two-dimensional component of any system through which a planar cut is represented by two circles, as for example for two parallel nanowires.) As the nanoparticles are brought closer together, the proximity of these charges to the molecule can be made arbitrarily close and hence the capacitive field sensed by the molecule commensurately large. The mutual interaction of the two nanospheres (or nanowires) also leads to an increase in the magnitude of the dipole induced in each component of the two-component nanosystem. The dipole induced in each nano-object arises from the combined field of the incident light and the intense field of its partner, which, in this configuration, leads to an amplification of the polarization. If this problem is solved naively as one of simple electrostatics involving two polarizable nanoparticles, then for light polarized along the interparticle axis, the enhancement increases approximately as d^{-8} where d is the gap size between the two nanowires.

Contrariwise, when the light is polarized in the other direction, illustrated in Fig. 1 (top), that is, normal to the axis joining the centers of the two nanospheres, a molecule in the interstitial region will not benefit from proximity to the induced charges, however closely the nanoparticles are brought together. Nor is there any other location where the field benefits distinctly from the fact that one has a system of two particles rather than one. Additionally, the mutual polarization of each of the two nanoparticles as a result of the field emitted by the other is not favorable.

Bringing two particles together also brings about other effects that one needs to be mindful of, for example, the plasmon resonance splits into two polarization-sensitive components [32, 33, 51,52,53], one of which has a resonance that depends strongly on the separation of the two nanospheres. Hence, as the two nanoparticles are separated, the SERS enhancement will diminish and the resonance condition will simultaneously change.

Further aggregation into larger clusters will create opportunities for other "hot" interstitials each of which with its own characteristics of polarization and field strength. Likewise one can modify this effect further by, for example, aggregating core-shell particles.

Is there a geometrical configuration of nanoparticles (of a given metal and given size or multiple metals and several sizes) that leads to the maximal field enhancement? And if so, how much larger would that field be (for comparable sizes of the interstitial spaces) as compared to that in the particle dimer? Although this problem is not yet solved, instinctively one feels that although perhaps another order of magnitude may be achievable by constructing a cluster of optimal geometry, the dimer configuration already accounts for the lion's share of the extra enhancement due to aggregation.

Another question that is worth asking and has not yet been solved is the dependence of the enhancement per interstitial site in a cluster of N nanoparticles. That is if one brings together N nanoparticles by linking them with M ligands (where $N \sim M$), and ignoring for the moment the problem of ensuring that every one of the ligands can profit from the appropriate polarization, is the enhancement per ligand less than or greater than its value when a single ligand molecule links two nanospheres? For a periodic structure such as a system of nanoparticles arranged on a square lattice connected by a single ligand linking neighboring nanoparticles, that enhancement per ligand is likely to be lower than that in a dimer but one that converges rapidly with increasing cluster size. But the situation for the cluster or arbitrary geometry arranged specifically with the aim of maximizing the per-ligand enhancement has not been solved.

The precise structure of the nanoparticle cluster has yet another important aspect that one needs to be mindful of. For highly (geometrically) symmetric aggregates, the degeneracies of the normal modes describing the surface-plasmon excitations (the problem is, to first order, isomorphic with a normal mode analysis of coupled oscillators, except that in common vibrational problems one deals with nuclear motion, while here the oscillators involved collective motion of electrons) will be such that they will generally be characterized by relatively narrow excitation spectra; while clusters with lower geometrical symmetries will have much broader excitation spectra. This will strongly affect resonance conditions.

The enhancement in an important class of large clusters – fractals – needs special mention. Nanoparticle aggregates grown by self-assembly from monomers in solution often show scaling symmetry either as self-similar (or, when deposited on a substrate) self-affine systems [66, 67]. *Stockman* and *Shalaev* [34, 35, 36, 37] have shown through a long series of papers that, due to the symmetry breaking that arises from the fact that the exciting field does not possess scaling symmetry while the cluster does, a form of energy localization arises that can lead to the formation of electromagnetic hot spots with field strengths that are often of the same magnitude as those discussed above for the nanoparticle dimer. What is more, the lack of translational symmetry in the nanoparticle aggregate leads to very broad excitation spectra ensuring resonance over a very broad range of wavelengths (although, of course, each resonance will correspond to a different set of hot spots, with fairly adjacent wavelengths often corresponding to very disparate patterns of hot spots). The existence of such hot spots has been shown experimentally both for Rayleigh and Raman scattering [67, 68]. Figure 2 shows an example of hot spots measured on a self-affine system of silver nanoparticles allowed to condense gravitationally out of solution. The images were obtained using a near-field optical fiber probe to obtain Raman maps of three SERS bands of phthalazine (or its photochemical products) adsorbed on a self-affine silver film [67, 68]. Also shown is a topographic map taken using shear-force microscopy of the region of the film at which the Raman maps were measured. The irregular distribution of SERS intensity over the surface of the self-affine silver film is clearly visible. Moreover, the pattern varies according to the Raman band probed, each of which corresponds to a different absolute wavelength. Although tedious, since the signal was weak and many points had to be probed to construct the map, the measurements were reproducible.

There are other electromagnetic mechanisms that can augment the SERS enhancement (and also modify the wavelength response of the system). Prime among them is the creation of structures such as ellipsoids and nanowires with regions of very large curvature [26, 27, 28, 29, 30]. This is sometimes referred to as the lightning-rod effect.

The reason SERS is seldom reported from single isolated nanoparticles is therefore easy to understand. With dimers and larger nanoparticle clusters capable of producing enhancements 5 or 6 orders and magnitude larger than those of single isolated particles even a handful of particle dimers or small clusters (or structures that are equivalent to small clusters, such as closely spaced nanoroughness features at a metal surface) will overwhelm the SERS signal from many thousands of isolated particles. As a result, all reports of SERS spectra claiming to originate from single, isolated particles must be looked upon with suspicion, even if rather compelling TEM images are

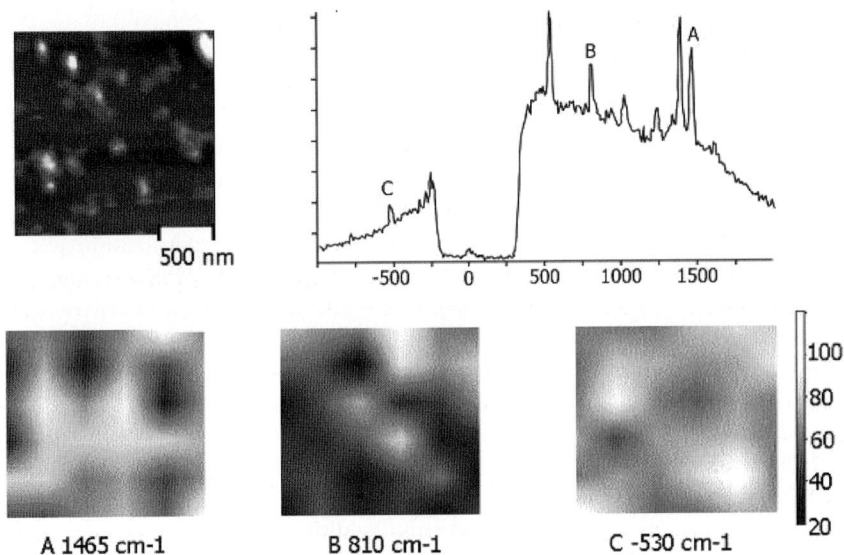

A 1465 cm-1   **B 810 cm-1**   **C -530 cm-1**

Fig. 2. *Top left:* Shear-force topographic image of a silver film produced by collapsing phthlazine-covered silver-nanoparticle aggregates gravitationally onto a pyrex cover slide. *Top right:* The near-field SERS spectrum of phthalazine measured on the film imaged in the previous panel. A, B, C: Near-field intensity maps of the SERS bands shown in the previous spectrum measured over a $3\,\mu m \times 3\,\mu m$ portion of the nanostructured silver film. Reprinted with permission from [69]

reproduced showing the presence of virtually only monomers. One cannot be certain that in the optical experiment the laser did not excite some of the very few clusters present in the sample.

It is also clear based on the foregoing, and the fact that essentially all SERS studies so far have utilized ensembles of strongly interacting metal particles such as aggregated metal nanoparticles or the closely spaced surface features present at cold-deposited metal films [70, 71], that in almost every instance, the average SERS enhancement $\sim 10^6$ is, in fact, an average over a very broad distribution of enhancements present within the portion of the SERS-active sample probed by the laser, which can range from values near unity to values exceeding 10^{11}.

A number of misconceptions exist regarding the em enhancement. A number of workers assume either that the em model implies the quasistatic approximation, in which the em fields are essentially those obtained by solving the Poisson equation, i.e., by assuming the problem is one of electrostatics, or that one needs to make the quasistatic approximation. Neither of these is correct, and in fact many studies, including even rather early studies, were carried out using electromagnetic formalisms that do not make the quasistatic approximation. Moreover, the recent interest in the excitation and propagation of optical signals generated or mediated by plasmons, referred to as plasmonics, or in photonic crystals, as well as in the near-field optics that often involve subwavelength geometrical features has resulted in many electromagnetic calculations of the fields in the vicinity of slits of various geometries, aperture arrays, and nanoparticle assemblies whose optical response cannot be described in terms the quasistatic approximation and are therefore treated by more general methods. Although not always presented as such, the results of those calculations are often pertinent to SERS of molecules residing in or near those structures. Because the quasistatic approximation leads to acceptably accurate results only for nanosystems with structural dimension much smaller than the wavelength of light (i.e., smaller than $\sim 40\,\mathrm{nm}$) this point is particularly important these days when engineered nanostructures [48,49,50], and other structures such as nanowires [72, 73] with somewhat larger structural dimensions are used.

It is clear that despite often misleading statements to the contrary, the em model is successful both in accounting for the major observations in SERS and as a prescriptive approach towards predicting the outcome of experiments. It is clearly not a complete account of all of the rich assortment of phenomena that can occur for molecules residing at a metal interface. To some extent these other effects are the most interesting since it is through them that the specificity of interaction between molecule and metal is expressed. Additionally, the time is ripe for the development of alternative expressions of the em model so as to take account of the quantum-mechanical nature of the problem, thereby better describing both the dynamics of the electrons in the metal and their interaction with the adsorbed molecule. This is not a new quest, but one that is not yet complete. Recent forays in that direction by *Brus* and coworkers [74, 75] and others point the way.

4 Possible Extensions of the Electromagnetic Model

Since most SERS signals come from fields that are concentrated in very small volumes of space it is also worth considering if we need to expand the basic approach to the calculation of field intensities in such restricted volumes due to the fact that the optical fields are rapidly varying both temporally as well as spatially. A preliminary gauge of the problem is demonstrated by the equation for the dipole moment induced in a particle placed in an inhomogeneous electromagnetic field:

$$\mu_\alpha = \alpha_{\alpha\beta} E_\beta + \tfrac{1}{3} A_{\alpha\beta\gamma} \frac{\partial E_\beta}{\partial \gamma} + G_{\alpha\beta} B_\beta + \cdots, \tag{4}$$

in which $\{\alpha,\ \beta,\ \gamma\}$ stand for $\{x,\ y,\ z\}$, the tensor convention is assumed on repeated subscripts, the tensors $\boldsymbol{\alpha}$, $\boldsymbol{A}$ and $\boldsymbol{G}$ are, respectively, the electric–dipole–electric–dipole, electric–dipole–electric–quadrupole, and electric–dipole–magnetic–dipole tensors and $\boldsymbol{E}$ and $\boldsymbol{B}$, are, respectively, the electric and magnetic optical fields [76]. In the far field, and for a particle small compared to the wavelength of light, only the first term of 1 contributes significantly because both the optical electric and magnetic fields vary slowly over the body of a small particle. However, for a particle or a molecule in the near field, especially in the field of a hot spot in the interstice between nanoparticles where the field amplitude varies greatly over a small distance, the contributions due to the second and third terms to the induced dipole could be significant. This sort of contribution will not only affect the spectroscopic selection rules of the molecules being examined (since the **A** and **G** tensors span different irreducible representation from these of the **α** tensor), but the "field-gradient" terms will also contribute to the dipolar plasmon fields radiated by the metal particles. That is, the electric dipole induced in a metal particle excited by the near field of a neighboring metal particle could have a significant contribution due to the interaction of the former's *quadrupole moment* with the gradient (or more correctly the dyadic tensor $\nabla \mathbf{E}$) of the field of the latter. Hence, even somewhat larger particles, with significant plasmon quadrupole moments that would normally not be implicated in the dipolar radiation by the particle, might under these circumstances produce sizable radiating dipolar plasmons both as a result of the more familiar first term of 1 as well as through one or both of the second and third terms.

Other formulations of the em theory of SERS might be attempted that would better conform to the structural details of the SERS-active system. Nowadays most of these consist of systems of interacting nanoparticles or other nanostructures displaying hot spots, especially systems in which one attempts to achieve a greater level of structural control than before. One can, for example, regard two closely spaced metal nanoparticles illuminated by an optical field as a radiating dipole antenna with a time-varying current traversing its gap that is driven by quantum tunneling whose intensity would be commensurately large as a result of excitation of the localized plasmon. Quantum states of molecules residing in the gap would constitute possible resonant energy-loss channels (as in inelastic tunneling spectroscopy [77, 78, 79]) that would, in turn, show their presence by the appearance of sidebands (i.e., Raman emissions) in the radiation field of the dipole antenna. Extending the antenna analogy to larger clusters, one could immediately determine the relative SERS enhancement and polarization properties of clusters of various shapes and sizes by determining which of the cluster's current modes are dipolar in character.

5 Conclusions

The em theory of SERS, despite its physical simplicity, can account for all major SERS observations including 1. the need for a nanostructured material as the SERS-active system, 2. the observation that some metals form good SERS-active systems while others do not, as well as predicting which are which, 3. the observation that strongly interacting metal nanoparticles result in very much more effective SERS-active systems than isolated single particles, and 4. the observed polarization sensitivity shown by nanoparticle aggregates. By extending the ideas inherent in the em model one can also understand the seminal features reported for single-molecule SERS, including the puzzling observation that only a few silver "particles" in an ensemble are "hot" (they are appropriately structured nanoparticle clusters), that for a hot particle once one is able to observe SERS, adding more adsorbate does not significantly alter the intensity (once the em hot spot is occupied, adding adsorbate to other sites on the nanoparticle cluster will not add greatly to the observed intensity). However, the em model does not account for all that is learned through SERS. Molecular resonances, charge-transfer transitions and other processes such as ballistic electrons transiently probing the region where the molecule resides [74,75] then modulating electronic processes of the metal as a result, certainly contribute to the rich information SERS reports; and by virtue of the fact that these contributions will vary from molecule to molecule, they will constitute among the most interesting aspects reported by SERS. But, the overall reason why SERS produces such inordinate enhancements is largely an electromagnetic property of nanostructures.

References

[1] M. Fleischman, P. J. Hendra, A. J. McQuillan: Chem. Phys. Lett. **26**, 123 (1974)
[2] D. L. Jeanmaire, R. P. Van Duyne: J. Electroanal. Chem. **84**, 1 (1977)
[3] M. G. Albrecht, J. A. Creighton: J. Am. Chem. Soc. **99**, 5215 (1977)
[4] M. Moskovits: J. Chem. Phys. **69**, 4159 (1978)
[5] K. Kneipp, H. Kneipp, I. Itzkan, R. R. Dasari, M. S. Feld: Chem. Phys. **247**, 155 (1999)
[6] K. Kneipp, H. Kneipp, R. Manoharan, E. B. Hanlon, I. Itzkan, R. R. Dasari, M. S. Feld: Appl. Spectrosc. **52**, 1493 (1998)
[7] K. Kneipp, H. Kneipp, V. B. Kartha, R. Manoharan, G. Deinum, I. Itzkan, R. R. Dasari, M. S. Feld: Phys. Rev. E **57**, R6281 (1998)
[8] K. Kneipp, Y. Wang, H. Kneipp, L. T. Perelman, I. Itzkan, R. R. Dasari, M. S. Feld: Phys. Rev. Lett. **78**, 1667 (1997)
[9] K. Kneipp, Y. Wang, H. Kneipp, I. Itzkan, R. R. Dasari, M. S. Feld: Phys. Rev. Lett. **76**, 2444 (1996)
[10] K. Kneipp, H. Kneipp, R. Manoharan, I. Itzkan, R. R. Dasari, M. S. Feld: J. Raman Spectrosc. **29**, 743 (1998)
[11] J. T. Krug, G. D. Wang, S. R. Emory, S. Nie: J. Am. Chem. Soc. **121**, 9208 (1999)
[12] S. R. Emory, W. E. Haskins, S. Nie: J. Am. Chem. Soc. **120**, 8009 (1998)
[13] W. A. Lyon, S. Nie: Anal. Chem. **69**, 3400 (1997)
[14] S. Nie, S. R. Emory: Science **275**, 1102 (1997)
[15] W. E. Doering, S. Nie: J. Phys. Chem. **106**, 311 (2002)
[16] D. J. Maxwell, S. R. Emory, S. Nie: Chem. Mater. **13**, 1082 (2001)

[17] M. Moskovits: Rev. Mod. Phys. **57**, 783 (1985)
[18] M. A. Paesler, P. J. Moyer: *Near-Field Optics: Theory, Instrumentation and Applications* (Wiley, New York 1996)
[19] N. Halas (Ed.): *Plasmonics: Metallic Nanostructures and Their Optical Properties* (SPIE 2003)
[20] J. I. Gersten: J. Chem. Phys. **72**, 5779 (1980)
[21] J. I. Gersten: J. Chem. Phys. **72**, 5780 (1980)
[22] J. I. Gersten, A. Nitzan: J. Chem. Phys. **73**, 3023 (1980)
[23] J. I. Gersten, A. Nitzan: J. Chem. Phys. **75**, 1139 (1981)
[24] S. L. McCall, P. M. Platzman: Phys. Rev. B **22**, 1660 (1980)
[25] S. L. McCall, P. M. Platzman, P. A. Wolff: Phys. Lett. **77A**, 381 (1980)
[26] M. Kerker: Appl. Opt. **18**, 1180 (1979)
[27] M. Kerker, O. Siiman, O. S. Wang: J. Phys. Chem. **88**, 3168 (1984)
[28] M. Kerker, D. S. Wang, H. Chew: Appl. Opt. **19**, 4159 (1980)
[29] D. S. Wang, M. Kerker: Phys. Rev. B **24**, 1777 (1981)
[30] D. S. Wang, M. Kerker: Phys. Rev. B **25**, 2433 (1982)
[31] J. D. Jackson: *Electromagnetic Theory*, 3rd ed. (Wiley, New York 1998)
[32] H. X. Xu, M. Käll: Chem. Phys. Chem. **4**, 1001 (2003)
[33] H. Xu, J. Aizpurua, M. Käll, P. Apell: Phys. Rev. B **62**, 4318 (2000)
[34] M. I. Stockman: Phys. Rev. E **56**, 6494 (1997)
[35] V. M. Shalaev, R. Botet, J. Mercer, E. B. Stechel: Phys. Rev. B **54**, 8235 (1996)
[36] V. M. Shalaev, A. K. Sarychev: Phys. Rev. B **57**, 13265 (1998)
[37] S. Gresillon, L. Aigouy, A. C. Boccara, J. C. Rivoal, X. Quelin, C. Desmarest, P. Gadenne, V. A. Shubin, A. K. Sarychev, V. M. Shalaev: Phys. Rev. Lett. **82**, 4520 (1999)
[38] see, for example, Sect. III D in [17]
[39] M. Moskovits, D. P. DiLella, K. Maynard: Langmuir **4**, 67 (1988)
[40] J. K. Sass, H. Neff, M. Moskovits, S. Holloway: J. Phys. Chem. **85**, 621 (1981)
[41] M. Moskovits: J. Chem. Phys. **77**, 4408 (1982)
[42] M. Moskovits, D. P. DiLella: J. Chem. Phys. **77**, 1655 (1982)
[43] E. J. Ayars, H. D. Hallen, C. L. Jahncke: Phys. Rev. Lett. **85**, 4180 (2000)
[44] H. D. Hallen, C. L. Jahncke: J. Raman Spectrosc. **34**, 655 (2003)
[45] U. Kreibig, C. Von Frags: Z. Physik **224**, 307 (1969)
[46] U. Kreibig, P. Zacharia: Z. Physik **231**, 128 (1970)
[47] M. J. Dignam, M. Moskovits: J. Chem. Soc. Faraday Trans. 2 **69**, 65 (1973)
[48] T. A. Kelf, Y. Sugawara, J. J. Baumberg, M. Abdelsalam, P. N. Bartlett: Phys. Rev. Lett. **95**, 116802 (2005)
[49] J. J. Baumberg, T. A. Kelf, Y. Sugawara, S. Pelfrey, M. Adelsalam, P. N. Bartlett, A. E. Russell: Nano Lett. **11**, 2262 (2005)
[50] N. M. B. Perney, J. J. Baumberg, M. E. Zoorob, M. D. B. Charlton, S. Mahnkopf, C. M. Netti: Opt. Express **14**, 847 (2006)
[51] P. K. Aravind, A. Nitzan, H. Metiu: Surf. Sci. **110**, 189 (1981)
[52] P. K. Aravind, H. Metiu: J. Phys. Chem. **86**, 5076 (1982)
[53] P. K. Aravind, H. Metiu: Surf. Sci. **124**, 506 (1983)
[54] N. Liver, A. Nitzan, J. I. Gersten: Chem. Phys. Lett. **111**, 449 (1984)
[55] M. Inoue, K. Ohtaka: J. Phys. Soc. Jpn. **52**, 1457 (1983)
[56] M. Inoue, K. Ohtaka: J. Phys. Soc. Jpn. **52**, 3853 (1983)
[57] M. L. Xu, M. J. Dignam: J. Chem. Phys. **100**, 197 (1994)

[58] M. L. Xu, M. J. Dignam: J. Chem. Phys. **99**, 2307 (1993)

[59] M. L. Xu, M. J. Dignam: J. Chem. Phys. **96**, 7758 (1992)

[60] C. E. Talley, J. B. Jackson, C. Oubre, N. K. Grady, C. W. Hollars, S. M. Lane, T. R. Huser, P. Nordlander, N. J. Halas: Nano Lett. **5**, 1569 (2005)

[61] C. Oubre, P. Nordlander: J. Phys. Chem. B **109**, 10042 (2005)

[62] S. J. Oldenburg, R. D. Averitt, S. L. Westcott, N. J. Halas: Chem. Phys. Lett. **288**, 243 (1998)

[63] J. Jackson, N. J. Halas: J. Phys. Chem. B **105**, 2743 (2001)

[64] S. J. Oldenburg, J. B. Jackson, S. L. Westcott, N. J. Halas: Appl. Phys. Lett. **111**, 2897 (1999)

[65] E. Prodan, C. Radloff, N. J. Halas, P. Nordlander: Science **302**, 419 (2003)

[66] V. M. Shalaev: *Nonlinear Optics of Random Media: Fractal Composites and Metal-Dielectric Films* (Springer, Heidelberg, Berlin, New York 2000)

[67] V. P. Drachev, S. V. Perminov, S. G. Rautian, V. P. Safonov: Nonlinear optical effects and selective photomodification of colloidal silver aggregates, in V. Shalaev (Ed.): *Optical Properties of Nanostructured Random Media* (Springer, Heidelberg, Berlin, New York 2002) pp. 113–147

[68] V. A. Markel, V. M. Shalaev, P. Zhang, W. Huynh, L. Tay, T. L. Haslett, M. Moskovits: Phys. Rev. B **59**, 10903 (1999)

[69] M. Moskovits, L. Tay, J. Yang, T. Haslett: SERS and the single molecule, in V. Shalaev (Ed.): *Optical Properties of Nanostructured Random Media* (Springer, Berlin Heidelberg New York 2002) pp. 215–226, copyright 2002 Springer Verlag

[70] C. Douketis, T. L. Haslett, Z. Wang, M. Moskovits, S. Iannotta: J. Chem. Phys. **113**, 11315 (2000)

[71] C. Douketis, Z. Wang, T. L. Haslett, M. Moskovits: Phys. Rev. B **51**, 11022 (1995)

[72] C. Douketis, Z. Wang, T. L. Haslett, M. Moskovits: J. Phys. Chem. B **108**, 12724 (2004)

[73] A. R. Tao, P. D. Yang: J. Phys. Chem. B **109**, 15687 (2005)

[74] A. M. Michaels, J. Jiang, L. Brus: J. Phys. Chem. B **104**, 11965 (2000)

[75] J. Jiang, K. Bosnick, M. Maillard, L. Brus: J. Phys. Chem. B **107**, 9964 (2003)

[76] A. D. Buckingham: Adv. Chem. Phys. **12**, 107 (1967)

[77] D. J. Scalapino, S. M. Marcus: Phys. Rev. Lett. **18**, 459 (1967)

[78] J. Lambe, R. C. Jaklevic: Phys. Rev. **165**, 821 (1968)

[79] M. G. Simonsen, R. V. Coleman, P. K. Hansma: J. Chem. Phys. **61**, 3789 (1974)

Index

Electromagnetic Mechanism of SERS

George C. Schatz, Matthew A. Young, and Richard P. Van Duyne

Department of Chemistry, Northwestern University, Evanston,
IL 60208-3113, USA
{schatz,vanduyne}@chem.northwestern.edu

1 Introduction

Surface-enhanced Raman spectroscopy (SERS) was originally discovered in the 1970s [1, 2] where it was found that submonolayers of small nonresonant organic molecules, such as pyridine, when adsorbed onto the surface of silver nanoparticles (and a few other metals such as copper and gold) would exhibit greatly enhanced Raman intensities (enhancements of 10^6). These results have always held the promise for using this technique to observe very low concentrations of molecules on nanoparticles and nanostructured surfaces, but only recently has this promise started to be fulfilled in a predictable way. Thanks to exciting advances in techniques for making nanoparticles (such as nanosphere and e-beam lithography), to characterizing the surfaces using electron and scanning probe microscopies, to functionalizing the surfaces of the particles using self-assembled monolayers with attached chemical receptors, and to the laser and optics technology associated with measuring the Raman spectra, a number of important applications have been reported recently [3]. Included in these results have been the determination of SERS excitation spectra for benzenethiol on lithographically fabricated silver surfaces (periodic particle arrays fabricated using nanosphere lihtography) and the first observation of glucose using SERS [4, 5].

Another area of great interest in the SERS community concerns the observation of single-molecule SERS (SMSERS) [6, 7, 8]. This technique, which was originally developed in 1997 [6, 7], has proven to be of fundamental interest due to the nominal enhancement factor of $> 10^{13}$ required for the observation of any signal, but it has so far proven to be elusive in providing widespread applicability, due to the limited numbers of molecules and substrates for which successful observations have been made. In particular, almost all of the observations to date have been with molecules that are resonant Raman scatterers (such as rhodamine 6G), which typically have larger Raman intensities for non-SERS applications than nonresonant molecules like pyridine by factors of 10^4 or more. In addition, the range of substrates that yield SMSERS has been limited mostly to colloidal aggregate structures, which presumably are efficient at producing particle junction structures (i.e., dimers and small clusters of particles in which a molecule is sandwiched between the particles) that

K. Kneipp, M. Moskovits, H. Kneipp (Eds.): Surface-Enhanced Raman Scattering – Physics and Applications, Topics Appl. Phys. **103**, 19–46 (2006)
© Springer-Verlag Berlin Heidelberg 2006

are needed to produce the extraordinarily high enhancement factors required for SERS.

As noted above, SERS has almost exclusively been associated with three metals, silver (by far the most important), gold and copper. The generalization to other metals and other materials has been explored since the discovery of SERS, but only in the last few years have the tools been available to generate well-characterized experiments where the reported enhancements are reliable. An interesting result of recent work is that there is evidence that SERS can be observed for metals like rhodium [9, 10, 11, 12] and platinum [13, 14, 15, 16, 17, 18, 19] that are of key interest to the chemical industry through their importance in catalysis. Other metals, such as ruthenium [11] and aluminum [20] have a smaller literature, but remain of great interest due to their technological applications. The connection of these results with SERS-enhancement mechanisms is of great interest, but not yet established.

The last few years have seen several studies [21, 22, 23] aimed at using theory, particularly computational electrodynamics to provide a realistic, and even quantitative description of SERS. Much of this work originated with studies of extinction and scattering spectra, which led to an understanding of the dependence of the optical properties of metal nanoparticles on size, shape, arrangement and dielectric environment [24, 25, 26, 27, 28, 29, 30, 31, 32, 33, 34, 35, 36]. In addition, this work builds on earlier theory [37, 38, 39, 40, 41, 42, 43, 44, 45] in which the basic mechanisms of SERS were first postulated. This earlier work has been reviewed several times, [46, 47, 48] so we will only highlight a few studies. Although the idea that plasmon excitation would lead to enhanced Raman intensities even predates the discovery of SERS [49], and its possible role was pointed out in early theory papers [43, 50], the first serious estimates of enhanced fields near metal nanoparticles were done in the late 1970s and early 1980s by *Gersten* and *Nitzan*, *Kerker* and others [40, 42, 51]. These calculations were based on simple models of the particle structures, such as spheres or spheroids, and some of the early estimates were not careful to distinguish the peak enhancement from average enhancement. A paper by *Zeman* and *Schatz* in 1987 [52] provided a detailed analysis of results for spheroids for many different metals. This work concluded that the average enhancement factor for silver was approximately 10^5 when optimized for particle size and shape. These conclusions were similar to those from the famous "posts" experiment at Bell labs [53], which for many years represented the only lithographically fabricated particles used in SERS experiments. However, the modeling of the posts was based on particle structures that were adjusted to match measured extinction spectra, so this did not represent an unparameterized comparison between theory and experiment.

In this Chapter we will review our recent studies of the electromagnetic mechanisms of SERS, including both experimental and computational studies. Our review begins with a discussion of the underlying theory of the electromagnetic mechanism along with a description of methods now used to calculate fields. Then we consider a variety of issues that are important to the development of an understanding of SMSERS, including an analysis of SMSERS intensities for isolated particles, SMSERS for junctions between metal particles, and field enhancements arising from long-range electromagnetic interactions. In addition, we describe recent electronic structure studies that are aimed at developing a first-principles understanding of SERS. Our review concludes with a description of recent experimental studies using wavelength-scanned SERS excitation spectroscopy (WS SERES) to probe the electromagnetic mechanism.

2 Electromagnetic Mechanism of SERS

Since Raman intensities scale as the product of the incident field intensity and polarizability derivative, it comes as no surprise that there are two commonly considered mechanisms for SERS, one of which involves enhancements in the field intensity as a result of plasmon resonance excitation, and the other the enhancement in polarizability due to chemical effects such as charge-transfer excited states [40, 44, 46, 47, 54, 55, 56, 57]. The first of these is the well-known electromagnetic enhancement mechanism, and in this mechanism the enhancement factor E at each molecule is (approximately) given by

$$\mathsf{E} = |\boldsymbol{E}(\omega)|^2 |\boldsymbol{E}(\omega')|^2, \tag{1}$$

where $\boldsymbol{E}(\omega)$ is the local electric-field enhancement factor at the incident frequency ω and $\boldsymbol{E}(\omega')$ is the corresponding factor at the Stokes-shifted frequency ω'. More rigorous expressions for the electromagnetic enhancement factor, that do not involve a product of fields at different frequencies, have been given by *Kerker* [58]. However, numerical values of the enhancements based on this more accurate expression are only slightly different, so (1) is almost exclusively used.

In conventional SERS, E is averaged over the surface area of the particles where molecules can adsorb to generate the observed enhancement factor $\langle \mathsf{E} \rangle$, while in single-molecule SERS (SMSERS) it is the maximum enhancement $\mathsf{E}_{\max}$ that is of interest. Note that $\mathsf{E}_{\max}$ can be orders of magnitude larger than $\langle \mathsf{E} \rangle$, so the distinction between these two enhancement estimates is important. Another point is that E is often approximated by assuming that $\boldsymbol{E}(\omega)$ and $\boldsymbol{E}(\omega')$ are the same, and hence $\mathsf{E} = |\boldsymbol{E}(\omega)|^4$. This approximation takes advantage of the fact that the plasmon width is often large compared to the Stokes shift. However, in studies of SERS on isolated homogeneous particles, this assumption leads to an overestimate of the enhancement factor by factors of 3 or more.

3 Numerical Methods for Calculating Electromagnetic Enhancement Factors

As noted in the introduction most of the early enhancement-factor estimates were based on analytical theory (either Mie theory for spheres or quasistatic approximations for spheroids). A number of theoretical estimates of SERS enhancement factors have been made in the last few years for nonresonant molecules on nanoparticle surfaces using computational electrodynamics methods such as the discrete dipole approximation (DDA) [59] and the finite difference time-domain (FDTD) [60, 61] method to solve Maxwell's equations to determine the local fields $E(\omega)$. In these methods the particle structures were represented using finite elements so it is not difficult to describe a particle of any shape, and sizes up to a few hundred nm are within standard computational capabilities. These theories can also be used to describe many particles, but ultimately they are limited by the total number of elements needed to converge the calculation. However, it is possible to couple many particles together using coupled multipole expansions.

In the DDA method, one represents the particle by a large number of cubes, each of which is assumed to have a polarizability that is determined by the dielectric constant associated with the cube. If there are N cubes whose positions and polarizabilities are denoted r_i and α_i, then the induced dipole P_i in each particle in the presence of an applied plane wave field is $P_i = \alpha_i E_{\mathrm{loc},i}(i = 1, 2, \ldots, N)$ where the local field $E_{\mathrm{loc}}(r_i)$ is the sum of the incident and retarded fields of the other $N - 1$ dipoles. For a given wavelength λ, this field is:

$$E_{\mathrm{loc},i} = E_{\mathrm{inc},i} + E_{\mathrm{dipole},i} = E_0 \exp(ik \cdot r_i) - \sum_{\substack{j=1 \\ j \neq i}}^{N} A_{ij} \cdot P_j \, i = 1, 2, \ldots, N \,, \quad (2)$$

where E_0 and $k = 2\pi/\lambda$ are the amplitude and wavevector of the incident wave, respectively. The dipole interaction matrix A is expressed as:

$$A_{ij} \cdot P_j = k^2 e^{ikr_{ij}} \frac{r_{ij} \times (r_{ij} \times P_j)}{r_{ij}^3}$$
$$+ e^{ikr_{ij}} (1 - ikr_{ij}) \frac{[r_{ij}^2 P_j - 3r_{ij}(r_{ij} \cdot P_j)]}{r_{ij}^5},$$
$$(i = 1, 2, \ldots, N, j = 1, 2, \ldots, N, j \neq i), \quad (3)$$

where r_{ij} is the vector from dipole i to dipole j. The polarization vectors are obtained by solving $3N$ linear equations of the form

$$\underset{\approx}{A'} \underset{\sim}{P} = \underset{\sim}{E} \,, \quad (4)$$

where the offdiagonal elements of the matrix, A'_{ij}, are the same as A_{ij}, and the diagonal elements of the matrix, A'_{ij}, are α_i^{-1}. Once the induced polarizations have been determined it is relatively easy to obtain fields, and therefore field enhancements, outside the particle using (2) but with all dipoles included in the sum.

The FDTD method is a time-propagation algorithm for solving Maxwell's curl equations on a cubic grid. For a metal in which the dielectric constant is represented via the Drude model

$$\varepsilon_r(\omega) = \varepsilon_\infty - \frac{\omega_\mathrm{p}^2}{\omega^2 + i\omega\gamma} \tag{5}$$

in terms of the plasmon frequency ω_p and width γ, these equations are:

$$\begin{array}{l} \frac{\partial H}{\partial t} = -\frac{1}{\mu_0}\nabla \times E \\ \frac{\partial E}{\partial t} = \frac{1}{\varepsilon_\mathrm{eff}}(\nabla \times H - J), \\ \frac{\partial J}{\partial t} = \varepsilon_0\omega_p^2 E - \gamma J \end{array} \tag{6}$$

where H, E, J, μ_0, and ε_0 are the magnetic field, electric field, current density, permeability of free space, and permittivity of free space, respectively. The effective dielectric constant is $\varepsilon_\mathrm{eff} = \varepsilon_0\varepsilon_\infty$ if the grid point is inside the metal, and $\varepsilon_\mathrm{eff} = \varepsilon_0$ outside. The finite differencing algorithm was developed by Yee and involves a staggered grid in which the electric- and magnetic-field components are evaluated at half-step intervals from one another both in space and in time. This leads to central-difference approximations to the derivatives that are second order in space and time.

The finite difference equations are solved subject to an initial wavepacket that represents the plane wave incident on the particle. Absorbing boundary conditions are imposed on the edges of the grid to avoid unphysical reflections. The propagation is continued until the wavepacket no longer interacts with the particle, and then the time-dependent field $\boldsymbol{E}(r, t)$ is Fourier transformed to determine the frequency-resolved field that is needed for the SERS enhancement-factor estimate.

4 Results of EM Calculations

Estimates of the SERS enhancements from electrodynamics calculations vary widely depending on what particle structures are used, but for the conventional SERS enhancement, most calculations find that $\langle \mathsf{E} \rangle$ is less than roughly 10^6 [62]. Since this value accounts for most of the observed enhancement factor in conventional SERS (10^6–10^8) this is one reason why it is often assumed that the chemical contribution to the nonresonant SERS enhancement factor is small (less than 10^2) [22]. The story for SMSERS is much less certain. A major problem in making meaningful estimates of SMSERS

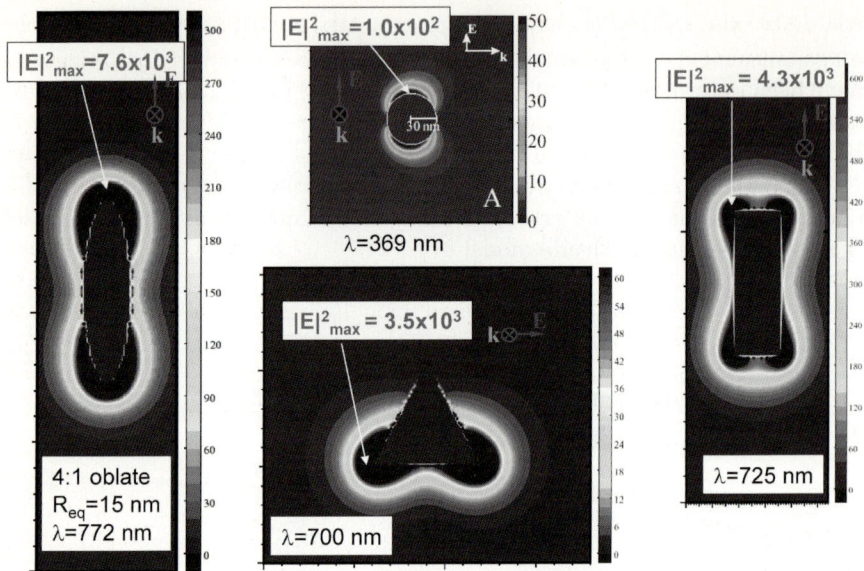

Fig. 1. Contours of the local field near silver particles at specified wavelengths, showing values of the peak field $|E|^2$

enhancement factors is that nearly all SMSERS studies have been done using colloidal aggregates so there are important questions about what aggregate structures are important, and if, as often is the case, the molecule is assumed to be sandwiched between two metal particles, then there are serious problems with the validity of continuum electrodynamics. For isolated particles, the peak $|E|^2$ values for particle sizes and shapes that are commonly studied in experiments are on the order of 10^4 or less, Fig. 1, which suggests enhancement factors of $E_{max} = |E|^4 = 10^8$, which is barely above $\langle E \rangle$. However, larger values, $|E|^4 = 10^{10}$–10^{11} can be obtained for dimers of silver particles, as demonstrated in Fig. 2 [21]. These values, which are associated with the gap between the two nanoparticles, are still below measured estimates of SMSERS enhancement factors (10^{14} or larger [6, 7]) by a factor of 10^3 or more, and this is one reason why it is often thought that SMSERS can only be observed for resonant Raman scatterers. However, this also raises the question as to whether other particle structures might have even larger enhancement factors. We will address this later.

In analyzing these enhancement-factor predictions it is important to note that the key parameter that controls the size of the enhancement factor for a dimer of nanoparticles is the size of the gap between the particles. In calculations by *Hao* and *Schatz* [21], it was found that E_{max} varies rapidly with gap size, and it is only for gaps on the order of 1 nm to 2 nm that one can obtain exceptionally large values such as $|E|^4 = 10^{11}$. Unfortunately, this is just where the reliability of the continuum electrodynamics calculations becomes a problem. One source of error is the use of a local dielectric response model, which is an issue that has occasionally been studied [63], and that always leads to smaller enhancement factors when factored in. In addition, for small particles, or even for large particles that have tips or other complex structures, the influence of scattering of the conduction electrons from the particle surfaces on the dielectric response can result in plasmon broadening that reduces enhancement factors [52]. Although corrections for this have been developed [64, 65, 66] these still use the continuum dielectric response model, so they ultimately fail for small enough particles. Recently, there has been some progress with the development of electronic structure methods for simulating SERS intensities [67, 68] that provides the promise of generating meaningful enhancement factors without using continuum electrodynamics. This will be described later.

5 Long-Range Electromagnetic Enhancement Effects

As noted above, the electromagnetic mechanism of SERS predicts enhancements of up to 10^{11} for a dimer of truncated tetrahedron-shaped particles [21] with a 1 nm gap. Other particle shapes give results that are comparable for optimized sizes, but what was not considered in this analysis is the possibility that large enhancements can be produced by combining the near-field effect considered in Fig. 2 with long-range coupling effects such as one can produce in an aggregate or array of widely spaced particles. There has been some interest in this point in the description of fractal clusters of particles, as here one finds local hot spots due to junction structures in the presence of a large structure that absorbs light efficiently at all wavelengths [8]. Arrays of particles can also be of interest [69, 70], and a particularly exciting possibility involves arrays of junction structures such as dimers of particles.

Figure 3 shows some of the motivation for how an array can influence plasmon excitation, in which we examine the extinction spectrum of a one-dimensional array of 50 nm (radius) spheres, here with the polarization vector and wavevector of the light both perpendicular to the array. The single-sphere extinction spectrum in this case shows a broad peak at about 400 nm that leads to very modest SMSERS enhancements ($E_{max} = 10^4$). We see in the figure that for spacings of 450 nm to 500 nm, there are very sharp peaks in the extinction spectrum that arise from long-range dipolar coherent dipolar interactions between the particles [71, 72]. Recent experiments have confirmed this result, although the experimental conditions that lead to the very sharp lines have yet to be achieved [34]. If we examine the SMSERS enhancement for the particles in this array at the resonant wavelengths, we find that $E_{max} = 10^6$, indicating that an extra factor of 10^2 is produced as a result of the long-range electrodynamic interactions [70].

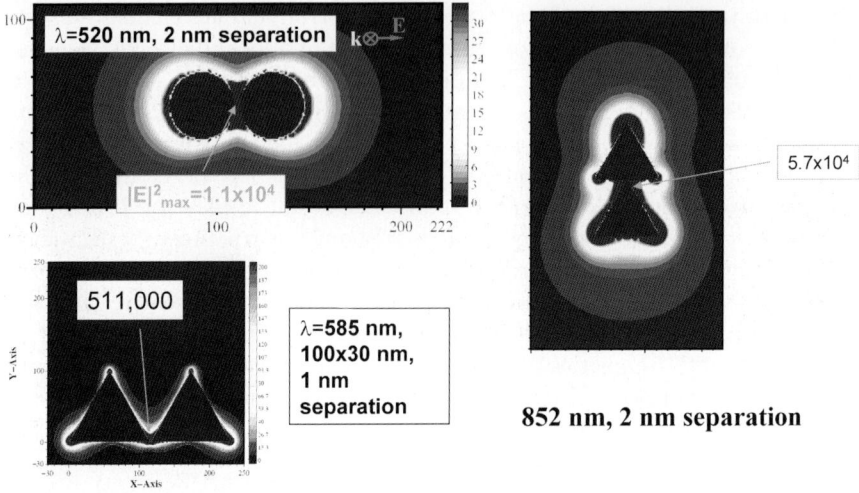

Fig. 2. Contours of the local field near dimers of silver particles at specified wavelengths, showing values of the peak field $|E|^2$

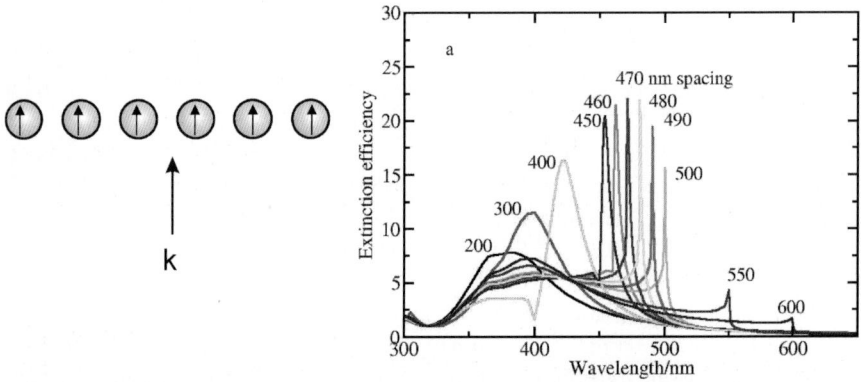

Fig. 3. Extinction spectra associated with an array of 400 silver particles, each with a diameter of 100 nm, for the case where the wavevector and polarization are perpendicular to the array axis. Results are presented for selected nanoparticle separations from 200 nm to 600 nm

Figure 4 shows a simple example of a structure that one might imagine fabricating that would combine the long-range effect just described with short-ranged interactions associated with a junction structure. Here we use an array of dimers of spheres with a spacing that is optimized to produce the highest possible field enhancement from near-field and long-range effects. This particular structure gives $E_{max} = 10^9$, but one can construct even better structures by using an array of spheres in which one sphere is replaced

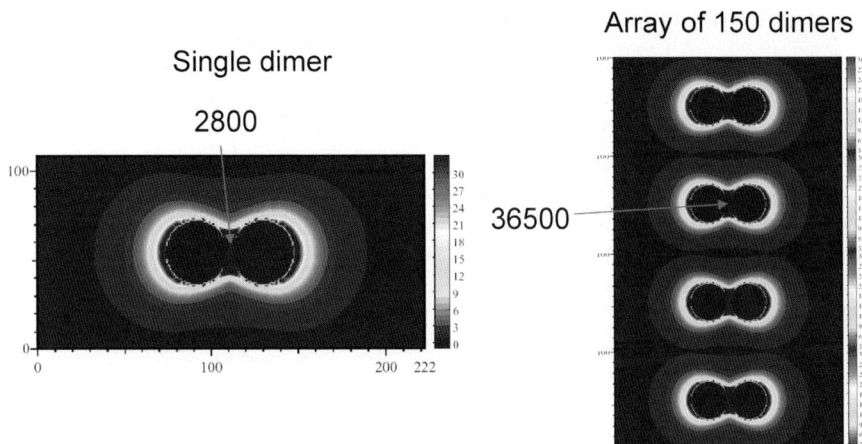

Fig. 4. Field contours for a single dimer of 100 nm diameter silver particles *(left)* at 561 nm (plasmon max), and for an array of 150 dimers *(right)* whose separation is 650 nm at 655 nm

by a dimer of truncated tetrahedrons, leading to SMSERS enhancement estimates of $E_{max} = 10^{13}$ [70]. This suggests that enhancements sufficient to generate SMSERS can be achieved without requiring any chemical enhancement. However, there are a number of problems with this analysis, including those mentioned previously concerning the validity of continuum electrodynamics for junctions. In addition, the structure that produces this result is hard to build, involving an array of a large number of spheres that are precisely placed, and with a dimer of specially fabricated particles located in the center of the array. It is also hard to build array structures that produce large electromagnetic enhancements at both the incident and Stokes-shifted wavelengths.

Given these complexities, it is desirable to use theory to identify simpler structures that will still achieve large SMSERS enhancements. One possibility is to generate array structures that allow for long-ranged interactions at multiple wavelengths, thereby removing the problem of generating large electromagnetic enhancements at two wavelengths. For example, the resonances in Fig. 3 that occur for a wavevector perpendicular to the array can also occur when the wavevector is parallel to the array, but for a somewhat different spacing. If one builds two-dimensional arrays that have one spacing in one direction and another spacing in the other direction, then exceptionally large electromagnetic enhancement factors should be possible. (Our preliminary calculations [73] suggest $E_{max} = 10^{13-14}$.) A second approach is to combine near-field enhancement from a dimer of metal particles with long-range enhancement that can be produced by dielectric structures. One version of this that has already been done is *Shalaev's* work on SERS from metal-particle aggregates on dielectric tubes in which whispering-gallery mode (WGM) resonances were excited [8]. However, WGM resonances tend to be extremely narrow, so this may only give enhancement at one wavelength. Indeed, the enhancement in Raman intensities from WGM excitation in water droplets is a well-known technique that can either be implemented at the incident or Stokes wavelengths, but not both [74, 75, 76, 77, 78, 79]. An alternative that we have recently modeled [80] is that the near-field region outside a micrometer-size dielectric sphere has a "hot spot" in the forward direction that arises from a simple focusing mechanism. Placing a dimer of metal particles at this hot spot leads to $E_{max} = 10^{13}$ under conditions that are relatively easy to achieve.

6 Electronic-Structure Studies

As mentioned earlier there are a number of problems with the use of classical electromagnetic theory to describe SMSERS that make it desirable to develop electronic-structure methods to calculate Raman intensities for molecules adsorbed onto metal particles. This possibility has been of interest for a long time, and there are a large number of electronic-structure studies of the static Raman intensities of molecules [81, 82, 83, 84, 85, 86, 87, 88, 89, 90, 91, 92, 93, 94, 95, 96, 97, 98, 99, 100, 101], in the presence of silver (or other metal) atoms. However, the calculation of frequency-dependent polarizability derivatives for molecules in the presence of metal particles for frequencies that are onresonance for excitation of the particles has only been considered in a few studies [23, 102, 103, 104, 105, 106], and only recently have the theoretical methods and computational resources become available to enable meaningful studies of this sort without serious approximations [68]. In these methods, molecule and metal are treated with the same electronic-structure theory, and electron correlation needs to be described well enough to include interband transition effects on the energies of plasmon excitations, and to determine molecular vibrational frequencies and transition moments accurately.

A promising method for doing this is time-dependent density functional theory (TDDFT) based on the Amsterdam density functional (ADF) code [68]. In this version, the finite lifetimes of the excited states of the metal particle are inserted as parameters in the determination of the induced polarization. These are adjusted to match measured extinction spectra or other properties. In a number of applications to metal particles, we find that lifetimes in the range 0.01 eV to 0.10 eV yield accurate absorption spectra.

Figure 5 shows some of the first results of our TDDFT calculations [67], in this case concerned with the resonance polarizability of a 20 atom (1 nm) silver cluster that is pictured in the upper left of the figure. The plot shows both the real and imaginary parts of the polarizability of the cluster as a function of photon energy. The imaginary part of the polarizability is proportional to the extinction cross section, and indeed this looks very much like what would come from a continuum electrodynamics calculation for the same

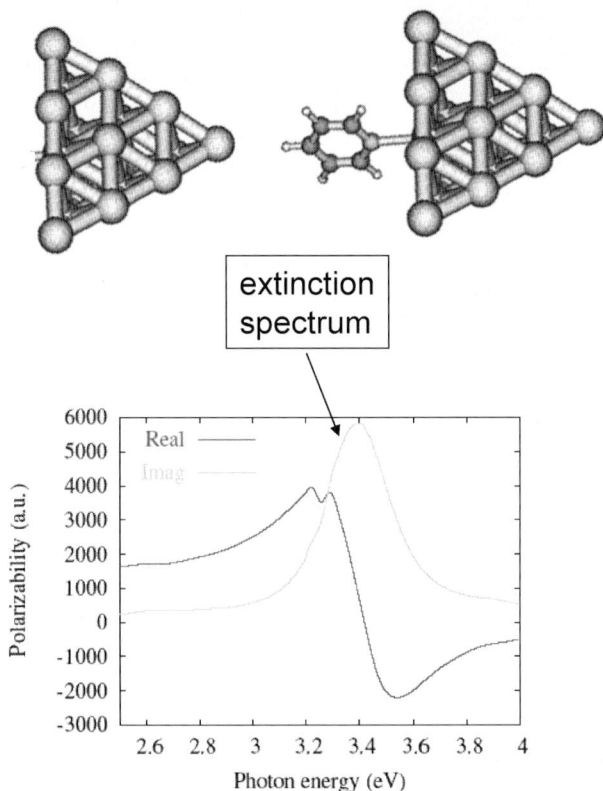

Fig. 5. Structure of 20 atom silver cluster *(upper left)*, and cluster with adsorbed pyridine *(upper right)*. Plot at *bottom* shows the real and imaginary parts of the Ag_{20} polarizability. The imaginary part is proportional to the extinction spectrum

particle using measured dielectric constants as input. In particular, the resonance wavelength is at about 365 nm, which is almost exactly that expected from classical electrodynamics. Of course the peak at 365 nm is not really plasmon excitation in the sense of a collective excitation of all the conduction electrons, however, we find that it consists of a superposition of many excited states, and these states involve many different kinds of excitations (one, two, three, etc., excitations) relative to the ground state.

Perhaps the most important feature of the resonance at 365 nm in Fig. 5 is that this leads to SERS enhancement for molecules interacting with the cluster, such as the pyridine pictured in the upper left of Fig. 5. In Fig. 6 we show TDDFT Raman spectra that we have calculated for pyridine, both as an isolated molecule (left panel) and on the cluster (right panel). The spectra are similar in appearance, but there is a factor of $E = 10^5$ enhancement in intensity due to the metal cluster. Thus, we find that enhanced Raman is possible for 1 nm clusters, although the effect is smaller than we estimated

Static and Resonance Raman Spectra of Pyridine-Ag_{20} Cluster

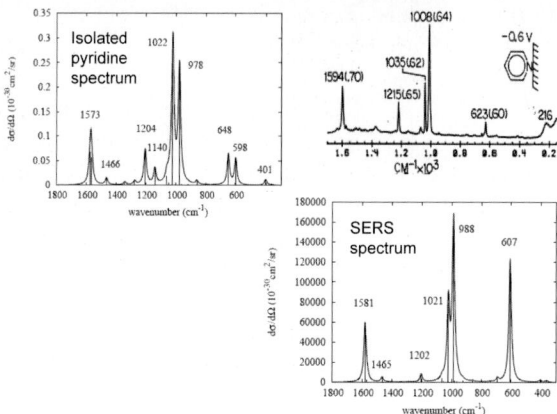

Fig. 6. TDDFT spectrum of pyridine *(left)*, and of the pyridine/Ag_{20} complex *(lower right)*. SERS results from experiment are *upper right*

earlier using classical electrodynamics for larger ($>10\,nm$) particles [22]. The upper trace in Fig. 5 shows a conventional SERS spectrum of pyridine, and we see that this is very close in appearance to what we have calculated.

The results in Fig. 6 can be used as the basis for decomposing the SERS enhancement into chemical and electromagnetic contributions. This work is incomplete, but preliminary results show [107] important electromagnetic and chemical mechanism contributions.

7 SERS Excitation Spectroscopy as a Probe of the Electromagnetic Mechanism

Wavelength-scanned SERES (WS SERES) involves the measurement of SERS enhancement for several laser excitation wavelengths λ_{ex}. This technique was recognized as a useful tool for probing the em mechanism immediately following the discovery of SERS. An obvious limitation of this technique is that the number of data points is determined by the tunability of the excitation laser and detection system. These substantial instrumental requirements have led to the majority of SERES publications suffering from low data point density and/or limited spectral coverage [108, 109, 110, 111, 112]. These limitations prevent the establishment of conclusive generalizations from SERES data. Additionally, most SERES experiments have been performed using surface-enhancing substrates with an unknown or poorly characterized distribution of roughness features. In the few cases where the surfaces are carefully characterized, it is shown that there is a wide distribution of roughness feature sizes [112, 113]. Other studies do not include characterization of the localized surface plasmon resonance (LSPR) of the substrate [108, 109], which prevents any direct comparison of the excitation profiles to the spectral location of the LSPR λ_{max}. The most common substrates historically employed in SERES experiments are Ag-island films and Ag-colloidal solutions. In these cases, the majority of the SERS excitation profiles peak at an excitation wavelength of maximum enhancement ($\lambda_{ex,max}$) near 500 nm to 600 nm [113, 114, 115, 116, 117]. The peaks of the excitation profiles have been shown to shift to longer wavelengths with increased aggregation [111, 113, 115, 116], which is a qualitative result predicted by the electromagnetic enhancement mechanism. With these substrates it is difficult to make a direct comparison between the LSPR of the substrate and the SERS excitation profile because the LSPR of the substrate is a superposition of a wide variety of LSPR wavelengths corresponding to the different roughness features.

Two exceptions to the above statements regarding roughness features are the well-known experiments by *Liao* and coworkers on microlithographically prepared Ag posts [53] and recent work by *Felidj* and coworkers on e-beam lithographically produced arrays of elongated gold nanoparticles [110]. The former ground-breaking work demonstrated excitation profiles where $\lambda_{ex,max}$ shifts to longer wavelengths with increased particle aspect ratio and with increased dielectric constant of the medium surrounding the particles. These results qualitatively agree with the em mechanism, but the LSPR of these substrates was not characterized for a direct comparison. In the latter work, the SERS enhancement was shown to peak at the midpoint between the excitation wavelength and the wavelength of the Raman scattered photon. This important experiment was the first observation of precisely what is predicted by the em mechanism. Unfortunately, this result was obtained on one sample with a profile consisting of three data points.

The limitation of laser and detection tunability has been circumvented by several researchers using a unique approach that involves investigating substrates with variations in the spectral location of the LSPR λ_{max} [118, 119, 120]. These variations allow investigation of the relationship between the LSPR and SERS enhancement using a single excitation wavelength. Our work using plasmon-sampled SERES (PS SERES) on well-defined arrays of nanoparticles was the first systematic study using this technique. PS SERES is a particularly attractive technique from a practical standpoint due to the fact that varying the LSPR λ_{max} of nanoparticles is typically far easier (and more cost effective) than accessing a variety of laser wavelengths. The conclusion of that study was that the condition for maximum enhancement occurred when the peak extinction wavelength of the LSPR, λ_{max}, is located near the midpoint between the energy of laser excitation and the energy of the Raman-scattered photons. This conclusion supports the em mechanism, which predicts that maximum SERS intensity is achieved when the LSPR strongly enhances both the incident and scattered photon intensities. WS SERES allows for a more thorough study of the relationship between the

LSPR and the Raman enhancement, and it affords the advantage of insuring that particle size and shape do not change throughout a given excitation spectrum.

The work described below utilizes a broadly tunable Raman system to measure excitation profiles with the greatest data point density ever demonstrated in a WS SERES experiment. A broadly tunable laser system, a versatile detection system, and a well-characterized surface-enhancing substrate were all employed in order to overcome the traditional shortcomings of WS SERES experiments. The use of a CW-modelocked Ti:sapphire and its harmonics allow for continuous tunability over the spectral ranges 350 nm to 500 nm and 700 nm to 1000 nm. The visible region not covered by the Ti:sapphire system was augmented with the use of a solid-state laser and a tunable dye laser. A triple spectrograph equipped with a CCD camera allows for rapid, multichannel spectral acquisition with efficient rejection of Rayleigh-scattered photons. The SERS substrates used in this work are triangular nanoparticle arrays fabricated by NSL. These substrates present a significant advantage over many of the traditional SERS substrates for SERES studies because NSL-fabricated triangular nanoparticles exhibit extremely narrow size distributions, making them an indispensable tool for probing the fundamental characteristics of SERS. Even though the surface coverage of these nanoparticles is $\sim 7\%$, strong SERS intensities are observed from analytes adsorbed to these substrates due to the strong enhancement ($E \sim 10^8$, vida infra) NSL-fabricated arrays exhibit [118].

For this detailed set of WS SERES experiments performed on optically and topographically characterized SERS substrates, the relative SERS enhancement of the substrates has been shown to vary by three orders of magnitude over the spectral range investigated. It is worth noting that this was not principally a study on the practical application of SERS for chemical analysis. Factors such as spectrograph throughput, detector efficiency, and the ν^4 scattering dependence of Raman photons play an important role in the practice of Raman spectroscopy. Instead, this study was performed in order to contribute fundamental insights into the origins of the SERS effect and to test various aspects of the em mechanism not previously studied. However, the conclusions reached are extremely important for the optimization of a surface-enhanced spectroscopy.

Figure 7 shows a schematic of the instrumentation used for the SERES experiments. All optical measurements were performed using an inverted microscope equipped with a $20\times$ objective (NA = 0.5). The light scattered by the samples was analyzed with a three-stage spectrograph equipped with a liquid-nitrogen-cooled, deep-depletion CCD detector. For the NSL-fabricated triangular nanoparticles, in-situ measurement of the LSPR spectrum was achieved by illuminating the sample with the microscope lamp and analyzing the transmitted light with a fiber-optically coupled miniature spectrometer. It is worthwhile to note that all illumination powers reported in this section were the laser powers incident on the microscope beamsplitter, not the power

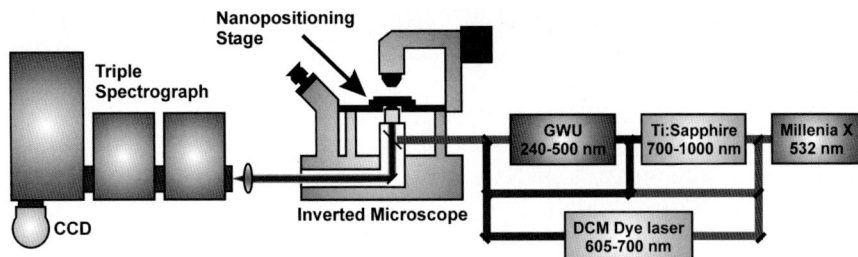

Fig. 7. Schematic diagram of the WS SERES apparatus

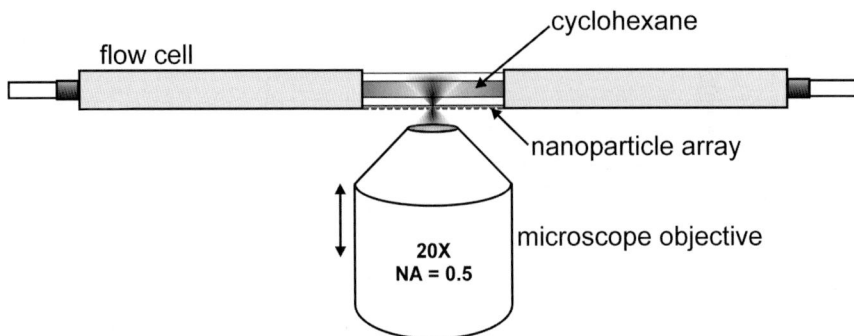

Fig. 8. Schematic diagram of flow cell containing cyclohexane for intensity standard measurements

incident on the sample. Based on experimental measurements, approximately 5 % to 10 % of the reported power is incident on the sample; however, because of the intensity standard the absolute power at the sample is not a critical measurement.

In order to correct for any variation of the SERS intensity not due to the enhancement by the substrate, the 1444 cm^{-1} normal Raman scattering band of neat cyclohexane was used as an intensity standard. This standard was used to correct for the inherent ν^4 behavior of Raman scattering, spectral dependence of the detection system, and differences in the illumination power. This was accomplished by mounting each sample face down as the bottom window of a transparent flow cell. When the flow cell was filled with cyclohexane, the nanoparticle array with an adsorbed benzenethiol monolayer was not in contact with the cyclohexane liquid. In this way, following each SERS acquisition, an intensity standard spectrum of cyclohexane could be taken by translating the inverted microscope objective $\sim 400\,\mu$m vertically. A schematic depiction of this setup is shown in Fig. 8.

A representative SERS spectrum of benzenethiol on a Ag-nanoparticle array is shown in Fig. 9. An AFM image of the sample from which this spectrum was taken is shown in the inset. This array was fabricated by depositing

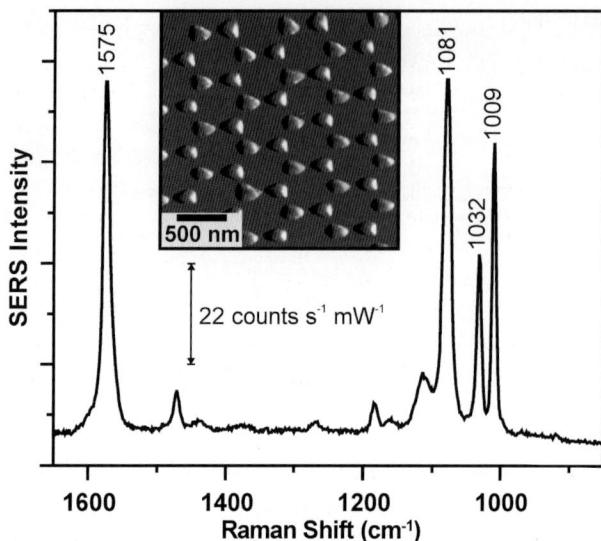

Fig. 9. Representative SERS spectrum of benzenethiol-dosed NSL substrate. λ_{ex} = 620 nm, $P = 3.0$ mW, acquisition time = 150 s. An atomic force micrograph of the sample is shown in the *inset*

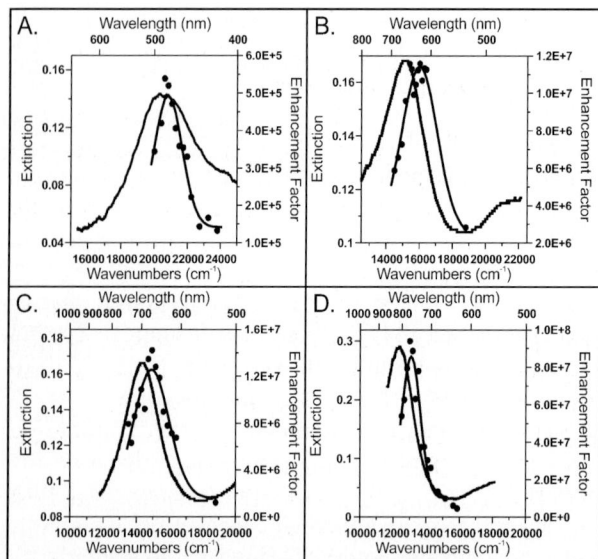

Fig. 10. Surface-enhanced Raman excitation spectra of the 1575 cm^{-1} peak of benzenethiol with cyclohexane as intensity standard. (**a**) Substrate annealed at 300 °C for 1 h. LSPR λ_{max} = 489 nm, profile fit maximum at $\lambda_{ex,max}$ = 480 nm. (**b**) LSPR λ_{max} = 663 nm, profile fit maximum at $\lambda_{ex,max}$ = 625 nm. (**c**) LSPR λ_{max} = 699 nm, profile fit maximum at $\lambda_{ex,max}$ = 671 nm. (**d**) LSPR λ_{max} = 810 nm, profile fit maximum at $\lambda_{ex,max}$ = 765 nm

55 nm of Ag through a mask formed with 450 nm diameter nanospheres. Figure 10 shows four excitation profiles for the 1575 cm^{-1} peak of benzenethiol, each with an LSPR λ_{max} at a distinctly different location. The SERES profile in Fig. 10a consists of 13 data points measured over the spectral range 420 nm to 500 nm. Because the formation of a monolayer of benzenethiol on these nanoparticle arrays results in a significant redshift in the position of the LSPR λ_{max}, it was necessary to anneal this sample at 300 °C for 1 h prior to benzenethiol addition in order to achieve a final LSPR λ_{max} at a wavelength shorter than 500 nm. It has been previously shown that annealing NSL-derived samples results in a large blueshift of the LSPR due to changing the shape of the nanoparticles [121]. The LSPR λ_{max} of this substrate was measured to be 489 nm (20 450 cm^{-1}). The largest SERS enhancement occurs at λ_{ex} = 485 nm. Fitting a Gaussian lineshape to the data reveals that the peak of the excitation profile, $\lambda_{ex,max}$, is 480 nm (20 833 cm^{-1}). The peak E value for this sample was calculated to be 5.5×10^5. This value is low in comparison to the values determined for the other samples because the shape of the nanoparticles is made more ellipsoidal by annealing. In addition to shifting the LSPR λ_{max} to shorter wavelengths, this change decreases the intensity of the electromagnetic fields at the nanoparticle surfaces.

The SERES profile in Fig. 10b consists of 14 data points measured over the spectral range 532 nm to 690 nm. The LSPR λ_{max} of this substrate was measured to be 663 nm (15 083 cm^{-1}). The largest SERS enhancement occurs for λ_{ex} = 625 nm. The maximum of a Gaussian lineshape fit to the data is 625 nm (16 000 cm^{-1}). The peak E value for this sample is 1.2×10^7. The SERES profile in Fig. 10c consists of 15 data points measured over the spectral range 532 nm to 740 nm. The LSPR λ_{max} of this substrate was measured to be 699 nm (14 306 cm^{-1}). The largest SERS enhancement occurs for λ_{ex} = 670 nm. The maximum of a Gaussian lineshape fit to the data is 671 nm (14 903 cm^{-1}). The peak E value for this sample is 1.4×10^7. The SERES profile in Fig. 10d consists of 15 data points measured over the spectral range 630 nm to 800 nm. The LSPR λ_{max} of this substrate was measured to be 810 nm (12 346 cm^{-1}). The largest SERS enhancement occurs for λ_{ex} = 770 nm. The maximum of a Gaussian lineshape fit to the data is 765 nm (13 072 cm^{-1}). The peak E value for this sample is 9.3×10^7.

In order to verify that this behavior can be generalized, two SERES experiments were undertaken in which a different benzenethiol band (1081 cm^{-1}) and intensity standard were monitored. In this case, the intensity standard was the 520 cm^{-1} phonon mode of silicon. The wavelength-dependent absorptivity of silicon requires that the measured Raman intensities must be corrected for differences in laser penetration depth. The penetration depth was calculated at all of the excitation wavelengths using the silicon absorptivities measured by *Aspnes* and *Studna* [122]. The silicon spectra were then normalized so that the intensities were representative of equivalent probe volumes. In addition, a correction was performed to account for the fact that the 520 cm^{-1} band of Si scatters at a significantly different wavelength from the $1081\,\mathrm{cm}^{-1}$ band of benzenethiol, particularly at redder excitation wavelengths. No correction was performed to account for variation in the Raman-scattering cross section of silicon because over the range of excitation wavelengths utilized in this work, the differences in the experimentally determined values of the polarizability of silicon are negligible [123]. The excitation spectra are shown in Fig. 11. The SERES profile in Fig. 11a consists of 13 data points measured over the spectral range 475 nm to 700 nm. The LSPR λ_{max} of this substrate was measured to be 690 nm ($14\,493\,\mathrm{cm}^{-1}$). The largest SERS enhancement occurs for $\lambda_{\mathrm{ex}} = 660$ nm. The maximum of a Gaussian lineshape fit to the data is 662 nm ($15\,106\,\mathrm{cm}^{-1}$). The peak E value for this sample is 1.9×10^{7}. The SERES profile in Fig. 11b consists of 17 data points measured over the spectral range 630 nm to 790 nm. The LSPR λ_{max} of this substrate was measured to be 744 nm ($13\,441\,\mathrm{cm}^{-1}$). The largest relative SERS intensity occurs for $\lambda_{\mathrm{ex}} = 700$ nm. The maximum of a Gaussian lineshape fit to the data is 715 nm ($13\,986\,\mathrm{cm}^{-1}$). The peak E value for this sample is 1.8×10^{7}.

Each substrate exhibits a SERES profile that has a similar lineshape to the extinction spectrum of the substrate. Also, the $\lambda_{\mathrm{ex,max}}$ for the NSL-fabricated substrates is consistently shorter than the LSPR λ_{max}. In all cases, the maximum SERS enhancement occurs when the substrate LSPR λ_{max} is located between λ_{ex} and λ_{vib}. Under these conditions, both the incident and scattered photons experience enhancement by the LSPR. This data is in accordance with the em mechanism of SERS and the experimental work performed previously using PS SERES.

If the peak in the SERS enhancement occurs when the LSPR λ_{max} of the sample is equal to $(\lambda_{\mathrm{ex}} + \lambda_{\mathrm{vib}})/2$, then $\lambda_{\mathrm{ex,max}}$ should be different for the various Raman bands of benzenethiol on a single sample. It is expected that $\lambda_{\mathrm{ex,max}}$ will have a larger separation from the LSPR λ_{max} for a large Raman shift than for a small shift. Excitation profiles for three benzenethiol peaks on a single substrate are shown in Fig. 12. For this substrate, the LSPR λ_{max} is 729 nm. Figure 12a shows the SERS excitation profile for the $1575\,\mathrm{cm}^{-1}$ peak of benzenethiol, normalized to the $1444\,\mathrm{cm}^{-1}$ peak of liquid cyclohexane. The separation in wave numbers between the LSPR λ_{max} and $\lambda_{\mathrm{ex,max}}$ is $734\,\mathrm{cm}^{-1}$. In Fig. 12b, the excitation profile for the $1081\,\mathrm{cm}^{-1}$ benzenethiol peak (normalized to the $1028\,\mathrm{cm}^{-1}$ peak of cyclohexane) is shown. The separation in wave numbers between the LSPR λ_{max} and $\lambda_{\mathrm{ex,max}}$ is $569\,\mathrm{cm}^{-1}$. Finally, in Fig. 12c, the excitation profile for the $1009\,\mathrm{cm}^{-1}$ benzenethiol peak (normalized to the $1028\,\mathrm{cm}^{-1}$ peak of cyclohexane) is shown, and the separation in wave numbers between the LSPR λ_{max} and $\lambda_{\mathrm{ex,max}}$ is $488\,\mathrm{cm}^{-1}$. This data demonstrates the qualitative trend whereby the $\lambda_{\mathrm{ex,max}}$ in the excitation spectra of larger Raman-shifted bands yield a larger separation from the LSPR λ_{max} than those of smaller Raman-shifted bands, and this once again lends support to the electromagnetic mechanism.

Previous work has demonstrated that the spectral location of the LSPR is extremely sensitive to the presence of molecular adsorbates [26, 124, 125].

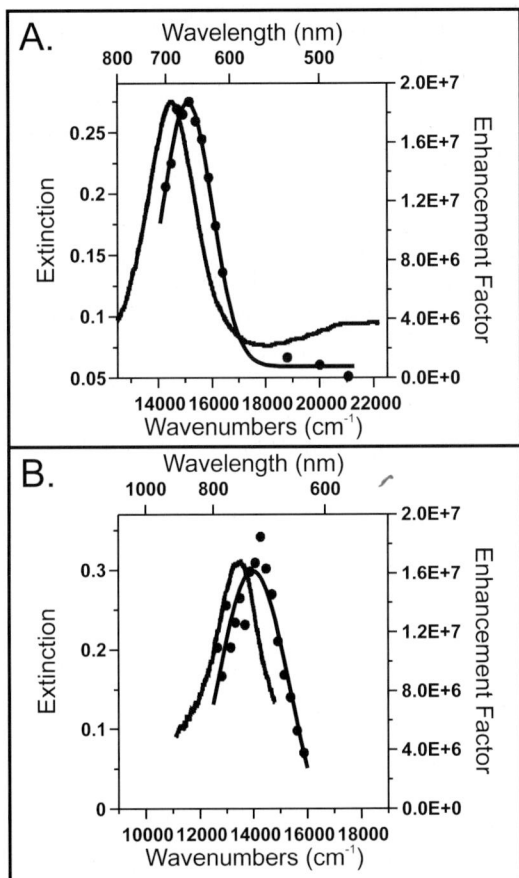

Fig. 11. Surface-enhanced Raman excitation spectra of the $1081\,\mathrm{cm}^{-1}$ peak of benzenethiol with Si as intensity standard. **(a)** LSPR $\lambda_{\mathrm{max}} = 690\,\mathrm{nm}$, profile fit maximum at $\lambda_{\mathrm{ex,max}} = 662\,\mathrm{nm}$. **(b)** LSPR $\lambda_{\mathrm{max}} = 744\,\mathrm{nm}$, profile fit maximum at $\lambda_{\mathrm{ex,max}} = 715\,\mathrm{nm}$

Therefore, it is important to note that the relationship between the LSPR spectra and SERES profiles depicted in Fig. 10, Fig. 11 and Fig. 12 pertain to LSPR spectra measured after adsorption of the analyte molecule. Figure 13 demonstrates the importance of considering this point. For a bare nanoparticle array, the LSPR λ_{max} was measured to be 672 nm. After incubation in 1 mM benzenethiol for $> 3\,\mathrm{h}$, thorough rinsing with methanol, and drying, the LSPR λ_{max} was observed to have redshifted by 57 nm to 729 nm. Measurement of the WS SERES profile yields $\lambda_{\mathrm{ex,max}} = 692\,\mathrm{nm}$. This is blueshifted with respect to the LSPR λ_{max} of the adsorbate-covered sample, as observed for the other samples used in this study, but redshifted with respect to the LSPR λ_{max} of the bare nanoparticle array. This demonstrates that it is criti-

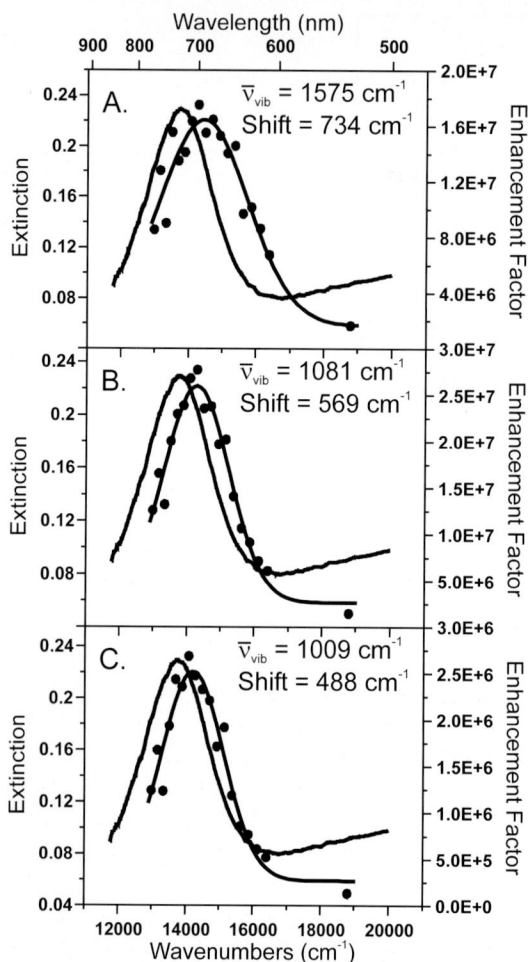

Fig. 12. Effect of Stokes Raman shift. (**a**) Profile of the $1575\,\mathrm{cm}^{-1}$ vibrational mode of benzenethiol. Distance between LSPR λ_{max} and excitation profile fit line $\lambda_{\mathrm{ex,max}} = 734\,\mathrm{cm}^{-1}$. $\mathsf{E} = 1.8 \times 10^7$. (**b**) $1081\,\mathrm{cm}^{-1}$ vibrational mode, shift $= 569\,\mathrm{cm}^{-1}$, $\mathsf{E} = 2.8 \times 10^7$. (**c**) $1009\,\mathrm{cm}^{-1}$ vibrational mode, shift $= 488\,\mathrm{cm}^{-1}$, $\mathsf{E} = 2.7 \times 10^6$

cal to characterize the LSPR of a SERS substrate after analyte adsorption in order to choose the appropriate laser excitation wavelength for maximizing E or to draw any conclusions about the fundamental mechanism of the SERS effect.

This work demonstrates the most thorough WS SERES experiments ever performed on optically and topographically characterized SERS substrates. The experimental apparatus utilized has proven effective for the measurement of relative SERS enhancements that vary by three orders of magnitude. This

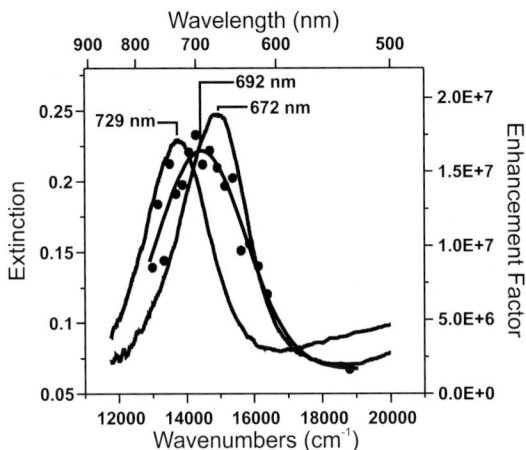

Fig. 13. LSPR shift and SERES profile for the $1575\,\text{cm}^{-1}$ peak of benzenethiol. The line with $\lambda_{\max} = 672\,\text{nm}$ is the LSPR extinction of the bare nanoparticle array. The line with $\lambda_{\max} = 729\,\text{nm}$ is the LSPR extinction of the nanoparticle array with an adsorbed monolayer of benzenethiol. The line with $\lambda_{\text{ex},\max} = 692\,\text{nm}$ is the best fit to the SERES data points

work demonstrates that the relationship between the substrate LSPR and the SERES profile for size-homogenous nanoparticles is consistent throughout the visible range. In all cases the experimentally observed behavior is consistent with that predicted by the em mechanism. Specifically, the strongest SERS enhancement occurs under conditions where the incident and Raman scattered photons are both strongly enhanced. The largest E measured was $\sim 10^8$ for the triangular nanoparticle arrays studied. Ultimately, refinement of the experimental apparatus and optimization of SERS enhancement will allow SERES to be performed using single-nanoparticle substrates. This level provides the best possible case in terms of reducing sample heterogeneity. These experiments are expected to provide key information to validate the em mechanism of SERS and will present an additional technique that can be used to study the SMSERS effect.

8 Conclusions

Using electrodynamics calculations, predicted E values as large as 10^{11} have been shown for nanoparticle dimers with small ($1\,\text{nm}$ to $2\,\text{nm}$) gaps. An E value of 10^{13} can be achieved with an array of dimers of truncated tetrahedra. This suggests that it is at least theoretically possible to realize an enhancement sufficient to generate SMSERS without appealing to any chemical enhancement. Because the structures required to produce the largest enhancements are difficult to build and unlikely to be what is occurring in experimental SMSERS, other simpler structures have also been explored.

To overcome the difficulties associated with using classical electromagnetic theory to model SERS for small nanoparticles with very small gaps, electronic structure methods are being developed for use in this area. Preliminary TDDFT calculations have shown good agreement with experiment for the SERS spectrum of pyridine. This method can be used to decompose the SERS enhancement into chemical and electromagnetic contributions, and the first results have shown important contributions from both sources.

The WS SERES results reviewed herein are the most detailed ever performed on optically and topographically characterized SERS substrates. The SERS enhancement factor has been shown to vary significantly as a function of laser wavelength for substrates with LSPR λ_{max} values throughout the visible spectrum. The consistent conclusion, verifying the electromagnetic mechanism prediction, is that the SERS enhancement factor is maximized when both the incident laser and Raman-scattered photons are strongly enhanced. This occurs when the incident laser is on the higher-energy side of λ_{max}, and the Raman shift is on the lower-energy side. Additionally, the electromagnetic mechanism prediction concerning the effect of the Stokes shift on SERES has been verified. Specifically, the data demonstrates that a peak with a smaller Raman shift shows a maximum enhancement closer in energy to the LSPR λ_{max} than a peak with a larger Raman shift. The conditions for maximum SERS enhancement are determined after the addition of the analyte molecule because of the significant shift in the LSPR λ_{max} caused by analyte adsorption. The adsorbate-induced LSPR shift and SERES Stokes shift are in opposite directions. The largest E measured was $\sim 10^8$ for the triangular nanoparticle arrays studied, which is in accordance with theory and previous experimental work.

Acknowledgements

The authors gratefully acknowledge support from the Air Force Office of Scientific Research MURI program (Grant F49620-02-1-0381) and the National Science Foundation (EEC-0118025, DMR-0076097, CHE-0414554, DGE-0114429).

References

[1] M. Fleischmann, P. J. Hendra, A. J. Mcquillan: Chem. Phys. Lett. **26**, 163 (1974)
[2] D. L. Jeanmaire, R. P. Van Duyne: J. Electroanal. Chem. Interf. Electrochem. **84**, 1 (1977)
[3] C. L. Haynes, A. D. McFarland, R. P. Van Duyne: Anal. Chem. **77**, 338A (2005)

[4] C. R. Yonzon, C. L. Haynes, X. Zhang, J. T. Walsh, Jr., R. P. Van Duyne: Anal. Chem. **76**, 78 (2004)

[5] O. Lyandres, N. C. Shah, C. R. Yonzon, J. T. Walsh Jr., M. R. Glucksberg, R. P. Van Duyne: Anal. Chem. **77**, 6134 (2005)

[6] K. Kneipp, Y. Wang, H. Kneipp, L. T. Perelman, I. Itzkan, R. R. Dasari, M. S. Feld: Phys. Rev. Lett. **78**, 1667 (1997)

[7] S. Nie, S. R. Emory: Science **275**, 1102 (1997)

[8] W. Kim, V. P. Safonov, V. M. Shalaev, R. L. Armstrong: Phys. Rev. Lett. **82**, 4811 (1999)

[9] L. Cui, Z. Liu, S. Duan, D.-Y. Wu, B. Ren, Z.-Q. Tian, S.-Z. Zou: J. Phys. Chem. B **109**, 17597 (2005)

[10] X.-F. Lin, B. Ren, Z.-Q. Tian: J. Phys. Chem. B **108**, 981 (2004)

[11] B. Ren, X.-F. Lin, Z.-L. Yang, G.-K. Liu, R. F. Aroca, B.-W. Mao, Z.-Q. Tian: J. Am. Chem. Soc. **125**, 9598 (2003)

[12] B. Ren, X.-F. Lin, J.-W. Yan, B.-W. Mao, Z.-Q. Tian: J. Phys. Chem. B **107**, 899 (2003)

[13] J.-Z. Zheng, B. Ren, D.-Y. Wu, Z.-Q. Tian: J. Electroanal. Chem. **574**, 285 (2005)

[14] S. A. Bilmes, J. C. Rubim, A. Otto, A. J. Arvia: Chem. Phys. Lett. **159**, 89 (1989)

[15] M. A. Bryant, S. L. Joa, J. E. Pemberton: Langmuir **8**, 753 (1992)

[16] P. Cao, Y. Sun, R. Gu: J. Phys. Chem. B **108**, 4716 (2004)

[17] R. Gomez, J. Solla-Gullon, J. M. Perez, A. Aldaz: Chem. Phys. Chem. **6**, 2017 (2005)

[18] Y. Kim, R. C. Johnson, J. T. Hupp: Nano Lett. **1**, 165 (2001)

[19] J. Miragliotta, T. E. Furtak: Mater. Res. Soc. Symp. Proc. **83**, 123 (1987)

[20] M. Muniz-Miranda: J. Raman Spectrosc. **27**, 435 (1996)

[21] E. Hao, G. C. Schatz: J. Chem. Phys. **120**, 357 (2004)

[22] G. C. Schatz, R. P. Van Duyne: Electromagnetic mechanism of surface enhanced spectroscopy, in J. M. Chalmers, P. R. Griffiths (Eds.): *Handbook of Vibrational Spectroscopy* (Wiley, Chichester 2002) p. 759

[23] S. Corni, J. Tomasi: Chem. Phys. Lett. **342**, 135 (2001)

[24] T. R. Jensen, M. L. Duval, K. L. Kelly, A. A. Lazarides, G. C. Schatz, R. P. Van Duyne: J. Phys. Chem. B **103**, 9846 (1999)

[25] T. R. Jensen, G. C. Schatz, R. P. Van Duyne: J. Phys. Chem. B **103**, 2394 (1999)

[26] M. D. Malinsky, K. L. Kelly, G. C. Schatz, R. P. Van Duyne: J. Am. Chem. Soc. **123**, 1471 (2001)

[27] C. L. Haynes, A. D. McFarland, L. Zhao, R. P. Van Duyne, G. C. Schatz, L. Gunnarsson, J. Prikulis, B. Kasemo, M. Kaell: J. Phys. Chem. B **107**, 7337 (2003)

[28] A. J. Haes, S. Zou, G. C. Schatz, R. P. Van Duyne: J. Phys. Chem. B **108**, 6961 (2004)

[29] A. J. Haes, S. Zou, G. C. Schatz, R. P. Van Duyne: J. Phys. Chem. B **108**, 109 (2004)

[30] C. R. Yonzon, E. Jeoung, S. Zou, G. C. Schatz, M. Mrksich, R. P. Van Duyne: J. Am. Chem. Soc. **126**, 12669 (2004)

[31] A. J. Haes, C. L. Haynes, A. D. McFarland, G. C. Schatz, R. P. van Duyne, S. Zou: MRS Bull. **30**, 368 (2005)

[32] A. J. Haes, J. Zhao, S. Zou, C. S. Own, L. D. Marks, G. C. Schatz, R. P. Van Duyne: J. Phys. Chem. B **109**, 11158 (2005)

[33] E. M. Hicks, X. Zhang, S. Zou, O. Lyandres, K. G. Spears, G. C. Schatz, R. P. Van Duyne: J. Phys. Chem. B **109**, 22351 (2005)

[34] E. M. Hicks, S. Zou, G. C. Schatz, K. G. Spears, R. P. Van Duyne, L. Gunnarsson, T. Rindzevicius, B. Kasemo, M. Kaell: Nano Lett. **5**, 1065 (2005)

[35] L. J. Sherry, S.-H. Chang, G. C. Schatz, R. P. Van Duyne, B. J. Wiley, Y. Xia: Nano Lett. **5**, 2034 (2005)

[36] X. Zhang, E. M. Hicks, J. Zhao, G. C. Schatz, R. P. Van Duyne: Nano Lett. **5**, 1503 (2005)

[37] P. K. Aravind, A. Nitzan, H. Metiu: Surf. Science **110**, 189 (1981)

[38] S. Efrima: J. Phys. Chem. **89**, 2843 (1985)

[39] P. J. Feibelman: Phys. rev. b, Condens. Matter Mater. Phys. **22**, 3654 (1980)

[40] J. Gersten, A. Nitzan: J. Chem. Phys. **73**, 3023 (1980)

[41] R. M. Hexter, M. G. Albrecht: Spectrochim. acta, part a, Molec. Biomol. Spectrosc. **233**, 35A (1979)

[42] M. Kerker, O. Siiman, L. A. Bumm, D. S. Wang: Appl. Opt. **19**, 3253 (1980)

[43] F. W. King, R. P. Van Duyne, G. C. Schatz: J. Chem. Phys. **69**, 4472 (1978)

[44] M. Moskovits: Solid State Commun. **32**, 59 (1979)

[45] A. Wokaun, J. P. Gordon, P. F. Liao: Phys. Rev. Lett. **48**, 957 (1982)

[46] H. Metiu, P. Das: Ann. Rev. Phys. Chem. **35**, 507 (1984)

[47] G. C. Schatz: Acc. Chem. Res. **17**, 370 (1984)

[48] M. Moskovits: Rev. Mod. Phys. **57**, 783 (1985)

[49] M. R. Philpott: J. Chem. Phys. **61**, 5306 (1974)

[50] M. Moskovits: J. Chem. Phys. **69**, 4159 (1978)

[51] D. S. Wang, H. Chew, M. Kerker: Appl. Opt. **19**, 2256 (1980)

[52] E. J. Zeman, G. C. Schatz: J. Phys. Chem. **91**, 634 (1987)

[53] P. F. Liao, J. G. Bergman, D. S. Chemla, A. Wokaun, J. Melngailis, A. M. Hawryluk, N. P. Economou: Chem. Phys. Lett. **82**, 355 (1981)

[54] M. Kerker: Studies Phys. Theoret. Chem. **45**, 3 (1987)

[55] M. R. Philpott: Colloque, J. Phys. **C10**, 295 (1983)

[56] A. Wokaun: Solid State Phys. **38**, 223 (1984)

[57] A. Wokaun: Molec. Phys. **56**, 1 (1985)

[58] M. Kerker: J. Colloid Interf. Sci. **118**, 417 (1987)

[59] W.-H. Yang, G. C. Schatz, R. P. Van Duyne: J. Chem. Phys. **103**, 869 (1995)

[60] A. Taflove, S. C. Hagness: *The Finite-Difference Time-Domain Method* (Artech House Inc., Norwood, MA 2005)

[61] L. Yin, V. K. Vlasko-Vlasov, A. Rydh, J. Pearson, U. Welp, S.-H. Chang, S. K. Gray, G. C. Schatz, D. E. Brown, C. W. Kimball: Appl. Phys. Lett. **85**, 467 (2004) Los Alamos National Laboratory, Preprint Archive

[62] K. L. Kelly, T. R. Jensen, A. A. Lazarides, G. C. Schatz: Modeling metal nanoparticles optical properties, in D. Feldheim, C. Foss (Eds.): *Metal Nanoparticles: Synthesis, Characterization and Applications* (Marcel-Dekker, New York 2002) p. 89

[63] P. T. Leung, W. S. Tse: Solid State Commun. **95**, 39 (1995)

[64] E. A. Coronado, G. C. Schatz: J. Chem. Phys. **119**, 3926 (2003)

[65] A. Hilger, T. von Hofe, U. Kreibig: Nova Acta Leopoldina **92**, 9 (2005)

[66] A. Reinholdt, R. Pecenka, A. Pinchuk, S. Runte, A. L. Stepanov, T. E. Weirich, U. Kreibig: European Phys. J. D, At. Molec. Opt. Phys. **31**, 69 (2004)

[67] L. Zhao, L. Jensen, G. C. Schatz: J. Am. Chem. Soc. **128**, 2911 (2006)

[68] L. Jensen, J. Autschbach, G. C. Schatz: J. Chem. Phys. **122**, 224115/1 (2005)

[69] D. A. Genov, A. K. Sarychev, V. M. Shalaev, A. Wei: Nano Lett. **4**, 153 (2004)

[70] S. Zou, G. C. Schatz: Chem. Phys. Lett. **403**, 62 (2005)

[71] S. Zou, N. Janel, G. C. Schatz: J. Chem. Phys. **120**, 10871 (2004)

[72] S. Zou, G. C. Schatz: J. Chem. Phys. **122**, 097102/1 (2005)

[73] S. Zou, G. C. Schatz: Coupled plasmonic plasmon/photonic resonance effects in sers, in K. Kneipp, M. Moskovitz, H. Kneipp (Eds.): *Surface Enhanced Raman Scattering: Physics and Applications* (Springer, Berlin 2006) p. 67

[74] R. Symes, R. J. J. Gilham, R. M. Sayer, J. P. Reid: Phys. Chem. Chem. Phys. **7**, 1414 (2005)

[75] R. Symes, R. M. Sayer, J. P. Reid: Phys. Chem. Chem. Phys. **6**, 474 (2004)

[76] J. F. Widmann, C. L. Aardahl, E. J. Davis: TrAC, Trends Anal. Chem. **17**, 339 (1998)

[77] E. J. Davis, C. L. Aardahl, J. F. Widmann: J. Dispers. Sci. Technol. **19**, 293 (1998)

[78] C. L. Aardahl, W. R. Foss, E. J. Davis: J. Aerosol Sci. **27**, 1015 (1996)

[79] K. Schaschek, J. Popp, W. Kiefer: J. Raman Spectrosc. **24**, 69 (1993)

[80] S. Zou, G. C. Schatz: Israel J. Chem. (2006)

[81] W.-H. Yang, J. Hulteen, G. C. Schatz, R. P. Van Duyne: J. Chem. Phys. **104**, 4313 (1996)

[82] R. F. Aroca, R. E. Clavijo, M. D. Halls, H. B. Schlegel: J. Phys. Chem. A **104**, 9500 (2000)

[83] T. Iliescu, M. Bolboaca, R. Pacurariu, D. Maniu, W. Kiefer: J. Raman Spectrosc. **34**, 705 (2003)

[84] P. E. Schoen, R. G. Priest, J. P. Sheridan, J. M. Schnur: Nature **270**, 412 (1977)

[85] E. A. Carrasco Flores, M. M. Campos Vallette, R. E. C. Clavijo, P. Leyton, G. Diaz F, R. Koch: Vibrat. Spectrosc. **37**, 153 (2005)

[86] K.-H. Cho, J. Choo, S.-W. Joo: J. Molec. Struct. **738**, 9 (2005)

[87] T. Iliescu, D. Maniu, V. Chis, F. D. Irimie, C. Paizs, M. Tosa: Chem. Phys. **310**, 189 (2005)

[88] A. V. Szeghalmi, L. Leopold, S. Pinzaru, V. Chis, I. Silaghi-Dumitrescu, M. Schmitt, J. Popp, W. Kiefer: Biopolymers **78**, 298 (2005)

[89] S. Thomas, N. Biswas, S. Venkateswaran, S. Kapoor, R. D'Cunha, T. Mukherjee: Chem. Phys. Lett. **402**, 361 (2005)

[90] S. Naumov, S. Kapoor, S. Thomas, S. Venkateswaran, T. Mukherjee: Theochem. **685**, 127 (2004)

[91] M. Baia, L. Baia, W. Kiefer, J. Popp: J. Phys. Chem. B **108**, 17491 (2004)

[92] G. Cardini, M. Muniz-Miranda, V. Schettino: J. Phys. Chem. B **108**, 17007 (2004)

[93] S. D. Silaghi, G. Salvan, M. Friedrich, T. U. Kampen, R. Scholz, D. R. T. Zahn: Appl. Surf. Sci. **235**, 73 (2004)

[94] S. Cinta Pinzaru, N. Leopold, I. Pavel, W. Kiefer: Spectrochim. acta, part a, Molec. Biomolec. Spectrosc. **60A**, 2021 (2004)

44   George C. Schatz et al.

[95] T. Iliescu, M. Baia, I. Pavel: J. Raman Spect. **37**, 318 (2006)
[96] B. Pergolese, M. Muniz-Miranda, G. Sbrana, A. Bigotto: Far. Disc. **132**, 111 (2006)
[97] T. Tanaka, A. Nakajima, A. Watanabe, T. Ohno, Y. Ozaki: Vibrat. Spectrosc. **34**, 157 (2004)
[98] T. Tanaka, A. Nakajima, A. Watanabe, T. Ohno, Y. Ozaki: J. Molec. Struct. **661–662**, 437 (2003)
[99] B. Giese, D. McNaughton: Biopolymers **72**, 472 (2003)
[100] D.-Y. Wu, B. Ren, X. Xu, G.-K. Liu, Z.-L. Yang, Z.-Q. Tian: J. Chem. Phys. **119**, 1701 (2003)
[101] M. Bolboaca, T. Iliescu, C. Paizs, F. D. Irimie, W. Kiefer: J. Phys. Chem. A **107**, 5144 (2003)
[102] P. K. K. Pandey, G. C. Schatz: J. Chem. Phys. **80**, 2959 (1984)
[103] H. Nakai, H. Nakatsuji: J. Chem. Phys. **103**, 2286 (1995)
[104] S. Corni, J. Tomasi: J. Chem. Phys. **114**, 3739 (2001)
[105] S. Corni, J. Tomasi: Chem. Phys. Lett. **365**, 552 (2002)
[106] S. Corni, J. Tomasi: J. Chem. Phys. **116**, 1156 (2002)
[107] L. Zhao, L. Jensen, G. C. Schatz: Nano Lettres **6** (2006) in press
[108] B. Vlckova, X. J. Gu, M. Moskovits: J. Phys. Chem. B **101**, 1588 (1997)
[109] B. W. Gregory, B. K. Clark, J. M. Standard, A. Avila: J. Phys. Chem. B **105**, 4684 (2001)
[110] N. Felidj, J. Aubard, G. Levi, J. R. Krenn, A. Hohenau, G. Schider, A. Leitner, F. R. Aussenegg: Appl. Phys. Lett. **82**, 3095 (2003)
[111] C. G. Blatchford, J. R. Campbell, J. A. Creighton: Surf. Sci. **120**, 435 (1982)
[112] R. P. Van Duyne, J. C. Hulteen, D. A. Treichel: J. Chem. Phys. **99**, 2101 (1993)
[113] K. U. Von Raben, R. K. Chang, B. L. Laube, P. W. Barber: J. Phys. Chem. **88**, 5290 (1984)
[114] D. A. Weitz, S. Garoff, T. J. Gramila: Opt. Lett. **7**, 168 (1982)
[115] M. Kerker, O. Siiman, D. S. Wang: J. Phys. Chem. **88**, 3168 (1984)
[116] D. Fornasiero, F. Grieser: J. Chem. Phys. **87**, 3213 (1987)
[117] H. Feilchenfeld, O. Siiman: J. Phys. Chem. **90**, 2163 (1986)
[118] C. L. Haynes, R. P. Van Duyne: J. Phys. Chem. B **107**, 7426 (2003)
[119] W. A. Weimer, M. J. Dyer: Appl. Phys. Lett. **79**, 3164 (2001)
[120] S. J. Oldenburg, S. L. Westcott, R. D. Averitt, N. J. Halas: J. Chem. Phys. **111**, 4729 (1999)
[121] T. R. Jensen, M. D. Malinsky, C. L. Haynes, R. P. van Duyne: J. Phys. Chem. B **104**, 10549 (2000)
[122] D. E. Aspnes, A. A. Studna: Phys. Rev. B, Condens Matter Mater. Phys. **27**, 985 (1983)
[123] M. Grimsditch, M. Cardona: Basic Research, Phys. Stat. Solidi B **102**, 155 (1980)
[124] A. D. McFarland, R. P. Van Duyne: Nano Lett. **3**, 1057 (2003)
[125] A. J. Haes, R. P. Van Duyne: J. Am. Chem. Soc. **124**, 10596 (2002)

Index

Electromagnetic Theory of SERS

Mark I. Stockman

Department of Physics and Astronomy, Georgia State University,
Atlanta, Georgia 30340, USA
mstockman@gsu.edu

1 Introduction

Surface-enhanced Raman scattering (SERS) is one of the strongest and most enigmatic effects in physics and optics, see the excellent review [1]. It was discovered approximately thirty years ago [2, 3, 4]. The SERS is manifested as an enhancement by many orders of magnitude of the intensity of Raman radiation by molecules bound to nanorough metal surfaces and nanostructured metal systems such as colloidal clusters of noble metals.

Theories of SERS that immediately followed its discovery considered molecules bound to metal spheroids [5, 6, 7, 8]. Hemispheroids on flat surfaces was a model emulating rough surfaces [9]. A result of general interest of this paper is a formula expressing the SERS intensity enhancement for a molecule bound at a position r_0 in terms of local field $\boldsymbol{E}(r)$ at this location,

$$g^{\mathrm{R}} = \left(\frac{|\boldsymbol{E}(r)|^2}{|\boldsymbol{E}_0|^2} \right)^2 , \tag{1}$$

where $|\boldsymbol{E}_0|^2$ is the excitation electric optical-field amplitude. In this expression a factor of

$$I(r) \equiv \frac{|\boldsymbol{E}(r)|^2}{|\boldsymbol{E}_0|^2} \tag{2}$$

takes into account the enhancement of the excitation rate due to the local field intensity increased by $I(r)$ times; yet another such factor describes the enhancement of the *Raman radiation* due to the nanosystem working as a resonant transmitting nanoantenna. Remarkably, the enhancement of local fields and the antenna transmission gain are the same.[1] We will return to this formula and compare it to our results repeatedly in this Chapter in conjunction with (44), (45), (46) and (50). To say in advance, it is not precisely reproduced by our exact result (44) but can serve as a resonable order-of-magnitude estimate.

---

[1] Obviously, the second factor I should be at the Raman-shifted frequency. However, this shift is normally small and can be neglected compared to the plasmonic resonance width in metal nanosystems.

K. Kneipp, M. Moskovits, H. Kneipp (Eds.): Surface-Enhanced Raman Scattering – Physics and Applications, Topics Appl. Phys. **103**, 47–66 (2006)
© Springer-Verlag Berlin Heidelberg 2006

The common problem with SERS models based on spheroids is that the enhancement, in accord with (1), is a pronounced plasmon-resonance peak, while experimentally when the SERS enhancement is giant, $g^R \gtrsim 10^5$, its spectral maximum is broad and shifted to the red (near-infrared) region [1]. To explain such a behavior, it has been proposed that it is due to aggregates of nanoparticles. First, the simplest of such aggregates, a dimer of nanospheres, was considered in [10]. There was a significant red shift of the SERS enhancement predicted in such dimers when the surface-to-surface distance between the individual spheres becomes small relative to their sizes. However, the enhancement spectral peaks still were too narrow, and the magnitude of enhancement was not as large as in experiments.

We have proposed that fractal clusters, which are experimentally the most efficient enhancers of SERS [1, 11], have specific properties conducive to the giant SERS enhancement [12, 13, 14]. One of these properties is the giant fluctuations of the local fields in fractals [15]. Because the SERS enhancement is quadratic in the local-field intensity enhancement, it is increased by the giant fluctuations. The spectrum of the surface-plasmon resonances in fractals does allow one to explain the broad spectrum of the SERS enhancement and its increase toward the red spectral region [13], see also Fig. 1 and its corresponsing discussion below in this Chapter.

The other reason for the SERS efficiency of fractals is their self-similarity, which is the defining property of fractals. The self-similarity leads to the transfer of the excitation down the scale of sizes enhancing its intensity [14], see Fig. 5 and its corresponding discussion.

One of the most remarkable and important developments in the field of SERS has been the discovery of the single-molecule SERS [16, 17] that has later been confirmed and studied in a body of experiments, see [11] in particular. In the case of the single-molecule SERS, the enhancement is truly giant, $g^R \gtrsim 10^{12}$ to 10^{14}. Such a giant enhancement can be explained in the framework of the electromagnetic theory only by combination of the enhancement factors: resonant enhancement, giant fluctuatons, and transfer of excitation down the ladder of spatial scales. We simulate such processes within the model of self-similar nanolenses built as an aggregate of a few nanospheres [14]. A comprehensive theory of the single-molecule SERS in fractal clusters and other complex systems is still to be developed.

We note that in this Chapter we do not consider the chemical mechanism of enhancement that is due to the formation of chemical bonds and hybridization of electrons between the molecular orbitals and metal Bloch functions. This mechanism, though quite possible, should be highly specific and sensitive to the purity, chemical modification, and physisorption at the metal surfaces. In contrast, the present electrodynamic mechanism is relatively insensitive to these factors and quite universal.

In this Chapter, we will also leave out the great application potential of the SERS. This is discussed abundantly in other Chapters of this book.

We point out only that the SERS is widely used in ultrasensitive sensors of chemical and biological objects, see, e.g., [18, 19, 20]

This Chapter begins with Sect. 2 devoted to the theory of SERS enhancement. This section contains Sect. 2.1 presenting Green's function spectral theory of optical responses of nanosystems. Next, Sect. 2.2 develops a general theory of the SERS enhancement based on Green's function method. Finally, Sect. 3 is devoted to numerical computations, results, and illustrations. Every effort has been made to make this theory as self-contained as possible. Its presentation is moderately paced and rather detailed, so it is designed for both theorists and experimentalists in the field. The Green's function method employed has allowed us to give a unified and, actually, simple presentation of the results where, in particular, a closed general expression for the SERS enhancement (44) is obtained.

2 Spectral Theory of SERS Enhancement

2.1 Spectral Expansion of Local Fields and Green's Function

We will consider a metal nanosystem embedded into a dielectric host medium. The metal is characterized by permittivity (dielectric function) $\varepsilon_m(\omega)$ that depends on the optical frequency ω. The permittivity of the dielectric is denoted as ε_d. We assume that the entire size of the nanosysten is much smaller than the light wavelength $\lambda = 2\pi c/\omega$, where c is the speed of light. In this case, we can use the quasistatic approximation where the electrostatic potential $\varphi(\boldsymbol{r})$ satisfies the continuity equation

$$\frac{\partial}{\partial \boldsymbol{r}} \varepsilon(\boldsymbol{r}) \frac{\partial}{\partial \boldsymbol{r}} \varphi(\boldsymbol{r}) = 0 \,, \tag{3}$$

and the dielectric function varying in space is expressed as

$$\varepsilon(\boldsymbol{r}) = \varepsilon_m(\omega)\Theta(\boldsymbol{r}) + \varepsilon_d[1 - \Theta(\boldsymbol{r})] \,. \tag{4}$$

Here $\Theta(\boldsymbol{r})$ is the so-called characteristic function of the system, which is equal to 1 when $\boldsymbol{r}$ belongs to the metal and 0 otherwise. We will be solving this equation following the spectral theory developed in [21, 22, 23].

Consider a system excited by an external field with potential $\varphi_0(\boldsymbol{r})$ at an optical frequency ω. This potential is created by external charges and, therefore, satisfies the Laplace equation within the system,

$$\frac{\partial^2}{\partial \boldsymbol{r}^2} \varphi_0(\boldsymbol{r}) = 0 \,. \tag{5}$$

We present the field potential as

$$\varphi(\boldsymbol{r}) = \varphi_0(\boldsymbol{r}) + \varphi_1(\boldsymbol{r}) \,, \tag{6}$$

where $\varphi_1(\boldsymbol{r})$ is the local field that satisfies homogeneous *Dirichlet* or *Neumann* conditions at a surface S surrounding the system, which we can set as

$$\varphi_1(\boldsymbol{r})\boldsymbol{n}(\boldsymbol{r})\frac{\partial}{\partial\boldsymbol{r}}\varphi_1(\boldsymbol{r})\bigg|_{\boldsymbol{r}\in S} = 0 \,, \tag{7}$$

with $\boldsymbol{n}(\boldsymbol{r})$ denoting a normal to surface S at a point of $\boldsymbol{r}$. While the boundary conditions (7) are essential and necessary to define the eigenproblem, we will also assume that the metal nanosystem is completely within the boundary S and does not touch it at any point. This leads to the following auxiliary boundary conditions

$$\Theta(\boldsymbol{r})\bigg|_{\boldsymbol{r}\in S} = 0, \quad \frac{\partial\Theta(\boldsymbol{r})}{\partial\boldsymbol{r}}\bigg|_{\boldsymbol{r}\in S} = 0 \,. \tag{8}$$

These auxiliary conditions are not principal and can always be satisfied for a nanosystem by a proper choice of the boundary surface outside of the metal subsystem, but significantly simplify the following formalism.

Substituting (6) into (3) and taking (4) and (5) into account, we obtain a second-order elliptic equation with the right-hand side that describes the external excitation source,

$$\frac{\partial}{\partial\boldsymbol{r}}\Theta(\boldsymbol{r})\frac{\partial}{\partial\boldsymbol{r}}\varphi_1(\boldsymbol{r}) - s(\omega)\frac{\partial^2}{\partial\boldsymbol{r}^2}\varphi_1(\boldsymbol{r}) = -\frac{\partial}{\partial\boldsymbol{r}}\Theta(\boldsymbol{r})\frac{\partial}{\partial\boldsymbol{r}}\varphi_0(\boldsymbol{r}) \,, \tag{9}$$

where we introduced spectral parameter [21]

$$s(\omega) \equiv \frac{\varepsilon_{\mathrm{d}}}{\varepsilon_{\mathrm{d}} - \varepsilon_{\mathrm{m}}(\omega)} \,. \tag{10}$$

To find the Green's function of this equation, we introduce eigenmodes $\varphi_n(\boldsymbol{r})$ and the corresponding eigenvalues s_n as a solution of the following generalized eigenproblem,

$$\frac{\partial}{\partial\boldsymbol{r}}\Theta(\boldsymbol{r})\frac{\partial}{\partial\boldsymbol{r}}\varphi_n(\boldsymbol{r}) - s_n\frac{\partial^2}{\partial\boldsymbol{r}^2}\varphi_n(\boldsymbol{r}) = 0 \,, \tag{11}$$

where the eigenfunctions $\varphi_n(\boldsymbol{r})$ satisfy the homogeneous *Dirichlet–Neumann* boundary conditions (7). Importantly, this eigenproblem depends only on the geometry of the system, but not on its material composition. Therefore, for a given geometry it can be solved only once, providing a basis for finding the optical responses of the system for all frequencies and arbitrary material compositions.

The physical eigenmodes defined by this equation are surface plasmons whose complex frequencies $\omega_n + \mathrm{i}\gamma_n$ are found from the complex equation [see also (19) and the corresponding discussion]

$$s(\omega_n + \mathrm{i}\gamma_n) = s_n \,. \tag{12}$$

For weak relaxation, $\gamma_n \ll \omega_n$, one finds that the real part of the surface-plasmon frequency ω_n satisfies the equation

$$\mathrm{Re}[s(\omega_n)] = s_n , \tag{13}$$

and that the surface-plasmon spectral width, γ_n, is expressed as

$$\gamma_n = \frac{\mathrm{Im}[s(\omega_n)]}{s'_n} , \quad s'_n \equiv \left. \frac{\partial \mathrm{Re}[s(\omega)]}{\partial \omega} \right|_{\omega=\omega_n} . \tag{14}$$

Note that these classical surface-plasmons have been quantized in [24] in connection with the prediction of a spaser, a nanoscale counterpart of a laser.

The Kramers–Kronig dispersion relations result in the following inequalities for the dielectric function of the metal [25]

$$\mathrm{Re}\frac{\partial \varepsilon_\mathrm{m}(\omega)}{\partial \omega} > 0 , \quad \mathrm{Re}\frac{\partial \omega \varepsilon_\mathrm{m}(\omega)}{\partial \omega} > 1 , \quad \mathrm{Re}\frac{\partial \omega^2 [\varepsilon_\mathrm{m}(\omega) - 1]}{\partial \omega} > 0 . \tag{15}$$

From the first of these inequalities and the fact that $\mathrm{Im}\varepsilon_\mathrm{m}(\omega) > 0$, it follows that

$$s'_n > 0 , \quad \gamma_n > 0 . \tag{16}$$

This inequality guarantees decay of the physical surface-plasmons with time and, consequently, causality.

Consider two functions, $\psi_1(\boldsymbol{r})$, $\psi_2(\boldsymbol{r})$ that are differentiable. For such functions, we can define a scalar product as the following operation:

$$(\psi_1|\psi_2) = \int_V \left[\frac{\partial}{\partial \boldsymbol{r}} \psi_2^*(\boldsymbol{r}) \right] \left[\frac{\partial}{\partial \boldsymbol{r}} \psi_1(\boldsymbol{r}) \right] \mathrm{d}^3 r , \tag{17}$$

where V is the volume of the system. This construction possesses all the necessary and sufficient properties of a scalar product: it is a binary, Hermitian self-adjoined, and positive-defined operation. From (7) and (11) it follows that these eigenfunctions are orthogonal with respect to the scalar product of (17) and can be normalized:

$$(\varphi_n|\varphi_m) = \delta_{nm} , \tag{18}$$

all eigenvalues s_n are real, and all eigenfunctions $\varphi_n(\boldsymbol{r})$ can be chosen to be real. It can also be shown that all eigenvalues are real numbers and $1 \geq s_n \geq 0$. These eigenfunctions form a complete basis in the space of all twice-differentiable functions satisfying the boundary conditions of (7), which we will denote $\mathbb{R}$.

The retarded Green's function $\bar{G}^\mathrm{r}(\boldsymbol{r}, \boldsymbol{r}'; \omega)$ of (9) satisfies the same equation with the δ-function right-hand side,

$$\left[\frac{\partial}{\partial \boldsymbol{r}} \Theta(\boldsymbol{r}) \frac{\partial}{\partial \boldsymbol{r}} - s(\omega) \frac{\partial^2}{\partial r^2} \right] \bar{G}^\mathrm{r}(\boldsymbol{r}, \boldsymbol{r}'; \omega) = \delta(\boldsymbol{r} - \boldsymbol{r}') , \tag{19}$$

Requiring that this Green's function belongs to $\mathbb{R}$, we expand it over the eigenfunctions φ_n with some unknown coefficients a_n,

$$\bar{G}^{\mathrm{r}}(\boldsymbol{r},\boldsymbol{r}';\omega) = \sum_n a_n(\boldsymbol{r}';\omega)\varphi_n(\boldsymbol{r}) \,. \tag{20}$$

Substituting this expansion into (19), multiplying by $\varphi_n(\boldsymbol{r})$ and integrating over V, we find a_n and, correspondingly, the expansion for the Green functions as

$$\bar{G}^{\mathrm{r}}(\boldsymbol{r},\boldsymbol{r}';\omega) = \sum_n \frac{\varphi_n(\boldsymbol{r})\,\varphi_n(\boldsymbol{r}')^*}{s(\omega) - s_n} \,. \tag{21}$$

This expression for the Green's function is exact (within the quasistatic approximation) and contains the maximum information on the linear responses of a nanosystem to an arbitrary excitation field at any frequency. It satisfies all the general properties of Green's functions due to the analytical form of (21) as an expansion over the eigenmodes (surface plasmons). Each such eigenmode is described by the corresponding pole of the Green's function. The frequency and spectral width of an n-th pole is given by (13) and (14). Because of inequality (16), all these poles are situated in the lower complex half-plane of ω. Consequently, the corresponding time-dependent Green's function, which describes ultrafast temporal dynamics of plasmonic systems, [23, 26, 27, 28] expressed as

$$\bar{G}^{\mathrm{r}}(\boldsymbol{r},\boldsymbol{r}';t) = \int \bar{G}^{\mathrm{r}}(\boldsymbol{r},\boldsymbol{r}';\omega)\exp\left(-i\omega t\right)\frac{d\omega}{2\pi} \tag{22}$$

is causal (retarded), i.e., $\bar{G}^{\mathrm{r}}(\boldsymbol{r},\boldsymbol{r}';t) = 0$ for $t < 0$. Due to this causality, it satisfies the *Kramers–Kronig* dispersion relations [25]. By the mere form of the spectral expansion (21), this Green's function satisfies all other exact analytical properties. This guarantees that in numerical simulation, it will possess these properties *irrespective of the numerical precision with which the eigenproblem is solved*. This insures an exceptional numerical stability of computational Green's function approaches (see Sect. 3).

It is useful to write an expression for this Green's function that is asymptotically valid near its poles, which can be obtained from (13) and (14) as

$$\bar{G}^{\mathrm{r}}(\boldsymbol{r},\boldsymbol{r}';\omega) = \frac{1}{s'(\omega)}\sum_n \frac{\varphi_n(\boldsymbol{r})\varphi_n(\boldsymbol{r}')^*}{\omega - \omega_n + i\gamma_n} \,. \tag{23}$$

For a frequency close to an n-th pole, the Green's function increases by a large factor of

$$Q_n = \frac{\omega_n}{\gamma_n} \,, \tag{24}$$

which is called the resonance (surface-plasmon) quality factor. To estimate this quality factor, we turn to the last of inequalities (15) where we take into account that, instead of the dielectric function, one should use the *relative* dielectric function of the metal in the dielectric host, i.e., replace $\varepsilon_{\mathrm{m}} \to \varepsilon_{\mathrm{m}}/\varepsilon_{\mathrm{d}}$. In such a way, we find,

$$Q_n \geq Q(\omega_n), \quad Q(\omega) = \frac{2\varepsilon_{\mathrm{d}}}{\mathrm{Res}(\omega)\mathrm{Im}\varepsilon_{\mathrm{m}}(\omega)} \gg 1 \,. \tag{25}$$

The lower-limit estimate of the quality factor, $Q(\omega)$, is function of frequency only (it does not depend on the surface-plasmon eigenfunctions). The inequality $Q(\omega) \gg 1$ is valid in the visible to near-IR spectral region for silver and other coinage metals [29]; it is responsible for the large enhancement of local optical fields and optical processes of plasmonic nanostructures.

Another very important property of the spectral Green's function method is the complete separation of the geometrical and material properties of the system. The eigenfunctions φ_n and eigenvalues s_n, which enter (21), depend only on geometry of the system and are completely independent of its material composition and excitation frequency. Therefore, the eigenproblem (11) should be solved only once for any given geometry. The material composition and excitation frequency enter the Green's function expression (21) only through the spectral parameter $s(\omega)$ that is a known function of the frequency if the dielectric permittivities of the system materials are known.

Once the Green's function is found, using (9) and (21), the local optical field potential is found as a contraction of this Green's function with the excitation potential $\varphi_0(\boldsymbol{r})$ as

$$\varphi_1(\boldsymbol{r}) = -\int_V \bar{G}^{\mathrm{r}}(\boldsymbol{r}, \boldsymbol{r}'; \omega) \frac{\partial}{\partial \boldsymbol{r}'} \Theta(\boldsymbol{r}') \frac{\partial}{\partial \boldsymbol{r}'} \varphi_0(\boldsymbol{r}') \, \mathrm{d}^3 r' \,. \tag{26}$$

From (6) and (26) using (8) and Gauss theorem, we obtain an expression for the field potential $\varphi(\boldsymbol{r})$ as a functional of the external (excitation) potential $\varphi_0(\boldsymbol{r})$,

$$\varphi(\boldsymbol{r}) = \varphi_0(\boldsymbol{r}) - \int_V \varphi_0(\boldsymbol{r}') \frac{\partial}{\partial \boldsymbol{r}'} \Theta(\boldsymbol{r}') \frac{\partial}{\partial \boldsymbol{r}'} \bar{G}^{\mathrm{r}}(\boldsymbol{r}, \boldsymbol{r}'; \omega) \, \mathrm{d}^3 r' \,. \tag{27}$$

Finally, differentiating this, we obtain a closed expression for the optical electric field $\boldsymbol{E}(\boldsymbol{r})$ as a functional of the excitation (external) field $\boldsymbol{E}^{(0)}(\boldsymbol{r})$ as

$$E_\alpha(\boldsymbol{r}) = E_\alpha^{(0)}(\boldsymbol{r}) + \int_V G_{\alpha\beta}^{\mathrm{r}}(\boldsymbol{r}, \boldsymbol{r}'; \omega) \Theta(\boldsymbol{r}') E_\beta^{(0)}(\boldsymbol{r}') \, \mathrm{d}^3 r' \,, \tag{28}$$

where $\alpha, \beta, \ldots$ are Euclidean vector indices ($\alpha, \beta, \ldots = x, y, z$) with summation over repeated indices implied; the fields are

$$\boldsymbol{E}(\boldsymbol{r}) = -\frac{\partial \varphi(\boldsymbol{r})}{\partial \boldsymbol{r}} \,, \quad \boldsymbol{E}^{(0)}(\boldsymbol{r}) = -\frac{\partial \varphi_0(\boldsymbol{r})}{\partial \boldsymbol{r}} \,, \tag{29}$$

and the tensor (dyadic) retarded Green's function is defined as

$$G^{\mathrm{r}}_{\alpha\beta}(\boldsymbol{r},\boldsymbol{r}';\omega) = \frac{\partial^2}{\partial r_\alpha \partial r_\beta}\bar{G}^{\mathrm{r}}(\boldsymbol{r},\boldsymbol{r}';\omega)\,. \tag{30}$$

One of the exact properties of this Green's function is its Hermitian symmetry,

$$G^{\mathrm{r}}_{\alpha\beta}(\boldsymbol{r},\boldsymbol{r}';\omega) = G^{\mathrm{r}}_{\beta\alpha}(\boldsymbol{r}',\boldsymbol{r};-\omega)^*\,. \tag{31}$$

Additionally, taking into account that the eigenproblem (11) is real and symmetric, its eigenvalues s_n are always real, and the eigenfunctions φ_n can always be chosen real. Therefore, Green's function satisfies also a simpler, permutation symmetry

$$G^{\mathrm{r}}_{\alpha\beta}(\boldsymbol{r},\boldsymbol{r}';\omega) = G^{\mathrm{r}}_{\beta\alpha}(\boldsymbol{r}',\boldsymbol{r};\omega)\,. \tag{32}$$

If the excitation is an optical wave, its wavefront is flat on the scale of the nanosystem, i.e., $\boldsymbol{E}^{(0)} = \text{const}$. Then from (28) we get

$$E_\alpha(\boldsymbol{r}) = [\delta_{\alpha\beta} + g_{\alpha\beta}(\boldsymbol{r},\omega)]E^{(0)}_\beta\,, \tag{33}$$

where the local field enhancement (tensorial) factor is a contraction of the retarded dyadic Green's function,

$$g_{\alpha\beta}(\boldsymbol{r},\omega) = \int_V G^{\mathrm{r}}_{\alpha\beta}(\boldsymbol{r},\boldsymbol{r}';\omega)\Theta(\boldsymbol{r}')\,\mathrm{d}^3 r'\,. \tag{34}$$

2.2 SERS Enhancement Factor in Green's Function Theory

Raman scattering is an incoherent process that we will describe by Raman polarization $\boldsymbol{P}^{\mathrm{R}}(\boldsymbol{r})$. The corresponding quasielectrostatic equation for the Raman-field potential $\varphi^{\mathrm{R}}(\boldsymbol{r})$ at the Raman frequency ω^{R} has the form

$$\frac{\partial}{\partial\boldsymbol{r}}\Theta(\boldsymbol{r})\frac{\partial}{\partial\boldsymbol{r}}\varphi^{\mathrm{R}}(\boldsymbol{r}) - s(\omega^{\mathrm{R}})\frac{\partial^2}{\partial r^2}\varphi^{\mathrm{R}}(\boldsymbol{r}) = \frac{4\pi}{\varepsilon_{\mathrm{d}}}\frac{\partial}{\partial\boldsymbol{r}}P^{\mathrm{R}}(\boldsymbol{r})\,. \tag{35}$$

Employing Green's function approach of Sect. 2.1, we obtain an integral solution of this equation

$$\varphi^{\mathrm{R}}(\boldsymbol{r}) = \frac{4\pi}{\varepsilon_{\mathrm{d}}}\int_V \bar{G}^{\mathrm{r}}(\boldsymbol{r},\boldsymbol{r}';\omega)\frac{\partial}{\partial\boldsymbol{r}'}P^{\mathrm{R}}(\boldsymbol{r}')\,\mathrm{d}^3 r'\,. \tag{36}$$

Assuming that $\boldsymbol{P}^{\mathrm{R}}(\boldsymbol{r}')$ vanishes at the boundary S, which is consistent with (8), and using the Gauss theorem (integration by parts), we obtain a Green's function expression for the Raman-frequency optical electric-field vector

$$E^{\mathrm{R}}_\alpha(\boldsymbol{r}) = \frac{4\pi}{\varepsilon_{\mathrm{d}}}\int_V G^{\mathrm{r}}_{\alpha\beta}(\boldsymbol{r},\boldsymbol{r}';\omega^{\mathrm{R}})P^{\mathrm{R}}_\beta(\boldsymbol{r}')\,\mathrm{d}^3 r'\,. \tag{37}$$

We consider Raman scattering by molecules attached to the surface of the metal nanosystem. Because Raman scattering is an incoherent process, i.e., $P_\beta^{\mathrm{R}}(\boldsymbol{r})$ randomly changes its phase from one Raman-emitting molecule to another, the *intensity* of the Raman radiation, and not its amplitude, is additive for different molecules. Therefore, it is sufficient to find the intensity of the Raman scattering from a single molecule positioned, say, at a point $\boldsymbol{r}_0$. That will resolve the problem of the single-molecule Raman scattering directly. If there are many such molecules, then the corresponding Raman intensity can always be found by appropriate summation over $\boldsymbol{r}_0$.

Assuming that the size of the Raman molecule is negligible with respect to the nanosystem, we can substitute

$$\boldsymbol{P}^{\mathrm{R}}(\boldsymbol{r}) = \delta(\boldsymbol{r} - \boldsymbol{r}_0)\boldsymbol{d}^{\mathrm{R}}(\boldsymbol{r}_0)\,, \tag{38}$$

where $\boldsymbol{d}^{\mathrm{R}}(\boldsymbol{r}_0)$ is the Raman transitional dipole (oscillating at frequency ω^{R}) for a molecule at a point of $\boldsymbol{r}_0$. Then from (37), we obtain

$$E_\alpha^{\mathrm{R}}(\boldsymbol{r}) = \frac{4\pi}{\varepsilon_{\mathrm{d}}} G_{\alpha\beta}^{\mathrm{r}}(\boldsymbol{r}, \boldsymbol{r}_0; \omega^{\mathrm{R}}) d_\beta^{\mathrm{R}}(\boldsymbol{r}_0)\,. \tag{39}$$

This electric field acts on the metal component whose dielectric susceptibility is

$$\chi_{\mathrm{m}}(\omega^{\mathrm{R}}) = \frac{\varepsilon_{\mathrm{m}}(\omega^{\mathrm{R}}) - \varepsilon_{\mathrm{d}}}{4\pi} = -\frac{\varepsilon_{\mathrm{d}}}{4\pi s(\omega_{\mathrm{R}})}\,. \tag{40}$$

This induces the total Raman dipole moment

$$\boldsymbol{D}^{\mathrm{R}} = \boldsymbol{d}^{\mathrm{R}}(\boldsymbol{r}_0) - \frac{\varepsilon_{\mathrm{d}}}{4\pi s(\omega_{\mathrm{R}})} \int_V \Theta(\boldsymbol{r}) \boldsymbol{E}^{\mathrm{R}}(\boldsymbol{r})\, \mathrm{d}^3 r\,, \tag{41}$$

where factor $\Theta(\boldsymbol{r})$ limits the integration volume to the metal component only.

Substituting the expression of (39) and taking into account the symmetry (32) of the Green's function, we obtain from (41) for the total radiating (transitional) Raman dipole,

$$D_\alpha^{\mathrm{R}} = \left[\delta_{\alpha\beta} - \frac{1}{s(\omega_{\mathrm{R}})} g_{\beta\alpha}(\boldsymbol{r}_0; \omega^{\mathrm{R}})\right] d_\beta^{\mathrm{R}}(\boldsymbol{r}_0)\,. \tag{42}$$

Here, the first term in the square brackets corresponds to the Raman dipole induced by the local optical field, and the second describes the antenna effect of the entire nanostructure that renormalizes and enhances this radiating Raman dipole.

The Raman dipole $\boldsymbol{d}^{\mathrm{R}}$ is proportional to the local field $\boldsymbol{E}(\boldsymbol{r}_0)$ that excites it with some Raman polarizability $\alpha^{\mathrm{R}}(\omega)$, which we for simplicity consider isotropic (scalar). Then, using (33) for the local optical field in combination with (42), we obtain the total radiating Raman dipole in terms of the excitation field,

$$D_\alpha^{\mathrm{R}} = \alpha^{\mathrm{R}}(\omega) \left[\delta_{\alpha\beta} - \frac{1}{s(\omega_{\mathrm{R}})} g_{\beta\alpha}(\boldsymbol{r}_0; \omega^{\mathrm{R}})\right] [\delta_{\beta\gamma} + g_{\beta\gamma}(\boldsymbol{r}_0, \omega)] E_\gamma^{(0)}\,. \tag{43}$$

For certainty, we assume that the excitation field $\boldsymbol{E}^{(0)}$ is linearly z-polarized. Then, enhancement coefficient g^{R} for the Raman radiation intensity is found from (43) as

$$g^{\mathrm{R}} = \sum_{\alpha=x,y,z} \left| \sum_{\beta=x,y,z} \left[\delta_{\alpha\beta} - \frac{1}{s(\omega_{\mathrm{R}})} g_{\beta\alpha}(\boldsymbol{r}_0; \omega^{\mathrm{R}}) \right] \left[\delta_{\beta z} + g_{\beta z}(\boldsymbol{r}_0, \omega) \right] \right|^2, \quad (44)$$

where the summations over vector indices are shown explicitly; of course, there is no summation over the z vector index.

This is the general expression for the enhancement of Raman radiation of a molecule bound to a metal nanostructure, which is our principal analytical result. Note that the assumption of the z-polarization of the excitation radiation has been made for simplicity only, is not principal, and can easily be eliminated, which would lead to a somewhat more cumbersome expression. Note that in (44), $g_{\beta z}(\boldsymbol{r}_0; \omega^{\mathrm{R}})$ describes the enhancement of the local field, while $g_{\beta\alpha}(\boldsymbol{r}_0; \omega^{\mathrm{R}})/s(\omega^{\mathrm{R}})$ describes the enhancement of the Raman dipole due to the antenna effect of the nanostructure. For a strong enhancement case, these factors are large, and (44) simplifies to

$$g^{\mathrm{R}} = \frac{1}{|s(\omega_R)|^2} \sum_{\alpha=x,y,z} \left| \sum_{\beta=x,y,z} g_{\beta\alpha}(\boldsymbol{r}_0; \omega^{\mathrm{R}}) g_{\beta z}(\boldsymbol{r}_0, \omega) \right|^2. \quad (45)$$

Often, a further simplified formula is used (as in [30]) that expresses the Raman enhancement as the product of the local intensity enhancement factors at the excitation and Raman-shifted frequencies at the position of the Raman-emitting molecule. As we noted in the Introduction, proposed in [9], this formula is given by (1). In the present notations, such a further-simplified expression is

$$g^{\mathrm{R}} = \sum_{\alpha=x,y,z} |g_{\alpha z}(\boldsymbol{r}_0; \omega^{\mathrm{R}})|^2 \sum_{\alpha=x,y,z} |g_{\alpha z}(\boldsymbol{r}_0; \omega)|^2. \quad (46)$$

This expression is close to (45) in the sense that it correctly takes into account the two multiplicative sources of the enhancement: the enhanced local fields inducing the Raman polarization, and the secondary enhancement due to the resonant antenna effect of the metal nanostructure. However, (46) has a different tensorial structure and lacks a prefactor in comparison with the correct approximation (45) and general expression (44).

3 Numerical Computations and Results

An order-of-magnitude estimate for the SERS enhancement factor (45) can be obtained by considering a resonance contribution to the Green's function as follows from (23) and (24),

$$g^{\mathrm{R}} \sim Q^4. \quad (47)$$

This estimate is frequently used, but a word of caution is appropriate here. This estimate takes into account only the energy denominator in (23) and completely disregards the effects of the surface-plasmon localization as described by the eigenfunctions in the numerator of (23). Therefore this formula can be used only for initial, very crude estimates that may not be valid even by the order of magnitude. Nevertheless, to get an initial idea about what can be expected we start with it and then will consider results of the qualitative theory.

The quality factor of all metals is highest for silver, where it changes between ~ 10 in the visible to ~ 100 in the near-IR. Thus one can expect the single-molecule SERS enhancement up to $g^R \sim 10^8$. In the single-molecule SERS experiments [11, 16, 17], the estimated enhancement is $g^R \sim 10^{12}$ to 10^{14}, i.e., significantly greater. This certainly indicates the presence of other enhancement mechanisms. One of them is the chemical enhancement that is due to the hybridization of the molecular and metal electron orbitals. This mechanism is possible though it should be very sensitive to the physical and chemical state of the surface, unlike the electromagnetic mechanism. The discussion of the chemical mechanism lies outside the scope of this Chapter.

Apart from the above-mentioned resonant electromagnetic enhancement, there is another electromagnetic mechanism that is based on the structural enhancement. This will appear only in complex systems (such as fractal clusters of colloidal metals, specially designed nanolenses, rough surfaces, etc.) on which we will concentrate in this section.

Note that the site-averaged (or ensemble-averaged) enhancement can be estimated if one takes into account that a probability for a molecule to resonantly interact with a surface-plasmon localized at a given arbitrary site is small proportionally to the width of the plasmonic resonance. This removes one power of the quality factor, leading to an estimate

$$g^R \sim Q^3. \tag{48}$$

In this case, the estimate $g^R \lesssim 10^6$ appears to be in line with the experiments [1].

3.1 SERS Enhancement in Fractals in Dipolar Approximation

It is known experimentally that a large SERS enhancement is observed in fractal clusters of colloidal silver [1]. The colloidal clusters are large enough (thousands or more of metal–nanosphere monomers), so the numerical solution of the eigenmode problem of Sect. 2.1 is still beyond existing computer capabilities for reasonable resources used. Therefore, for fractals we used the spectral theory where only dipole fields of the monomers were taken into account [13].

The dependence of the enhancement factor g^R averaged over binding sites in individual clusters and over an ensemble of such clusters is shown in Fig. 1.

g^R

120000
100000
80000
60000
40000
20000
0

○ Theory
▪ Experiment

400   450   500   550   600

Wavelength λ (nm)

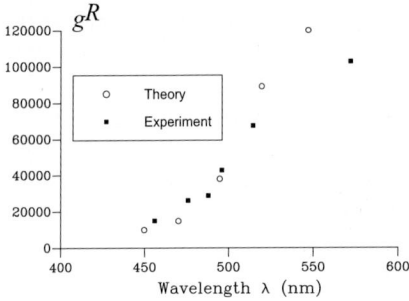

Fig. 1. SERS enhancement factor g^R averaged over an ensemble of fractal clusters of silver as a function of excitation wavelength λ. Adapted from [13]

We use the dielectric permittivity for silver from [29]. Clusters are simulated in the cluster–cluster aggregation model [31, 32]. As we see from Fig. 1, the SERS enhancement dramatically increases from the blue region where $g^R \lesssim 10^4$ to the red spectral region where it grows to $g^R \gtrsim 10^5$. This is an expected range and behavior of g^R that is due to the increase of the Q factor toward the red region.

Despite the evident success of the dipolar approximation for fractal clusters in explaining the many-molecule SERS, a word of caution should be given. The dipole approximation yields fields that diverge at small distances and does not converge well at larger scale. This means that it only describes well the intermediate region of scales in large fractal clusters. Therefore the local field and SERS enhancement that it can reliably describe are not extremely large [cf. Fig. 1]. In contrast, the single-molecule Raman scattering is characterized by much higher enhancement, $g^R \sim 10^{12} - 10^{14}$ and, therefore, is likely to originate at the minimum scale of the system where the local fields are very strong but the dipolar approximation is inapplicable.

3.2 Single-Molecule SERS in Random Systems

From these data, it is also clear that the phenomenon of the single-molecule SERS cannot be explained by the averaged behavior of g^R. Here, we will explore a cause of the SERS enhancement associated with the giant fluctuations of local fields existing in random, especially fractal, systems [15]. These fluctuations play an important role in single-molecule SERS because g^R is proportional to the square of the local field intensity enhancement [cf. (44), (45) and (47)] despite the fact that Raman scattering is an optically linear effect. As a result, the strong fluctuations give higher average g^R and a very high *maximum* g^R. One can say that the single-molecule SERS is a result of a giant positive fluctuation at the site where the Raman molecule is bound. This peculiarity has been well understood starting from our early paper [13]. Recently, a special role of the geometric local anisotropy has been elucidated [33]. This theory is in line with our understanding of the plasmonic eigenmode distribution and localization as being exclusively controlled by the geometry of a nanosystem [see above in Sect. 2.1].

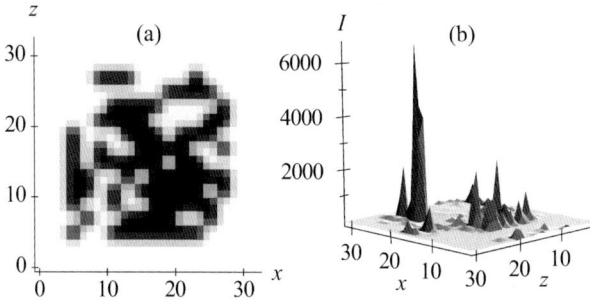

Fig. 2. (a) Topography of random planar composite (RPC). The grid used is $32 \times 32 \times 32$ cells. The monomer size is 2 grid steps. (b) Distribution of the intensity enhancement $I(\boldsymbol{r})$ in the plane of RPC at the surface of the metal for excitation frequency $\hbar\omega = 1.55\,\mathrm{eV}$

Unfortunately, many-multipole computations for large fractal clusters are still more complex than is possible to carry out with a realistic volume of computations at the present time. Therefore, we concentrate here on another model of random nanosystems, namely random planar composites (RPC). Such a composite is generated on a planar rectangular grid (say, in the xz-plane) by randomly filling grid cells with metal monomers with some probability until a given fill factor f is achieved. The monomers are supposed to be small metal cubes, to the system is in the xz-plane with small but finite thickness in the y-direction. An example of such a system, for which we have carried out computations presented below in this section, is shown in Fig. 2a. Note that the system is a thin plane layer of disordered metal but the field equations are three-dimensional as is their solution. To improve the numerical convergence and avoid singularities at the metal surface, the RPC is smoothed out with a Gaussian filter, as is shown in panel (a). The eigenmodes and Green's function are found numerically using a finite difference method of the third-order accuracy in the grid step, as described in [23]. In the quasistatic approximation, there is no dependence on the total size of the system, only on its form, since the wavelength λ is much larger than the system's size. Therefore we give all dimensions in grid steps. Realistically, the grid step is 1 nm to 2 nm.

We calculate the enhancement factor of the local field intensity (i.e., the local field intensity with the unit excitation field) as

$$I(\boldsymbol{r};\omega) = \sum_{\alpha=x,y,z} |g_{\alpha z}(\boldsymbol{r}_0;\omega)|^2 . \tag{49}$$

An example of the local field intensity for the RPC under consideration is shown in Fig. 2b. As one can see, it has a wide distribution of singular peaks

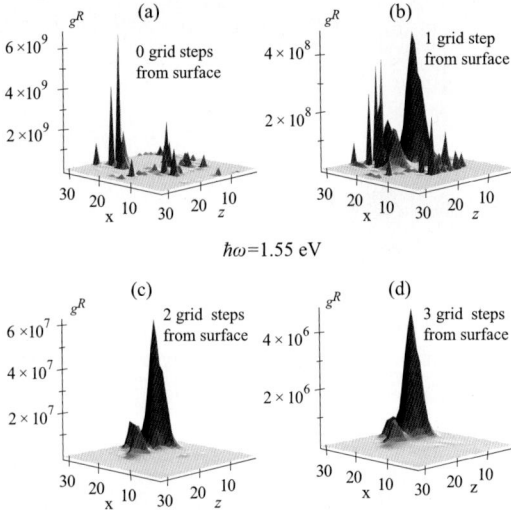

Fig. 3. SERS enhancement factor for single molecule situated at different planes above a metal surface as a function of the position in the plane. (**a**) The plane is at the metal surface, and (**b**)–(**d**) for 1–3 grid steps (1 nm to 6 nm) above this surface. The excitation frequency is $\hbar\omega = 1.55\,\text{eV}$, the Raman radiation frequency $\hbar\omega^{\text{R}} = 1.50\,\text{eV}$

(hot spots) where the energy of the local fields is localized.[2] The maximum SERS enhancement factor that can be estimated from this intensity according to (46) is $g^{\text{R}} \sim 10^7$ to 10^8.

We have calculated the enhancement factor of SERS from the exact formula (44) for the RPC system discussed above where the results are presented in Fig. 3 for g^{R} of a single molecule located at different planes above the metal surface. The frequency chosen of $1.55\,\text{eV}$ corresponds to $\lambda = 800\,\text{nm}$, i.e., to the radiation of a Ti:sapphire laser as one of the most common optical generators. As we see from panel (a), the enhancement is very large, up to $g^{\text{R}} \approx 6 \times 10^9$ for a molecule at the metal surface. This is a large value but still much less than $g^{\text{R}} \gtrsim 10^{12}$ needed for the observable single-molecule SERS. Note that this value is significantly greater than the simplified estimate obtained from (46) above in this section.

With the distance from the metal surface increased by just a few grid steps (note that the grid step defines the minimum scale of the system's roughness), the SERS enhancement is drastically, by orders of magnitude, suppressed. This reduction in g^{R} is stronger than exponential and is certainly due to very high multipoles of the surface plasmon eigenmodes involved. Similar

---

[2] To prevent possible misunderstandings, we note that the eigenmodes that are responsible for these peaks are singular but not Anderson-localized [22]. The statements in the literature about the Anderson localization of the plasmonic eigenmodes (see, e.g., [34]) are incorrect.

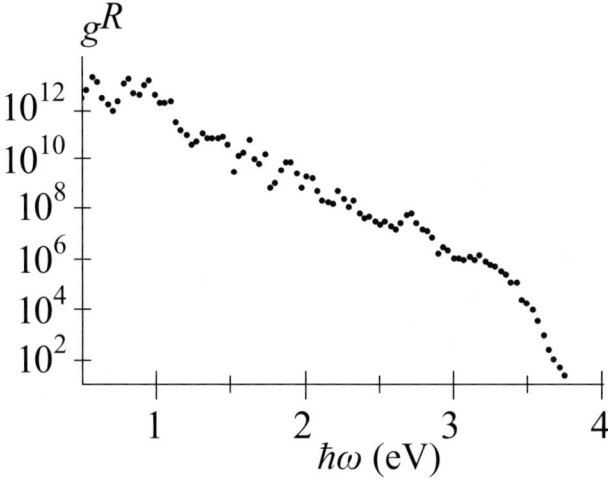

Fig. 4. SERS enhancement factor for a single molecule situated at "hottest spot" at a metal surface as a function of excitation frequency $\hbar\omega$. The Raman frequency shift $\hbar\Delta\omega^R = \hbar\omega - \hbar\omega^R = 0.05\,\text{eV}$

results have been obtained for other excitation frequencies (data not shown). Note that not only the maximum enhancement changes with the distance to the metal RPC surface, but also the spatial dependence of g^R experiences dramatic redistribution, i.e., the eigenmodes contributing at various distances are completely different.

For the same system, we show in Fig. 4 the frequency dependence of the maximum enhancenment factor of the single-molecule SERS. For each frequency, the Raman molecule is put at the position of the maximum enhancement (the Raman "hottest spot"). The Raman frequency shift $\Delta\omega^R = \hbar\omega - \hbar\omega^R = 0.05\,\text{eV}$ is constant for all these spectral points. As we see, there is a dramatic increase of the single-molecule SERS for the frequency changing from the near-ultraviolet to near-IR, where it reaches the values $g^R \sim 10^{12}$ typical for the single-molecule SERS. However, in the entire visible region, $g^R \lesssim 10^9$, significantly less than required to observe the SERS from a single molecule.

3.3 Single-Molecule SERS in Nanosphere Nanolens

The fact is that the fractals are the most efficient enhancers of the SERS [1, 11, 16]. This is likely to be related to two reasons: (i) self-similarity of fractals on average and (ii) giant fluctuations of local optical fields. As we argued, the self-similarity in random systems leads to giant fluctuations [15].

The enhancing effect of the self-similar geometry can be qualitatively understood from the following arguments. First, as soon as the size of the system is much less than λ, the plasmonic eigenfrequencies do not depend on size,

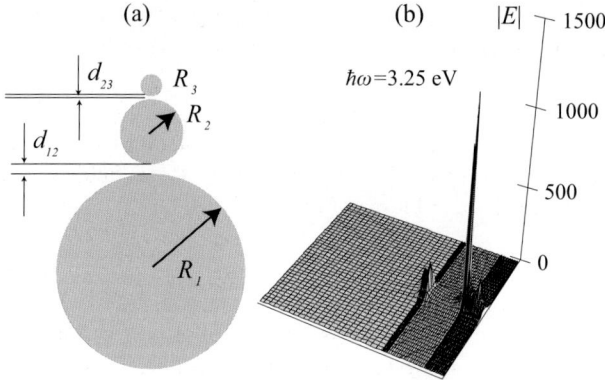

Fig. 5. (a) Geometry of a silver nanolens consisting of three silver nanospheres is shown in the cross section through its plane of symmetry. Nanosphere radii are: $R_1 = 45\,\text{nm}$, $R_2 = 15\,\text{nm}$, and $R_3 = 5\,\text{nm}$. The separations between the nanosphere surfaces are $d_{12} = 4.5\,\text{nm}$ and $d_{23} = 1.5\,\text{nm}$. (b) Local field-enhancement coefficient $I(\boldsymbol{r};\omega)$ is computed from (49) and shown as a function of the position in the cross section through the plane of symmetry. Adapted from [14]

rather they depend only on the form factor. Consider a fractal system at the largest scale within the self-similarity range. There are plasmonic eigenmodes localized at this maximum scale. Such eigenmodes resonant to the external radiation will generate enhanced local fields whose characteristic extension is also this maximum scale. Therefore, these large-scale local fields are almost uniform on some much smaller scale that is also present due to the self-similarity of the fractal. They play the role of the external excitation for the smaller-scale eigenmodes whose magnitude is enhanced by the Q-factor. These smaller-scale eigenmodes will respond in kind creating on their scale a local field enhanced by a factor of $\sim Q^2$. This transfer of excitation toward the progressively smaller scales enhances the local field by a factor of Q each time it takes place. This transfer is limited by the minimum scale of the fractal.

The above-described idea of the self-similar transfer of excitation down the spatial scales can be implemented in a simple, fully controllable, and reproducible system shown in Fig. 5a. This system is comprised by a linear aggregate of three silver nanospheres whose radii decrease by a factor of 1/3 from one sphere to another (from bottom to top). This specific reduction factor should be significantly less than 1, but its precise value is not of principal importance. The larger nanosphere's radius should be much smaller than the wavelength, and the minimum radius should be not too small to use the continuous electrodynamics. Our choice of three nanosphere radii is $R_1 = 45\,\text{nm}$, $R_2 = 15\,\text{nm}$, and $R_3 = 5\,\text{nm}$. The gaps between the surfaces of the nanosphere are chosen to be 0.3 of the radius of the smaller adjacent nanosphere, i.e., $d_{12} = 0.3R_2 = 4.5\,\text{nm}$, and $d_{23} = 0.3R_3 = 1.5\,\text{nm}$. This is a finite, deterministic, self-similar system that possesses the above-discussed mechanism of enhancement based on transfer down the spatial scale.

The external field with frequency close to the nanosphere surface-plasmon resonance excites the local field around the largest nanosphere enhanced by a factor of $\sim Q$. This local field is nearly uniform on the scale of the next, smaller nanosphere and plays the role of an external excitation field for it. This in turn creates the local field enhanced by $\sim Q^2$. Similarly, the local fields around the smallest nanosphere are enhanced by a factor of $Q^3 \sim 10^3$ for the realistic $Q \sim 10$ in the blue spectral region where the nanospheres have their surface-plasmon resonances. This process would stop when the next nanosphere is too small for the continuous electrodynamics to apply.

The problem has been solved in [14] in the quasistatic approximation using a multipolar expansion for Green's function. The local field shown in Fig. 5b is computed as $|E(\boldsymbol{r})| = [I(\boldsymbol{r}, \omega)]^{\frac{1}{2}}$, where $I(\boldsymbol{r}; \omega)$ is computed from (49). As we see, there is the hottest spot of local fields (nanofocus of this nanolens) that is situated in the smallest gap at the surface of the smallest nanosphere. This local field is enhanced by a factor of $|E| \approx 1200$, in excellent qualitative agreement with our order-of-magnitude estimate. From (45), we can estimate the SERS enhancement factor as

$$g^{\mathrm{R}} \sim \frac{1}{|s(\omega)|^2} |E|^4 \sim 10^{13}, \tag{50}$$

where we took into account that for the surface-plasmon resonance of a nanosphere $s(\omega) \approx 1/3$. Thus, for a molecule bound in the smallest gap, the nanolens is predicted to yield the enhancement factor that is sufficient to observe the single-molecule SERS. We are planning to carry out accurate computations of the enhancement for various nanolenses.

Acknowlededgments

This work was supported by grants from the Chemical Sciences, Biosciences and Geosciences Division of the Office of Basic Energy Sciences, Office of Science, US Department of Energy, a grant from National Science foundation, and a grant from the US–Israel Binational Science Foundation.

References

[1] M. Moskovits: Rev. Mod. Phys. **57**, 783 (1985)
[2] M. Fleischmann, P. J. Hendra, A. J. McQuillan: Chem. Phys. Lett. **26**, 163 (1974)
[3] D. L. Jeanmaire, R. P. Van Duyne: J. Electroanal. Chem. **84**, 1 (1977)
[4] M. G. Albrecht, J. A. Creighton: J. Am. Chem. Soc. **99**, 5215 (1977)
[5] J. I. Gersten: J. Chem. Phys. **72**, 5779 (1980)
[6] J. I. Gersten, A. Nitzan: J. Chem. Phys. **75**, 1139 (1981)

[7] S. L. McCall, P. M. Platzman, P. A. Wolff: Phys. Lett. A **77**, 381 (1980)

[8] D.-S. Wang, M. Kerker, H. W. Chew: Appl. Opt. **19**, 2315 (1980)

[9] R. Ruppin: Solid State Commun. **39**, 903 (1981)

[10] P. K. Aravind, A. Nitzan, H. Metiu: Surf. Sci. **110**, 189 (1981)

[11] Z. J. Wang, S. L. Pan, T. D. Krauss, H. Du, L. J. Rothberg: Proc. Nat. Acad. Sci. USA **100**, 8638 (2003)

[12] V. M. Shalaev, M. I. Stockman: Sov. Phys. JETP **65**, 287 (1987)

[13] M. I. Stockman, V. M. Shalaev, M. Moskovits, R. Botet, T. F. George: Phys. Rev. B **46**, 2821 (1992)

[14] K. Li, M. I. Stockman, D. J. Bergman: Phys. Rev. Lett. **91**, 227402 (2003)

[15] M. I. Stockman, L. N. Pandey, L. S. Muratov, T. F. George: Phys. Rev. Lett. **72**, 2486 (1994)

[16] K. Kneipp, Y. Wang, H. Kneipp, L. T. Perelman, I. Itzkan, R. R. Dasari, M. S. Feld: Phys. Rev. Lett. **78**, 1667 (1997)

[17] S. Nie, S. R. Emory: Science **275**, 1102 (1997)

[18] K. E. Shafer-Peltier, C. L. Haynes, M. R. Glucksberg, R. P. Van Duyne: J. Am. Chem. Soc. **125**, 588 (2003)

[19] C. R. Yonzon, C. L. Haynes, X. Y. Zhang, J. T. Walsh, R. P. Van Duyne: Anal. Chem. **76**, 78 (2004)

[20] X. Zhang, M. A. Young, O. Lyandres, R. P. Van Duyne: J. Am. Chem. Soc. **127**, 4484 (2005)

[21] D. J. Bergman, D. Stroud: Properties of macroscopically inhomogeneous media, in H. Ehrenreich, D. Turnbull (Eds.): *Solid State Physics*, vol. 46 (Academic, Boston 1992) pp. 148–270

[22] M. I. Stockman, S. V. Faleev, D. J. Bergman: Phys. Rev. Lett. **87**, 167401 (2001)

[23] M. I. Stockman, D. J. Bergman, T. Kobayashi: Phys. Rev. B **69**, 54202 (2004)

[24] D. J. Bergman, M. I. Stockman: Phys. Rev. Lett. **90**, 27402 (2003)

[25] L. D. Landau, E. M. Lifshitz: *Electrodynamics of Continuous Media* (Pergamon, Oxford and New York 1984)

[26] M. I. Stockman: Phys. Rev. Lett. **84**, 1011 (2000)

[27] M. I. Stockman: Phys. Rev. B **62**, 10494 (2000)

[28] M. I. Stockman, S. V. Faleev, D. J. Bergman: Phys. Rev. Lett. **88**, 67402 (2002)

[29] P. B. Johnson, R. W. Christy: Phys. Rev. B **6**, 4370 (1972)

[30] C. E. Talley, J. B. Jackson, C. Oubre, N. K. Grady, C. W. Hollars, S. M. Lane, T. R. Huser, P. Nordlander, N. J. Halas: Nano Lett. **5**, 1569 (2005)

[31] D. A. Weitz, M. Oliveria: Phys. Rev. Lett. **52**, 1433 (1984)

[32] M. Kolb, R. Botet, J. Julienne: Phys. Rev. Lett. **51**, 1123 (1983)

[33] S. V. Karpov, V. S. Gerasimov, I. L. Isaev, V. A. Markel: Phys. Rev. B **72**, 205425 (2005)

[34] A. K. Sarychev, V. A. Shubin, V. M. Shalaev: Phys. Rev. B **60**, 16389 (1999)

Index

Coupled Plasmonic Plasmon/Photonic Resonance Effects in SERS

Shengli Zou and George C. Schatz

Department of Chemistry, Northwestern University, 2145 Sheridan Road,
Evanston, Illinois, 60208-311
schatz@chem.northwestern.edu

1 Introduction

The observation of the surface-enhanced Raman scattering (SERS) [1,2,3] by analyte molecules near metal surfaces has stimulated tremendous interest in the optical properties of rough metal surfaces [4, 5, 6] and metal particles [7, 8, 9, 10, 11, 12, 13, 14, 15, 16, 17, 18, 19, 20, 21]. The most popular metal used in these studies has been silver due to intense plasmon resonance excitation that nanometer-size silver particles exhibit at visible wavelengths. This leads to large enhancement factors in surface-enhanced Raman scattering (SERS) studies on silver, and SERS is also detectable on gold and copper surfaces [22, 23,24]. Conventional SERS is often considered to have an enhancement factor of $10^5 - 10^6$, but the recent observation of single-molecule SERS [8, 25, 26] has produced enhancement estimates of 10^{12}–10^{15}.

The mechanisms leading to SERS still remain a matter of controversy [27], but two mechanisms are popularly mentioned in the literature. In the electromagnetic mechanism [15, 28, 29, 30, 31, 32], the local electric fields around metal-particle surfaces are enhanced as a result of plasmon excitation leading to more intense electronic transitions in molecules adsorbed near the particle surfaces, and intense Raman scattering. This mechanism leads to the strong sensitivity of SERS intensities to particle size, shape, dielectric environment and arrangement [33]. The second or chemical mechanism involves charge-transfer excitation [34,35] between analyte molecules and the metal particles to give a resonance Raman enhancement process.

Advances in nanotechnology have recently led to the preparation of metal particles with controllable sizes and shapes using wet-chemical methods, including rods [36, 37], prisms [38], disks [39], stars [40], and many others. Other particle shapes, such as truncated tetrahedra, have been fabricated with nanosphere lithography [41]. Particle arrays with precisely controlled size and shape, and a predesigned pattern can be produced using e-beam lithography [42] and nanosphere lithography [14, 41, 43, 44, 45, 46, 47, 48]. For particle arrays it is possible for the plasmon resonances in each particle to be coupled, leading to shifts in plasmon wavelengths, and changes in Raman intensities. These effects become especially interesting when the distances between the particles are close to the wavelength of light, for in this case one can combine plasmon-resonance effects in the particles with photonic

K. Kneipp, M. Moskovits, H. Kneipp (Eds.): Surface-Enhanced Raman Scattering – Physics and Applications, Topics Appl. Phys. **103**, 67–86 (2006)
© Springer-Verlag Berlin Heidelberg 2006

resonances associated with the array structure to produce new structures in extinction lineshapes, including sharp dips and peaks. The coupling of particles in arrays can also influence SERS intensities, leading to the possibility of photonic contributions to enhancement effects in combination with plasmonic enhancements.

The focus of this Chapter will be on the enhanced local electric fields around noble metal (silver) nanoparticles and their contribution to the electromagnetic mechanism in SERS. In particular after first examining the plasmon resonances of single particles and dimers, we study the plasmonic/photonic resonances of several kinds of particle arrays, to assess how long-range photonic interactions influence SERS intensities. These arrays include one-dimensional structures that are either perpendicular and parallel to the wavevector, as well as combinations of parallel or perpendicular arrays, as these simple array structures can be studied by analytical models, as well as computational methods, that provide good mechanistic insight. We show that exceptionally high local field enhancements occur for the combined parallel and perpendicular array structures, especially for structures in which a dimer of nanoparticles is placed in the arrays at the peak field location. These field enhancements are significantly larger than can be obtained for isolated nanoparticles, and they suggest that enhancement factors can be obtained that exceed anything that has so far been studied in the experiments.

2 Methods

For a single metal nanoparticle, the optical properties of the particle depend on its composition, size, shape and dielectric environment [33]. For a cluster of spherical particles, a T-matrix method that combines multipole expansions around each particle [49] provides an efficient approach for the analysis of the optical properties of the cluster [50, 51, 52, 53, 54, 55]. In this method, the scattered fields from each particle are expanded in terms of vector spherical harmonics (VSH) from Mie theory [56, 57], with the expansion coefficients determined by satisfying electromagnetic boundary conditions on the surfaces of each particle. As a result, exact solutions to Maxwell's equations may be generated as long as high enough expansion orders are used for each particle. When only dipole excitation is involved, the T-matrix method reduces to the coupled dipole (CD) approximation method that will be used extensively in this Chapter.

To study the properties of nanoparticles and arrays of nanoparticles that are not spherical, it is necessary to use a finite element-based approach. Among these methods, the discrete dipole approximation (DDA) method [33, 58, 59] is one of the most simple and efficient methods. The DDA method is similar numerically to the CD method, but in DDA one uses many dipoles to represent each particle, and all dipoles are located on a cubic lattice. Convergence of the results with respect to the lattice size leads to results that are usually close to the exact solution to Maxwell's equations. Other methods for treating nonspherical particles include the finite difference time-domain (FDTD) method [60, 61], the modified long-wavelength approximation (MLWA) [15, 33], and the multiple multipole (MMP) method [62, 63].

We have used T-matrix theory [50, 51, 52, 54, 55], the CD approximation [64] and a semianalytical formula based on the CD method to investigate the extinction spectra and electric fields of isolated spherical particles, dimers of spheres, spheres in one-dimensional arrays, and in more sophisticated configurations. For particles in arrays, when the particles are well separated, the CD method provides a quantitative description of the optical properties of the arrays, and it is much faster to use than T-matrix theory. However T-matrix theory is still useful, as it provides an efficient method for modeling the electrodynamics of spherical particles with arbitrary radii and separations, enabling us to check the accuracy of the CD results and model the systems where the particle separations are small. All of our studies of nonspherical particles refer to truncated tetrahedrons, and for these we have used the DDA method to determine local fields. The original DDA method treats isolated particles, but we have modified it to treat periodic arrays in some of the calculations we present.

The T-matrix method is described in detail by *Mackowski* and *Mishchenko* [50, 51, 52, 53, 54, 55]. We briefly introduce the CD method here. In the CD method, we consider an array of N particles whose positions and polarizabilities are denoted r_i and α_i. The induced polarization $\boldsymbol{P}_i$ in each particle in the presence of an applied plane wave field is $\boldsymbol{P}_i = \alpha_i \boldsymbol{E}_{\mathrm{loc},i} (i = 1, 2, \ldots, N)$ where the local field $\boldsymbol{E}_{\mathrm{loc},i}$ is the sum of the incident and retarded fields of the other $N - 1$ dipoles. For a given wavelength λ, this field is:

$$\boldsymbol{E}_{\mathrm{loc},i} = \boldsymbol{E}_{\mathrm{inc},i} + \boldsymbol{E}_{\mathrm{dipole},i} = \boldsymbol{E}_0 \exp(i\boldsymbol{k} \cdot \boldsymbol{r}_i) - \sum_{\substack{j=1 \\ j \neq i}}^{N} \boldsymbol{A}_{ij} \cdot \boldsymbol{P}_j \, i = 1, 2, \ldots, N \,,$$

$$(1)$$

where E_0 and $k = 2\pi/\lambda$ are the amplitude and wavevector of the incident wave, respectively. The dipole-interaction matrix $\boldsymbol{A}$ is expressed as:

$$\boldsymbol{A}_{ij} \cdot \boldsymbol{P}_j = k^2 e^{ikr_{ij}} \frac{\boldsymbol{r}_{ij} \times (\boldsymbol{r}_{ij} \times \boldsymbol{P}_j)}{r_{ij}^3}$$

$$+ e^{ikr_{ij}} (1 - ikr_{ij}) \frac{[r_{ij}^2 \boldsymbol{P}_j - 3\boldsymbol{r}_{ij}(\boldsymbol{r}_{ij} \cdot \boldsymbol{P}_j)]}{r_{ij}^5},$$

$$(i = 1, 2, \ldots, N, j = 1, 2, \ldots, N, j \neq i), \quad (2)$$

where $\boldsymbol{r}_{ij}$ is the vector from dipole i to dipole j. Note that the first term in (2) has a $1/r$ dependence on interparticle spacing, while the second has $1/r^2$ and $1/r^3$ variations. As a result, the first term, which is associated with radiative dipolar interactions between the particles, is often dominant for large array spacings.

The polarization vectors are obtained by solving $3N$ linear equations of the form

$$\underset{\sim}{A'}\underset{\sim}{P} = \underset{\sim}{E}, \tag{3}$$

where the offdiagonal elements of the matrix $\underset{\sim}{A'}$, A'_{ij} are the same as A_{ij} in (2), and the diagonal elements A'_{ii} are α_i^{-1}. After obtaining the polarization vectors, we can calculate the extinction cross section using:

$$C_{\text{ext}} = \frac{4\pi k}{|E_0|^2} \sum_{j=1}^{N} \text{Im}(E^*_{\text{inc},j} \cdot P_j). \tag{4}$$

A simple analytical solution [64] to (3) can be obtained in the case of an infinite array of particles for the case where the wavevector is perpendicular to the array axis (or plane) by assuming that the induced polarization in each array element is the same. This leads to the following expression for the polarization of each particle:

$$P = \frac{\alpha_s E_0}{1 - \alpha_s S} = \frac{E_0}{1/\alpha_s - S} \tag{5}$$

and the extinction cross section for each particle

$$C_{\text{ext}} = 4\pi k \Im\left(\frac{P}{E_0}\right) = 4\pi k \text{Im}\left(\frac{1}{1/\alpha_s - S}\right), \tag{6}$$

where S is the retarded dipole sum

$$S = \sum_{j \neq i} \left[\frac{(1 - ikr_{ij})(3\cos^2\theta_{ij} - 1)e^{ikr_{ij}}}{r_{ij}^3} + \frac{k^2 \sin^2\theta_{ij} e^{ikr_{ij}}}{r_{ij}} \right]. \tag{7}$$

The polarizability in (6) is

$$\alpha_s = \frac{3a_1}{2k^3}, \tag{8}$$

where a_1 is the expansion coefficient in Mie theory associated with the scattered electric fields around the particles. This is given by

$$a_1 = \frac{\mu m^2 j_1(m\rho)[\rho j_1(\rho)]' - \mu_1 j_1(\rho)[m\rho j_1(m\rho)]'}{\mu m^2 j_1(m\rho)[\rho h_1(\rho)]' - \mu_1 h_1^1(\rho)[m\rho j_1(m\rho)]'}. \tag{9}$$

In this expression, $\rho = kr$, where r is the radius of the particle, m is the ratio of the indices of refraction inside and outside the particle, μ_i and μ are the magnetic permeabilities inside and outside, and j_1 and h_1 are the usual spherical Bessel functions. In our earlier work, we used (5) to show how coupling between the particles leads to a sharp peak in the plasmon-resonance lineshape [16, 19, 64].

Now let us generalize this derivation to consider a three-dimensional periodic array. Here (5) still applies, but the dipole sum is modulated by an extra term e^{ikz} (where z measures the distance along the wavevector). If we take the polarization to be in the x-direction, then (7) is replaced by:

$$
\begin{aligned}
S &= \sum_{j\neq i} e^{ikz_{ij}} \left[\frac{(1 - ikr_{ij})(3x_{ij}^2 - r_{ij}^2)e^{ikr_{ij}}}{r_{ij}^5} + \frac{k^2(y_{ij}^2 + z_{ij}^2)e^{ikr_{ij}}}{r_{ij}^3} \right] \\
&= \sum_{j\neq i} e^{ik(r_{ij}+z_{ij})} \left[\frac{(1 - ikr_{ij})(3x_{ij}^2 - r_{ij}^2)}{r_{ij}^5} + \frac{k^2(y_{ij}^2 + z_{ij}^2)}{r_{ij}^3} \right],
\end{aligned} \tag{10}
$$

where z_{ij} is the z-axis coordinate of particle j relative to particle i, etc.

In (10), the first term in the square bracket corresponds to the short-range dipole interaction. This leads to the strongest interactions between particles for the case where the array is a chain along the polarization direction [48,65]. The second term represents the long-range interaction. In this case, if the chain is parallel to the polarization direction (the x-axis), the long-range term is zero, and narrow lines are not produced [16]. The best structures for generating long-range interactions are chains in the yz-plane (where z specifies the wavevector direction), especially one-dimensional chains along the y- or the z-axes [16, 19, 66].

3 Single Nanoparticles and Dimers

Comprehensive investigations of the extinction spectra of single nanoparticles have been reported by *Kelly* et al. [33]. Electromagnetic-field calculations for single and dimer silver particles with different shapes, arrangements and interparticle distances have been extensively studied by *Hao* et al. [65]. In this section, we consider field enhancements for particle shapes that we consider later in our array studies, to examine how these field enhancements vary with structure, taking the particles to be silver and the surrounding medium to be vacuum. Spherical silver particles normally have resonance peaks at short wavelengths (i.e., 365 nm). The resonance peaks redshift when the particles become prolate spheroids along the polarization direction or possess sharp tips. Oblate spheroids with a symmetry axis along the polarization direction are expected to generate resonance peaks at shorter wavelengths. The enhanced local electric fields around particles are very sensitive to changes in the particle shapes especially for particles with sharp tips.

The enhanced local electric-fields $|\boldsymbol{E}|^2$ of silver particles with spherical and tetrahedral shapes are shown in Fig. 1. The radii of the spherical particles are 30 nm and 50 nm, respectively. The truncated tetrahedron (regular

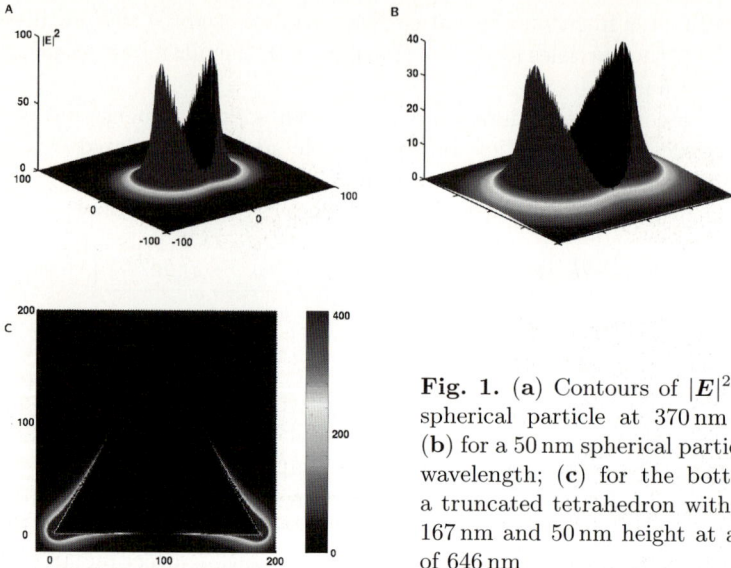

Fig. 1. (a) Contours of $|\boldsymbol{E}|^2$ for a 30 nm spherical particle at 370 nm wavelength; (b) for a 50 nm spherical particle at 405 nm wavelength; (c) for the bottom plane of a truncated tetrahedron with inplane size 167 nm and 50 nm height at a wavelength of 646 nm

tetrahedron with one tip removed) has a 50 nm height and a 167 nm base (where the base is the bisector of the equilateral triangle that forms the base of the tetrahedron). This makes its volume equivalent to that of a spherical particle with a 50 nm radius. The silver dielectric constants for all calculations are from *Palik* [67]. For the electric-field calculations, we used Mie theory for the spheres, which provides exact $|\boldsymbol{E}|^2$ values that we evaluate right at the particle surfaces. The truncated tetrahedron calculations use the DDA method, and we follow the earlier work of *Hao* and *Schatz* [16] where the field is calculated half a grid point outside the surface boundary. The grid size in the current calculations is 1 nm.

Figure 1 shows that the peak $|\boldsymbol{E}|^2$ for an isolated 30 nm spherical particle is 90 (relative to the applied asymptotic field) at the plasmon-resonance wavelength of 370 nm. The peak value for a 50 nm particle declines to 40 at the resonance wavelength of 405 nm due to radiative damping effects. The resonance wavelength redshifts to 646 nm for the truncated tetrahedral particle. The peak $|\boldsymbol{E}|^2$ around the particle tips is 47 000 which is 1000 times higher than that for an isolated spherical particle.

As noted in many studies [65, 68, 69, 70], the electric fields produced by a particle dimer are significantly higher than those around an isolated particle. Electric-field contour plots for spherical particle dimers with 30 nm and 50 nm radii are shown in Fig. 2. The electric fields near a truncated tetrahedron dimer equivalent in volume to a spherical particle with a 50 nm radius are also included. In the calculations, the wavevector is perpendicular to the dimer axis, which is defined as the axis linking the centers of the two particles,

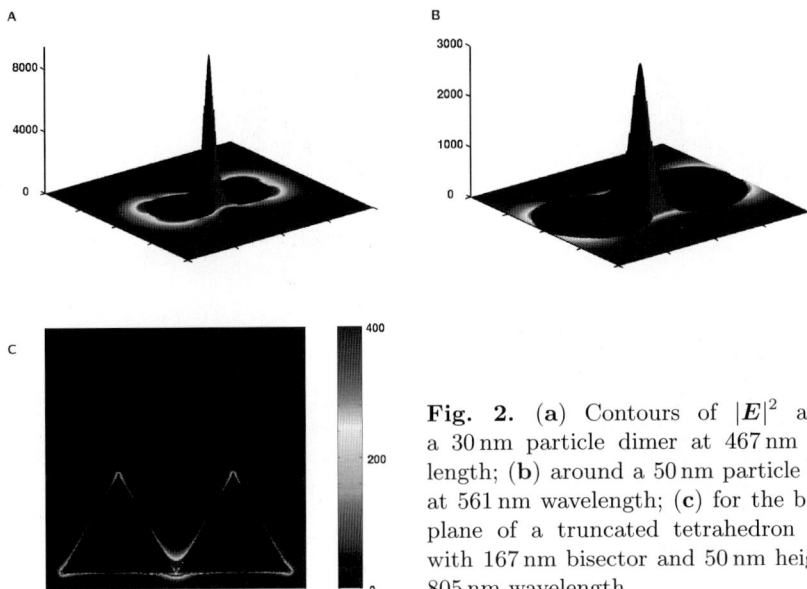

Fig. 2. (**a**) Contours of $|\boldsymbol{E}|^2$ around a 30 nm particle dimer at 467 nm wavelength; (**b**) around a 50 nm particle dimer at 561 nm wavelength; (**c**) for the bottom plane of a truncated tetrahedron dimer with 167 nm bisector and 50 nm height at 805 nm wavelength

and the polarization vector is parallel to the axis. The distances between the particles in each dimer are taken to be 1 nm.

Figure 2 shows that the highest electric field between a 30 nm spherical particle dimer is 9 500. This occurs at the resonance wavelength of 467 nm. The peak field falls off to 2 800 for the 50 nm particle dimer at its resonance wavelength of 561 nm. The decline of the electric fields in going from 30 nm to 50 nm particle dimers is due to radiative damping (i.e., larger particles emit radiation so efficiently that it is hard to induce a large polarization in them). When the two spherical particles are replaced by truncated tetrahedrons with an equivalent volume of a 50 nm spherical particle, the peak value of the electric fields between the particles skyrockets to 1.9×10^6.

In the electromagnetic mechanism of SERS, the signal enhancement is proportional (as demonstrated by *Kerker* et al. [28]) to the average over the surface of the metal particle of the local field $|\boldsymbol{E}(\omega)|^2$ at the pump frequency ω multiplied by $|\boldsymbol{E}(\omega')|^2$ at the Stokes-shifted frequency ω'. This $|\boldsymbol{E}(\omega)|^2|\boldsymbol{E}(\omega')|^2$ factor is often approximated by $|\boldsymbol{E}(\omega)|^4$ as the plasmon-resonance peaks are broad enough that the change in $|\boldsymbol{E}(\omega)|^2$ between the incident and Stokes-shifted frequency is small (typically less than a factor of three). For single-molecule SERS, the average electric fields are replaced by their peak values.

For the truncated tetrahedron results in Fig. 2, the peak enhanced electric field indicates a 4×10^{12} enhancement factor for single-molecule SERS, which is close to values estimated from experimental measurements. However, we should inject a note of caution into the meaning of this, as this is based on a

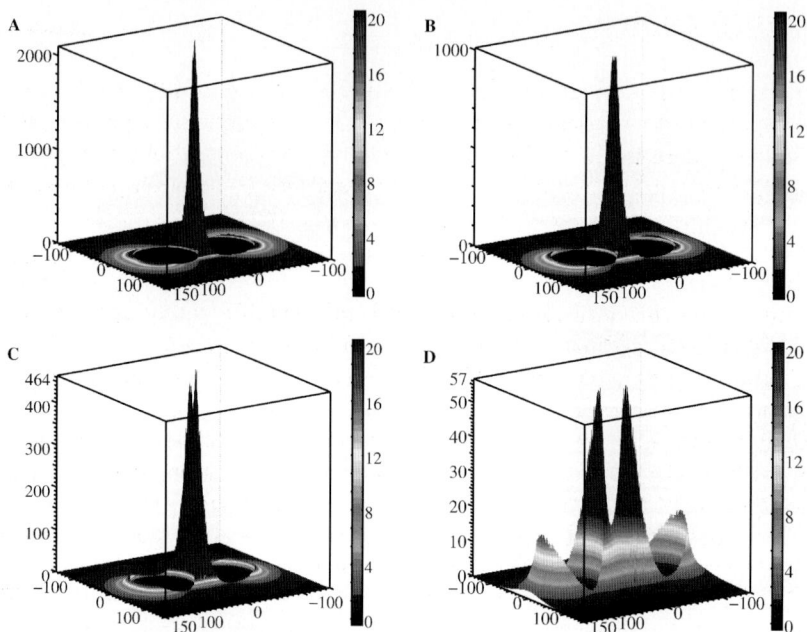

Fig. 3. Contours of $|E|^2$ around an isolated 50 nm particle dimer with different interparticle distances and wavelengths. (a) 2 nm separation at 555 nm wavelength; (b) 5 nm separation at 520 nm wavelength; (c) 10 nm separation at 495 nm wavelength; (D) 50 nm separation at 430 nm wavelength

continuum electrodynamics calculation, and the accuracy of such calculations for interparticle distances of 1 nm is not known. While the 1 nm spacing has been used in other studies [26], there is evidence based on studies that used nonlocal dielectric constants [71] that continuum electrodynamics based on local dielectric constants overestimates the field enhancements.

The electric field between the dimer particles is very sensitive to the interparticle distance. To study this, in Fig. 3 we present electric-field contour plots for 50 nm sphere dimers with different gaps. Here, we see that the peak electric field drops from 2 800 for a particle separation of 1 nm (resonance wavelength = 561 nm) to 2 000 for a 2 nm gap (resonance wavelength = 555 nm). The field further slips to 1 000 for a 5 nm spacing (resonance wavelength = 520 nm), and to 464 for a 10 nm gap (resonance wavelength = 495 nm). When the distance between two particles reaches 50 nm, the coupling between the two particles is sufficiently weak that the highest field region is split into two parts, with each part having a peak at 57 (resonance wavelength = 430 nm). The 50 nm results are reasonably close to those for an isolated particle, for which the peak field is 43 and the resonance wavelength is 405 nm.

4 Particle Arrays

Advances in e-beam lithography [42] and nanosphere lithography [43,72] have made it possible to fabricate particle (and hole) arrays with a predesigned size, shape and pattern, leading to interesting questions about the use of these structures in SERS applications. The interactions between particles in one- or two-dimensional arrays have been investigated extensively [16, 17, 18, 19, 20, 64, 73, 74, 75], but so far very little in the way of experimental studies have been done.

Many of the computational studies of local fields for many-particle systems have focused on dimers, small clusters and fractal aggregates [65,76,77]. Additional work has been reported by *Ebbesen's* group [78] who calculated the local fields in periodic arrays of holes that are separated by half the incident wavelength, and by *Porto* et al. [79] who reported transmission resonances on a metallic grating by excitation of coupled surface-plasmon polaritons and by the coupling of the incident wave with waveguide resonances.

In our work [16, 19], we found that remarkably narrow resonance peaks can be achieved in the extinction spectra of one- and two-dimensional silver particle arrays, and that these resonances lead to local electric fields around the particles that are over an order of magnitude larger than is obtained when the same particles are isolated. The narrow resonances arise because of coupling between plasmons in the metal particles and the photonic states of the particle arrays. Two factors are crucial in generating these resonances. First, there is a minimum particle radius of about 30 nm, as the scattered fields need to be large enough to collectively perturb the local field at each particle sufficiently to produce a measureable change in extinction (i.e., smaller particles lead to such narrow resonances that particle inhomogeneities and dephased excitation average their influence to zero). Another important factor is that the medium surrounding the particles needs to be homogeneous in order to produce sufficient coherent coupling between the particles. Thus, if the particles are exposed partially to vacuum and partially to a substrate, for example glass, the dipolar interactions between the particles are partially dephased, and the sharp resonances are lost.

We found that one-dimensional arrays give the strongest coherent coupling between particles, leading to the highest enhanced local electric fields [17]. In a one-dimensional array, the polarization direction needs to be perpendicular to the chain to achieve optimized coherent interactions between the scattered electric fields from particles. The wavevector can either be perpendicular or parallel to the chain, but the resonance properties are quite different for these two situations. When the wavevector is perpendicular to the chain, the shortest wavelength associated with the sharp resonance wavelengths is close to the interparticle distance. The results for the parallel wavevector case are more complex as the resonance wavelength does not converge to a fixed number as the number of particles in the chain is increased, however, for chains with a few hundred silver particles, there are two important res-

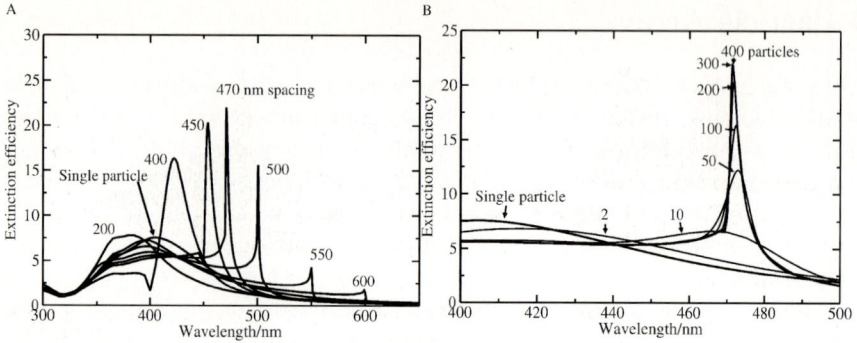

Fig. 4. (a) Extinction spectra of a one-dimensional chain of 50 nm silver particles with different interparticle distances. The chain is arranged to be perpendicular to the wavevector direction. (b) Extinction spectra for a one-dimensional chain of 50 nm particles. Chain perpendicular to the wavevector direction for a spacing of 470 nm with chain size varying from 1 to 400 particles

onances in the visible, one with wavelength equal to the particle separation and one with wavelength equal to twice the separation.

4.1 Chains Perpendicular to the Wavevector

The extinction spectra of silver particles in one-dimensional arrays are shown in Fig. 4 for the parallel wavevector case. The spectra in Fig. 4a are from chains with 400 particles, and the quantity plotted is the extinction efficiency (ratio of extinction cross section to geometrical area of the particles in the array) versus wavelength. This shows a broad peak at around 375 nm wavelength when the particle distance is 200 nm. The peak wavelengths redshift as the distance between the particles increases and the resonance widths narrow. When the interparticle distance is increased to 400 nm, a relatively narrow peak occurs at around 420 nm wavelength. The highest peak intensity is found at 471.4 nm wavelength when the particle spacing is 470 nm. Still larger separations lead to even smaller resonance widths, but the intensities of the resonance peaks fall off very quickly when the particle spacing is increased above 500 nm. This fall-off of the peak intensities indicates weak interactions between particles when the interparticle distances are over 500 nm. Note that the narrow peaks only fall within the envelope of wavelengths that is longer than the single-particle resonance wavelength at 400 nm. A detailed interpretation of this phenomenon has been presented [66].

The influence of the number of particles in the chain on the extinction spectra is presented in Fig. 4b. Results for 1 to 400-particles are considered, for a fixed particle spacing of 470 nm. This shows that a broad peak appears near 470 nm for short chains, and grows and narrows as the number of par-

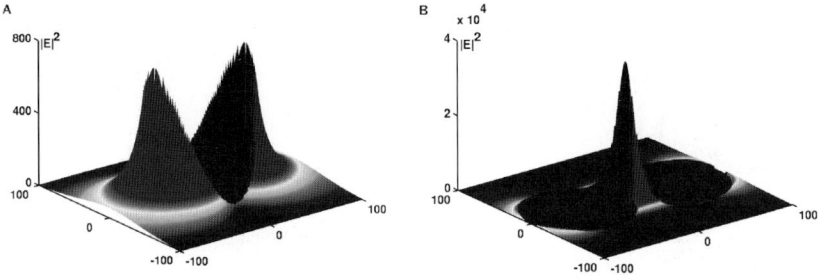

Fig. 5. (a) Contours of $|E|^2$ for the central particle in an array at a wavelength of 471.4 nm; (b) Fields around the central dimer in a one-dimensional array with a spacing of 650 nm at a wavelength of 655 nm. Contour refers to a plane that passes through the center of the particle and is perpendicular to the wavevector

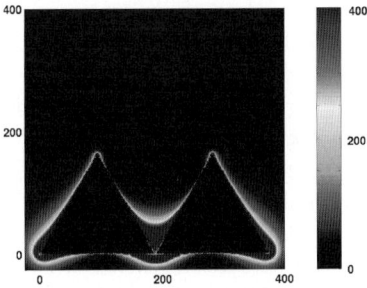

Fig. 6. Contours of $|E|^2$ for the bottom plane of the central dimer in a truncated tetrahedron dimer array at 1000.1 nm, the truncated tetrahedron is 50 nm high with 167 nm bisector

ticles increases, leading to a narrow peak for 50 particles, and a result that is independent of particle number for more than 300 particles.

To examine the application of this narrow resonance phenomenon to SERS, we have calculated the local electric fields around the particles in the one-dimensional arrays. The results are presented in Fig. 5 for a chain of 400-particles and also for a chain of 150 dimers of particles. The 400 particle chain results are in Fig. 5a for a spacing of 470 nm and a wavelength of 471.4 nm (the peak in Fig. 4). The contour plane passes through the center of the central particle in the array and is perpendicular to the wavevector. The figure shows that the peak value of $|E|^2$ is 700, which is 18 times that for an isolated particle. The dimer array results are in Fig. 5b, taking the distance between particles in each dimer to be 1 nm. Here, the polarization direction is parallel to the dimer axis and perpendicular to the chain direction. The spacing between dimers is taken to be 650 nm as this generates the most intense resonance peak at 655 nm wavelength. The peak value of the electric fields around the central dimer in the array is 3.7×10^4, which is 13 times that of the isolated dimer.

As already discussed in the previous paragraphs, particles with sharp tips generate higher enhanced electric fields around them than spheres. Contours

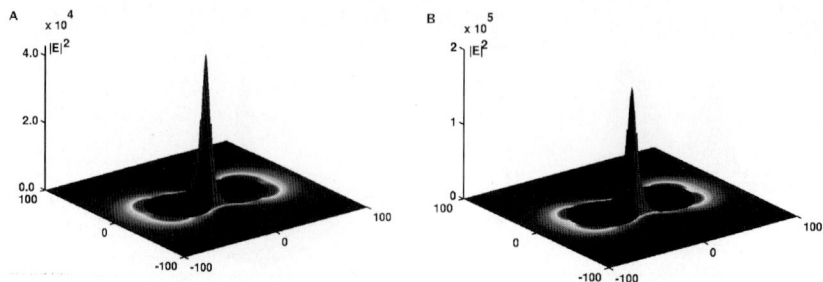

Fig. 7. (a) Contours of $|E|^2$ around the central dimer for 30 nm particles in a one-dimensional array at 490 nm wavelength; (**b**) around the central dimer in a one-dimensional array composed mainly of 50 nm single particles at 471.4 nm wavelength

of the electric field around the central dimer in a one-dimensional array of truncated tetrahedrons are shown in Fig. 6. To calculate these contours we used an extended DDA program [20] with periodic boundary conditions. The interdimer distance is taken to be 1000 nm. The central dimer in the array shows a peak electric field of 9.1×10^6 at a resonance wavelength of 1000.1 nm. This $|E|^2$ corresponds to an $|E|^4$ SERS enhancement factor of nearly 10^{14}, which is similar to estimates from a variety of experiments [8, 25, 26]. Of course, there are a variety of factors that could make this estimate an upper bound to what might actually be observed, including the use of continuum electrodynamics with a local dielectric constants for particles spaced by 1 nm as noted earlier. In addition, there is a problem with getting a large enhancement factor at both the incident and Stokes-shifted wavelengths that we consider later.

Even though coherent interactions between particles in an array provide an additional enhancement on top of the plasmon enhancement to local fields that occur for isolated particles, the resonance wavelengths of the isolated particles or dimers are quite different from those associated with the photonic states of the one-dimensional arrays. However, if we can generate structures of single particles or dimers that have nearly the same peak wavelength as the one-dimensional arrays, then the enhanced local electric fields are even larger. To show how this works, we consider a dimer of 30 nm spherical particles with a 1 nm separation. As discussed earlier, this has a plasmon maximum at 467 nm. This is very close to the narrow resonance wavelength of 471.4 nm that is found for a one-dimensional array of 50 nm particles with 470 nm spacing. If the 30 nm particle dimer is inserted in a one-dimensional array of the 50 nm particles with the 470 nm spacing, an even higher field-enhancement factor is found. Here we take the dimer axis to be parallel to the applied field that is perpendicular to the chain.

The electric fields of the central dimer in a one-dimensional array composed of 30 nm particle dimers (all with 1 nm separation) are first calculated and presented in Fig. 7a. 150 pairs of particles are included in the calculations.

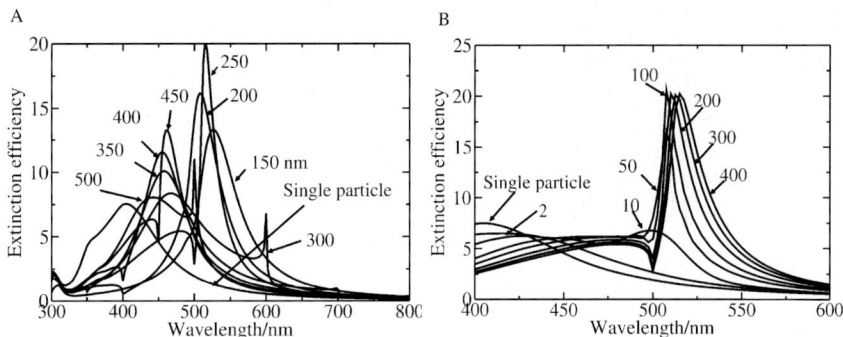

Fig. 8. (a) Extinction spectra of one-dimensional chains of 50 nm silver particles with different interparticle distances parallel to the wavevector direction; (b) Extinction spectra of one-dimensional chains of 50 nm particles parallel to the wavevector direction at a spacing of 250 nm with chain size varying from 1 to 400 particles

This shows that the peak value of the electric fields around the central dimer is 4.3×10^4 at 490 nm wavelength when the interdimer spacing is 480 nm. The enhancement factor is 5 times that of a single dimer, which is 9 500. Figure 7b shows that when a single 30 nm dimer is located at the center of a 50 nm particle array, the peak field is $|E|^2 = 1.6 \times 10^5$ at 471.4 nm wavelength. This is 17 times what is obtained from an isolated dimer and four times that obtained from the array of dimers.

4.2 Chains Parallel to the Wavevector

As noted above, a disadvantage of chains that are perpendicular to the wavevector for SERS applications is that the resonance peaks associated with dipole coupling between the particles are so narrow that it is not possible to excite the resonance significantly at both the incident and Stokes-shifted wavelengths. As a result, the SERS enhancement factor is significantly smaller than the $|E|^4 \sim 10^{14}$ estimate that we gave earlier. To fix this problem, we may fabricate chains with two distinct spacings such that resonances at both wavelengths can be excited [17]. Another way is to use chains that are parallel to the wavevector direction as these can produce resonances where the resonance widths are more tunable than they are for perpendicular wavevectors [66].

Figure 8 presents the extinction spectra for chains parallel to the wavevector. 400 spheres are included in the results in Fig. 8a, and we see that when the distance between the particles is 150 nm, a broad peak appears at around 525 nm. When the particle distance increases to 200 nm, the resonance wavelength blueshifts to 508 nm and the resonance width narrows. This is quite different from the resonance peak in Fig. 4 for the chains perpendicular to the wavevector, which for a 200 nm spacing gives a resonance wavelength close

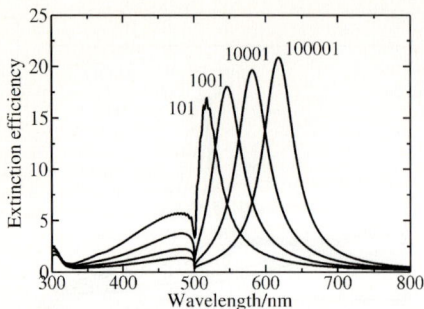

Fig. 9. Extinction spectra of one-dimensional chains of 50 nm particles at a spacing of 250 nm based on the semianalytical formula. The chain size varies from 101 to 100 001 particles

to 360 nm. The resonance wavelength in Fig. 8a redshifts to 515 nm when the spacing between particles is increased to 250 nm, while the extinction efficiency peak intensity grows and the width narrows to 20 nm. With further increase in the particle distances, a narrow resonance continues to occur with a wavelength close to twice the interparticle spacing, however, the peak intensities fall off. For example, the resonance wavelength at 515 nm when the spacing is 250 nm, shifts to 600 nm when the particle spacing is extended to 300 nm. However, new peaks appear near 450 nm for these larger spacings. These blueshift and narrow when the interparticle distances are increased from 300 nm to 400 nm, and then for larger spacings there is a sharp resonance whose wavelength is close to the interparticle spacing. For example, the resonance wavelength is at about 500 nm when the particle spacing is 500 nm. From these results we conclude that the parallel configuration leads to a series of resonances whose wavelength is roughly twice the spacing divided by an integer $(1, 2, \ldots)$, with the restriction that the wavelength needs to be longer than about 500 nm in order for a narrow resonance to appear.

The extinction spectra of chains having different numbers of particles are presented in Fig. 8 for a fixed interparticle distance of 250 nm. Figure 8b shows that the narrow peaks become significant when the particle number is larger than about 20. However, the peak wavelength does not converge with increasing particle number. To study this behavior further, in Fig. 9 we present extinction spectra for much longer chains, this time using the semianalytical formula (10). This shows that increasing the particle number from 10^2 to 10^5 leads to continuous broadening and redshifting of the resonance [66]. This demonstrates the extremely long-range nature of the dipole sums for this configuration. It is also noteworthy that the broad and tunable resonances that are produced are useful for SERS applications as we now show.

To study the use of parallel chains for SERS, in Fig. 10 we present contours of the electric fields around the central particle in the 400-particle chain

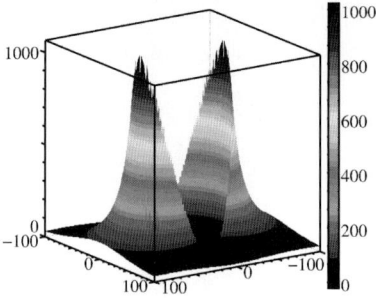

Fig. 10. Contours of $|\boldsymbol{E}|^2$ around the central particle in a one-dimensional array composed of 50 nm particles at 250 nm spacing and 513 nm wavelength

considered in Fig. 8, again for a separation of 250 nm. The wavelength for this figure is chosen to be the resonance wavelength at 513 nm. The figure shows that the peak electric field is $|\boldsymbol{E}|^2 = 1070$, which is higher than the 700 value obtained for a one-dimensional chain with the same particle number and wavevector perpendicular to the chain axes at its optimized spacing of 470 nm and the resonance wavelength of 471.4 nm.

4.3 Configurations that Combine Parallel and Perpendicular Chains

We have also examined arrays composed of a one-dimensional chain perpendicular to the wavevector that crosses one that is parallel to the wavevector (forming a "+" shape). By suitably choosing the spacings of the two arrays, it is possible to make the plasmonic/photonic resonances in each array overlap. We start with a one-dimensional chain of 50 nm particles perpendicular to the wavevector that was used in Fig. 7b, and add another chain having the same size particles and spacing that is parallel to the wavevector. To yield large field enhancements, we replace the particle at the crossing of the two arrays with a dimer composed of 30 nm particles with a 1 nm separation.

Contours of the electric fields around the central particles in this structure are shown in Fig. 11. In the calculations, 302 particles are included. 150 particles are arranged parallel to the wavevector while another 150 are perpendicular to it. The interparticle distance is taken to be 470 nm and the electric fields are calculated at the resonance wavelength of 471.4 nm. The central dimer is taken to be parallel to the applied polarization that is perpendicular to both chains. Figure 11 shows that the peak value of the electric field at the central particle dimer is 3.9×10^5, which is twice the value for the central dimer in a single chain with perpendicular wavevector.

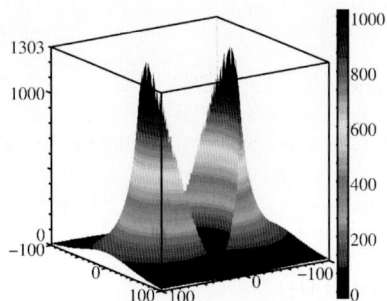

Fig. 11. Contours of $|\boldsymbol{E}|^2$ around the central particle dimer for particles of the "+" array configuration

5 Conclusion

The present studies show that array structures can be used to produce local field enhancements that are considerably larger than is found for isolated particles or dimers. These enhancements are large enough, in fact, that single-molecule SERS measurements may be possible. The largest enhancements were obtained using a structure of two one-dimensional arrays, one parallel to the wavevector direction and one perpendicular to it. The particle at the crossing point of the two arrays is replaced by a dimer of particles, so that the local field between the particles is enhanced both by short-range near-field behavior of the dimer and also by long-range photonic interactions associated with resonances in the two arrays. This structure produced a field enhancement of 3.9×10^5 for a dimer of spheres, which is twice what we have been able to generate for a dimer in a single chain. Computational limitations make it impossible for us to study a dimer of nonspherical particles for this array structure, but similar studies for an isolated dimer of truncated tetrahedrons produced an extra factor of over 10^2 in the enhancement over the equivalent sphere result, suggesting that $|\boldsymbol{E}|^2$ values of over 10^7 are possible for the "+" structure with a truncated tetrahedron dimer at the center. Furthermore, the resonance structures in this case are broad enough that significant enhancements at both the incoming and outgoing wavelengths is possible. Thus we conclude that this structure holds promise for single-molecule SERS measurements.

Acknowledgements

This work was supported by the National Science Foundation through the Nanotechnology Science and Engineering Center (NSEC), and by the Air Force Office of Scientific Research MURI program (F49620-02-1-0381).

References

[1] M. G. Albrecht, J. A. Greghton: J. Am. Chem. Soc. **99**, 5215 (1977)

[2] D. L. Jeanmaire, R. P. Van Duyne: J. Electroanal. Chem. **84**, 1 (1977)

[3] R. K. Chang, T. Furtak: in *Surface-Enhanced Raman Scattering* (Plenum, New York 1981)

[4] S. G. Schultz, M. Janik-Czachor, R. P. Van Duyne: Surf. Sci. **104**, 419 (1981)

[5] U. Laor, G. C. Schatz: J. Chem. Phys. **76**, 2889 (1982)

[6] I. Pockrand: J. Electron. Spectrosc. Relat. Phenom. **29**, 257 (1983)

[7] G. W. Robinson: Chem. Phys. Lett. **80**, 404 (1981)

[8] S. Nie, S. R. Emory: Science **275**, 1102 (1997)

[9] J. Grand, M. L. de la Chapelle, J. L. Bijeon, P. M. Adam, A. Vial, P. Royer: Phys. Rev. B **72**, 33407 (2005)

[10] D. A. Genov, A. K. Sarychev, V. M. Shalaev, A. Wei: Nano Lett. **4**, 153 (2004)

[11] A. J. Haes, C. L. Haynes, A. D. McFarland, G. C. Schatz, R. P. Van Duyne, S. Zou: MRS Bull. **30**, 368 (2005)

[12] A. D. McFarland, M. A. Young, J. A. Dieringer, R. P. Van Duyne: J. Phys. Chem. B **109**, 11279 (2005)

[13] L. A. Dick, A. D. McFarland, C. L. Haynes, R. P. Van Duyne: J. Phys. Chem. B **106**, 853 (2002)

[14] E. M. Hicks, S. Zou, G. C. Schatz, K. G. Spears, R. P. Van Duyne, L. Gunnarsson, T. Rindzevicius, B. Kasemo, M. Kall: Nano Lett. **5**, 1065 (2005)

[15] E. J. Zeman, G. C. Schatz: J. Phys. Chem. **91**, 634 (1987)

[16] S. Zou, N. Janel, G. C. Schatz: J. Chem. Phys. **120**, 10871 (2004)

[17] S. Zou, G. C. Schatz: Chem. Phys. Lett. **403**, 62 (2004)

[18] S. Zou, G. C. Schatz: SPIE Proc. **5513**, 22 (2004)

[19] S. Zou, G. C. Schatz: J. Chem. Phys. **121**, 12606 (2004)

[20] S. Zou, L. Zhao, G. C. Schatz: Proc. SPIE **5221**, 174 (2003)

[21] D. S. Citrin: Nano Lett. **5**, 985 (2005)

[22] H. D. Ladouceur, D. E. Tevault, R. R. Smardzewski: J. Chem. Phys. **78**, 980 (1983)

[23] G. Niaura, A. K. Gaigalas, V. L. Vilker: J. Phys. Chem. B **101**, 9250 (1997)

[24] C. A. Jennings, G. J. Kovacs, R. Aroca: Langmuir **9**, 2151 (1993)

[25] K. Kneipp, Y. Wang, H. Kneipp, L. T. Perelman, I. Etzkan, R. R. Dasari, M. S. Feld: Phys. Rev. Lett. **78**, 1667 (1977)

[26] H. Xu, E. J. Bjerneld, M. Kall, L. Borjesson: Phys. Rev. Lett. **83**, 4357 (1999)

[27] R. Dornhaus, M. B. Long, R. E. Benner, R. K. Chang: Surf. Sci. **93**, 240 (1980)

[28] M. Kerker, D. S. Wang, H. Chew: Appl. Opt. **19**, 3373 (1980)

[29] G. C. Schatz: Acc. Chem. Res. **17**, 370 (1984)

[30] H. Metiu, P. Das: Annu. Rev. Phys. Chem. **35**, 507 (1984)

[31] H. Xu, J. Aizpurua, M. Kall, P. Apell: Phys. Rev. E **62**, 4318 (2000)

[32] G. C. Schatz, R. P. Van Duyne: *Handbook of Vibrational Spectroscopy*, vol. 1 (Wiley, New York 2002) p. 759

[33] K. L. Kelly, E. Coronado, L. Zhao, G. C. Schatz: J. Phys. Chem. B **107**, 668 (2003)

[34] F. J. Adrian: J. Chem. Phys. **77**, 5302 (1982)

[35] H. Yamada, H. Nagata, K. Toba, Y. Nakao: Surf. Sci. **182**, 269 (1987)

[36] K. K. Caswell, C. M. Bender, C. J. Murphy: Nano Lett. **3**, 667 (2003)

[37] N. R. Jana, L. Gearheart, C. J. Murphy: Chem. Commun. **7**, 617 (2001)

[38] R. C. Jin, Y. W. Cao, C. A. Mirkin, K. L. Kelly, G. C. Schatz, J. G. Zheng: Science **294**, 1901 (2001)

[39] E. Hao, K. L. Kelly, J. T. Hupp, G. C. Schatz: J. Am. Chem. Soc. **124**, 15182 (2002)

[40] E. Hao, R. C. Bailey, G. C. Schatz, J. T. Hupp, S. Li: Nano Lett. **4**, 327 (2004)

[41] C. L. Haynes, R. P. Van Duyne: J. Phys. Chem. B **105**, 5599 (2001)

[42] M. Kahl, E. Voges, S. Kostrewa, C. Viets, W. Hill: Sens. Actuators B **51**, 285 (1998)

[43] R. P. Van Duyne, J. C. Hulteen, D. A. Treichel: J. Chem. Phys. **99**, 2101 (1993)

[44] C. L. Haynes, A. D. McFarland, L. Zhao, R. P. Van Duyne, G. C. Schatz, L. Gunnarsson, J. Prikulis, B. Kasemo, M. Kall: J. Phys. Chem. B **107**, 7337 (2003)

[45] A. Haes, E. Voges, R. P. Van Duyne: J. Am. Chem. Soc. **124**, 10596 (2002)

[46] A. Haes, S. Zou, G. C. Schatz, R. P. Van Duyne: J. Phys. Chem. B **108**, 109 (2004)

[47] A. Haes, S. Zou, G. C. Schatz, R. P. Van Duyne: J. Phys. Chem. B **108**, 6961 (2004)

[48] L. Gunnarsson, T. Rindzevicius, J. Prikulis, B. Kasemo, M. Kall, S. Zou, G. C. Schatz: J. Phys. Chem. B **109**, 1079 (2005)

[49] P. C. Waterman: Phys. Rev. D **3**, 825 (1971)

[50] M. I. Mishchenko, D. W. Mackowski, L. D. Travis: Appl. Opt. **34**, 4589 (1995)

[51] M. I. Mishchenko, L. D. Travis: Opt. Commun. **109**, 16 (1994)

[52] M. I. Mishchenko, L. D. Travis, D. W. Mackowski: J. Quant. Spectrosc. Radiat. Transfer **55**, 535 (1996)

[53] D. W. Mackowski, M. I. Mishchenko: J. Opt. Soc. Am. A **13**, 2266 (1996)

[54] D. W. Mackowski: J. Quant. Spectrosc. Radiat. Transfer. **70**, 441 (2001)

[55] D. W. Mackowski: J. Opt. Soc. Am. A **11**, 2851 (1994)

[56] G. Mie: Ann. Phys. **25**, 377 (1908)

[57] C. F. Bohren, D. R. Huffman: *Absorption and Scattering of Light by Small Particles* (Wiley, New York 1983)

[58] B. T. Draine, P. J. Flatau: *User Guide for the Discrete Dipole Approximation Code DDSCAT.6.0* (2003) URL http://arxiv.org/ags.astro-ph/0309069

[59] B. T. Draine, P. J. Flatau: J. Opt. Soc. Am. A **11**, 1491 (1994)

[60] A. Taflove: *Advances in Computational Electrodynamics: The Finite-Difference Time-Domain Method* (Artech House, Boston 1995) p. 599

[61] R. X. Bian, R. C. Dunn, X. S. Xie, P. T. Leung: Phys. Rev. Lett. **75**, 4772 (1995)

[62] L. Novotny, R. X. Bian, X. S. Xie: Phys. Rev. Lett. **79**, 645 (1997)

[63] E. Moreno, D. Erni, C. Hafner, R. Vahldieck: J. Opt. Soc. Am. A **19**, 101 (2002)

[64] L. Zhao, K. L. Kelly, G. C. Schatz: J. Phys. Chem. B **107**, 7343 (2003)

[65] E. Hao, G. C. Schatz: J. Chem. Phys. **120**, 357 (2004)

[66] S. Zou, G. C. Schatz: Nanotechnology (2006) in press

[67] D. W. Lynch, W. R. Hunter: Optical constants of metals, in E. D. Palik (Ed.): *Handbook of Optical Constants of Solids* (Academic, New York 1985) pp. 350–356

[68] H. Xu, M. Kall: Phys. Rev. Lett. **89**, 246802 (2002)

[69] P. K. Aravind, H. Metiu: Surf. Sci. **124**, 506 (1983)
[70] M. Micic, N. Klymyshyn, Y. D. Suh, H. P. Lu: J. Phys. Chem. **107**, 1574 (2003)
[71] P. T. Leung, W. S. Tse: Solid State Commun. **95**, 39 (1995)
[72] J. C. Hulteen, R. P. Van Duyne: J. Vac. Sci. Technol. A **13**, 1553 (1995)
[73] K. T. Carron, W. Fluhr, M. Meier, A. Wokaun, H. W. Lehmann: J. Opt. Soc. Am. B **3**, 430 (1986)
[74] V. A. Markel: J. Mod. Opt. **40**, 2281 (1993)
[75] S. Zou, G. C. Schatz: J. Chem. Phys. **12**, 97102 (2005)
[76] W. D. Bragg, V. A. Markel, W. T. Kim, K. Banerjee, M. R. Young, J. G. Zhu, R. L. Armstrong, V. M. Shalaev, Z. C. Ying: J. Opt. Soc. Am. B **18**, 698 (2001)
[77] K. Li, M. I. Stockman, D. J. Bergman: Phys. Rev. Lett. **91**, 227402 (2003)
[78] W. L. Barnes, A. Dereux, T. W. Ebbesen: Nature **424**, 824 (2003)
[79] J. A. Porto, F. J. Garca-Vidal, J. B. Pendry: Phys. Rev. Lett. **83**, 2845 (1999)

Index

Estimating SERS Properties of Silver-Particle Aggregates through Generalized Mie Theory

Hongxing Xu[1,2] and Mikael Käll[3]

[1] Institute of Physics, Chinese Academy of Sciences, P.O. Box 603-146, Beijing, 100080, P. R. China
[2] Division of Solid State Physics, Lund University, Box 118, 221 00, Lund, Sweden
[3] Applied Physics, Chalmers University of Technology, 412 96, Göteborg, Sweden
`kall@fy.chalmers.se`

1 Introduction

It has long been known that the classical electromagnetic (em) enhancement mechanism is by far the most important contributor to SERS and related surface-enhanced spectroscopies [1]. The em theory provides a quantitative understanding of all the principal characteristics of SERS, including variation in enhancement with wavelength, polarization, nanostructure morphology and type of metal. When applied to aggregates of particles, it also makes it understandable why SERS can provide single-molecule sensitivity [2, 3]. Unfortunately, realistic estimates of em enhancement effects in nanostructured materials are in general not straightforward and invariably involves approximations. In this Chapter, we briefly summarize our recent work on em-enhancement phenomena in nanoparticle aggregates, focusing on theoretical results obtained through generalized Mie theory [2, 3, 4, 5, 6, 7, 8, 9, 10, 11]. The fundamental approximation of the model is thus that the metal nanoparticles are treated as spheres. Although this might be seen as a gross oversimplification, the advantage of the Mie approach is that it provides a complete and analytic solution to the full Maxwell's equations, including, for example, retardation effects. The Chapter is organized as follows: After introducing the basic elements of the em enhancement effect and generalized Mie theory (GMT), we briefly summarize the calculation technique in Sect. 2. Section 3 then discusses the main theory results, focusing on the concept of "hot sites" in nanoparticle junctions, their polarization dependence and the relation to far-field optical properties. We also briefly bring up the importance of optical forces in SERS and an extension of the em SERS theory that includes a quantum optical treatment of the molecular response.

1.1 Electromagnetic Enhancement

Consider a molecule located at a point r in the vicinity of a nanostructure and let the molecule–nanostructure system be illuminated by an external incident field E_i with wavevector k. Because of the electromagnetic response of the system, the local field E_l at r will be different from E_i by a factor

K. Kneipp, M. Moskovits, H. Kneipp (Eds.): Surface-Enhanced Raman Scattering – Physics and Applications, Topics Appl. Phys. **103**, 87–104 (2006)
© Springer-Verlag Berlin Heidelberg 2006

$M = |\boldsymbol{E}_l|/|\boldsymbol{E}_i|$. This field-enhancement factor can be above or below unity, and its magnitude will be largest if the system supports internal electromagnetic resonances at the illumination frequency ω_0. The local field will induce a dipole moment in the molecule, and part of the molecular dipole scattering will appear at Raman-shifted frequencies $\omega_0 \pm \omega_{\mathrm{vib}}$. Because of electromagnetic reciprocity, that part of the dipole far-field that scatters into the $-\boldsymbol{k}$ direction and has the same polarization as the incident field will also be enhanced by the factor M, although the factor should now be evaluated at the Raman-shifted frequency of interest. Hence, the effective molecular Raman polarizability will be enhanced by a factor M^2 and the effective Raman cross section by a factor M^4. This power-of-four dependence makes it understandable why SERS is such a pronounced effect – even relatively modest field-enhancement factors, of the order 10–30, result in a huge Raman enhancement of the order $\sim 10^4$ to $\sim 10^6$. Similar types of arguments can be applied also to other types of surface-enhanced spectroscopies. In the case of fluorescence, the M^4 factor has to be multiplied by a factor $1/M_{\mathrm{d}}^2$, which compensates for the enhancement of the excited-state decay rate. If the fluorophore is far from the metal surface, we have that $M_{\mathrm{d}} \approx M$, which implies that the effective fluorescence cross section scales as M^2. For shorter distances (a few nanometers), $M_{\mathrm{d}} \gg M$ in general, and the fluorescence is quenched [9, 10, 11].

As noted above, a large field-enhancement factor requires some kind of electromagnetic resonance in the nanostructure, the most efficient and well known being localized surface plasmons (LSPs) of various types. This explains why silver, which can support sharp LSP modes over the entire visible to near-infrared wavelength range, and gold, which is useful at frequencies below the interband region ($\lambda > \sim 550\,\mathrm{nm}$), are the optimal materials for construction of SERS-active nanostructures. However, other kinds of resonances, for example diffractive modes in particle arrays or other types of spatially extended nanostructures, and purely electrostatic effects, such as the lightning-rod effect, can also contribute substantially to the field-enhancement factor. Naturally, the largest SERS signal is expected when several types of electrodynamic and electrostatic enhancement effects work in unison. This is the case in the single-molecule SERS experiments illustrated in Fig. 1 [2]. Here, hemoglobin (Hb) was incubated together with a heterogeneous Ag colloidal solution in a ratio of approximately one protein molecule per three Ag nanoparticles. The proteins, together with a low concentration of NaCl, cause a slight aggregation of the colloid, resulting in the formation of particle dimers and a few larger clusters. These small aggregates turned out to be highly SERS active, whereas the single isolated particles did not produce any measurable signal. Moreover, the Raman signal was maximal for dimers that were oriented parallel to the incident polarization, for which the SERS enhancement factor was experimentally estimated to be of the order 10^{10}. As will be discussed below, these observations can be understood from general-

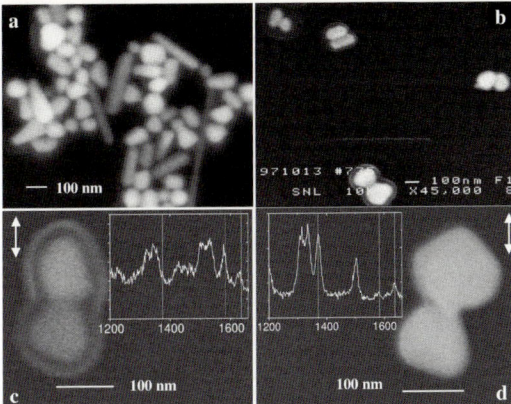

Fig. 1. (Reproduced from [2]), SEM images of immobilized Ag-particles. The pictures show (**a**) overview of Ag-particle shapes and sizes, (**b**) Ag-particle dimers observed after incubation with hemoglobin, and (**c**), (**d**) hot dimers and corresponding single-molecule Raman spectra. The *double arrows* in (**c**) and (**d**) indicate the polarization of the incident laser field

ized Mie theory (GMT) calculations of field-enhancement factors in the gap between Ag nanoparticles [2, 3].

1.2 Generalized Mie Theory

Mie theory [12] is a method to solve the boundary condition problem involved in the scattering and extinction of light by a single sphere situated in a homogeneous medium. This can be done through an expansion of the incident and scattered fields into vector spherical harmonics (VSHs). Mie theory has formed the basis for a vast range of techniques that can also be used to understand more complex scattering problems, for example a sphere above a flat surface, spheres composed of multiple concentric shells and ensembles of interacting spheres, which is the focus of the present chapter. Although any formal definition seems meaningless, we group these methods under a common heading, i.e., generalized Mie theory (GMT). Many of the GMT methods have proved instrumental for the development of a theoretical basis for surface-enhanced spectroscopy, plasmonics, and nano-optics in general. GMT also provides a convenient analytical comparison for a range of more novel grid-based methods in computational electromagnetics, such as the discrete dipole approximation (DDA) and the finite difference time-domain (FTDT) methods. Pioneering Mie theory contributions on SERS from aggregates include those of [13, 14, 15, 16, 17, 18].

A crucial component of GMT is the transition matrix (T-matrix) technique, introduced by *Waterman* [19, 20]. The T-matrix can be used to relate the VSH expansion coefficients for different spheres in an ensemble based on the addition theorem of *Stein* [21] and *Cruzan* [22] and the total scattered field can then be obtained directly by solving a linear system of equations [23]. Various techniques to calculate the T-matrix of multisphere systems have been reported [20, 23, 24]. However, many of these methods are tedious when the distances between spheres are small or when the number of spheres is large. A complementary approach is the order-of-scattering (OS) method, which expresses the total scattered field from an ensemble as a sum of different scattering orders. By tracing the light paths in the multiple scattering processes, the contribution to the total field from each scattering event can be obtained by applying the boundary conditions for each single scatterer. *Fuller* has given the solution for the case of two spheres [25]. Recently, we developed a technique to calculate the scattering from multisphere systems using a recursive OS approach [7, 8]. This method is described briefly in the following paragraph.

2 The Recursive Order-of-Scattering Method

Similar to the original Mie theory [12], the incident electric field and the scattered electric field of an ensemble of L spheres are expanded in vector spherical harmonics (VSHs) as:

$$
{}^{i}\boldsymbol{E}_l = \sum_{n=1}^{\infty} \sum_{m=-n}^{n} \sum_{p=1}^{2} {}^{i}C_{mnp}^{l} \, |mn1p\rangle
$$

$$
{}^{s}\boldsymbol{E} = \sum_{l=1}^{\oplus L} {}^{s}\boldsymbol{E}_l = \sum_{l=1}^{\oplus L} \sum_{n=1}^{\infty} \sum_{m=-n}^{n} \sum_{p=1}^{2} {}^{s}C_{mnp}^{l} \, |mn3p\rangle, \tag{1}
$$

where ${}^{i}C_{mnp}^{l}$ and ${}^{s}C_{mnp}^{l}$ are the expansion coefficients for the VSH $|mnjp\rangle$ centered at the l-th sphere, with $p = 1$ for $\boldsymbol{M}_{mn}^{j}$ and $p = 2$ for $\boldsymbol{N}_{mn}^{j}$, respectively. The index $j = 1, 2, 3, 4$ corresponds to spherical Bessel and Hankel functions j_n, y_n, $h_n^{(1)}$, $h_n^{(2)}$, respectively [26], and the symbol $\oplus$ means that the sum should be performed in Cartesian coordinates. The scattering coefficients ${}^{s}C_{mnp}^{l}$ are functions of the incident coefficients ${}^{i}C_{mnp}^{l}$, the corresponding Lorenz–Mie coefficients a_n^l and b_n^l, and the translation coefficients ${}^{lh}A_{mn}^{\mu\nu}$ and ${}^{lh}B_{mn}^{\mu\nu}$, relating sphere l to sphere h [21, 22], i.e:

$$
{}^{s}C_{mnp}^{l} = {}^{L}T_l({}^{i}C_{\mu\nu q}^{h}, a_\nu^h, b_\nu^h, {}^{lh}A_{mn}^{\mu\nu}, {}^{lh}B_{mn}^{\mu\nu}). \tag{2}
$$

Similarly, the corresponding magnetic fields can be expanded as:

$$
\boldsymbol{H}_i^l = \frac{k}{i\omega\mu} \sum_{n=1}^{\infty} \sum_{m=-n}^{n} \sum_{p=1\neq p'}^{2} {}^{i}C_{mnp}^{l} \, |mn1p'\rangle
$$

$$
\boldsymbol{H}_s = \frac{k}{i\omega\mu} \sum_{l=1}^{\oplus L} \boldsymbol{H}_s^l = \frac{k}{i\omega\mu} \sum_{l=1}^{\oplus L} \sum_{n=1}^{\infty} \sum_{m=-n}^{n} \sum_{p=1\neq p'}^{2} {}^{s}C_{mnp}^{l} \, |mn3p'\rangle \tag{3}
$$

due to the relations $\boldsymbol{H} = \frac{1}{i\omega\mu}\nabla \times \boldsymbol{E}$, $\boldsymbol{N} = \frac{1}{k}\nabla \times \boldsymbol{M}$, $\boldsymbol{M} = \frac{1}{k}\nabla \times \boldsymbol{N}$.

The recursive OS method utilizes a matrix representation of T in (2), which is described as a product of the matrices G, which contain the expansion coefficients of the incident field for each sphere, and the response matrix Ψ of the L-sphere system, which transfer the incident coefficients to the scattering coefficients (for details see [8]). One can thus write

$$[^L T_1, {}^L T_2, {}^L T_3 \cdots {}^L T_L] = [G_1, G_2, G_3 \cdots G_L]\Psi^{(L)},\qquad(4)$$

where

$$\Psi_{LL}^{(L)} = S_L \sum_{i=0}^{N_{os}} (\Omega^{(L-1)}\Psi^{(L-1)}\Omega'^{(L-1)}S_L),$$

$$\Psi_{pL}^{(L)} = \sum_{j=1}^{L-1} \Psi_{pj}^{(L-1)}\Omega_{j,L}\Psi_{LL}^{(L)}\qquad\qquad p = 1,\ldots,L-1,$$

$$\Psi_{Lq}^{(L)} = \Psi_{LL}^{(L)} \sum_{j=1}^{L-1} \Omega_{L,j}\Psi_{jq}^{(L-1)}\qquad\qquad q = 1,\ldots,L-1,$$

$$\Psi_{pq}^{(L)} = \Psi_{pq}^{(L-1)} + \Psi_{pL}^{(L)} \sum_{j=1}^{L-1} \Omega_{L,j}\Psi_{jq}^{(L-1)}\qquad p,q = 1,\ldots,L-1.\qquad(5)$$

Here, Ω_{pq} are matrices that contain the transfer coefficients relating VSHs originating in the p-th sphere to those originating in the q-th sphere, S_p are the matrices that contain the Mie scattering coefficients, while $\Omega^{(L-1)} = [\Omega_{L1}, \Omega_{L2}, \ldots, \Omega_{L,L-1}]$ and $\Omega'^{(L-1)} = [\Omega_{L1}, \Omega_{L2}, \ldots, \Omega_{L,L-1}]^T$ [8]. Moreover, it can be shown that the summation over different scattering orders $i \in [0, N_{OS}]$ up to infinite order can be obtained through matrix inversion instead of summation, according to:

$$\lim_{N_{OS}\Rightarrow\infty} \sum_{i=0}^{N_{os}} (\Omega^{(L-1)}\Psi^{(L-1)}\Omega'^{(L-1)}S_L)$$
$$= \frac{1}{1 - \Omega^{(L-1)}\Psi^{(L-1)}\Omega'^{(L-1)}S_L}.\qquad(6)$$

In order to find the response matrix $\Psi^{(L)}$ for L spheres, we first find the response matrix $\Psi^{(L-1)}$ for $(L-1)$ spheres, and so on. Finally, the response matrix for the first sphere is found as $\Psi^{(1)} = S_1$. Using this recursive method, the response matrix $\Psi^{(L)}$ of L-spheres can be obtained according to (1)–(6).

Scattering problems that involve an arbitrary number of interacting spheres can in principle be solved exactly by the technique described briefly above. In practice, however, one needs to calculate a very large number of complex functions and translation coefficients that include a large number of multipoles n in order to ensure convergence, in particular for closely spaced particles. If we let n range from 1 to N, the total number of VSHs will be $N \times (2N + 1) \times 2 \times 2$ for a two-sphere system, i.e., more than eighty thousand for $N = 100$. The total number of translation coefficients will be $\left[N \times (2N + 1) \times 2 \times 2 \right]^2$, i.e., more than 6 billion. Moreover, if we have L spheres, the number of translation coefficients increases to $\frac{L}{2}(L - 1) \times \left[N \times (2N + 1) \times 2 \times 2 \right]^2$. Considering the complexity of calculating each translation coefficient, it is not surprising that GMT calculations of strongly interacting particles has turned out to be a challenge, both in terms of mathematical methodology and the computer capacity needed.

3 Examples of GMT Calculations for Ag-Particle Aggregates

3.1 "Hot Sites" Between Metal Particles

Figure 2 illustrates the effect of aggregation of metal particles in SERS. Using GMT, we have calculated the local field at two positions around two large (diameter $D = 200$ nm) silver spheres separated by a gap of length d. The two spheres are illuminated by a plane wave polarized parallel to the dimer axis. For position A, located 0.5 nm from one of the spheres along the dimer axis, the factor M^4 increases dramatically for decreasing interparticle distances and reaches $\sim 10^9$ for $d = 1$ nm. At position B, which is located 0.5 nm from the surface in the direction perpendicular to the dimer axis, there is no substantial enhancement at any separation. These effects can, in fact, be well accounted for by a simple electrostatic model. Imagine that the two particles were perfectly conducting and situated in a uniform electrostatic field E_i oriented in the direction of the dimer axis. Since the field will be excluded from the perfectly conducting spheres, the electrostatic potential drop will be concentrated to the interparticle region. From the geometry of the problem, this results in a local field between the spheres of strength $E_l = E_i(D + d)/d$. A molecule situated between the particles thus experiences a SERS enhancement factor of the order $M^4 = (D/d + 1)^4$, while molecules outside the interparticle region will experience no net enhancement ($M^4 \equiv 1$). The comparison between the simple electrostatic model and the full electrodynamic calculation in Fig. 2 clearly shows that the former captures much of the essential physics of the interparticle coupling problem.

Figure 3 illustrates the wavelength dependence of the SERS enhancement for dimers composed of identical Ag spheres of varying diameter D but with a fixed surface-to-surface separation $d = 1$ nm. The prominent peaks that are evident in those spectra originate in plasmon modes of the dimer system. In particular, the longest-wavelength peak in each spectrum can be assigned to a mode dominated by the dipolar plasmons of each sphere coupled in phase. This mode has recently been identified in scattering spectra of nanofabricated Ag-particle dimers, and its dispersion with d and D has been shown

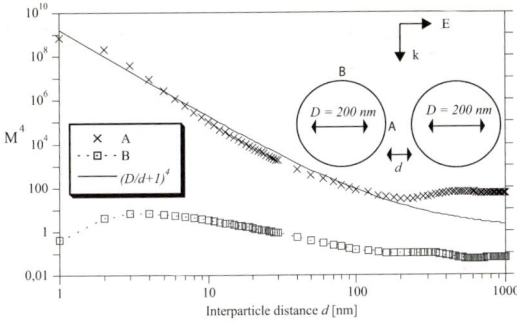

Fig. 2. (Adapted from [4]), Illustration of SERS enhancement, approximated as the fourth power of the field enhancement factor M, at two positions around a pair of silver spheres of diameter $D = 200$ nm for varying surface-to-surface separation d. The wavelength is $\lambda = 514.5$ nm and dielectric data for silver from [27] have been used in the GMT calculations. The *thin solid line* shows the electrostatic estimate of the enhancement effect. Note the logarithmic scales

to be in good agreement with electrodynamic calculations [28]. However, the prominence of the coupled dipole mode in the spectra of Fig. 3 does not imply that the field-enhancement effects can be understood within the dipole approximation. On the contrary, in order to build up the electrostatic component of the field enhancement between particles, a large number of multipolar resonances have to be taken into account. This is illustrated in Fig. 4, which shows the intensity enhancement factor M^2 versus the number of multipoles N used in a GMT calculation for a dimer of $D = 90$ nm spheres excited with light polarized parallel to the dimer axis. We see that for a separation distance of around $d = 10$ nm, it is necessary to include multipoles up to order $N = 10$ whereas for distances down to 1 nm more than 50 multipolar terms are needed to reach convergence. The latter d value is probably close to the smallest interparticle distance for which classical electrodynamic calculations can be trusted. For even smaller separations, one approaches the spill-out region of the Ag s electrons [29], which should lead to interparticle currents that reduce the field strength in the gap region. Moreover, for smaller distances or for gaps between sharper structures, the nonlocal dielectric response of the metal has to be taken into account. Nonlocal effects can be expected to reduce the contribution from high-order multipoles [30], although to what extent is a matter of debate.

From GMT calculations such as those in Fig. 2 and Fig. 3, we find that the maximum SERS enhancement factor M^4 at "hot sites", i.e., in gaps, in Ag-dimer systems peaks at around 10^{11} to 10^{12} for gap distances of the order 1 nm. Taking into account the Stokes shift in the Raman process and the possibility that the imaginary part of the Ag dielectric function may be larger than in the experimental values used [27], a more conservative value for the maximum em SERS enhancement might be 10^{10}. Although no formal proof

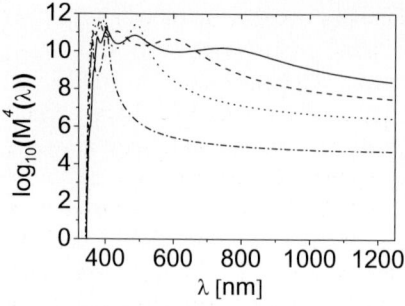

Fig. 3. (Adapted from [3]), SERS enhancement factor M^4 as function of wavelength for the midpoint between two identical Ag spheres with surface-to-surface separation $d = 1\,\mathrm{nm}$ and different diameters: $D = 140\,\mathrm{nm}$, $100\,\mathrm{nm}$, $60\,\mathrm{nm}$ and $20\,\mathrm{nm}$, from *top* to *bottom*. The dimer is illuminated with light polarized parallel to the dimer axis and with incident $\boldsymbol{k}$ vector normal to this axis

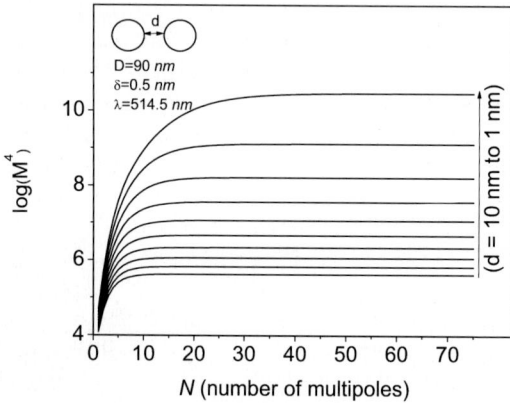

Fig. 4. (Adapted from [4]), Calculated intensity enhancement M^2 for a position along the dimer axis $0.5\,\mathrm{nm}$ from the surface of one of the spheres as a function of the number of multipoles N included in the GMT calculation. The diameters of the spheres are $90\,\mathrm{nm}$, and the incident polarization is parallel to the dimer axis. The calculations are performed for different surface-to-surface distances d, from $1\,\mathrm{nm}$ to $10\,\mathrm{nm}$

exists, it seems likely that this is also close to the maximum em enhancement factor in any type of Ag or Au nanostructure that contains gaps or crevices of similar dimensions.

3.2 Polarization Anisotropy

From the GMT results above, we have seen that the SERS enhancement in small nanoparticle aggregates is expected to be a highly localized phenomenon. Due to the combination of electrostatic effects and electrodynamic LSP modes, the calculated M^4 factor can reach above 10^{10} in the gap regions between particles, and at the same time be at or below unity outside these areas. The gaps between particles thus constitute "hot sites" for SERS and the molecules at those sites are expected to dominate the net SERS signal from an aggregate, even if the whole aggregate is covered by molecules. However, the gap regions are not "hot" for all incident polarizations. If a dimer is excited with light polarized perpendicular to the dimer axis, the field is instead excluded from the gap region, resulting in an enhancement below unity. We thus expect the SERS signal from dimers and other small aggregates to be highly polarization dependent. Figure 5 gives an experimental example that illustrates this effect. The samples in this study were prepared in a similar way to the ones in Fig. 1, i.e., by first incubating hemoglobin and colloidal Ag particles and then immobilizing the resulting Hb/Ag aggregates for Raman and electron microscopy investigations. However, the Hb:Ag-particle ratio was ~ 100 times higher in the present case. Figure 5a shows a SEM image of a number of Hb/Ag aggregates within one investigated area. The corresponding polarized Raman images for different angles between the incident polarization and the coordinate system of the sample are shown in Fig. 5b. Six bright spots, marked as A to F, could be clearly identified as different aggregates in the SEM, and the intensities of these spots varied to different degrees with the incident polarization. Figure 5c shows a polar plot of this variation for spot A, which turned out to be a dimer with its dimer axis rotated $\alpha_0 \approx 125°$ relative to the vertical axis in Fig. 5a. It is clear that the Raman intensity has a maximum when the incident polarization is parallel to the dimer axis – in fact, the angular variation can be quite well described by a $\cos^4(\alpha - \alpha_0)$ dependence, which is the expected variation if a "hot" gap site completely dominates the Raman response. For aggregates composed of more than two particles, the polarization dependence turned out to be in general more isotropic than for isolated dimers. Figure 5d shows a polar plot of the Raman intensity from spot C, which was composed of five Ag particles. In this case, a large signal was observed for all polarization angles, but with two noticeable anisotropic intensity peaks for polarization parallel to the $80°/260°$ and $160°/340°$ directions. We interpret such complicated polarization dependencies as a result of an entangled electromagnetic coupling between several particles. In this particular case, the angular variation indicates an interpretation in terms of a superposed signal from two dominating but perpendicular dimers within the five-particle aggregate. Figure 6 shows a GMT calculation that illustrates the variation in enhancement with polarization for a five-particle system (*not* chosen to mimic cluster C in Fig. 5). As expected, one finds that each polarization directions tend to "select" those gap sites for which the respective dimer axes overlap most with the incident polarization. However, the variation is not as clear-cut as for a symmetric dimer, simply because the total field at each gap site in a large low-symmetry cluster also involves retarded dipolar fields from distant particles that add to the local "dimer field" with varying phase factors.

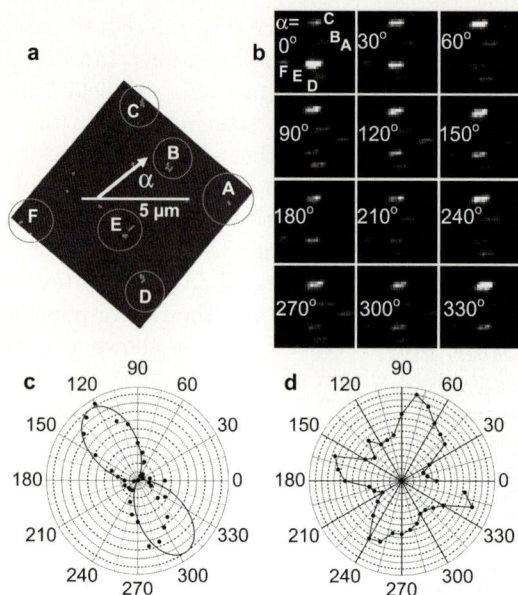

Fig. 5. (Reproduced from [6]), (**a**) SEM image of Hb/Ag clusters and (**b**) the corresponding polarized Raman images ($11 \times 11 \, \mu\mathrm{m}^2$) for different incident polarizations ($\lambda_\mathrm{I} = 514.5 \, \mathrm{nm}$, $I_0 \approx 0.1 \, \mathrm{W}/\mu\mathrm{m}^2$, Raman intensity integrated over a Stokes shift from $700 \, \mathrm{cm}^{-1}$ to $2200 \, \mathrm{cm}^{-1}$). The polar plots show the angular variation of the Raman intensity for (**c**) spot A and (**d**) spot C. The *full line* in (**c**) is a fit to a $\cos^4(\alpha - \alpha_0)$ dependence

3.3 Comparisons between Near-Field and Far-Field Spectra

In the preceding discussion, we have focused on the nano-optical phenomenon of greatest relevance to SERS, i.e., the near-field enhancement effect in Ag-particle aggregates. However, far-field optical properties, in particular extinction spectra, are often measured in conjunction with SERS experiments and it is therefore interesting to briefly compare the calculated near-field and far-field properties of Ag-particle aggregates. In Fig. 7, we do this for two different cases, a dimer and a low-symmetry pentamer. All gap sizes in both systems equals $d = 1 \, \mathrm{nm}$ but the spheres have different sizes in order to simulate the "nonideal" morphology of colloidal aggregates. In addition, the spheres have been positioned such that their lower boundaries form a plane, in order to simulate aggregates immobilized on a surface. As before, we compute near-field spectra for gap sites only.

In the case of the dimer, we first note the prominence of the aforementioned long-wavelength dipolar peak for polarization parallel to the dimer axis. This mode dominates both the intensity enhancement spectrum for the gap site and the far-field extinction spectrum, which in turn is composed mainly of elastic scattering. In addition, the short-wavelength region con-

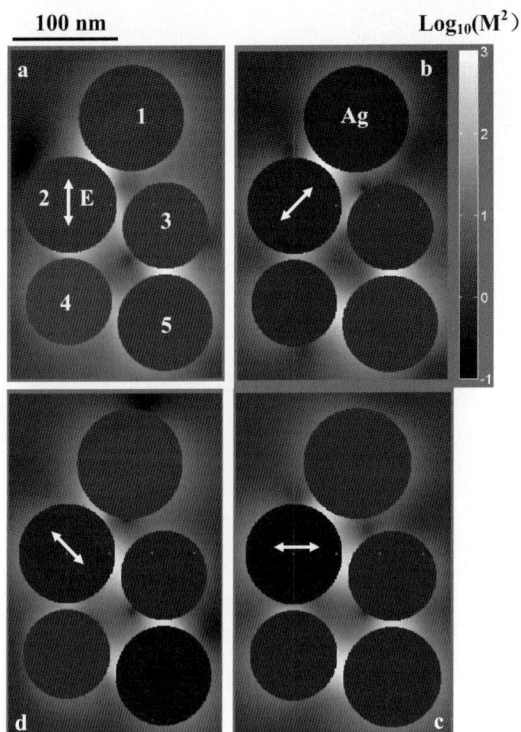

Fig. 6. (Adapted from [8]), Intensity enhancement factor M^2 (logarithmic scale) at 514.5 nm in a plane through the centers of five different Ag spheres with diameters $D_1 = 110$ nm, $D_2 = D_5 = 100$ nm and $D_3 = D_4 = 90$ nm. The gap dimensions are $d_{12} = d_{24} = d_{35} = 5.5$ nm and $d_{23} = d_{45} = 7.4$ nm. The incident k vector is normal to the plane of the paper and arrows mark the different incident polarization vectors. The calculation is based on $N = 16$ multipoles and includes scattering orders up to $N_{os} = 200$

tains some well-defined modes that mainly originate in various hybridized quadrupolar resonances. When the polarization is turned perpendicular to the dimer axis, the enhancement at the gap site more-or-less vanishes, as before. The extinction spectrum is still dominated by a coupled dipolar mode, but its energy in this polarization configuration is such that it overlaps with higher-order multipolar peaks, and it does not contribute an enhancement at the gap site. All in all, there is clearly a reasonably good and understandable correspondence between far-field and near-field properties of the dimer. In the case of the pentamer, however, the situation is quite different. The extinction spectrum is dominated by a broad long-wavelength peak of dipolar origin, qualitatively similar to what is observed for strongly aggregated colloids. The near-field intensities in the junctions, in contrast, exhibit resonances that are much sharper than for the dimer and extremely polarization dependent. It is

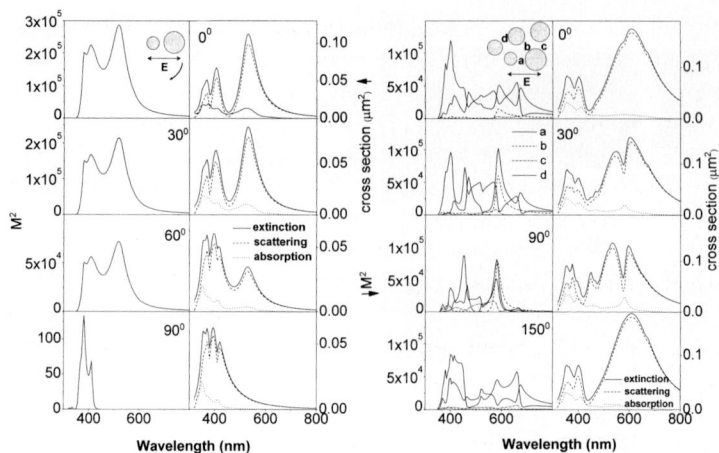

Fig. 7. Near-field intensity enhancement $M^2(\lambda)$ at "hot" gap sites and far-field cross sections (extinction, scattering, absorption) for a dimer ($D = 60$ nm and $D = 100$ nm) an a pentamer ($D = 60$ nm, 70 nm, 80 nm, 90 nm and 100 nm) in vacuum with gap dimensions $d = 1$ nm throughout. The spheres are situated on a planar virtual surface and the incident $\boldsymbol{k}$ vector is normal to this surface with polarization as indicated in the insets. The calculations included multipoles up to order $N = 30$ and scattering processes up to infinite order through the matrix-inversion technique

possible that these resonances signal the onset of diffractive coupling effects caused by the coherent addition of fields emanating from distant particles. We note that similar effects, leading to high field enhancements, have been predicted and observed for both linear arrays of Ag particles [31, 32] and for large Au-particle clusters of fractal dimensions [33, 34]. Although the results require further analysis, the pentamer case thus illustrates that the near-field and the far-field properties of the same aggregate can be distinctly different. The results also indicate that the concept of "hot" junction sites that dominate the SERS response remains valid also for large particle aggregates of low symmetry. This is in qualitative agreement with experimental SERS studies of dense Ag particle layers, which exhibit single-molecule fluctuations even when analyte molecules are present at high concentrations [35, 36].

3.4 Including Molecular Quantum Dynamics into the EM SERS Theory

The electromagnetic mechanism of SERS discussed above treats the molecule as a "passive observer" of the field-enhancement effects. Recently, however, an extension of the em theory that includes a quantum optical description of a model molecule was put forward [9, 10, 11]. The model molecule is described by two electronic states (HOMO/LUMO) and one vibrational mode, and the molecular dynamics is calculated using a density matrix approach, which allows one to compute absolute Raman, fluorescence and absorption cross sections and spectra for arbitrary irradiation levels. The parameters of the model were chosen so as to simulate the absorption spectrum of the fluorophore rhodamine 6G, which has become the work-horse of single-molecule SERS and fluorescence experiments (see, e.g., [36, 37]). The corresponding calculated energy integrated fluorescence cross section was found to be $\approx 10^{-16} \, \text{cm}^2$, which agrees with experimental values for many fluorophores. The model molecule was then placed at a "hot" gap site with $d = 1 \, \text{nm}$ and $M^4 \approx 10^{10}$ and the combined Raman and fluorescence spectrum was calculated. The integrated SERS cross section was found to be $\approx 10^{-14} \, \text{cm}^2$, i.e., two orders-of-magnitude higher than the fluorescence cross section of the free model-molecule and ~ 15 orders-of-magnitude higher than a "typical" Raman cross section of a nonresonant molecule ($\sigma_R \approx 10^{-29} \, \text{cm}^2$). These values are in agreement with recent single-molecule SERRS (surface-enhanced *resonance* Raman scattering) studies of rhodamine 6G [36,37], and suggest that no additional enhancement mechanisms beyond the "electromagnetic SERS theory" are needed in order to understand the single-molecule SERS phenomenon. Interestingly, the calculations also showed that a substantial enhanced fluorescence similar to the intense "SERS background" seen in many single-molecule SERRS experiments [36, 37] remained for a molecule in a "hot" gap site, whereas the familiar fluorescence-quenching effect dominated for molecules located near single particles.

3.5 Optical Forces

Optical forces has been a largely neglected aspect of SERS and nanoplasmonics in general, although it seems likely that such forces can be of both practical and fundamental importance. The total optical force that acts on a single particle or molecule situated in an inhomogeneous field can in general be resolved into two components: the gradient force and the dissipative force. In MKS units, we have:

$$\boldsymbol{F}_{\text{grad}}(\boldsymbol{r}) = \frac{n_{\text{m}}}{2c} \Re\left\{\alpha(\lambda)\right\} \nabla I(\boldsymbol{r})$$

$$\boldsymbol{F}_{\text{diss}}(\boldsymbol{r}) = \boldsymbol{k}\frac{n_{\text{m}}}{c} \Im\left\{\alpha(\lambda)\right\} I(\boldsymbol{r}). \tag{7}$$

Here, $\alpha(\lambda)$ is the wavelength-dependent dipole polarizability of the particle or molecule, n_{m} is the refractive index of the surrounding medium, c is the speed of light in vacuum and $I(\boldsymbol{r})$ is the optical intensity profile.

From (7), we see that the gradient force is repulsive when $\Re\left\{\alpha\right\} < 0$. This situation can occur for an incident wavelength on the blue side of a dipole resonance, such as a localized plasmon or molecular absorption band. For illumination on the red side of a resonance, on the other hand, the object is pulled towards regions of high local intensity, which is why dielectric objects can be easily trapped using laser tweezers. The dissipative force, in contrast, is proportional to the extinction cross section $\sigma_{\text{ext}} = k\Im\{\alpha\}$ and therefore always positive, i.e., the object is pushed in the direction along the incident $\boldsymbol{k}$ vector.

From the discussions in preceding sections, we have seen that nanoparticle aggregates exhibit regions of ultrahigh field enhancement and at the same time regions where the optical field is not enhanced at all, or even excluded (see, e.g., the dark spots in Fig. 6). In [5], we examined whether the resulting intensity gradients could be large enough to actually pull molecules into the em "hot spots". Figure 8a shows the theoretical result for a linear trimer of Ag spheres, which exhibits its largest intensity enhancement $M^2 \approx 9.2 \times 10^5$ in the near-infrared ($\lambda_{\text{LSP}} \approx 760\,\text{nm}$). The molecular polarizability was chosen to mimic rhodamine 6G, which is resonant in the green ($\lambda_{\text{HOMO}\rightarrow\text{LUMO}} \approx 530\,\text{nm}$). The figure shows the conservative gradient force at $\lambda \approx 760\,\text{nm}$ in terms of the optical potential U in units of $k_{\text{B}}T$ at $T = 300\,\text{K}$ and for an incident intensity of $10\,\text{mW}/\mu\text{m}^2$, which is a rather typical excitation intensity in NIR SERS experiments. For the specific parameters used, the optical potential in the gap turns out to be $U \approx 6k_{\text{B}}T$. Thus, the optical force will be able to compete effectively with the random thermal motion of the molecule despite an irradiation wavelength far from the molecular resonance. Although no experimental verifications exists to date, it thus seems reasonable to believe that nanoscale optical forces acting on the molecular level could be of fundamental importance and, perhaps, be used to actively drive molecules to specific "hot spots" in plasmonic nanostructures. This is, however, not the only optical force effect that may be of relevance to SERS. In a recent experimental study, it was shown that the SERS intensity obtained from a colloidal solution of Ag particles covered by thiophenol could be increased dramatically if a high-intensity near-infrared laser beam was made to overlap with the green excitation laser in the Raman microscope [38]. This phenomenon was interpreted as the combined result of two effects: an accumulation of nanoparticles in the dual laser focus, i.e., the optical tweezers effect, and an optical aggregation of those nanoparticles due to electromagnetic interactions. The latter effect has been investigated theoretically in [5,39] and it is likely that it contributes to the plasmon renormalization recently observed when Ag particle dimers were created through optical manipulation [40].

The optical forces that act on a given particle due to the scattered fields from neighboring particles can be computed from Maxwell's stress tensor and represented in terms of an optical potential. An example is shown in Fig. 8b. The calculation is performed for a Ag-particle dimer illuminated by a beam of $\lambda = 514.5\,\text{nm}$ light polarized along the dimer axis and with a unit incident intensity of $1\,\text{mW}/\mu\text{m}^2$. The interparticle optical potential is shown as a function of gap size d for a few different sphere radii. For small distances, we find a strong attractive interaction, far greater than $k_{\text{B}}T$, that has a fairly long range, in particular for the larger particles. For longer sep-

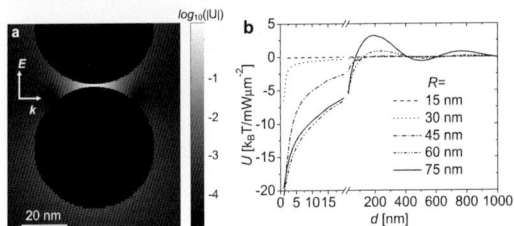

Fig. 8. (Adapted from [5]), (**a**) Spatial variation of the optical potential U in units of k_BT ($T = 300\,K$) around a three-sphere system ($D = 50\,nm$, $d = 1\,nm$, $I_0 = 10\,mW/\mu m^2$) in water excited at $760\,nm$. The molecular polarizability was chosen to simulate rhodamine 6G. (**b**) Interparticle optical potential $U(d; R)$ for dimers in water in units of k_BT ($T = 300\,K$) for an illumination intensity of $1\,mW/\mu m^2$, a wavelength of $514.5\,nm$ and a polarization along the dimer axis

arations ($d \approx \lambda/2$) the interaction turns repulsive because the dipole fields of the individual particles are no longer in phase. The interaction is also repulsive for short separations if the incident field is polarized perpendicular to the dimer axis. The interparticle optical force can thus be expected to pull properly oriented particles together, creating "hot" gap sites, while aggregates that are oriented perpendicular to the incident polarization should be pushed apart. Experimental investigations in this direction are underway and will be reported elsewhere.

4 Summary

Using generalized Mie theory, it is possible to estimate field-enhancement effects in strongly interacting nanoparticle systems, including dense aggregates of several nanoparticles. Apart from the inherent approximation of a spherical particle shape, the method is a complete solution to Maxwell's equations. Based on GMT, we have shown that Ag nanoparticle aggregates exhibit polarization-dependent "hot" gap sites that will dominate the SERS response. This result is in agreement with the majority of SERS experiments and in line with previous theory results from the 1980s. The "hot" gap sites can be expected to yield SERS enhancements factors of the order of 10^{10} or above. When combined with a proper description of the quantum optical response of the molecule, this factor quantitatively explains SERS from single resonant molecules. We have also discussed optical forces in SERS and shown that molecules could be attracted to sites with high surface enhancement while properly aligned particles could be pulled closer together, even for relatively modest irradiation levels. These effects could be of both fundamental and practical importance.

References

[1] M. Moskovits: Rev. Mod. Phys. **57**, 783 (1985)
[2] H. X. Xu, et al.: Phys. Rev. Lett. **83**, 4357 (1999)
[3] H. X. Xu, et al.: Phys. Rev. E **62**, 4318 (2000)
[4] H. X. Xu, et al.: Proc. SPIE **4258**, 35 (2001)
[5] H. X. Xu, M. Käll: Phys. Rev. Lett. **89**, 246802 (2002)
[6] H. Xu, M. Käll: Chem. Phys. Chem. **4**, 1001 (2003)
[7] H. X. Xu: Phys. Lett. A **312**, 411 (2003)
[8] H. X. Xu: J. Opt. Soc. Am. **21**, 804 (2004)
[9] H. X. Xu, et al.: Phys. Rev. Lett. **93**, 243002 (2004)
[10] P. Johansson, H. X. Xu, M. Käll: Phys. Rev. B **72**, 035427 (2005)
[11] M. Käll, H. X. Xu, P. Johansson: J. Raman Spectrosc. **36**, 510 (2005)
[12] G. Mie: Ann. Phys. **25**, 377 (1908)
[13] P. K. Aravind, A. Nitzan, H. Metiu: Surf. Sci. **110**, 189 (1981)
[14] P. K. Aravind, H. Metiu: Surf. Sci. **124**, 506 (1983)
[15] P. K. Aravind, H. Metiu: J. Phys. Chem. **86**, 5076 (1982)
[16] J. I. Gersten: J. Chem. Phys. **72**, 5779 (1980)
[17] J. I. Gersten, A. Nitzan: J. Chem. Phys. **73**, 3023 (1980)
[18] M. Inoue, K. Othtaka: J. Phys. Soc. Jpn. **52**, 3853 (1989)
[19] P. C. Waterman: Phys. Rev. D **3**, 825 (1971)
[20] M. I. Mischenko, et al.: J. Quantum Spectrosc. Radiat. Transfer **88**, 357 (2004)
[21] S. Stein: Quart. Appl. Math. **19**, 15 (1961)
[22] O. R. Cruzan: Quart. Appl. Math. **20**, 33 (1962)
[23] A. K. Hamid, I. R. Ciric, M. Hamid: IEEE Proc. H **138**, 565 (1991)
[24] Y. M. Wang, W. C. Chew: IEEE Trans. Antennas Propag. **41**, 1633 (1993)
[25] K. A. Fuller: Appl. Opt. **30**, 4716 (1991)
[26] J. A. Stratton: *Electromagnetic Theory* (McGraw-Hill, New York 1941)
[27] P. B. Johnson, R. W. Christy: Phys. Rev. B **6**, 4370 (1972)
[28] L. Gunnarsson, et al.: J. Phys. Chem. B **109**, 1079 (2005)
[29] A. Liebsch: *Electronics Excitations at Metal Surfaces* (Plenum, New York 1997)
[30] A. Pack, M. Hietschold, R. Wannemacher: Opt. Comm. **194**, 277 (2001)
[31] S. Zou, N. Janel, G. C. Schatz: J. Chem. Phys. **120**, 10871 (2004)
[32] E. M. Hicks, et al.: Nano Lett. **5**, 1065 (2005)
[33] V. P. Drachev, et al.: *Nonlinear Optical Effects and Selective Photomodification of Colloidal Silver Aggregates* (Springer, Berlin, Heidelberg, New York 2002) p. 113
[34] E. J. Bjerneld, P. Johansson, M. Käll: Single Mol. **1**, 239 (2000)
[35] E. J. Bjerneld, et al.: J. Phys. Chem. A **108**, 4187 (2004)
[36] S. M. Nie, S. R. Emory: Science **275**, 1102 (1997)
[37] K. A. Bosnick, J. Jiang, L. E. Brus: J. Phys. Chem. B **106**, 8096 (2002)
[38] F. Svedberg, M. Käll: Faraday Discuss. **132**, 35 (2006)
[39] A. J. Hallock, P. L. Redmond, L. E. Brus: Proc. Natl. Acad. Sci. **102**, 1280 (2005)
[40] J. Prikulis, et al.: Nano Lett. **4**, 115 (2004)

Index

Studying SERS from Metal Nanoparticles and Nanoparticles Aggregates with Continuum Models

Stefano Corni[1] and Jacopo Tomasi[2]

[1] INFM-CNR-S3 National Research Center on nanoStructures and bioSystems at
   Surfaces, via Campi 213/A, 41100, Modena, Italy
   corni.stefano@unimore.it
[2] Dipartimento di Chimica e Chimica Industriale, Università di Pisa,
   via Risorgimento 35, 56126, Pisa, Italy
   tomasi@dcci.unipi.it

1 Introduction

In this Chapter we shall give an overview of the models that we have developed in recent years to study surface-enhanced (SE) phenomena, i.e., the phenomena related to the interaction of a molecule with an electromagnetic (em) field that are amplified by the presence of a nearby metal specimen. The most famous among them is surely SERS, surface-enhanced Raman scattering. Generalities on SERS are extensively and authoritatively discussed elsewhere in this book, and we shall not repeat them here. We prefer instead to introduce some general considerations and then focus on our own approach to SERS and related phenomena.

The models used for calculating SERS quantities can be divided into two main classes: one collects those favoring the em-related aspects of the metal/molecule system ("em models") and the other groups the models focused on the chemical enhancement ("chemical models"). The existing em models (several examples can be found in [1, 2]) exploit the point dipole approximation for the molecule, and usually neglect the effects of adsorption and the existence of charge-transfer states. However, the description of the metal can reach quite accurate levels, for both its response [3, 4] and its shape [5]. Since the em enhancement is the most important contribution to the total amplification of the Raman signal in SERS, the em models have been generally used to estimate the enhancement obtainable from a given metallic system. However, the assumption of a point dipole molecule is quite serious also in the framework of em models, since the molecular details and the chemical nature of the adsorbed molecule are completely neglected.

On the other hand, the chemical models cannot say much regarding the total enhancement, but since the description of the molecule and of the metal–molecule local interactions are quite refined, they are more suitable for the study of the appearance of Raman spectra (vibrational frequencies and relative intensities) [6, 7, 8, 9, 10, 11, 12]. However, these models disregard the

K. Kneipp, M. Moskovits, H. Kneipp (Eds.): Surface-Enhanced Raman Scattering – Physics and
Applications, Topics Appl. Phys. **103**, 105–124 (2006)
© Springer-Verlag Berlin Heidelberg 2006

effects on the molecule due to the details of the electric field in proximity of the metallic surface. In addition, there were no methods describing the SERS spectrum appearance for systems in which the metal and the molecule are immersed in a solvent. This brief discussion shows the need in SERS studies for a model that could merge the good description of the metal-induced em field enhancement given by the em models with the detailed description of the molecule nature given by the chemical models. In [13, 14, 15] we proposed a model addressing this point, and also capable of describing other optical phenomena, such as surface-enhanced infrared absorption (SEIRA) and metal-affected molecular luminescence. This Chapter gives an account of our studies based on this model. In particular, in Sect. 2 we describe how to treat SERS from molecules adsorbed on a metallic nanoparticle with a complex shape; in Sect. 3 the refinements needed to treat aggregates of nanoparticles are presented; in Sect. 4 the application of our model to phenomena different from SERS is introduced; finally, in Sect. 5 an overview of the possible developments of our continuum model is given. In presenting our results, we chose to reserve more space for the unpublished ones, while brief accounts and proper citations are given for the already published results.

2 Description of the Model: A Molecule Close to a Complex-Shaped Nanoparticle

Our model for SE phenomena was elaborated after the observation that, if we limit ourselves to the study of molecules physisorbed on a metal, i.e., molecules that are not chemically bound to the metal surface, the problem of a proper description of such systems strongly resembles the one for solvated molecules. In fact, a molecule in solution weakly interacts with a great number of other electrons and nuclei, feeling all the fields of external as well as internal origin, as a molecule physisorbed on a metal surface does. The experiences of our group in describing optical effects of chromophores dispersed into a homogeneous dielectric phase [16, 17] through the use of the polarizable continuum model (PCM) [18] made such an approach natural also for the investigation of the properties of chromophores located near a metal surface. Basically, in PCM the solvent is described as a continuum dielectric medium in which a molecularly shaped cavity housing the solute is built [19]. The latter is described at a high QM level. Exploiting the analogy mentioned above, in our model for SERS the metal is described as a continuous body characterized by electric response properties only (perfect conductor for static fields and dielectric for time-dependent fields).

Keeping with the analogy, the chromophore (we have explicitly considered a single molecule only) has been treated at the QM level (Hartree–Fock or density functional theory so far), in the determination of both its ground state and its properties. This model for the molecule represents a remarkable progress in the accuracy of the description compared to polarizable point dipole models.[1] Solvation effects have been described with the PCM.

No chemisorption effects have been included in the versions of the model elaborated so far and no charge-transfer (CT) states have been considered. Thus, our model belongs to the em class, with the important specification that the molecule is not reduced to a point dipole. The idea of working on optical properties of metal/molecule systems with an em model mainly arises from two considerations: 1. The em contribution is the most important one in the determination of SERS intensity. An error of an order of magnitude in the estimate of the em enhancement (which, we believe, is a cautious estimate for the available em models) can completely hide the chemical enhancement. 2. While the em mechanism is common to several SE optical properties, the nature of the chemical enhancement is much more property dependent. Thus, if one aims at a general model for such kind of properties, an em model seems a good idea. In conclusion, the only interactions considered among the various portions of the complex system (the molecule, the metal, and, possibly, the solvent) are electrostatic in origin. Such electrostatic interactions has been treated through the integral equation formalism (IEF) [21], numerically solved by a BEM (boundary element method). The IEF is based on the possibility of writing the electrostatic potential inside a given region of space in terms of a fictitious surface density of charge placed on the region boundary. This density is obtained by solving an integral equation defined on such boundary. The BEM consists in translating the integral equation into a matrix equation by discretizing the boundary.

The em interaction with a probing field and among the various oscillating densities of charge in the system is treated in the quasistatic approximation, and the resulting Poisson problems are once again solved with an IEF/BEM methodology. More details on the IEF/BEM implementation can be found in [13, 22].

2.1 The Calculation of Surface-Enhanced Raman Spectra

The ingredients needed to calculate vibrational Raman spectra (and thus enhancement factors) for a given molecule are basically *vibrational frequencies*, *scattering cross sections* and *band-shape functions* for each transition in the frequency range of interest. As for band shape, a simple Lorentzian form is often a reasonable approximation. Harmonic vibrational frequencies can be calculated in the framework of our model. Moderate effects (frequency shifts in the range of $10 \, \text{cm}^{-1}$, comparable to solvent-induced ones) have been found (see [14, 22]) and thus we shall not address this problem further. The calculation of scattering cross sections is summarized in the following.

---

[1] A similar model was used by *Hilton* and *Oxtoby* [20] to calculate the static polarizability of an hydrogen atom (the "chromophore") placed on a planar infinite perfect conductor.

2.1.1 Enhanced Scattering Cross Section

Let us begin by considering an isolated molecule (in the ground state). The radiant intensity $I_R(\boldsymbol{k})$ of the vibrational scattering along the direction of the wavevector $\boldsymbol{k}$ with polarization unit vector $\hat{e}$ is [23]:

$$
I_R(\boldsymbol{k}) = I_0 k' k^3 \left| \sum_r \frac{\langle m|\boldsymbol{\mu}\cdot\hat{e}|r\rangle\langle r|\boldsymbol{\mu}\cdot\hat{e}'|0\rangle}{\hbar(\omega_{r0}-\omega'-i\gamma_r)} + \frac{\langle m|\boldsymbol{\mu}\cdot\hat{e}'|r\rangle\langle r|\boldsymbol{\mu}\cdot\hat{e}|0\rangle}{\hbar(\omega_{r0}+\omega+i\gamma_r)} \right|^2 , \quad (1)
$$

where I_0 is the intensity of the incident field, $\boldsymbol{\mu}$ is the dipole moment operator, $\hat{e}'$ is the unit vector of the incident radiation electric field and k' the modulus of the incident light wavevector. $\hbar\omega_{r0}$ is the energy difference between the vibronic state r and 0 (the ground state) and ω', ω are the frequencies of the incident and scattered radiation, respectively. $2\gamma_r$ is the inverse lifetime of the vibronic state r. In order to treat a molecule interacting with a metal particle and, possibly, in solution, we have to reconsider the theoretical basis on which (1) is based. Such an equation can be obtained by using a second-order perturbation theory in the dipolar approximation, which assumes the perturbation to be written as $-\boldsymbol{\mu}\cdot\boldsymbol{E}$, where $\boldsymbol{E}$ is the radiation electric field. In solution $\boldsymbol{E}$ has to be replaced with the Maxwell electric field in the medium $\boldsymbol{E}^M$. In addition, it is necessary to sum to the dipole $\boldsymbol{\mu}$ the dipole induced by the molecule in the metal and in the solvent (called $\tilde{\boldsymbol{\mu}}$). The radiant intensity for the molecule in the composite system is thus:

$$
I_R^{com}(\boldsymbol{k}) = I_0 k' k^3 \left| \sum_r \frac{\langle m^{com}|[\boldsymbol{\mu}+\tilde{\boldsymbol{\mu}}(\omega)]\cdot\hat{e}|r^{com}\rangle\langle r^{com}|[\boldsymbol{\mu}+\tilde{\boldsymbol{\mu}}(\omega')]\cdot\hat{e}'|0^{com}\rangle}{\hbar(\omega_{r0}^{com}-\omega'-i\gamma_r^{com})} \right.
$$
$$
\left. + \frac{\langle m^{com}|[\boldsymbol{\mu}+\tilde{\boldsymbol{\mu}}(\omega')]\cdot\hat{e}'|r^{com}\rangle\langle r^{com}|[\boldsymbol{\mu}+\tilde{\boldsymbol{\mu}}(\omega)]\cdot\hat{e}|0^{com}\rangle}{\hbar(\omega_{r0}^{com}+\omega+i\gamma_r^{com})} \right|^2 .
$$
$$(2)$$

In (2), we have explicitly indicated through the superscript com (it stands for *composite*) the quantities that changed with respect to the isolated molecule. Assuming the Born–Oppenheimer approximation and following *Placzek* [24], Raman intensities can be related to the derivatives of the electronic polarizability. If the incident light is linearly polarized, the scattered light is collected perpendicularly to the direction of the incident radiation and of the incident electric field, and the molecules are randomly oriented, it is possible to obtain:

$$
I_R^{com} = \frac{\hbar k^4}{2\omega_{01}} I_0 \frac{45\alpha_{com}'^2 + 7\gamma_{com}'^2 + 5\delta_{com}'^2}{45} , \quad (3)
$$

where ω_{01} is the frequency of the vibrational transition and $\alpha_{com}'^2$, $\gamma_{com}'^2$ and $\delta_{com}'^2$ are defined, in the double harmonic approximation, as functions of polarizability element derivatives with respect to the Q normal coordinate relative to the excited vibration ($\partial \alpha_{ij}^{\text{com}} / \partial Q$):

$$\alpha_{\text{com}}'^2 = \frac{1}{9} \left| \sum_i \frac{\partial \alpha_{ii}^{\text{com}}}{\partial Q} \right|^2 \qquad \gamma_{\text{com}}'^2 = \frac{3}{8} \sum_{ij} \left| \frac{\partial \alpha_{ij}^{\text{com}}}{\partial Q} + \frac{\partial \alpha_{ji}^{\text{com}}}{\partial Q} \right|^2 - \frac{9}{2} \alpha_{\text{com}}'^2 \qquad (4)$$

$$\delta_{\text{com}}'^2 = \frac{3}{8} \sum_{ij} \left| \frac{\partial \alpha_{ij}^{\text{com}}}{\partial Q} - \frac{\partial \alpha_{ji}^{\text{com}}}{\partial Q} \right|^2 . \qquad (5)$$

We have taken into account that polarizability derivatives can be complex numbers since the contribution from $\tilde{\boldsymbol{\mu}}$ depends on the metal polarization and the dielectric constant of the metal has a non-negligible imaginary part. In (3) the square of the antisymmetric anisotropy δ' [defined in (5)] appears. Such an invariant of the derived polarizability tensor is different from zero if this tensor is not symmetric. For isolated molecules in the Placzek approximation (in particular, far from electronic resonances) such a tensor is symmetric and $\delta' = 0$. If we instead consider the whole metal–molecule system, due to the fact that the metal response is different at the incident and scattered frequencies, the derived polarizability tensor is nonsymmetric and thus the contribution of δ' to the scattered light intensity should be taken into account. Polarizability derivatives that appear in (4) to (5) are evaluated numerically by calculating the polarizability at the exciting frequency for two molecular geometries obtained by making small displacements from the equilibrium geometry along a given normal coordinate. The required polarizability α^{com} has been calculated in the time-dependent Hartree–Fock (TDHF) framework in [14, 15]. We shall not detail here the procedure to get α_{ij}^{com} (see [13, 22]); suffice it to say that it includes the effects on the molecule of both the "reflected" field, i.e., the field due to the polarization of the metal induced by the applied field and the "image field", due to the polarization induced in the metal by the oscillating density of charge of the molecule. In addition, α^{com} naturally includes the enhancement, due to the metal, of the field scattered by the molecule, evaluated at the proper frequency. In particular, all these molecule–field interactions are treated via a BEM procedure that permits taking into account terms in the multipolar expansion of such interactions also higher than the dipolar one. Finally, the procedure also takes into account the presence of a solvent described with the PCM.

In [14] we applied our model to pyridine adsorbed on elongated nanoparticles obtained by the coalescence of spherical nanoparticles. We found that the most convenient molecule location for high enhancement is the crevice that forms at the junction between the nanoparticles, in agreement with the findings of *Xu* et al. [5]. In addition, we also described the differences in the appearance of the pyridine SERS spectrum calculated via our model and by simply imposing a surface-induced selection rule. One point that we did not analyze quantitatively in our published studies was the different enhancements between Stokes and anti-Stokes signals [25]. In fact, the amplification due to the metal specimen of the em field scattered by the molecule depends on the frequency of the scattered field (this feature is taken into account in our model), which is obviously different for Stokes and anti-Stokes branches of the Raman spectrum. In particular, for the $1000 \, cm^{-1}$ band of pyridine adsorbed in the crevice between two spherical Ag nanoparticles (radius: 20 nm, distance between the particle centers: 30 nm), we found that the anti-Stokes/Stokes enhancement ratio may be different from 1 and depends on the excitation frequency. Our model yielded anti-Stokes/Stokes enhancement ratios of 0.92, 0.89, 4.2, and 5.0 for excitation wavelengths of 514.5 nm, 632.8 nm, 780 nm, 830 nm, respectively.

3 Extending the Model to Metal-Particle Aggregates

It is now well established that single nanoparticles with a complex shape can give sizable enhancement, in particular when the molecule is placed in crevices [26]. Larger aggregates of metallic nanoparticles have been used as SERS substrates for a long time. Recently, it has been shown that they can give enhancement factors high enough to allow single-molecule detection, very likely without any (molecular) resonance Raman contribution [27, 28, 29]. These aggregates have a fractal nature (nothing special has to be done to obtain such a feature, it is simply a result of the aggregate-growing mechanism). It has been demonstrated, both theoretically and experimentally, that fractal metal-particle aggregates are capable of inhomogeneously redistributing in their volume the intensity of an em field acting on them, thus creating regions (comprising some metal particles) of very high electric field, named *hot spots* [30, 31, 32, 33, 34, 35].

In SERS of a molecule adsorbed on a fractal aggregate, one should distinguish between two em-enhancement mechanisms: on the one hand, the structure of the aggregate in close proximity of the chromophore provides an enhancement due, for example, to crevices, very similar to what happens for isolated particle dimers; on the other hand the aggregate as a whole is responsible for the existence of hot spots involving different metal particles. A method capable to separately calculate such contributions and to combine them in a total enhancement factor is highly desirable, since it can help in shedding light on the origin of the measured huge enhancements. In [15] we proposed a method having this capability. Briefly, we have treated the problem of SERS due to a single molecule adsorbed on a metal particle cluster by partitioning the system in three "layers", exploiting different models for the different layers. The *first layer* is composed of the whole cluster. It has been treated as a collection of spherical metallic particles described as polarizable dipoles [30, 31, 32, 33, 34]. From such a description, it is possible to recognize where the hot spots are and how strong the electric field locally acting on them is. The *second layer* groups the metal particles in a given hot spot. They are considered as a unique complex-shaped particle, treated with the model presented in Sect. 2, which takes into account effects on the electric field due to the complex shape well beyond the polarizable dipole approximation. Finally, the molecule forms the *third layer*. Coherently with Sect. 2, it has been described at the ab-initio level, taking into account the electrostatic interaction with the metal particle and properly treating the electric field acting on it.

The procedure that we proposed for the calculation of Raman intensities of the adsorbed molecule is composed of 4 steps:

1. The building of the metal-particle aggregate is done by exploiting a cluster–cluster aggregation (CCA) model [36]. These CCA clusters provide good simulations of empirically observed fractal aggregates of metal particles in solution [30, 37]. Input: the aggregate parameters (initial concentration of the unaggregated nanoparticles, number of particles per aggregate, seed of the random step generator that, given the simulation procedure and the other parameters, uniquely identifies the aggregate). Output: the geometry of the aggregate.

2. The resolution of the em problem for a field impinging on the whole aggregate is performed by solving the coupled-dipole (or discrete-dipole) equations, which are introduced elsewhere in this book. Input: the geometry of the aggregate (from step 1), the properties of the incident light (frequency, polarization, direction, intensity), the dielectric properties of the metal (including corrections for the limited electron mean free path), the size of the metal nanoparticles, the dielectric constant of the solvent. Output: the electric field acting on each single particle of the aggregate and thus the hot-spot positions and size.

3. The evaluation of Raman scattering intensities for a molecule physisorbed in the hot spot. Let us call E_{hot} the field acting on the molecule and the few particles in the hot spot due solely to the other particles in the aggregate (obtained from 2). We found in [15] that the Raman-scattered light $I_R^{com}(k)$ of (2) can be identify with the light emitted from a dipole P_{hot} oscillating at the scattering frequency with expression:

$$P_{hot} = \sqrt{\frac{h}{4\pi\omega_{10}}} \frac{\partial\alpha^{com}}{\partial Q} \cdot E_{hot}, \qquad (6)$$

where the derivative $\partial\alpha^{com}/\partial Q$ that appears in (6) is calculated as sketched in Sect. 2.1. Input: E_{hot} (from step 2), the physical parameters of the molecule (chemical nature, geometry, position in the hot spot), and the dielectric/geometrical parameters of the metallic particles in the hot spot. Output: the oscillating dipole P_{hot}.

4. The field emitted by the dipole P_{hot} is further scattered by the other particles of the aggregate to finally give the total scattered field. The coupled dipole equations have been adapted to this problem [15], and the intensity of the scattered light in a given direction as a function of the incident one is finally obtained. Input: the aggregate parameters (as in step 2), the position of the hot spot from step 2, the oscillating dipole P_{hot} from step 3 and

the direction and polarization of the collected scattered light. Output: the intensity of the scattered light far from the aggregate I_{tot}^{sc} as a function of the incident light, or, in other words, the total Raman scattering cross section.

When the total em enhancement Y is of interest, it can be calculated by taking the ratio

$$Y = \frac{I_{tot}^{sc}}{I_{mol}^{sc}}, \tag{7}$$

where I_{mol}^{sc} is a reference intensity (e.g., the Raman intensity for the molecule without any metal specimen for the same exciting light intensity used to calculate I_{tot}^{sc}).

In conclusion, we feed our method with: 1. The parameters of the unaggregated metallic nanoparticles (concentration, size of the particles, dielectric properties of the metal); 2. The dielectric constant of the solvent; 3. The properties of the incident light (frequency, polarization, direction, intensity); 4. The direction and polarization of the light collection; and we get as the main results: 1. The intensity of the Raman scattering light as a function of the incident one, and thus the Raman enhancement factor Y, once a reference intensity is defined; 2. The position and the nature of the hot spots; 3. em quantities related to the different origins of the total em enhancements, such as the electric field acting in the hot spot for a given incident field, and the scattering properties of the molecule–hot-particle system.

Clearly, besides the above-mentioned physical parameters as input quantities, we have also to choose calculation parameters, such as basis sets and convergence thresholds for the QM calculations and tessellation details for the BEM. The model has been implemented by modifying and interfacing two pre-existing codes, DDSCAT [38] for the coupled dipole equation resolution and GAMESS [39] for the SERS of the molecule–hot-spot system. Simulation of nanoparticle aggregate formation via CCA has been implemented ex novo.

3.1 Decomposition of the Total Enhancement

As said above, an enticing feature of our model is the possibility of decomposing the total enhancement in two factors: one related to the overall behavior of the aggregate, the other related to the local topology of the few metal particles closest to the molecule. In [15] we took advantage of this capability and defined three contributions to the enhancement (Y_1, Y_2, Y_3), with $Y = Y_1 \cdot Y_2 \cdot Y_3$. In particular, Y_1 represents the enhancement of the incident light intensity achieved in the few metal particles hot spot, Y_2 is the contribution obtained by considering the molecule in the hot spot and neglecting the rest of the aggregate, and Y_3 is the intensity enhancement factor arising from the interaction with the rest of the aggregate of the field scattered by the molecule in the hot spot. Thus, $Y_{loc} = Y_2$ represents the contribution from the local topology of the metal particles, while $Y_{agg} = Y_1 \cdot Y_3$ originates from the overall (fractal) structure of the aggregate. The main conclusion of [15] concerning the relative importance of Y_{loc} and Y_{agg} was that $Y_{loc} \gg Y_{agg}$, i.e., the main source of enhancement is the local structure (in particular the junction between contiguous nanoparticles). However, this finding does not mean, as is sometimes stated, that the rest of the aggregate is useless for the enhancement: in typical cases, with parameters close to the ones of Kneipp's experiments, we found that $Y_{loc} \approx 10^7 – 10^9$ while $Y_{agg} \approx 10^4$. Assuming that the total enhancement (including em, chemical, and molecular resonance contributions) needed to measure single-molecule SERS is around $10^{12} – 10^{14}$, we noted that Y_{loc} alone is not enough: it should be accompanied by a resonance Raman contribution (as in the metallic dimers experiments of Käll and coworkers) or by the help (relatively little but essential) of the fractal aggregate (as likely in Kneipp's experiments).

3.2 Hot Spots and Aggregate Size

Although we have seen in the previous section that most of the enhancement is due to the local (nonfractal) structure of a two-particle system, the hot-spot enhancement can give a sizable contribution, sizable enough to make the total enhancement sufficient for single-molecule detection. In this section, we shall investigate how the number of hot spots and their average morphology depends on the size of the aggregate. One of the aims of this investigation is to establish, from the point of view of the hot-spot properties, at which size an aggregate starts to present the large-aggregate behavior. In fact, when the cluster is formed by a small number of particles (such as 2–3), it is clear that it cannot present the (fractal) structure responsible for the inhomogeneous field distribution and thus of the arising of hot spots. First, we have to give an operative definition of hot spot. To do this, we choose a threshold for the electric-field enhancement. If the modulus of the electric field acting on a given particle is above such a threshold, the particle is defined as *hot*. Then, after having identified all the hot particles, one checks which ones are connected together. A group of connected hot particles is called a *hot spot*.[2] We have applied this definition by assuming that a particle should be considered hot if the modulus of the electric field acting on it is 20 times larger than the incident field one. This enhancement threshold would be enough, according to the results of [15], to obtain single-molecule sensitivity. Particles of Ag with radius equal to 20 nm and immersed in water have been

---

[2] We remark that in the literature the term "hot spot" is used with somewhat different meanings. For example, in the works by *Shalaev* (see [30] for a comprehensive report) hot spot has been defined in a way similar to our qualitative definition, i.e., to indicate regions of a metal-particle aggregate where an incident electric field is strongly enhanced. On the contrary, the term "hot spot" is sometimes used for the molecular-sized region where the high field is achieved (e.g., between two nanoparticles). The two meanings refer to regions of space differing in size by some orders of magnitude.

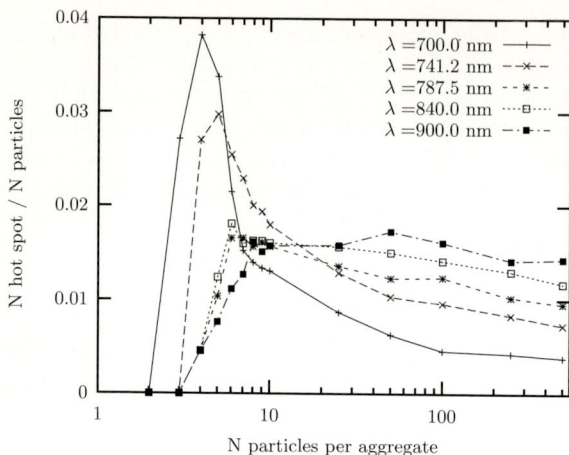

Fig. 1. Number of hot spots vs. number of particles as a function of the size of the aggregate. Particles of Ag with radius equal to 20 nm and immersed in water have been considered

considered. In these studies, we have neglected the effects of the damping due to the reduced free path of the metal electrons. No rotational averaging has been performed.

For each aggregate size considered, an ensemble of randomly chosen aggregates with the given size has been built by exploiting the CCA model. The total number of particles in each ensemble (i.e., the number of particles per aggregate multiplied by the number of aggregates) is approx. 50 000. Coupled-dipole equations are solved for each aggregate. The total number of hot spots and the mean number of particles per hot spot are thus determined.

The number of hot spots per aggregate vs. the number of particles in the aggregate as a function of the size of the aggregate for five different excitation wavelengths is depicted in Fig. 1.

Two well-definite peaks are visible in the trends: one for $\lambda = 700$ nm for aggregates containing 4 particles and the other for $\lambda = 741.2$ nm and 5 particles. Both can be attributed to the presence, for these wavelengths and these aggregate sizes, of resonances of the surface plasmons. Such resonances are the usual ones found for elongated metal nanoparticles [1, 2] and are not related to a fractal structure since the number of particles is too small.

For $\lambda = 840$ nm (close to the λ experimentally used by *Kneipp* et al. [27, 28, 29], 830 nm) the number of hot spots does not depend appreciably on the aggregate size for aggregates larger than 6–7 particles. This is in agreement with the experimental findings of *Kneipp* et al. [27, 28, 29], who observed an almost constant SERS activity for aggregates containing more than 6–10 particles and no measurable activity for smaller aggregates.

The trends reported in the figure become settled for a number of particles per aggregate between 20 and 30. These results suggest that to fully exploit

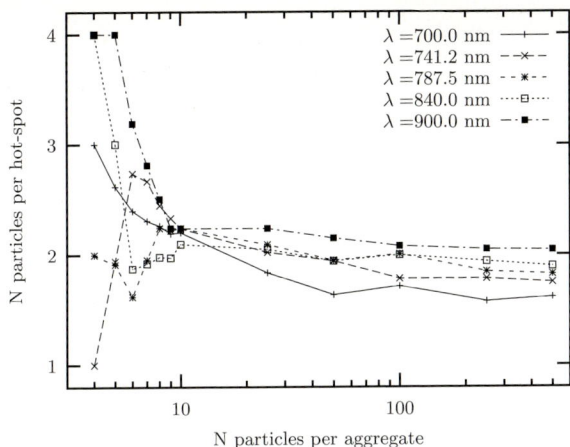

Fig. 2. Mean number of particles per hot spot as a function of the size of the aggregate. Particles of Ag with radius equal to 20 nm and immersed in water have been considered

the fractal nature of colloidal particle clusters for SERS, $\approx$ 30 particle aggregates are enough. The slightly asymptotic decrease of the number of hot particles observed on increasing the size is probably due to electrodynamics effects: by increasing the aggregate size the quantity of radiation scattered by the system increases and thus the em energy available to create hot spots decreases. Hence, there is no point in experimentally using metal clusters that are too large.

Together with the number of hot spots, we have also extracted from the simulations information regarding their shape and size. We have verified that almost all the hot spots are linear in shape, with the major axis parallel to the polarization direction of the incident field. The mean number of particles contained in each hot spot as a function of the aggregate size is reported in Fig. 2. The behavior for aggregates containing less than 10 particles is quite complex. For larger aggregates, the average number of particles per hot spot becomes stable at around 2. As has been noted for Fig. 1, also for Fig. 2, the asymptotic behavior seems to be fully shown for an aggregate size as small as 20–30 particles.

4 Beyond SERS

Although SERS is the best known and the most studied among the surface-enhanced properties, several other molecular properties are affected by the presence of a curved metal specimen. Our model for SERS has been extended to some of these other properties. In particular, here we shall give a brief account of our works on surface-enhanced infrared absorption (SEIRA) and molecular fluorescence.

4.1 SEIRA

SEIRA has been reported, e.g., for molecules on electrode/electrolyte interfaces, metal-coated IR transparent substrate, island metal films, metal colloids spread on an inert substrate (see, for example, [40] and references therein). The enhancements usually measured are smaller than those for SERS (10–1000 for SEIRA to 10^5–10^6 for SERS) and no single-molecule detection has been reported so far. As for SERS, the IR absorption enhancement can also be decomposed into an em and a chemical contribution. The smaller total enhancement is generally assigned to a smaller em contribution (in SERS the metal enhances both the incident and the scattered field, whereas in SEIRA there is no scattered field at all) and to the absence of some chemical enhancement mechanisms (e.g., resonant metal–molecule charge-transfer states).

The quantity that characterizes the intensity of vibrational absorption is the integrated absorption coefficient A. For a gas-phase molecule, A^{gas} can be derived from the radiation-induced transition probability $W^{\mathrm{gas}}_{1 \leftarrow 0}$ between two vibrational states 0 and 1 in the dipolar approximation:

$$W^{\mathrm{gas}}_{1 \leftarrow 0} = \frac{2\pi}{\hbar} \delta(\mathcal{E}_1 - \mathcal{E}_0 - \hbar\omega)|\langle 1|\boldsymbol{\mu} \cdot \boldsymbol{E}|0\rangle|^2 \,, \tag{8}$$

where $\mathcal{E}_0$, $\mathcal{E}_1$ are the energies of the states 0 and 1 and $\boldsymbol{E}$ is the radiation electric field acting on the molecule. Neglecting anharmonicity effects, from (8) it is possible to obtain A^{gas} for the transition between states associated at the i mass-weighted normal coordinate Q_i as:

$$A^{\mathrm{gas}} = \frac{\pi N_{\mathrm{A}}}{3c^2} \left| \frac{\partial \boldsymbol{\mu}}{\partial Q_i} \right|^2 \,, \tag{9}$$

where an isotropic orientation of the molecule has been implicitly assumed. In (9) N_{A} is Avogadro's number and c is the velocity of light in vacuo. For molecules in close proximity of a metal particle and possibly in the presence of a solvent, (8) has to be modified: if a solvent is present, $\boldsymbol{E}$ is replaced by the electric field of the radiation in the medium $\boldsymbol{E}^{\mathrm{M}}$; the energies $\mathcal{E}$ of the vibrational states of the isolated molecule become free energies of the molecule-near-to-the-metal states; in addition, it is necessary to sum to the dipole $\boldsymbol{\mu}$ the dipole induced by the molecule on the metal particle and on the solvent. This additional term is called $\tilde{\boldsymbol{\mu}}$ as in SERS and its calculation for the case of a molecule in an isotropic solution is described in [41]. The absorption intensity for the molecule in the metal/solvent composite environment (A^{com}) is thus:

$$A^{\mathrm{com}} = \frac{\pi N_{\mathrm{A}}}{3n_{\mathrm{s}}c^2} \left| \frac{\partial(\boldsymbol{\mu} + \tilde{\boldsymbol{\mu}})}{\partial Q_i} \right|^2 \,. \tag{10}$$

In (10) n_s is the refractive index of the medium in which the metal particle and the molecule are immersed. The calculation of the term $\partial\boldsymbol{\mu}/\partial Q_i$ at the HF or DFT level is a simple application of coupled perturbed HF(DFT). The procedure in the presence of a metal is explained in detail in [22]. $\partial\tilde{\boldsymbol{\mu}}/\partial Q_i$ is instead obtained by the IEF/BEM procedure.

We have applied our model to the calculation of SEIRA spectra and enhancement factors for the p-nitrobenzoate ion adsorbed on spherical silver nanoparticles. IR spectra of monolayers of p-nitrobenzoate ion adsorbed on silver-island films of different thickness (ranging from 4 nm to 14 nm) have been experimentally determined by *Osawa* and *Ikeda* [40]. In particular, the UV spectrum of the thinnest film resembles that of isolated silver particles in a dielectric environment, so it is reasonable to assume that for this film thickness silver islands are sufficiently apart from each other to neglect em interactions between them. In addition, since the peak is in the visible region of the spectrum, it is possible to exclude the presence of metal electron resonances in the IR. We have thus considered as a model for this system a molecule of p-nitrobenzoate placed in close proximity to a perfect conductor particle with a radius of 2 nm. The interactions between different adsorbates have been neglected and the metal–molecule system has been considered in vacuo. The p-nitrobenzoate ion is described at the DFT level with the B3LYP exchange-correlation functional and the 6-31G** basis set (the code for the calculation of frequency, normal coordinates and IR intensities has been implemented in a development version of the Gaussian [42] package of programs). The carboxylic group is the one nearest to the metal (it has been experimentally observed that p-nitrobenzoic acid adsorbs on the island film as p-nitrobenzoate); two limiting orientations are considered: one in which the main axis of the molecule is perpendicular to the metal surface and the other in which it is tangential. In Fig. 3 and Fig. 4 the calculated spectra for the radial (perpendicular) and tangential (parallel) orientations, respectively, are shown together with the experimental spectrum taken from [40] (the absorption intensity scale for the calculated spectra is not the same in the two figures).

The comparison between the experimental spectrum and the calculated ones show a quite good agreement, better for the spectrum given in Fig. 3 than for the one of Fig. 4. In fact, the strong $895\,\mathrm{cm}^{-1}$ band seen only in Fig. 4 is not present in the experimental spectrum. Such a peak can be assigned to the symmetric wagging of the aromatic-ring hydrogen atoms. The corresponding transition dipole is perpendicular to the metal surface if the ion is tangentially oriented, while it is tangential if the ion is perpendicularly oriented. Taking into account these different orientations of the molecular dipole moment derivative, we can explain qualitatively the differences in intensity by considering $\tilde{\boldsymbol{\mu}}$, the dipole moment induced by the molecule in the metal: a simple image-charge model applied to a point-like dipole gives that the dipole moment induced on a conductor surface by an external dipole is parallel to it if the external dipole is perpendicular to the surface and antipar-

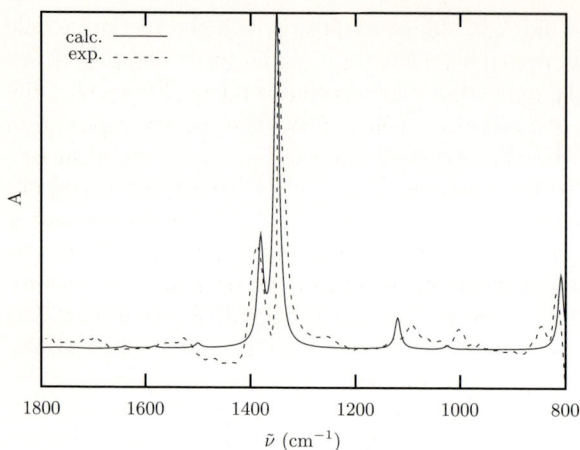

Fig. 3. Calculated ("calc.") IR absorption spectrum for p-nitrobenzoate adsorbed on a conductor spherical particle with the main axis perpendicular to the metal surface. The bandwidth in the calculated spectrum is chosen arbitrarily. Experimental spectrum ("exp.") for p-nitrobenzoate adsorbed on a 4 nm thick silver-island film is also reported (after [40])

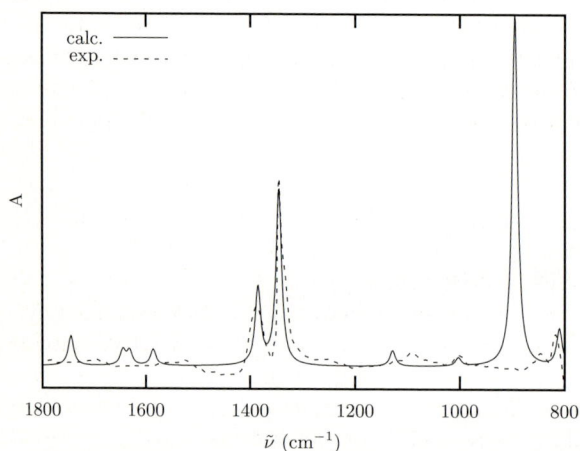

Fig. 4. Calculated ("calc.") IR absorption spectrum for p-nitrobenzoate adsorbed on a conductor spherical particle with the main axis tangential to the metal surface. The bandwidth in the calculated spectrum is chosen arbitrarily. Experimental spectrum ("exp.") for p-nitrobenzoate adsorbed on a 4 nm thick silver-island film is also reported (after [40])

allel if the external dipole is parallel to the surface [1]. Thus, the total dipole moment derivatives associated with the 895 cm^{-1} vibration is enhanced if the molecule is placed tangentially to the metal surface and is decreased if the molecule is perpendicular to the metal surface. From the spectra reported in Fig. 3 and Fig. 4 it seems thus that the molecule adsorbs on the metal surface almost perpendicularly. Actually this conclusion does not take into account the differences in absolute intensities: in fact, we found that the intensities for the tangential orientation are much smaller than those for the perpendicular one. Thus, we can only say that the molecules in the perpendicular (or almost perpendicular) orientation are responsible for the SEIRA spectrum, but we cannot rule out the possibility that a non-negligible number of molecules prefer other (more or less tangential) orientations.

The maximum enhancement (about 9) is achieved for the OCO stretching band in the perpendicular orientation. In [40] the enhancement with respect to the 1350 cm^{-1} band of p-nitrobenzoic acid is estimated to be ≈ 30. A direct comparison between our theoretical result and the experimental one is hindered by the different reference intensity used (isolated p-nitrobenzoate ion in the calculation and solid p-nitrobenzoic acid in [40]), but at least the two numbers have the same order of magnitude.

Finally, we note that our result deviates from the one obtained with the point-dipole model for the molecule (data not shown, see [22]). In particular, the OCO and the ONO stretching vibrations present different enhancement factors in our model, due to the different distances of these moieties from the metal. Clearly, a point-like model for the molecule cannot generate such effects.

4.2 Molecular Fluorescence

Our model for SERS has been extended to treat fluorescence for a molecule close to a metal nanoparticle [43, 44]. In particular, we calculated the radiative decay rate Γ^{r} and the nonradiative decay rate Γ^{nr} due to the energy transfer from the molecule to the metal. Details on the calculations are given in [43, 45]. Briefly, the model of the metal–particle system is that presented in Sect. 2. The radiative decay rate is obtained as:

$$\Gamma^{\mathrm{r}} = \frac{4n_{\mathrm{s}}\omega^3}{3\hbar c^3}|\boldsymbol{\mu}_{K0} + \tilde{\boldsymbol{\mu}}_{K0}|^2 , \tag{11}$$

where n_{s} is the refractive index of the solvent, $\boldsymbol{\mu}_{K0}$ is the transition dipole for the $K \to 0$ transition of the molecule interacting with the metal particle and $\tilde{\boldsymbol{\mu}}_{K0}$ is the dipole induced in the metal and in the solvent from the transition density $K \to 0$ of the molecule. $\boldsymbol{\mu}_{K0}$ is calculated with TDHF (or time-dependent density functional theory, TDDFT) in the random phase approximation (RPA)-like formulation, while $\tilde{\boldsymbol{\mu}}_{K0}$ is calculated exploiting IEF/BEM (see [44] for more details). Γ^{nr} has been obtained from the imaginary part of the excitation energies of the molecules interacting with the metal

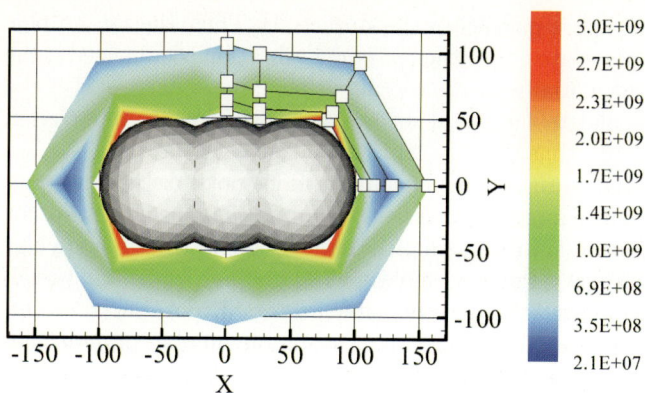

Fig. 5. Radiative decay rate (in s^{-1}) for the coumarin of [44] near a silver aggregate composed by three identical interlocking spheres. The *white squares* indicate the molecule positions for which calculations have been performed. Cartesian axes are in Å. Reprinted with permission from [44]. Copyright 2004, American Institute of Physics

nanoparticle, as described in [22, 45]. Our group (the authors of the present Chapter in collaboration with Oliviero Andreussi and Benedetta Mennucci) has investigated the behavior of these quantities for different systems (planar metal surfaces, rough or smooth, spherical nanoparticles, complex-shaped nanoparticles) and as a function of different parameters (metal–molecule distance, metal-particle size, complexity of the nanoparticle shape, position of the molecule on the nanoparticle surface). All these studies are documented in [22, 43, 44]. Here, we report only one of the findings of [44]: for a coumarine chromophore close to a metal nanoparticle, it has been found that a kind of *cold spot* exists for Γ^r (see Fig. 5), i.e., a region where Γ^r is depressed with respect to the value for the isolated molecule.

Such a cold spot is the result of $\tilde{\boldsymbol{\mu}}_{K0}$ being out-of-phase with $\boldsymbol{\mu}_{K0}$ in (11) that happens close to the resonance of a metal-particle plasmon.

Notably, in [43] we took into account the nonlocal nature of the metal response by using a modified Lindhard–Mermin dielectric constant. In that work, only planar surfaces were considered. However, this opened up the way to the inclusion of such nonlocal effects also for the other SE phenomena.

5 Summary and Perspectives

In this Chapter we have briefly presented the model that we proposed, in analogy with continuum solvation models, to study enhanced optical properties of molecules close to metal nanoparticles and nanoparticle aggregates. In particular, we focused on some results that were not previously published, and we showed that the model is not limited to SERS, but can be extended to other surface-enhanced (or, more generally, affected) phenomena such as SEIRA and molecular luminescence. We end this Chapter by discussing the possible new directions of our work. The main lines of development are twofold:

1. Refining the model;
2. Extending the model to other properties.

Concerning 1, we find that the description of the molecule at the HF/DFT level is at present satisfactory. The main limitation is probably the size of molecules that can be investigated with these methods: They are usually too demanding to treat common dyes such as Cy5 or rhodamine 6G. For this reason, members of our group (Marco Caricato and Oliviero Andreussi) are currently working to extend the models to the ZINDO level, a semiempirical method suitable for spectroscopic calculations. As for the description of the metal, we think that it may be improved. In particular, we have already included the nonlocal effects of the metal dielectric response in our studies of the molecular fluorescence close to planar metal surface. The extension to more complex shapes is not difficult (it has been already done, in another context [46]). For the smallest nanoparticles, when quantum size effects are at work, the continuum dielectric description could be replaced by a jellium one [47]. On the other hand, the quasistatic approximation used so far in our model must be superseded for larger nanoparticles. Actually, BEM (the method that we use to solve the electrostatic equations) has been already formulated to solve the full electrodynamics problem [48], and this extension can be incorporated in our model. Finally, to complete the model for SERS, we should also be able to treat chemical enhancement effects. A simple tentative approach would be to include in the portion of the system that is described at the QM level some metal atoms.

As for 2, the extension of the model to other properties, we have several possibilities. Different reports of "enhanced fluorescence" for molecules close to a metal specimen recently appeared in the literature [49]. Although we have the main ingredients to calculate such enhancement, we still have to suitably combine them to properly model this phenomenon. SE properties related to nonlinear optical response of the molecule has been also reported, such as SE hyper-Raman scattering and SE two-photon absorption. The calculations of these quantities in the framework of our model do not require any new idea. Simply, the proper equation must be implemented. Finally, other properties more related to plasmonics are in the scope of our model: we refer in particular to the fluorescence resonance energy transfer between two chromophores in the presence of a metal specimen. Recently, it has been experimentally shown that the metal can dramatically affect this process [50]. In conclusion, we believe that our model, although relatively simple, has a good chance to produce other interesting results in the framework of SE phenomena. Surely, it can highly benefit from the continuous interplay with its elderly brothers, the continuum solvation models.

References

[1] M. Moskovits: Rev. Mod. Phys. **57**, 783 (1985)
[2] G. C. Schatz, R. P. Van Duyne: *Handbook of Vibrational Spectroscopy* (Wiley, New York 2002)
[3] G. Korzeniewski, T. Maniv, H. Metiu: Chem. Phys. Lett. **73**, 212 (1980)
[4] T. Maniv, H. Metiu: J. Chem. Phys. **72**, 1996 (1980)
[5] H. Xu, J. Aizpurua, M. Käll, P. Apell: Phys. Rev. E **62**, 4318 (2000)
[6] W.-H. Yang, G. C. Schatz: J. Chem. Phys. **97**, 3831 (1992)
[7] W.-H. Yang, J. Hulteen, G. C. Schatz, R. P. Van Duyne: J. Chem. Phys. **104**, 4313 (1996)
[8] J. Arenas, M. Woolley, I. Tocón, J. Otero, J. I. Marcos: J. Chem. Phys. **112**, 7669 (2000)
[9] R. F. Aroca, R. E. Clavijo, M. D. Halls, H. B. Schlegel: J. Phys. Chem. A **104**, 9500 (2000)
[10] T. Tanaka, A. Nakajima, A. Watanabe, T. Ohno, Y. Ozaki: Vib. Spectrosc. **34**, 157 (2004)
[11] D. Y. Wu, M. Hayashi, S. H. Lin, Z. Q. Tian: Spectrochim. Acta A **60**, 137 (2004)
[12] M. Muniz-Miranda, G. Cardini, V. Schettino: Theor. Chem. Acc. **111**, 264 (2004)
[13] S. Corni, J. Tomasi: J. Chem. Phys. **114**, 3739 (2001)
[14] S. Corni, J. Tomasi: Chem. Phys. Lett. **342**, 135 (2001)
[15] S. Corni, J. Tomasi: J. Chem. Phys. **116**, 1156 (2002)
[16] R. Cammi, B. Mennucci, J. Tomasi: J. Am. Chem. Soc. **120**, 8834 (1998)
[17] R. Cammi, M. Cossi, B. Mennucci, J. Tomasi: J. Chem. Phys. **105**, 10556 (1996)
[18] S. Miertuš, E. Scrocco, J. Tomasi: Chem. Phys. **55**, 117 (1981)
[19] J. Tomasi, B. Mennucci, R. Cammi: Chem. Rev. **105**, 2999 (2005)
[20] P. R. Hilton, D. W. Oxtoby: J. Chem. Phys. **72**, 6346 (1980)
[21] W. Hackbusch: *Integral Equations – Theory and Numerical Treatment* (Birkhäuser, Basel 1995)
[22] S. Corni: *Continuum Models for the Study of Optical Properties of Metal-Molecules Complex Systems*, Ph.D. thesis, Scuola Normale Superiore, Pisa (2003)
[23] D. A. Long: *The Raman Effect* (Wiley, Chichester 2002)
[24] G. Placzek: *Handbuch der Radiologie*, vol. VI, part 2 (Akademische Verlagsgesellschaft, Leipzig 1934)
[25] A. G. Brolo, A. C. Sanderson, A. P. Smith: Phys. Rev. B **69**, 045424 (2004)
[26] H. Xu, E. J. Bjerneld, M. Käll, L. Börjesson: Phys. Rev. Lett. **83**, 4357 (1999)
[27] K. Kneipp, Y. Wang, H. Kneipp, I. Itzkan, R. R. Dasari, M. S. Feld: Phys. Rev. Lett. **78**, 1667 (1997)
[28] K. Kneipp, H. Kneipp, G. Deinum, I. Itzkan, R. R. Dasari, M. S. Feld: Appl. Spectrosc. **52**, 175 (1998)
[29] K. Kneipp, H. Kneipp, R. Manoharan, E. Hanlon, I. Itzkan, R. R. Dasari, M. S. Feld: Appl. Spectrosc. **52**, 1493 (1998)
[30] V. M. Shalaev: *Nonlinear Optics of Random Media*, vol. 158, Springer Tracts in Modern Physics (Springer, Berlin, Heidelberg 2000)

[31] M. I. Stockman, V. M. Shalaev, M. Moskovits, R. Botet, T. F. George: Phys. Rev. B **46**, 2821 (1992)

[32] V. A. Markel, V. M. Shalaev, E. B. Stechel: Phys. Rev. B **53**, 2425 (1996)

[33] V. M. Shalaev, E. Y. Poliakov, V. A. Markel: Phys. Rev. B **53**, 2437 (1996)

[34] V. A. Markel, L. S. Muratov, M. I. Stockman, T. F. George: Phys. Rev. B **43**, 8183 (1991)

[35] Z. Wang, S. Pan, T. D. Krauss, H. Du, L. J. Rothberg: Proc. Natl. Acad. Sci **100**, 8638 (2003)

[36] T. Vicsek: *Fractal Growth Phenomena* (World Scientific, Singapore 1989)

[37] D. Weitz, M. Oliveria: Phys. Rev. Lett. **52**, 1433 (1984)

[38] B. T. Draine, P. J. Flatau: *User Guide to the Discrete Dipole Approximation Code DDSCAT (Version 5a10)* (2000)
URL http://arxiv.org/abs/astro-ph/0008151v4

[39] M. W. Schmidt, et al.: J. Comp. Chem. **14**, 1347 (1993)

[40] M. Osawa, M. Ikeda: J. Phys. Chem. **95**, 9914 (1991)

[41] R. Cammi, C. Cappelli, S. Corni, J. Tomasi: J. Phys. Chem. A **104**, 9874 (2000)

[42] M. J. Frisch, et al.: *Gaussian 99, Development Version (Revision A.10+)* (Gaussian, Inc., Pittsburgh PA 1998)

[43] S. Corni, J. Tomasi: J. Chem. Phys. **118**, 6481 (2003)

[44] O. Andreussi, S. Corni, B. Mennucci, J. Tomasi: J. Chem. Phys. **121**, 10190 (2004)

[45] S. Corni, J. Tomasi: J. Chem. Phys. **117**, 7266 (2002)

[46] S. Corni: J. Phys. Chem. B **109**, 3423 (2005)

[47] W. Ekardt, Z. Penzar: Phys. Rev. B **34**, 8444 (1986)

[48] N. Ida: *Numerical Modeling for Electromagnetic Non-Destructive Evaluation* (Chapman Hall, New York 1995)

[49] J. R. Lakowicz, et al.: J. Fluorescence **14**, 425 (2004)

[50] P. Andrew, W. L. Barnes: Science **306**, 1002 (2004)

Index

SERS From Transition Metals and Excited by Ultraviolet Light

Zhong-Qun Tian, Zhi-Lin Yang, Bin Ren, and De-Yin Wu

State Key Laboratory of Physical Chemistry of Solid Surfaces and Department of Chemistry, College of Chemistry and Chemical Engineering, Xiamen University, Xiamen, 361005, China
zqtian@xmu.edu.cn

1 Introduction

Since the mid-1990s, surface-enhanced Raman scattering (SERS) has greatly advanced and gained wider application and renewal interest than in the previous two decades [1, 2, 3, 4, 5, 6, 7, 8, 9]. There have been several new and creative developments, e.g., SERS of single molecules, nanostructures and transition metals, tip-enhanced Raman scattering (TERS), surface-enhanced hyper-Raman scattering (SEHRS), ultraviolet-excited SERS (UV-SERS), surface-enhanced resonance Raman scattering (SERRS). In this book, this Chapter is probably the only one describing metals other than Ag, Au and Cu. It may be necessary to first give a brief introduction on the development of extending SERS study to transition metals.

The major obstacle hampering the generality of SERS is that only the three coinage metals (Au, Ag, and Cu) and some alkali metals (Na [10], K [11], Li [12]) can provide the huge enhancement because they belong to free-electron metals whose surface-plasmon resonance can be efficiently excited by the visible light, and only Au, Ag and Cu can be used for practical applications. This fact severely limited the SERS application in other materials. Whereas transition metals have a much wider application in modern industries and technologies, such as electrochemistry, corrosion and catalysis, they had been commonly considered as non-SERS-active substrates by the communities of surface science and spectroscopy [13]. In fact, we do not find any SERS review articles definitely stating that transition metals exhibit SERS before 1998. In the past three decades, the development of SERS into a powerful spectroscopic tool for a wide variety of materials had been slower than expected.

However, researchers never gave up their efforts to extend SERS to the study of metallic and nonmetallic surfaces other than coinage metals [14, 15, 16, 17, 18, 19, 20, 21, 22, 23, 24, 25, 26, 27, 28, 29, 30, 31, 32, 33, 34, 35, 36, 37, 38, 39, 40, 41, 42, 43, 44, 45]. In the 1980s, a strategy based on "borrowing SERS" was proposed, either by depositing a SERS-active metals onto non-SERS-active substrates including semiconductors [14, 15] or by depositing non-SERS-active materials over SERS-active substrates [16, 17, 18, 19, 20, 21]. For example, a SERS-active silver layer was deposited over a n-GaAS sur-

K. Kneipp, M. Moskovits, H. Kneipp (Eds.): Surface-Enhanced Raman Scattering – Physics and Applications, Topics Appl. Phys. **103**, 125–146 (2006)
© Springer-Verlag Berlin Heidelberg 2006

face and the resonance Raman signal of dye molecules was obtained [14, 15]. Alternatively, Ni, Co, Fe, Pt, Pd, Rh and Ru ultrathin films were electrochemically deposited over SERS-active Ag or Au electrodes to obtain signals from these films. With the aid of the long-range effect of the electromagnetic (em) enhancement created by the SERS-active substrate underneath, weak SERS spectra of various adsorbates on the transition-metal overlayers have been obtained [16, 17, 18, 19, 20, 21].

It should be noted that the strong electromagnetic field generated on the SERS-active substrate will be attenuated exponentially with the increasing of thickness of the coated film so that the film has to be ultrathin, normally 3–10 atomic layers, to achieve a reasonable signal. However, it was very difficult to completely cover randomly rough substrates with such a ultrathin layer. Thus the "pinhole" in the overlayer made it extremely difficult to eliminate entirely the contribution of the giant SERS of the substrate. Weaver and coworkers made significant progress in overcoming this problem. They reported a series of work on "pinhole-free" transition metals over the SERS-active Au surface by electrochemical atomic-layer epitaxy using constant-current deposition at a low current density or by redox replacement of underpotential-deposited metals on Au [22, 23, 24, 25]. This method is very promising if one can prepare a "pinhole-free" ultrathin film of different materials with good stability in a wide range of potential or/and temperature. It makes SERS a versatile tool in studying various material surfaces of practical importance. In addition to studying the surface adsorption and reaction, the overlayer method can also be used to characterize the fine structure of the ultrathin film itself. This includes oxides, semiconductors and polymers [26, 27, 28, 29, 30]. Its advantage of high sensitivity enables one to probe ultrathin films with a few atomic monolayers.

Another totally different strategy is to generate SERS directly from transition metals [33, 34, 35, 36, 37, 38, 39, 40, 41, 42, 43, 44, 45]. There is no doubt that the surface preparation is more straightforward and the stability is higher than the substrate coated with an ultrathin film. This strategy is much more challenging as it contradicts the commonly accepted notion that transition metals were not SERS-active. This approach appeared to be the most difficult or even impossible one, and has been demonstrated since the early days of SERS. Several groups attempted to obtain unenhanced [33, 34] and enhanced [35, 36, 37, 38, 39, 40, 41] Raman signals from adsorbates on either roughened or mechanically polished Pt, In, and Rh electrodes, or porous Ni, Pd, Pt, Ti and Co films [42, 43]. In all of these studies, the reported surface spectra were obtained only under optimal conditions or by data manipulation using spectral-subtraction methods [33, 34, 35, 36]. Surface Raman signals were typically too weak to be investigated as a function of the electrode potential or temperature. In practical applications, however, such a type of measurement is quite essential. Furthermore, some results could not be repeated by other groups. The reported results were not strongly supportive of SERS studies on transition metals, and pointed to a gloomy future in this direction. Indeed, only a few papers [38, 41, 44, 45], among many theoretical and experimental works, claimed that transition metals might have relatively weak SERS activity in comparison with the coinage metals of Au, Ag and Cu, which had not been recognized by the scientific community. As a consequence, the activity in the SERS field declined substantially from the late-1980s to the mid-1990s.

The situation has changed dramatically since the late-1990s [46, 47, 48, 49, 50, 51, 52, 53, 54] with advances in Raman instrumentation, i.e., the advent of the confocal microscope and the holographic notch filter [55, 56]. The Raman experiment, which normally employed high-dispersion double or triple monochromators to filter out the elastically scattered laser radiation, can now be performed simply with a single spectrograph together with a holographic notch filter. The throughput of a single-grating system is far higher than, for example, a triple monochromator. The optical configuration of the confocal Raman microscope was found to be very helpful for obtaining the very weak signal of the surface species without the interference of the strong signal from the bulk phase. These new developments have resulted in unprecedented sensitivity that was vitally important for making the several important advances in this field in the late-1990s.

One of the progresses is confirmation of SERS directly from many pure massive transition metals [46, 47, 48, 49, 50, 51, 52, 53, 54]. Since 1996 our group has reported our efforts to optimize the confocal Raman microscope in order to obtain the highest sensitivity, and has developed various surface-preparation procedures for different transition metals to obtain surface spectra with better signal to noise ratio. Generally, two types of substrates have often been employed in SERS studies: roughened metal electrodes and metal nanoparticle (colloid) sols. In our studies, six SERS activation procedures have been performed and/or developed for different transition metals and for different studies, i.e., potential-controlled oxidation and reduction cycle(s) (ORC) [46], current-controlled ORC [57], chemical etching [52, 53], electrode-position [51], template synthesis [58, 59], and nanoparticles-on-electrode [60].

So far we have been able to obtain surface Raman spectra of good quality not only on the electrode surfaces made of many transition metals, such as Pt, Ni, Ru, Rh, Pd, Co, Fe and their alloys, but also over a very wide potential region (e.g., $-2.0\,\mathrm{V}$ to $+1.4\,\mathrm{V}$) [9, 13, 46, 47, 48, 49, 50, 51, 52, 53, 54]. The molecular-level investigation of diverse adsorbates at various transition-metal electrodes can be realized by Raman spectroscopy [46, 47, 48, 49, 50, 51, 52, 53, 54]. The important adsorbates already studied include CO, H, O, Cl^-, Br^-, SCN^-, CN^-, pyridine, thiourea, benzotriazole and pyrazine [46, 47, 48, 49, 50, 51, 52, 53, 54, 61, 62, 63, 64, 65]. It has been demonstrated that SERS can be widely used to study electrochemical interfaces including many transition-metal-based systems, such as electrochemical adsorption and reaction, electrocatalysis, corrosion and fuel cells [63, 64, 65, 66].

It has been found that the increase in the surface signal intensity is not simply proportional to the surface-roughness factor (R). Therefore, one should consider the existence of the SERS effect on these transition-metal surfaces. It should be pointed out that in the past about three decades, among many theoretical studies, only two papers [44, 45] claimed that transition metals might have relatively weak SERS activity in comparison with the coinage metals. In the mid-1980s *Chang* and coworkers reported a very interesting electrodynamics calculation on spheroids of Pt, Rh, Ni and Pd [44], but these important papers had not been recognized by the scientific community because transition metals were not generally considered as effective SERS substrates. Since we have already a large surface Raman signal from transition metals, it might be possible for us to clarify whether there are SERS on transition metals. We proposed a method to accurately calculate the surface-enhancement factor (G) after considering the special feature of the confocal Raman system [48, 56]. We obtained G of five roughened transition metals Fe, Co, Ni, Rh, Pd and Pt experimentally ranging from 10^2 to 10^4, in comparison with 10^6 to 10^7 for Ag, using pyridine as the probe molecule and the 632.8 nm excitation.

On the basis of the work on SERS of transition-metal surfaces, we will divide the rest of this chapter into two parts. First, we will discuss the enhancement mechanism, mainly from the physical aspect, for transition metals in comparison with the coinage metals. A three-dimensional finite difference time-domain (3D-FDTD) method was used to calculate and evaluate the local electromagnetic field that is critically dependent on the surface geometry and the size, shape and interparticle coupling of nanoparticles. Second, a new approach of using ultraviolet laser to study UV-SERS is presented, showing that the transition metals have a higher SERS activity than the coinage metals in the UV region.

2 The Physics behind SERS of Transition Metals

2.1 The Electronic Structure and Dielectric Constants of Transition Metals

Since we have successfully and systematically obtained SERS from transition metals and demonstrated that there is a SERS effect on many transition metals, it is interesting to get a clear picture of the role of the electromagnetic (em) field enhancement in the SERS of transition-metal systems. It has been well known that the em enhancement is mainly contributed by surface-plasmon resonance at geometrically defined metal nanoparticles or nanostrutures [67, 68, 69, 70, 71, 72]. The em field of the light at the surface can be greatly enhanced under conditions of collective electron resonance for "free-electron" metals classified as the coinage metals (1B group) and alkali

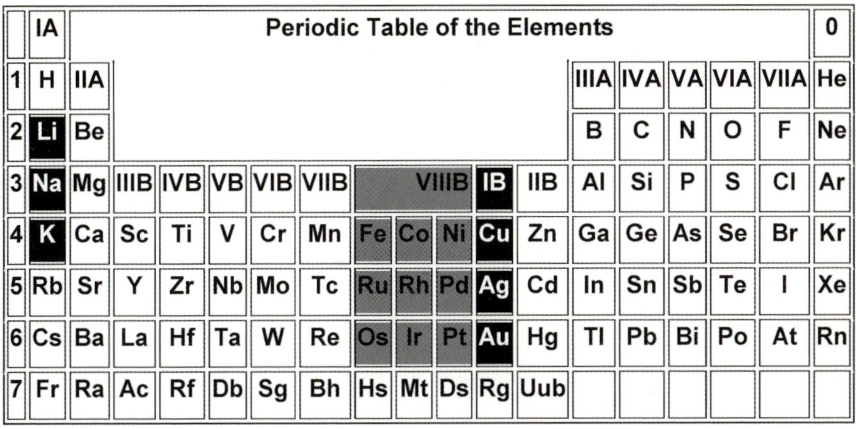

	IA		Periodic Table of the Elements														0
1	H	IIA									IIIA	IVA	VA	VIA	VIIA	He
2	Li	Be										B	C	N	O	F	Ne
3	Na	Mg	IIIB	IVB	VB	VIB	VIIB		VIIIB		IB	IIB	Al	Si	P	S	Cl	Ar
4	K	Ca	Sc	Ti	V	Cr	Mn	Fe	Co	Ni	Cu	Zn	Ga	Ge	As	Se	Br	Kr
5	Rb	Sr	Y	Zr	Nb	Mo	Tc	Ru	Rh	Pd	Ag	Cd	In	Sn	Sb	Te	I	Xe
6	Cs	Ba	La	Hf	Ta	W	Re	Os	Ir	Pt	Au	Hg	Tl	Pb	Bi	Po	At	Rn
7	Fr	Ra	Ac	Rf	Db	Sg	Bh	Hs	Mt	Ds	Rg	Uub						

Fig. 1. A simplified periodic table of the elements. The part marked in *black* are "free-electron" metals and those marked in *gray* are transition metals

metals (1 A group) in the periodic table (see Fig. 1). In fact, all metals including transition metals can enhance the field more or less, mainly depending on the ability of metals to sustain the surface plasmon of high resonance quality. In comparison with the coinage and alkali metals, transition metals (VIIIB group) have very different electronic structures [73]. The transition metals cannot be well described by the free-electron model. The $3d$, $4d$ and $5d$ shells are always strongly mixed with the $4s$, $5s$ and $6s$ states. From the physical point of view, the Fermi level of transition metals locates at the d band. Therefore, the interband excitation occurs very possibly in the visible light region [74,75]. The coupling between the high density of states and interband electronic transitions depresses the quality of the surface-plasmon resonance of transition metals considerably [68, 71].

To meet the conditions of good surface-plasmon resonance, the metal usually needs a small value of the imaginary component of the dielectric constant [76]. However, this is impossible for the transition metals because of their large values of the imaginary part of dielectric constants in the visible light region (see Fig. 2a). The dielectric constants in the figure are derived from the experimentally determined optical constants [77, 78] through the relationships of $\varepsilon_r = n^2 - k^2$, $\varepsilon_i = 2nk$.

Figure 2b illustrates dramatic differences in the SPR character between Pt and Ag elliptical nanoparticles with an aspect ratio of 2 : 1. The calculation is based on a two-dimensional array model proposed by *Chu* and *Wang* [79]. It can be seen that the enhancement factor of the coinage metal increases sharply when the frequency of the excitation light is close to the frequency of the surface-plasmon resonance [72, 76]. In contrast, the curves are much broader for Pt. SERS of Pt does not show any distinct character of surface-plasmon resonance, and the enhancement factors just show a slight variation

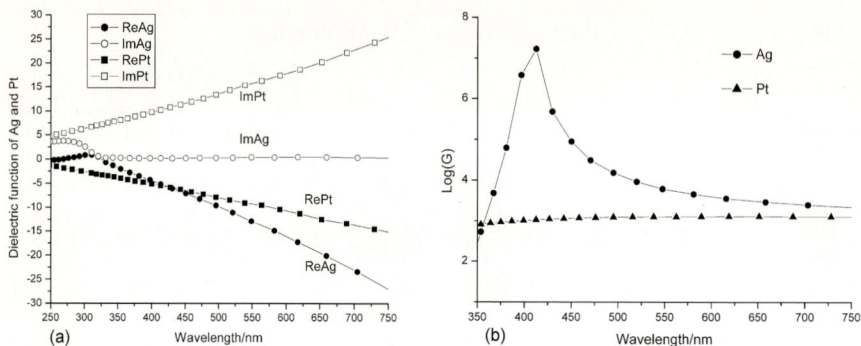

Fig. 2. (a) Dielectric constants of Ag and Pt; (b) Dependence of the surface enhancement factor (G) for two-dimensional Ag and Pt nanospheroids array on the wavelength of the incident plane wave. The semiminor axis and aspect ratio of spheroids are 20 nm and 2 : 1, respectively. The probe molecule is assumed to be located at the tip of spheroids

with the incident photon energy. This means that the SERS intensity of Pt will not vary much with the excitation wavelength in the visible light region. Our experimental results have confirmed this prediction. For instance, Pt shows almost the same SERS enhancement if we change the excitation laser from 632.8 nm (1.96 eV) to 514.5 nm (2.41 eV) while keeping all other conditions unchanged. Although the intrinsic optical property of transition metals prevents them from exhibiting intense SERS, the em enhancement still exists in the transition-metal systems with different characteristics.

2.2 Theoretical Simulation of the Local Electric Field by the Finite Difference Time-Domain Method

Several methods have been utilized to perform electrodynamics calculation of light scattering from particles of an arbitrary shape, including the finite element method (FEM), discrete dipole approximation (DDA), finite difference time-domain (FDTD) method and T-matrix method [80,81,82]. Among them, FDTD is a powerful tool to simulate the distribution of the electromagnetic field around the illuminated nanoparticles or substrate with an arbitrary geometry by numerically solving Maxwell's equations. Here, we will simulate the local electric field on transition-metal surfaces using the three-dimensional FDTD (3D-FDTD) method. To obtain the necessary parameters like static permittivity, infinite frequency permittivity, conductivity and the relaxation time, we fit them to the complex permittivity of metals [78] for the optical frequency range under investigation. Their relationship can be described by the general Drude model as follows [82,83],

$$\varepsilon(\omega) = \varepsilon_\infty + \frac{\varepsilon_s - \varepsilon_\infty}{1 + i\omega\tau} + \frac{\sigma}{i\omega\varepsilon_0},$$

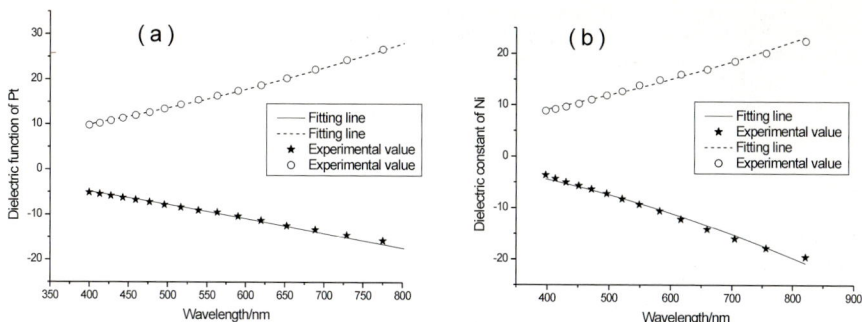

Fig. 3. Complex permittivity for Pt (**a**) and Ni (**b**) solid: *fitting line* using the general Drude model; *scatter*: experimental data

where ε_s, ε_∞, σ, τ represent the static permittivity, the infinite frequency permittivity, the conductivity and the relaxation time, respectively. ω is the angular frequency of the excitation light and ε_0 is the permittivity of free space. The four parameters ε_s, ε_∞, σ, τ can be adjusted through curve-fitting techniques to correctly match the experimental complex permittivity. Figure 3a shows a good fitting curve for the complex permittivity for platinum in the visible light region where ε_s, ε_∞, σ, τ are -62, 1, 1.18×10^6 S/m and 6.6×10^{-16} s. Similarly, a good fitting curve can be obtained for Ni with the static permittivity, the infinite frequency permittivity, the conductivity and the relaxation time of -315, 1, 2.1×10^6 S/m, and 1.6×10^{-15} s, respectively (see Fig. 3b). A more detailed description of the theoretical treatment will be given elsewhere [60].

Based on the previous theoretical works [72], we have carried out a comparative theoretical study of transition metals and coinage metals. We first used a model of a Ni spheroid with different aspect ratios to calculate the local optical electric field contributing to SERS without taking into account the coupling between particles. The 3D-FDTD result indicates that the Raman signal from a single nanoparticle critically depends on the shape of the nanoparticle and the dielectric constant of the metal as shown in Fig. 4. In order to accurately simulate the detailed nanostructure of nanoparticles, the Yee cell size used in our calculation was set to be $1 \times 1 \times 1$ nm^3, which was much smaller than the size of relevant nanoparticle features. The step size of time was set as 1.73 as, and the number of periods of the incident sinusoidal plane wave was set to be 8 to guarantee the calculation convergence, which could be judged by checking whether the near-zone electric field values reach steady state. The amplitude of the sinusoidal plane wave was set to be 1 V/m in the calculation, and the excitation wavelength 632.8 nm.

For a single 60 nm Ni sphere under 632.8 nm excitation, the electric-field enhancement at the tip is just about 4 times the incident optical electric field. Since the SERS enhancement can be estimated by $G = [E_L(\omega_i)/E_{in}(\omega_i)]^2 [E_L(\omega_s)/E_S(\omega_s)]^2$ [68,69], in this case, the SERS enhance-

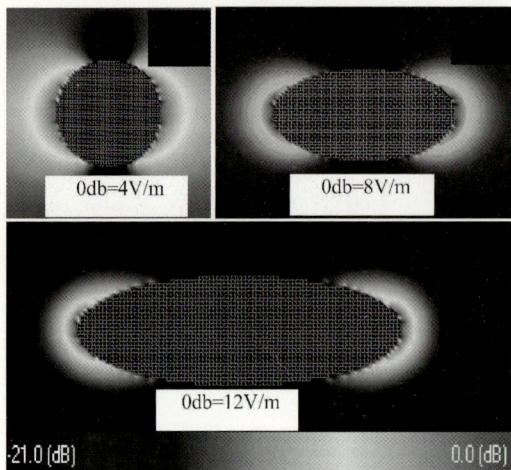

Fig. 4. FDTD simulated electric field distribution of nickel nanospheroid with various aspect ratios, (**a**) 1 : 1, (**b**) 2 : 1, (**c**) 3 : 1. The excitation polarization is along the major axis

ment is about 2.6×10^2. In this approximation, we neglected the influence of the frequency shift of the Raman scattering light on the field enhancement. With the increase of the aspect ratio of the nanoparticle, the maximum *E-field* enhancement increases rapidly. For example, for a spheroid with aspect ratios of 2 : 1 or 3 : 1, the maximum E-field enhancement increases by more than 8 and 12 times, respectively (corresponding to 4×10^3 and 2×10^4 SERS enhancement at the tip, respectively). However, if the aspect ratio of the spheroid becomes 5 : 1 or higher, the maximum electric-field enhancement will decrease slowly. This tendency of the enhancement factor with increasing aspect ratio of the nanoparticle is in agreement with the results of *Schatz* and *Van Duyne* who used electrodynamics corrections to the spheroid model [72]. The above FDTD result is also very close to our previous results using an analytical method for the two-dimensional array model [84].

2.3 3D-FDTD Simulation of the Electromagnetic Field Distribution over a Cauliflower-Like Nanostructure

We found several surfaces with a cauliflower-like morphology, as shown in Fig. 5, to exhibit the highest SERS activity. To explain these experimental observations, the electric-field enhancement of SERS-active Rh systems was estimated by using the 3D-FDTD method. The cauliflower-like nanoparticle was modeled by a 120 nm sphere studded with many 20 nm semispheres, as shown in Fig. 6.

Shown in Fig. 6 is the calculated electric-field distribution on the surface of a Rh sphere with a diameter of 120 nm and covered with 20 nm semi-

(a)

(b)

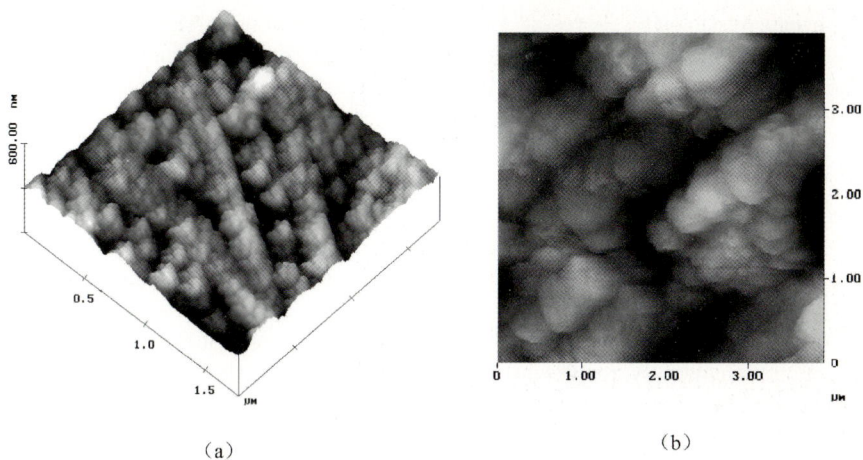

Fig. 5. AFM images of a roughened cauliflower-like Rh electrode (**a**) and a Co electrode (**b**)

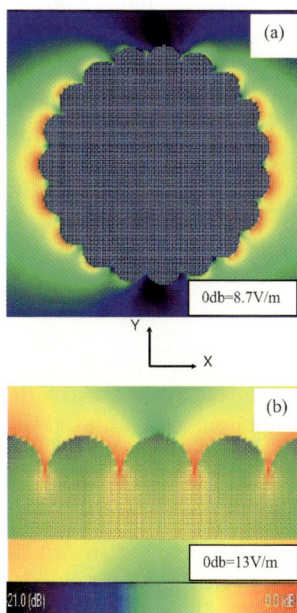

Fig. 6. FDTD simulated electric-field distribution for cauliflower-like Rh nanoparticles. The incident beam illuminates along the y-direction with x-polarization

spherical particles to represent the cauliflower-like nanoparticle (Fig. 6a). It can be seen that the magnitude of the maximally enhanced electric field on the cauliflower nanoparticle is about 8–9 times higher than that of the incident light, and the highest enhancement usually appears at the apex of small semispheres in the "cauliflower". It corresponds to an ca. 6×10^3-fold SERS enhancement on these sites. Note that the magnitude of the maximum electric-field enhancement is just about 4 times larger than the incident light for the smooth sphere under the same excitation condition [60], which means that the maximal enhancement factor for the cauliflower-like nanoparticle is just about 17 times larger than that for the smooth sphere. This different field enhancement could be probably understood on the basis of the lightning-rod effect, which usually results in a relatively large electric-field enhancement near high curvature sites on the surface [76].

It is noted that the field enhancement for the cauliflower-like nanoparticle is not enhanced significantly at the junction point as expected due to the coupling effect between small semispheres. However, if the cauliflower is detached and extended into a plane grating-like structure, the field maximum is shifted to the crevices, as shown in Fig. 6b. The symmetric nature of the nanoparticle may play a role in limiting the electromagnetic modes when the nanostrucutre is coupled with an incident plane wave. Our results seem to indicate that the em enhancement is very sensitive not only to the wavelength and polarization of the exciting light, the electronic property of the metal and the surface morphology, but also to the symmetric nature of the SERS nanostructures.

2.4 3D-FDTD Simulation of the Electromagnetic Field Distribution over a Nanocube Dimer

In our experiment, we found that the SERS activity from Pt nanocubes assembled on a glassy carbon surface is much higher than that from the roughened Pt electrodes. In order to understand this phenomenon, we evaluated the interparticle coupling effects since these Pt nanocube are closely connected to each other at the surface, as can be seen in [60]. When two Pt nanocubes approach each other, the classical electromagnetic theory predicts that a coherent interference of the enhanced field around each particle will result in a dramatic increase in the E-field in the junction between them. Again, the 3D-FDTD method was adopted to simulate the interparticle coupling of Pt nanoparticle dimers.

Figure 7 represents the geometric model in our calculation. On the basis of the TEM image of cubic Pt nanoparticles, the edge length of nanocubes is set to be 20 nm, and the gap between two nanocubes at 1 nm that is just enough to accommodate a layer of molecules. The incident vector is depicted in the figure and the polarization is parallel to the axis along the particle pair (x-axis).

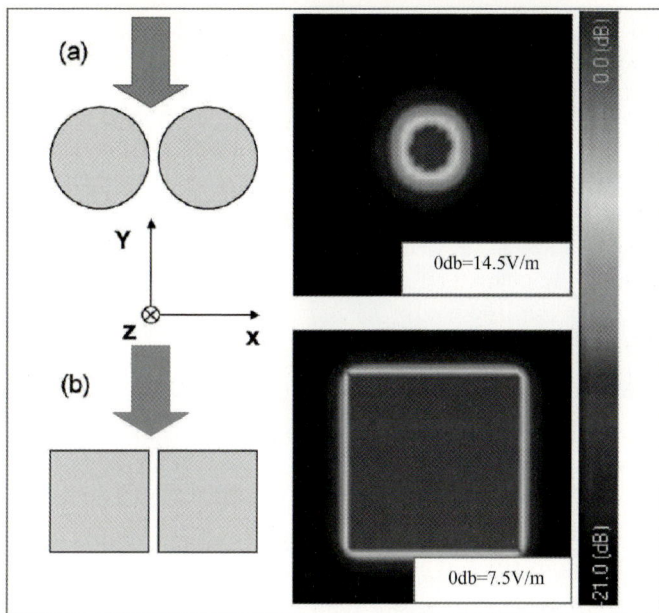

Fig. 7. FDTD simulations of the electric-field distribution in the yz-plane at $x = 0$. The excitation polarization is along the x-axis

It can be seen clearly from the calculation that the electric field reaches the maximum in the gap region in both the sphere (Fig. 7a) and cube models (Fig. 7b). This is in good agreement with the work of *Käll* and coworkers [85] and *Brus* and coworkers [86] that a nanoparticle gap or junction can produce a very large enhancement. The theoretical calculations by *Käll* and coworkers [87, 88] and *Martin* and coworkers [89] can adequately explain the experimental phenomena.

In the present study, the electric-field enhancement is about 14.5 in the gap of the two Pt spheres, while it is about 7.4 in the gap of cubes. This corresponds to SERS enhancement factors of 4.4×10^4 and 3.0×10^3 in the gap for the two shapes, respectively. However, it should be noted that the effective volume in the junction that can offer the largest SERS enhancement is very limited in the case of spheres compared with that of cubes. This fact results in a comparable, even stronger, total SERS signal from nanocubes than from nanospheres.

2.5 3D-FDTD Simulation of Electromagnetic-Field Enhancement of Core-Shell Nanoparticles

To make use of the electromagnetic field enhancement of SERS-active metal to study the non-SERS active overlayers has been a subject of interest for

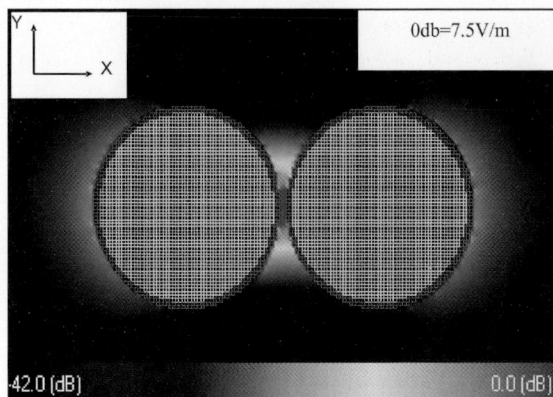

Fig. 8. FDTD simulated electric-field distribution for Au@Pd nanoshell dimers. The excitation polarization is along the x-axis

about 20 years. Recently, we also used the same strategy, however, to prepare Au core Pd shell(Au@Pd) nanoparticles. We found that the SERS activity decays exponentially with increasing shell thickness, which is similar to our previous observation from the Rh layers electrodeposited on a SERS-active Au substrate [90]. The UV-vis data (not shown here) also indicate that the surface-plasmon band of Au is gradually damped with increasing thickness. Most importantly, the highest SERS signal obtained on these Au@Pd nanoparticles can reach up to about 1000 counts per second (cps), which is remarkably higher than that of a roughened Pd electrode (5 cps) and monometallic Pd nanoparticles (60 cps). It clearly demonstrates that a major contribution to the phenomena comes from the long-range effect of the strong electromagnetic field generated by the Au core. To understand the effect of the core, we also simulated the system by the FDTD method.

Figure 8 shows the FDTD-simulated electric-field distribution of the Au@Pd nanoparticle dimers to take account of the coupling effect. In the calculation, the thickness of the Pd shell is set to be 2 nm, the size of the Au core is set to be 55 nm, and the interparticle distance is 1 nm. The 632.8 nm excitation laser is polarized along the axis connecting the two particles. The simulated result shows the maximum electric-field enhancement is also in the junction, which is about 50, corresponding to a SERS enhancement of 6×10^6. It may be necessary to point out the dielectric constants of the shell were directly taken from the bulk value since the large value of the imaginary component of the dielectric constant of transition metals will just result in a relatively small change of the dielectric constant compared with coinage metals in the visible light region. If the thickness is less than 2 nm, the size-dependent effect of the dielectric function of the Pd shell should be taken into account, but we cannot find the necessary parameters, such as the electron mean-free-path for Pd, in the literature. Nevertheless, in our preliminary calculation, we noticed that the em enhancement increases rapidly with decreasing Pd shell thickness to 1 nm (not shown here) when the bulk dielectric constant was used. The calculation and the experimental data are in good agreement. However, the size-dependent dielectric constants should be used in order to perform more quantitatively simulation on the SERS properties of core-shell nanoparticles when the thickness is around 1 nm.

If the shell thickness exceeds 6 nm, the SERS activity of Au@Pd nanoparticles will show no difference from pure Pd nanoparticles. It should be emphasized here that although the em enhancement is very high in the junction of the nanoshell dimers, the surface-averaged SERS enhancement factor is much lower than 6 orders of magnitude since the number of hottest SERS-active sites, located in the junction, is very limited.

Finally, it may be necessary to note that many molecules and ions adsorb more strongly at transition metals because of the empty d orbital(s) in comparison with the coinage-metals systems. In surface chemistry, most molecules studied can strongly interact with (bind to) the surface and surface chemists cannot ignore the interaction of the surface species with the substrates [13, 91, 92, 93, 94]. It is important for them to describe a clear picture about the molecule–surface interaction, adsorption orientation, surface configuration and coverage, etc. and to find the answer to the phenomenological behavior by analyzing the SERS spectra in much more detail [95, 96]. Therefore, although the electromagnetic effect is the dominant mechanism for most SERS systems, it is necessary, at least for surface chemists, to understand the chemical enhancement, and especially the influence of chemical enhancement on the spectral feature of surface species. Furthermore, when the strong chemical bond is formed, this chemisorption not only changes the electronic structure of the adsorbate itself, but also influences to some extent the surface electronic structure. This may cause the shift of the surface-plasmon resonance frequency and lead to the change of the local electric field at the metal surface [97].

3 SERS From Transition Metals with Ultraviolet Excitation

In all previous SERS studies, light from the visible to the near-infrared region were used for excitation. Extending SERS studies to the UV region may initiate some new approaches. First, it may allow investigation of molecules and nanostructures under new experimental condition including UV-SERS active materials that may or may not be SERS active in the visible or near-infrared region of the em spectrum. Second, since the UV energies are in preresonance or full resonance with electronic absorption of many molecular systems, macromolecules and biomolecules, if some substrates can support the em enhancement in the UV region, the UV laser can excite both the resonance Raman and SERS effect, and the possible applications of surface-enhanced resonance Raman scattering (UV-SERRS) could be boundless. Moreover, the relevant theoretical approach could provide a deep physical insight into the SERS/SERRS phenomenon. However, many groups including ours have tried but failed to observe SERS from the coinage metals using the ultraviolet (UV) excitation. This is mainly due to the inherent difficulties in generating observable surface enhancement for coinage metals in the UV spectral region, due to the damping of the surface-plasmon resonance by the interband transition absorption when the excitation line is moved to the UV region. Since the optical properties of transition metals are very different from coinage metals, it is worth testing the UV-SERS activity. Recently, we have carried out some preliminary studies using the UV excitation. The results show that UV-SERS can be obtained from a number of adsorbates on different transition-metal surfaces [98, 99].

3.1 Potential-Dependent UV-Raman Spectra from Transition Metals

The surface UV-Raman studies were performed using a He–Cd laser with the wavelength of 325 nm for excitation [98, 99]. Figure 9a shows the potential-dependent spectra of pyridine adsorbed on a roughened Rh electrode surface with a cauliflower-like nanostructure. In the present study, the objective used is of 15× magnification, therefore the interference of the solution signal to the surface signal is quite severe. In order to extract the signal of the surface species, all the surface Raman spectra shown have been subtracted with the corresponding solution spectrum with the intensity of the 933 cm^{-1} band (total symmetric stretching of ClO_4^-) as an internal reference. As can be seen from Fig. 9a, the frequency and the relative intensity of the major bands of pyridine show a clear potential dependence. The redshift of these bands with the positive movement of the electrode potential is clear evidence of surface species whose surface orientation and coverage are dependent on the applied potential.

Compared with pyridine, SCN^- is a smaller Raman scatterer. However, despite this fact, good-quality spectra were obtained for SCN^- adsorbed on a roughened Rh surface. The experiment was performed with 0.1 mol·l^{-1} $NaClO_4$ + 0.01 mol·l^{-1} NaSCN solution as the electrolyte, where the ClO_4^- in the solution was again used as the internal reference for subtracting the solution signal. The surface Raman spectra are presented in Fig. 9b, which shows that the intensity of the CN stretching band increases with the negative movement of the electrode potential. In addition, the band frequency redshifts significantly showing an electrochemical Stark effect with a tuning rate of around 73 cm^{-1}·V^{-1} of excellent linearity [99].

The good-quality spectra of adsorbed SCN^- can also be observed from Ru and Co electrode surfaces [98, 99] as well as Pt nanocube-on-electrodes [100]. The potential-dependent spectra of the adsorbed SCN^- on a Ru-coated glassy carbon electrode and on a roughened Co massive electrode are shown in

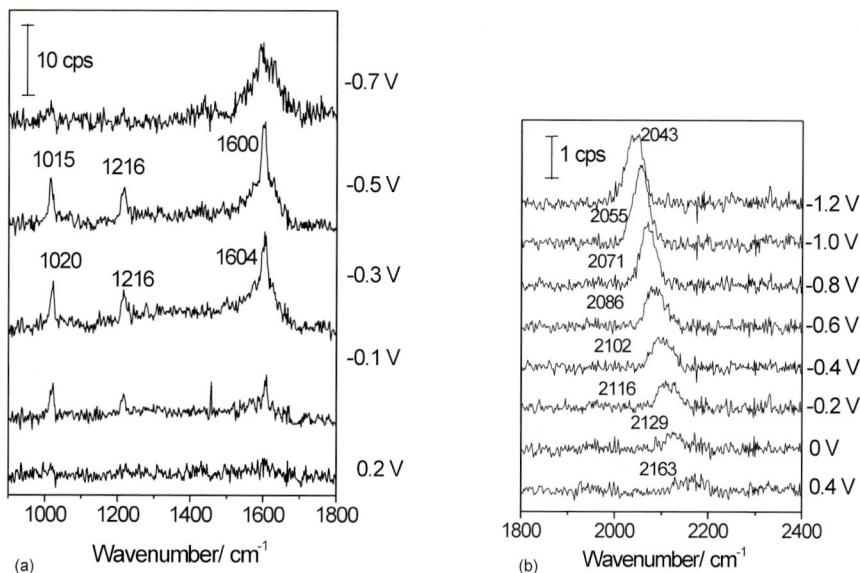

Fig. 9. (a) UV-SER spectra of pyridine adsorbed on a Rh electrode at different potentials; (b) UV-SER spectra of SCN^- adsorbed on a roughened Rh electrode in $0.1 \, mol \cdot l^{-1}$ $NaClO_4$ + $0.01 \, mol \cdot l^{-1}$ NaSCN at different potentials. Excitation line: 325 nm

Figs. 10a and 10b. On these two electrodes, only one peak was observed in the negative potential region and the v_{CN} blueshifts with the positive movement of the potential. Another peak appears at the high-frequency side of the original peak from -0.1 V for Ru electrode and -0.5 V for Co electrode, indicating that there is another type of surface SCN^- species in the positive-potential region.

It may be necessary to note that photodecomposition on the surface is often observed and it may limit to some extent the application of UV-SERS. However, the successful observation of a surface UV-Raman signal from adsorbed species indicates a promising future for this approach. Furthermore, according to our calibration, the signal of silicon ($520.6 \, cm^{-1}$) obtained on the visible Raman system is about 40-fold higher than that on the UV-Raman system. This is almost opposite to that predicted by theory if we consider the fourth-power-law relationship of Raman intensity with that of the incident laser frequency. This situation makes the UV-Raman technique very promising since we still have a great chance to improve significantly the detection sensitivity of the UV-Raman instrument.

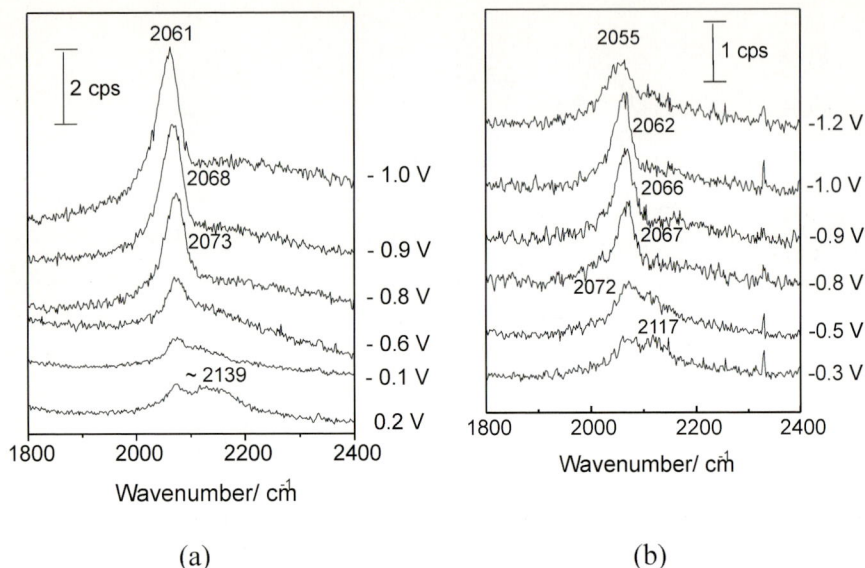

Fig. 10. UV-SER spectra SCN$^-$ adsorbed on a Ru-coated glass carbon electrode (a) and on a Co electrode (b) in $0.1\,\mathrm{mol}\cdot\mathrm{l}^{-1}$ NaClO$_4$ + $0.01\,\mathrm{mol}\cdot\mathrm{l}^{-1}$ NaSCN at different potentials. Excitation line: 325 nm, acquisition time: 200 s

3.2 Confirmation of UV-SERS Effect on Transition Metals

It can be seen that the UV-Raman signal of SCN$^-$ on the Ru coated glassy carbon electrode is stronger than that on the roughened Rh and Co surfaces. This may indicate the importance of proper matching of the incident light frequency with that of the optical properties of the surface nanostructures. It is necessary to clarify if there is the SERS effect with ultraviolet excitation and whether transition metals can exhibit UV-SERS activity. We performed a similar calculating method mentioned above and also considering the special feature of the UV Raman microscope [99]. From the experimental results shown in Fig. 9a, the surface enhancement factor for the adsorbed SCN$^-$ is calculated to be about 345, indicating that the surface enhancement factor of the Rh surface in the UV regions is at least 2 orders of magnitude. It should be pointed out that although this value is considerably lower compared with that of coinage metals in the visible region, it is indeed meaningful for studying the adsorbed surface species with UV excitation.

It is of special interest to work out why transition metal and not the typical SERS metals support SERS activity in the UV region. We performed a preliminarily theoretical calculation of the wavelength-dependent surface enhancement factor for Rh and Ag using the method described by *Zeman* and *Schatz* [101], based on the em model. The Rh and Ag surfaces were simulated with a Rh and a Ag ellipsoidal nanoparticles with 3 : 1 aspect ratio (semi-

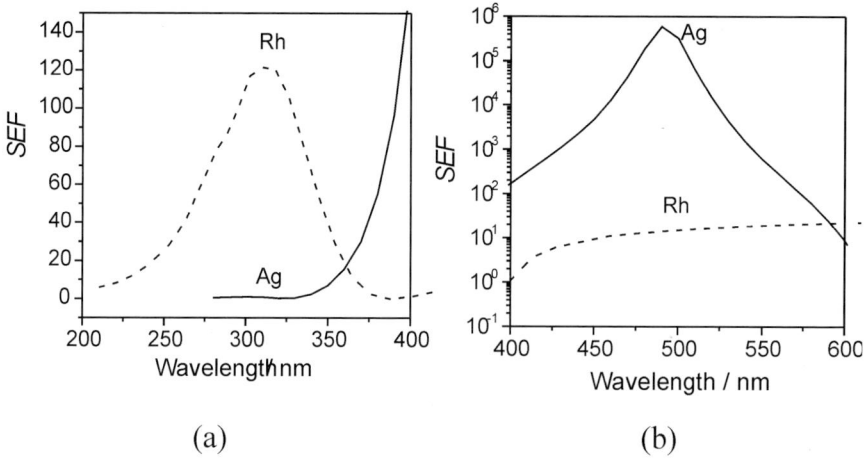

Fig. 11. Averaged enhancement factors for Ag and Rh ellipsoids with the aspect ratio of 3 : 1 in the UV spectral region (**a**) and the visible region (**b**) of the em spectrum

major axis $a = 45\,$nm and semiminor axis $b = 15\,$nm), respectively. As is well known, the field at the tip of the ellipsoid is much stronger than the other region of the particles when the electric-field vector of the incident light is polarized along the major axis. However, in a real SERS experiment, the signal is an average of the signal from all over the particle's surface. Therefore, only the averaged enhancement is meaningful for comparing experimental and theoretical results. The calculated averaged surface enhancement factors at different wavelengths for the above two systems are shown in Fig. 11. The averaged G for Ag in the visible region can be as high as 10^6, contributed by both surface-plasmon resonance and the lightning-rod effect. However, it decreases sharply when the excitation wavelength is moved into the UV spectral region and there is essentially no enhancement when the wavelength is shorter than $325\,$nm because this condition is then not appropriate to induce surface-plasmon resonance. The real dielectric constant of Ag is changed from the suitable negative value at about -10 at around $500\,$nm to the positive value of about 0.5 at around $325\,$nm. Accordingly, it seems to be easy to understand the diminishment of the enhancement in the UV region for Ag. In contrast, the G value of a Rh ellipsoidal does not change so significantly as wavelength varies in the visible region, but shows a small peak with a maximum of ca. 10^2 at around $325\,$nm. The real dielectric constant of Rh is changed from -18 at around $500\,$nm to about -10 at around $325\,$nm.

This preliminary theoretical calculation infers that some transition metals, rather than the typical SERS metals, may have relatively suitable optical properties to present observable SERS signal in the UV region. Although a more systematic study is required, our preliminary approach of UV-SERS not only demonstrates the importance of optical properties of the substrate in UV-SERS, but also provides some useful information on the physics behind SERS.

4 Conclusion

In this Chapter, we employed the 3D-FDTD method to calculate and evaluate the local electromagnetic field by investigating the effect of the surface geometry, the size, shape, and aggregation forms of nanoparticles on the SERS activity of transition metal systems. Our calculation on the cauliflower-structured nanoparticles shows that the em enhancement is sensitive not only to the polarization of the exciting light, the electronic property of metal and the surface morphology, but also to the symmetric nature of the SERS nanostructures. We demonstrated that the reason for the high enhancement in the nanocube system could be due to the large effective volume involved in the strong coupling between neighboring nanocubes. For the core-shell system, we show that the em enhancement decreases rapidly with increasing shell thickness and when the shell thickness exceeds 6 nm, the core no longer influences the SERS of the shell. The detection of a reasonably good UV-SERS signal on Pt, Rh, Ru and Co surfaces rather than Ag and Au is mainly due to the intrinsically suitable optical properties of transition metals in the UV region. Overall, the extension of SERS substrate to the transition-metal systems and use of UV excitation have improved the generality of the SERS technique and could be helpful for comprehensive understanding of SERS.

References

[1] Z. Q. Tian: Special issue on SERS, J. Raman Spectrosc. **36** (2005)
[2] A. Campion, P. Kambhampati: Chem. Soc. Rev. **27**, 241 (1998)
[3] K. Kneipp, H. Kneipp, I. Itzkan, R. R. Dasari, M. S. Feld: Chem. Rev. **99**, 780 (1999)
[4] M. Moskovits, L. L. Tay, J. Yang, T. Haslett: SERS and the single molecule, in V. Shalaev (Ed.): *Optical Properties of Nanostructured Random Media* (Springer, Berlin, Heidelberg, New York 2002) pp. 215–226
[5] K. L. Kelly, E. Coronado, L. L. Zhao, G. C. Schatz: J. Phys. Chem. B **107**, 668 (2003)
[6] J. Jiang, K. Bosnick, M. Maillard, L. Brus: J. Phys. Chem. B **107**, 9964 (2003)
[7] T. R. Jensen, M. D. Malinsky, C. L. Haynes, R. P. Van Duyne: J. Phys. Chem. B **104**, 10549 (2000)
[8] S. M. Nie, R. N. Zare: Ann. Rev. Biophys. Biomol. Struct. **26**, 567 (1997)
[9] Z. Q. Tian, B. Ren: Ann. Rev. Phys. Chem. **55**, 197 (2004)
[10] R. A. Lund, R. R. Smardzewski, R. E. Tevault: J. Chem. Phys. **88**, 1731 (1984)

[11] W. Schulze, K. P. Charle, U. Kloss: Surf. Sci. **156**, 822 (1985)

[12] D. P. DiLella, J. S. Suh, M. Moskovits: *Proceedings of the 8th International Conference on Raman Spectroscopy* (Wiley, Chichester, UK 1982) p. 63

[13] Z. Q. Tian, B. Ren, D. Y. Wu: J. Phys. Chem. B **106**, 9463 (2002)

[14] R. P. Van Duyne, J. P. Haushalter: J. Phys. Chem. **87**, 2999 (1983)

[15] J. C. Rubim, G. Kannen, D. Schumacher, J. Dunnwald, A. Otto: Appl. Surf. Sci. **37**, 233 (1989)

[16] M. Fleischmann, Z. Q. Tian: J. Electroanal. Chem. **217**, 385 (1987)

[17] M. Fleischmann, Z. Q. Tian, L. J. Li: J. Electroanal. Chem. **217**, 397 (1987)

[18] G. Mengoli, M. M. Musiani, M. Fleischmann, B. W. Mao, Z. Q. Tian: Electrochem. Acta **32**, 1239 (1987)

[19] L. W. H. Leung, M. J. Weaver: J. Am. Chem. Soc. **109**, 5113 (1987)

[20] L. W. H. Leung, M. J. Weaver: J. Electroanal. Chem. **217**, 367 (1987)

[21] Y. Zhang, X. Gao, M. J. Weaver: J. Phys. Chem. **97**, 8656 (1993)

[22] S. Zou, M. J. Weaver: Anal. Chem. **70**, 2387 (1998)

[23] S. Zou, C. T. Williams, E. K. Y. Chen, M. J. Weaver: J. Am. Chem. Soc. **120**, 3811 (1998)

[24] M. J. Weaver, S. Zou, H. Y. H. Chan: Anal. Chem. **72**, 38 A (2000)

[25] M. F. Mrozek, Y. Xie, M. J. Weaver: Anal. Chem. **73**, 5953 (2001)

[26] C. A. Melendres, M. Pankuch, Y. S. Li, R. L. Knight: Electrochim. Acta **37**, 2747 (1992)

[27] C. J. Zhong, Z. Q. Tian, Z. W. Tian: Sci. China B **33**, 656 (1990)

[28] G. Xue, Y. Lu, G. Shi: Polym. **35**, 2488 (1994)

[29] C. J. L. Constantino, T. Lemma, P. A. Antunes, R. Aroca: Anal. Chem. **73**, 3674 (2001)

[30] R. F. Aroca, C. J. L. Constantino: Langmuir **16**, 5425 (2000)

[31] C. A. Melendres, N. Camillone, T. Tipton: Electrochim. Acta **34**, 281 (1989)

[32] C. A. Melendres: *Spectroscopic and Diffraction Techniques in Interfacial Electrochemistry* (Kluwer Academic, Dordrecht 1990) p. 181

[33] R. P. Cooney, M. Fleischmann, P. J. Hendra: J. Chem. Soc. Chem. Commun. **7**, 235 (1977)

[34] M. Fleischmann, D. Sockalingum, M. M. Musiani: Spectrochim. Acta **46A**, 285 (1990)

[35] C. Jennings, R. Aroca, A. M. Hor, R. O. Loutfy: Anal. Chem. **56**, 2033 (1984)

[36] R. Aroca, C. Jennings, G. J. Kovacs, R. O. Loutfy, P. S. Vincett: J. Phys. Chem. **89**, 4051 (1985)

[37] T. Maeda, Y. Sasaki, C. Horie, M. Osawa: J. Electron. Spectrosc. Relat. Phenom. **64/65**, 381 (1993)

[38] S. A. Bilmes, J. C. Rubim, A. Otto, A. J. Arvia: Chem. Phys. Lett. **159**, 89 (1989)

[39] S. A. Bilmes: Chem. Phys. Lett. **171**, 141 (1990)

[40] C. Shannon, A. Campion: J. Phys. Chem. **92**, 1385 (1988)

[41] L. Guo, Q. J. Huang, X. Y. Li, S. H. Yang: Phys. Chem. Chem. Phys. **3**, 1661 (2001)

[42] H. Yamada, Y. Yamamoto: Chem. Phys. Lett. **77**, 520 (1981)

[43] H. Yamada, Y. Yamamoto: Surf. Sci. **134**, 71 (1983)

[44] M. P. Cline, P. W. Barber, R. K. Chang: J. Opt. Soc. Am. B **3**, 15 (1986)

[45] B. J. Messinger, K. U. von Raben, R. K. Chang, P. W. Barber: Phys. Rev. B **24**, 649 (1981)

[46] Z. Q. Tian, B. Ren, B. W. Mao: J. Phys. Chem. B **101**, 1338 (1997)
[47] Z. Q. Tian, J. S. Gao, X. Q. Li, B. Ren, Q. J. Huang, W. B. Cai, F. M. Liu, B. W. Mao: J. Raman Spectrosc. **29**, 703 (1998)
[48] W. B. Cai, B. Ren, X. Q. Li, C. X. She, F. M. Liu, X. W. Cai, Z. Q. Tian: Surf. Sci. **406**, 9 (1998)
[49] Q. J. Huang, X. Q. Li, J. L. Yao, B. Ren, W. B. Cai, J. S. Gao, B. W. Mao, Z. Q. Tian: Surf. Sci. **427/428**, 162 (1999)
[50] B. Ren, Q. J. Huang, W. B. Cai, B. W. Mao, F. M. Liu, Z. Q. Tian: J. Electroanal. Chem. **415**, 175 (1996)
[51] J. S. Gao, Z. Q. Tian: Spectrochim. Acta A **53**, 1595 (1997)
[52] Q. J. Huang, J. L. Yao, R. A. Gu, Z. Q. Tian: Chem. Phys. Lett. **271**, 101 (1997)
[53] P. G. Cao, J. L. Yao, B. Ren, B. W. Mao, R. A. Gu, Z. Q. Tian: Chem. Phys. Lett. **316**, 1 (2000)
[54] D. Y. Wu, Y. Xie, B. Ren, J. W. Yan, B. W. Mao, Z. Q. Tian: Phys. Chem. Comm. **18**, 1 (2001)
[55] B. Chase: Appl. Spectrosc. **48**, 14 A (1994)
[56] Z. Q. Tian, B. Ren: *Encyclopedia of Analytical Chemistry* (Wiley, New York 2000) p. 9162
[57] B. Ren, X. F. Lin, J. W. Yan, B. W. Mao, Z. Q. Tian: J. Phys. Chem. B **107**, 899 (2003)
[58] J. L. Yao, J. Tang, D. Y. Wu, D. M. Sun, K. H. Xue, B. Ren, B. W. Mao, Z. Q. Tian: Surf. Sci. **514**, 108 (2002)
[59] J. L. Yao, G. P. Pan, K. H. Xue, D. Y. Wu, B. Ren, D. M. Sun, J. Tang, X. Xu, Z. Q. Tian: Pure Appl. Chem. **72**, 221 (2000)
[60] Z. Q. Tian, Z. L. Yang, B. Ren, J. F. Li, Y. Zhang, X. F. Lin, J. W. Hu, D. Y. Wu: Faraday Discuss. **132**, 159 (2006)
[61] X. Xu, B. Ren, D. Y. Wu, H. Xian, X. Lu, P. Shi, Z. Q. Tian: Surf. Interf. Anal. **28**, 111 (1999)
[62] X. Xu, D. Y. Wu, B. Ren, H. Xian, Z. Q. Tian: Chem. Phys. Lett. **311**, 193 (1999)
[63] S. Z. Zou, C. T. Williams, E. K. Y. Chen, M. J. Weaver: J. Am. Chem. Soc. **120**, 3811 (1998)
[64] M. J. Weaver: Top. Catal. **8**, 65 (1999)
[65] V. M. Browne, S. G. Fox, P. Hollins: Catal. Today **9**, 1 (1991)
[66] B. Ren, X. Q. Li, C. X. She, D. Y. Wu, Z. Q. Tian: Electrochim. Acta **46**, 193 (2000)
[67] R. K. Chang, T. E. Furtak: *Surface Enhanced Raman Scattering* (Plenum, New York 1982)
[68] A. Otto: Surface-enhanced Raman-scattering – Classical and chemical origins, in M. Cardona, G. Guntherodt (Eds.): *Light Scattering in Solids* (Springer, Berlin 1984) pp. 289–418
[69] M. Kerker: Acct. Chem. Res. **17**, 271 (1984)
[70] H. Metiu, P. Das: Annu. Rev. Phys. Chem. **35**, 507 (1984)
[71] M. Moskovits: Rev. Mod. Phys. **57**, 783 (1985)
[72] G. C. Schatz, R. P. Van Duyne: *Handbook of Vibrational Spectroscopy* (Wiley, Chichester 2002) pp. 759–774
[73] H. Ebert: Rep. Prog. Phys. **59**, 1665 (1996)
[74] J. H. Weaver: Phys. Rev. B **11**, 1416 (1975)

[75] M. A. Ordal, R. J. Bell, R. W. Alexander, L. L. Long, M. R. Querry: Appl. Opt. **24**, 4493 (1985)

[76] J. Gersten, A. Nitzan: J. Chem. Phys. **73**, 3023 (1980)

[77] P. B. Johnson, R. W. Christy: Phys. Rev. B **6**, 4370 (1972)

[78] E. D. Palik: *Handbook of Optical Constants of Solids* (Academic, New York 1985)

[79] L. C. Chu, S. Wang: Phys. Rev. B **31**, 693 (1985)

[80] E. Hao, G. C. Schatz: J. Chem. Phys. **120**, 357 (2004)

[81] P. W. Barber, S. C. Hill: *Light Scattering by Particles: Computational Methods* (World Scientific, Singapore 1990)

[82] K. S. Kunz, R. J. Luebbers: *The Finite Difference Time Domain Method for Electromagnetics* (CRC and LLC, Boca Raton 1993)

[83] J. T. Krug II, E. J. Sanchez, X. S. Xie: J. Chem. Phys. **116**, 10895 (2002)

[84] Z. L. Yang, D. Y. Wu, J. L. Yao, J. Q. Hu, B. Ren, H. G. Zhou, Z. Q. Tian: Chin. Sci. Bull. **47**, 1983 (2002)

[85] H. X. Xu, E. J. Bjerneld, M. Käll, L. Borjesson: Phys. Rev. Lett. **83**, 4357 (1999)

[86] M. Michaels, M. Nirmal, L. E. Brus: J. Am. Chem. Soc. **121**, 9932 (1999)

[87] H. X. Xu, J. Aizpurua, M. Käll, P. Apell: Phys. Rev. E **62**, 4318 (2000)

[88] H. X. Xu, M. Käll: Phys. Rev. Lett. **89**, 246802 (2002)

[89] J. P. Kottmann, O. J. F. Martin, D. R. Smith, S. Schultz: Phys. Rev. B **64**, 235402 (2001)

[90] S. Zou, M. J. Weaver, X. Q. Li, B. Ren, Z. Q. Tian: J. Phys. Chem. B **103**, 4218 (1999)

[91] A. Otto, I. Mrozek, H. Grabhorn, W. Akemann: J. Phys. Condens. Matter **4**, 1143 (1992)

[92] D. Y. Wu, B. Ren, Y. X. Jiang, X. Xu, Z. Q. Tian: J. Phys. Chem. A **106**, 9042 (2002)

[93] D. Y. Wu, B. Ren, X. Xu, G. K. Liu, Z. L. Yang, Z. Q. Tian: J. Chem. Phys. **119**, 1701 (2003)

[94] D. Y. Wu, M. Hayashi, S. H. Lin, Z. Q. Tian: Spectrochim. Acta A **60**, 137 (2004)

[95] J. Billman, A. Otto: Surf. Sci. **138**, 1 (1984)

[96] J. F. Arenas, I. L. Tocon, J. C. Otero, J. I. Marcos: J. Phys. Chem. **100**, 9254 (1996)

[97] M. D. Malinsky, K. L. Kelly, G. C. Schatz, R. P. Van Duyne: J. Am. Chem. Soc. B **123**, 1471 (2001)

[98] B. Ren, X. F. Lin, Z. L. Yang, R. F. Aroca, B. W. Mao, Z. Q. Tian: J. Am. Chem. Soc. **125**, 9598 (2003)

[99] X. F. Lin, B. Ren, Z. L. Yang, G. K. Liu, Z. Q. Tian: J. Raman Spectrosc. **36**, 606 (2005)

[100] L. Cui, Y. Zhang, Z. L. Yang, B. Ren, Z. Q. Tian: in preparation

[101] E. J. Zeman, G. C. Schatz: J. Phys. Chem. **91**, 634 (1987)

Index

Electronic Mechanisms of SERS

Andreas Otto[1] and Masayuki Futamata[2]

[1] Institut für Physik der konensierten Materie, Heinrich-Heine Universität
Düsseldorf, D-40225 Düsseldorf, EU
`otto@uni-duesseldorf.de`
[2] Nanoarchitectonics Research Centre, National Institute of Advanced Industrial
Science and Technology, 1-1-1 Higashi, Tsukuba 305-8562, Japan
`m.futamata@aist.go.jp`

1 Introduction

Within the last decade, the field of Mie resonances in small metal particles [1]
and of surface-plasmon polaritons at plane metal surfaces [2] has matured into
the topic of plasmonics, see for example [3, 4]. This field is thriving, because
patterned structures can be produced by nanotechnology. Large enhancements of the local field at the laser frequency can be achieved, which support
surface-enhanced Raman scattering by adsorbed molecules [5, 6].

The special SERS intensities for the coinage and alkali metals [7, 8], and
less for transition metals (for instance [9]), the strong dependence on the
mesoscopic structure of the systems involved, the strong dependence of the
SERS intensity on the state of aggregation of colloidal nanoparticles (for
instance [10]) are essentially incomprehensible without invoking the electromagnetic (em) theory. Though a "chemical effect" has sometimes been postulated for smooth surfaces [11], detailed experiments and theory [12] have not
confirmed this opinion (with the exception of molecules without π^* states).
At a plane interface, enhanced Raman scattering has only been achieved by
tip-enhanced Raman scattering [13] or by exciting a surface-plasmon polariton by using a coupling prism above the surface [14].

Large em enhancement without interfering electronic ("chemical") effects
is the aim of analytical chemistry, where molecules must be recognized and
their concentration be measured irrespective of the chemical nature of the
adsorbed molecule [5].

The research in the electronic contribution to SERS is driven by the surface-science interest in the metal–adsorbate electronic and vibrational interaction, also in dependence of the atomistic structure of the surface, reactions
between different adsorbates and the hope to contribute to the understanding of important catalytic reactions, as Raman spectroscopy can bridge the
"pressure gap" or allows investigations at the solid/liquid interface.

The unavoidable electronic effect can spoil in some cases the great expectations based on the theory of the em enhancement. In the case of horseradish
peroxidase on nanopatterned silver structures, full enzymatic activity could
only be preserved by a primary reaction of the silver particles with hydrogen peroxide H_2O_2. In this case, the single-molecule SERS sensitivity was

K. Kneipp, M. Moskovits, H. Kneipp (Eds.): Surface-Enhanced Raman Scattering – Physics and
Applications, Topics Appl. Phys. **103**, 147–182 (2006)
© Springer-Verlag Berlin Heidelberg 2006

lost [15]. Maybe this has some relation to the so-called quenching of SERS by oxygen exposure under ultrahigh vacuum (UHV) [16].

For the beautiful space-resolved SERS spectra from the interior of living cells [17], by self-insertion of gold colloids forming aggregates the electronic effect is important. Here the absence of a strong SERS line of the adenine ring-breathing mode was not taken as an indication of adenine missing, but as an indication that the native DNA (double strand) and not the denatured DNA (single strand) contributed to the spectra. Adenine inside of the double helix has no direct contact to the Au surface and therefore is not observed [18], in spite of the theoretically enormous electromagnetic enhancement in fractal structures [19] (but see the Chapter by *Tian* et al. and [20]).

This Chapter starts with the historic development of "hard facts" of the "chemical effect" and ends with conjectures on the electronic contribution to single-molecule SERS, based on the experimental work of one of the authors and his collaborators (M. F.). The pioneers of single-molecule SERS are *Katrin Kneipp* and collaborators [21, 22] and *Shuming Nie* and collaborators [23]. Preceeding works pointing in the direction of SM-SERS are from *Hildebrand* and *Stockburger* [24] and *Pettinger* et al. [25, 26, 27]. Starting with Sect. 8 we will discuss to what extent electronic effects may contribute to single-molecule SERS.

Though analytical SERS is mainly performed at sample/gas, sample/ aqueous electrolyte, or sample/cell plasma interfaces, SERS of adsorbates at cold-deposited coinage metal surfaces in ultrahigh vacuum offers some advantages for mechanistic research on the SERS mechanisms. Coverage can be controlled by dosing the exposure and by thermodesorption spectroscopy, the electronic contact of the adsorbates with the metal is monitored by electrical surface resistance. The absence of water or water vapor allows for infrared reflectance and transmission spectroscopy. By controlled thermal annealing of the cold-deposited samples the structure on an atomic and mesoscopic scale is changed, and the existence of chemical active sites can be demonstrated. Besides optical reflection spectroscopy also inverse photoemission and electron energy-loss spectroscopy may be used to learn more about the excitations of substrate and adsorbates (in our case small molecules). Therefore the experimental examples in the next sections are mostly taken from cold-deposited films.

2 Long-Range Electromagnetic (em)-Enhancement G_{em} and "Chemical", "First-Layer"-Enhancement $G_{first\ layer}$ of Various Ag Samples

In order to understand the relevant enhancement mechanisms of SERS, the investigation of the spatial range off the metal surface of the enhancement is crucial. Two experiments, published in 1980 and 1981 [28, 29] shaped for

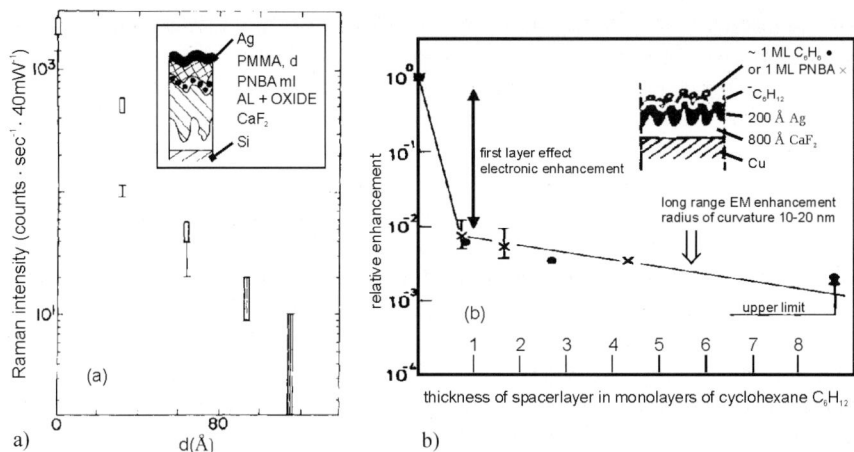

Fig. 1. (a) Spacer experiment with silver film of about 200 Å thickness (*inset* shows the geometry). The intensity of the 1595 cm^{-1} (*open symbols*) and 1425 cm^{-1} (*closed symbols*) Raman peaks of PNBA above the background (logarithmic scale) vs. the thickness d of the PMMA layer. Error bars are the size of the symbols and represent statistical accuracy only [29]. (b) Spacer experiment, see *inset*. Dependance of the relative enhancement of the CC ring-breathing mode of benzene (990 cm^{-1}; *points*) and the ν_3 mode of PNBA (1598 cm^{-1}; *crosses*) on precoverage by cyclohexane

about 10 years the prevalent opinion, that the long-range em enhancement is the nearly exclusive contribution to SERS.

Zwemer et al. [28], used an iodine-roughened silver surface in UHV at 120 K, first covered by a monolayer of pyridine (C$_5$D$_5$N), followed by condensation of deuterated pyridine (C$_5$D$_5$N). Since both species contributed to the Raman signal, the authors concluded that the enhancement extended well beyond the first monolayer and that there was no "first-layer" effect, discarding a previous result of *Smardzewski* et al. [30].

This experiment was repeated by *Mrozek* and *Otto* [31] with a silver-island film in UHV, albeit at a temperature of only 30 K. The SERS spectrum showed only the isotope species of pyridine chosen as the first layer, not from the alternative isotopic species in the subsequent layers. Only after warming the samples to 92 K and higher temperatures did a place exchange start and was a spectrum of both species observed. Hence, *Zwemer* et al. [28] had observed the first-layer effect of the scrambled pyridine isotopes.

Murray et al. [29] studied SERS from organic molecules, especially para-nitrobenzoic acid (PNBA), separated by polymer spacer layers from mesoscopically rough silver films. Thin-film multilayered structures Al/Aloxide/monolayer of PNBA/polymer spacer/silver film were prepared on a rough CaF$_2$ film on a Si wafer, see the inset of Fig. 1a, taken from [29].

The intensity versus the thickness d of the spacer layer showed clearly the long-range electromagnetic enhancement but no clear evidence for an extra enhancement without the spacer layer.

With a similar thin-film configuration (see inset of Fig. 1b), these experiments were repeated about 8 years later [32]. CaF$_2$ underlayers of about 80 nm thickness were thermally evaporated on a polished Cu block and transferred into UHV. A silver film of about 20 nm was evaporated at room temperature on this film. This film contours the CaF$_2$ film, corresponding to a root mean square (rms) roughness of about 3 nm to 4 nm and a roughness correlation length of about 20 nm [33]. After cooling to 30 K, the samples were first exposed to cyclohexane to form stable and tight spacer layers (for the calibration by monolayers see [31]) and directly afterwards to PNBA or benzene. The dependence of the relative enhancement for one monolayer of benzene (by observing the CC-ring-breathing mode) and PNBA (by observing the ν_3 mode at 1598 cm^{-1}) as a function of the number of cyclohexane spacer layers is shown in Fig. 1b. There is a clear extra enhancement for the directly adsorbed molecules ("first-layer effect") of about a factor of 100. This is of electronic origin, see below. Extrapolating the long-range signal to zero thickness of the spacer layer corresponds to the em enhancement for adsorbates G_{em}. In the case of the ring-breathing mode of benzene, the absolute enhancement for an adsorbed monolayer is 2×10^4, which corresponds to $G_{\mathrm{em}} \approx 200$. Using the spheroidal particle model, the decay of the long-range em enhancement, given by [34]

$$G_{\mathrm{em}}(d) = G_{\mathrm{em}}(d = 0) \left[\frac{r}{r + d} \right]^{10} , \qquad (1)$$

with r being the radius of local curvature and d the distance to the local surface, was fitted to the data. This yielded $r \approx 10$ nm to 20 nm, which is on the order of the correlation length of the silver-film roughness.

The discrepancy between Fig. 1a and Fig. 1b could be explained [32] by diffusion of PNBA through the polymer spacer layers and a photoreaction of PNBA at the silver surface, probably to azodibenzoate. The SERS spectra of [29] did resemble that of the photoreaction product of PNBA. Nevertheless, at the missing PMMA polymer spacer layer there is a considerable SERS signal for PNBA below (!) the 200 Å thick silver film. Light emission from such a Al–Aloxide–silver MIM (metal–insulator–metal)-junction (albeit without internal adsorbate) has been investigated and no light emission from the interior of the MIM junction was found [35, 36]. If one follows the theorem of optical reciprocity, a strong optical excitation within the MIM junction is not possible. Therefore the explanation of this interesting experiment [29] is still open.

The "first-layer effect" of SERS on all silver and copper samples in UHV is quenched by a weak exposure to oxygen gas, for the example of Fig. 1b see [31].

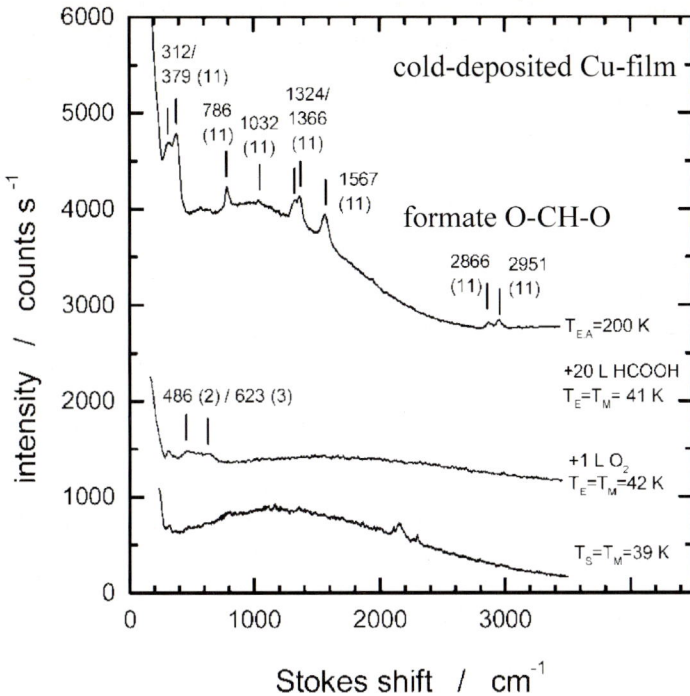

Fig. 2. SERS spectra of a cold-deposited copper film, deposited at $T_S = 39\,\text{K}$, exposed at $T_E = 42\,\text{K}$ to $1\,\text{L}\ O_2$ and subsequently to $20\,\text{L}$ HCOOH. The uppermost spectrum was recorded after annealing to $T_{EA} = 200\,\text{K}$

An example of oxygen quenching and reappearance of the SERS signal after getting oxygen off the surface by a chemical reaction [37] is shown in Fig. 2. In the lowest spectrum the typical Raman background is seen with some CO contamination. The Raman-background and CO signals are quenched by an exposure of $1\,\text{L}$ (10^{-6} Torr $\times$ s) of O_2, which is much less than a monolayer coverage given the porous structure of cold-deposited films. The quenching also prevents any signal of formic acid (HCOOH) after a $20\,\text{L}$ exposure (top spectrum). By warming the film to $200\,\text{K}$ the reaction of formic acid to formate

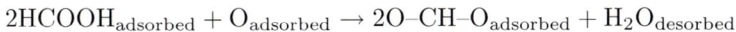

$$2\text{HCOOH}_{\text{adsorbed}} + \text{O}_{\text{adsorbed}} \rightarrow 2\text{O--CH--O}_{\text{adsorbed}} + \text{H}_2\text{O}_{\text{desorbed}}$$

strips all the oxygen from the surface and the background reappears strongly with all the vibrational bands of the formate [37].

Oxygen quenching allows the isolation of G_{em} from the full SERS enhancement for all silver samples listed in the first column of Fig. 3. The rightmost column gives the scattering configuration (backscattering (1) and $60°$ scattering (2)). The overall enhancement G has been measured for the

substrate	Reference substrate ($\frac{c}{s\,WL}$)	2	3	4 log G	scattering configuration
silver film on grating	sapphire 1.0 ± 0.5				①
silver film on rough CaF$_2$ film	rough CaF$_2$ film 1.35 ± 0.3			$d_m \sim 42$ Å	②
silver island film on sapphire	sapphire 0.67 ± 0.2	$d_m \sim 77\ 30\ 61$ Å			②
silver island film on carbonaceous substrates	carbonaceous substrate 0.6 ± 0.2	CSC C EG $d_m \sim 48$ Å C			②
silver islands on stochastic post structure	stochastic post structure 0.7 ± 0.2				①
cold-deposited silver film	sapphire 1.0 ± 0.5				①

1   2   3   4   5

log G

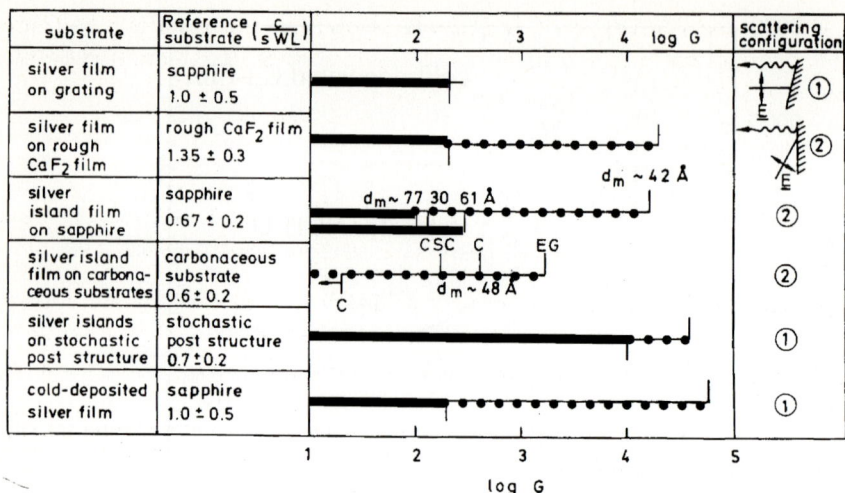

Fig. 3. Separation of long-range em enhancement (*solid lines*) and first-layer enhancement (*dotted lines*) contribution to the overall enhancement G, see text

CC-breathing vibration of benzene C_6H_6 at submonolayer coverage [16]. It is given on the logarithmic scale of G by the sum of the horizontal continuous and dotted bars. The dotted part of the enhancement is lost after oxygen quenching. This part is $\log(G_{\text{first layer}})$. The remaining part (continuous bar) is $\log(G_{\text{em}})$. Both values have been obtained by measuring the unenhanced Raman spectrum of a thick condensed benzene film on the substrates given in the second column of Fig. 3. The quantitative reference value is given in counts per second and watt of laser power and per L of the exposure needed to form a thick condensed film of benzene. In separate experiments it was checked carefully that the sticking coefficient of the gases used at the low temperatures used is one [16, 38]. This and the fact that the reference value is taken in the same laser beam and scattering configuration guarantees the accuracy of the enhancement values in Fig. 3.

Good and nearly pure electromagnetic enhancers are silver islands grown (like snow caps) on isolated posts (second lowest line in Fig. 3), whereas for cold-deposited Ag films, $G_{\text{em}} \approx G_{\text{first layer}}$. For island films on carbonaceous materials, the electromagnetic enhancement was below the threshold of the experimental sensitivity.

For island films at the percolation threshold (defined by the appearance of a DC current path from one end to the other end of the film) one can make good use of a theory of em intensity fluctuations in these films by *Shalaev* and collaborators [39]. The island film on sapphire in the middle line of Fig. 3 with the mass thickness of 6.1 nm is believed to be close to the percolation threshold. For the em enhancement factor for the first monolayer (ensemble SERS) the theorem of optical reciprocity yields for molecules with a Raman tensor $\alpha\delta_{ij}$ the exact result of SERS by local em field enhancement (the Ω are the directions and polarizations of the incident laser light and emitted Stokes light, the average, given by $\langle x \rangle$, is over all sites $\underline{r}$) [20]

$$\langle G_{\text{em}}(\underline{r}, \omega_{\text{laser}}, \omega_{\text{Stokes}}, \underline{\Omega}_{\text{laser}}, \underline{\Omega}_{\text{Stokes}}) \rangle$$

$$= \left\langle \frac{|\underline{E}(\underline{r}, \omega_{\text{laser}})|^2}{|\underline{E}_{\text{incident}}(\omega_{\text{laser}}, \underline{\Omega}_{\text{laser}})|^2} \frac{|\underline{E}(\underline{r}, \omega_{\text{Stokes}})|^2}{|\underline{E}_{\text{incident}}(\omega_{\text{Stokes}}, -\underline{\Omega}_{\text{Stokes}})|^2} \right\rangle. \quad (2)$$

In the island films the local em field enhancements at the laser and Stokes frequencies ω_{laser} and ω_{Stokes} are decoupled, (it is very rare that a hot spot at ω_{laser} is at the same site $\underline{r}$ as a hot spot at ω_{Stokes}). This is discussed in detail in [20].

The theoretical results are compared with the experimental results for the 3 island film on sapphire (Fig. 3) in Table 1.

Table 1. Electromagnetic enhancements G_{em} of silver island films on sapphire, comparsion experimental value and theoretical values for two different relaxations of silver

Sample	G_{em}(exp)	G_{em}(theory), relaxation $= 0.021\,\text{eV}$	G_{em}(theory), relaxation $= 3.5 \times 0.021\,\text{eV}$
Silver islands on sapphire	~ 100–274	4.08×10^4	~ 272

The relaxation value for silver in the middle of Table 1 is that of well-prepared continuous Ag films [40]. However, for small particles the relaxation rate is increased considerably, see pages 83, 311, 318 in [41]. According to a private communication by A. Pucci, the relaxation rate of silver films deposited in UHV just above the percolation limit can be increased by a factor up to 10. Assuming a factor of 3.5, one does achieve a very good agreement between experiment and Shalaev's theory [20], see Table 1.

3 The Electronic Origin of the "First-Layer Effect" of SERS

Though already in early times, there were conjectures of SERS-active sites in atomic surface roughness [42] (dubbed the "adatom model"), it took some time to proove this, see the next section. However, a clear demonstratation of a transitory charge-transfer (CT) mechanism in SERS from adsorbates at silver electrodes was already achieved in 1982 [43,44]. The CT mechanism led to SERS intensity versus electrode potential plots that depended on the laser wavelength. All this is discussed in detail in the review [45]. Because at the used aqueous electrolytes only the contact adsorbates are influenced by the electrode potential, this transitory charge-transfer mechanism is a "first-layer effect".

The chemical and vibrational specificity of SERS caused by the transitory CT mechanism is shown in Fig. 4. In the case of resonance, the CT transition is to the C–C antibonding π^* orbitals, which couples only to the "skeleton modes", but not to the C–H stretch modes, which are only excited by short excursions to the C–H antibonding σ^* orbitals (so-called impact scattering). These mechanisms are described at length and in detail in [12].

Therefore the first-layer enhancements of the CH stretch modes are on the order of a factor of 30 above the em enhancements, however, for adsorbates at all sites, whereas the CC stretch modes are at least a factor of 10 stronger, but originating only from SERS-active sites, see sections 5 and 6. Assuming for instance a suface density of SERS-active sites of 10 %, the ratio of the enhancement factors of CC to CH stretch vibrations of adsorbates at SERS-active sites would be about 300. If a molecule has no π^*-LUMO, the observed CC stretch signal strength is about a factor 10 less than for adsorbates with π^*-LUMO (see Fig. 4), but it is originating from all adsorption sites. The comparison of C_6H_6 and C_6F_6 is interesting. The differences between the enhancements of the C–H and C–F stretch modes are about 10^3. Due to the high electronegativity of F the C–F antibonding σ^* orbital comes in resonance with the laser photon energy. This difference cannot be explained by the so-called propensity rule [47].

4 The Raman-Continuum of Electron–Hole-Pair Excitations

The so-called inelastic background of roughened silver was first explicitly mentioned in [48]. The history of this phenomenon up to the year 1983 is well described in *Pockrand*'s review article [49]. In 1994 two articles [50, 51] proved that the background of cold-deposited silver films exists without adsorbates, and is genuine Raman scattering, because it shifts with the exciting laser frequency. Adsorption of benzene, pyridine and ethane did not measurably change the intensity and spectral shape of the background, with cyclohexane it increased slightly but reversibly (W. Akemann, unpublished). The background is not a bulk electronic phenomenon, because it is quenched by a very small coverage of atomic oxygen, see the example of a cold-deposited copper film in Fig. 2 [37].

The electronic Raman scattering on the Stokes side, measured at 40 K, of cold-deposited silver films annealed at 50 K, 200 K and 250 K is displayed in the lower part of Fig. 5 (from [50]). In the dashed part of the spectrum the phonon contribution [52] below $250\,\mathrm{cm^{-1}}$ has been deleted. The background extends beyond $-250\,\mathrm{cm^{-1}}$ on the anti-Stokes side (Fig. 6), where it depends strongly on the temperature during measurement.

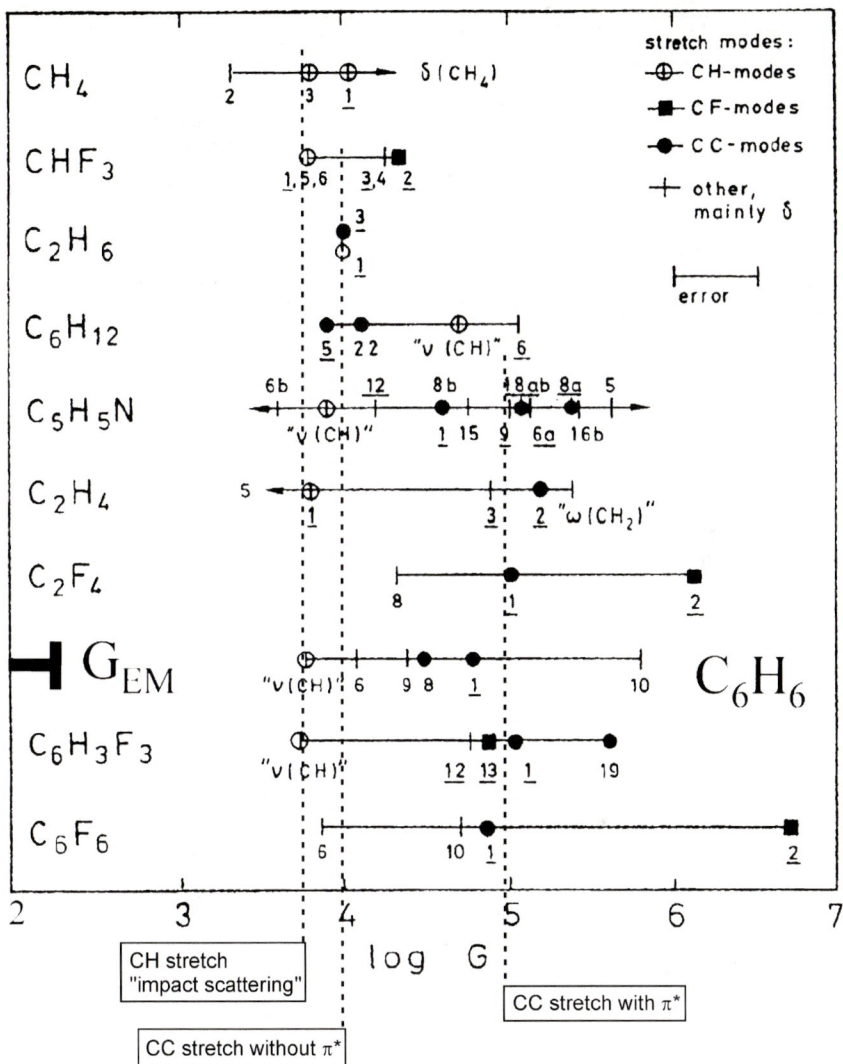

Fig. 4. Absolute enhancement G of the integrated SERS bands of various adsorbate vibrations at submonolayer coverage of silver films, cold deposited and exposed at 40 K. Laser wavelength 514.5 nm. Reference samples are condensed films on sapphire. Notations of the vibration of the aromatic molecules are according to Wilson, and of the other molecules according to Herzberg. *Arrows to the left* or *right* denote vibrations with no detectable SERS, respectively, normal Raman bands. "ν(C–H)" denotes the integration over all unassigned C–H stretch SERS bands normalized by the integral over all C–H stretch bands of the reference sample. "$\omega(CH_2)$" has the analogous meaning. Numbers of the totally symmetric modes are *underlined*. The em enhancement G_{em} was measured for the CC breathing vibration of benzene (from [46]) [12]

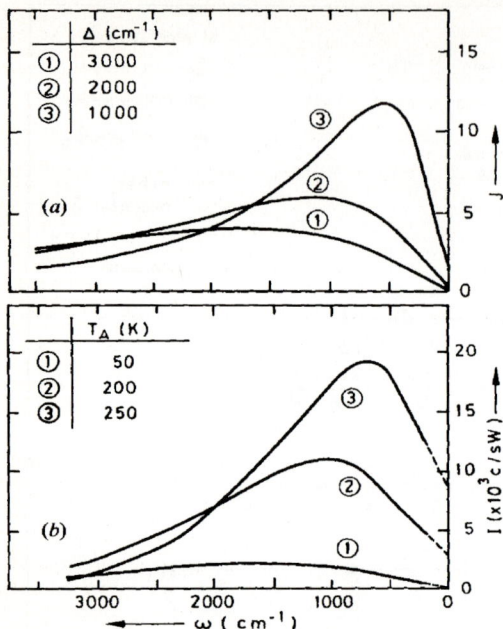

Fig. 5. *Top*: Calculated joint density of states $J(\omega)$ of intraband transitions within bands of Lorentzian density with half-width Δ and maximum weight at the Fermi level; arbitrary units. The temperature is taken to be 50 K. *Bottom*: Electronic Raman spectra at 40 K after annealing at T_A. Noise is omitted. Dotted extrapolation in the range of the observed phonon structure [50]

A very simple model allowed the background to be fitted as function of the annealing temperature and the measuring temperature, both on the Stokes and anti-Stokes side, see Fig. 5 and Fig. 6. The model assumes a density $g(E)$ of electron- and hole-states in a Lorentzian band

$$g(E) = \frac{\Delta}{(E - E_F)^2 + (\Delta/2)^2} \tag{3}$$

centered at the Fermi energy E_F. The width Δ of the Lorentzian is decreasing with increasing annealing temperature. Values of Δ are obtained from least-square fits to the experimental curves. The distribution of electrons and holes is given by the Fermi–Dirac distribution

$$f(E) = \frac{1}{e^{(E-E_F)/kT} + 1} . \tag{4}$$

The joint density of states $J(\hbar\omega)$ of electron–hole pairs at an energy of $\hbar\omega$ is given by

$$J(\hbar\omega) = \int g(E')f(E')g(E' + \hbar\omega)[1 - f(E' + \hbar\omega)]\, \mathrm{d}E' . \tag{5}$$

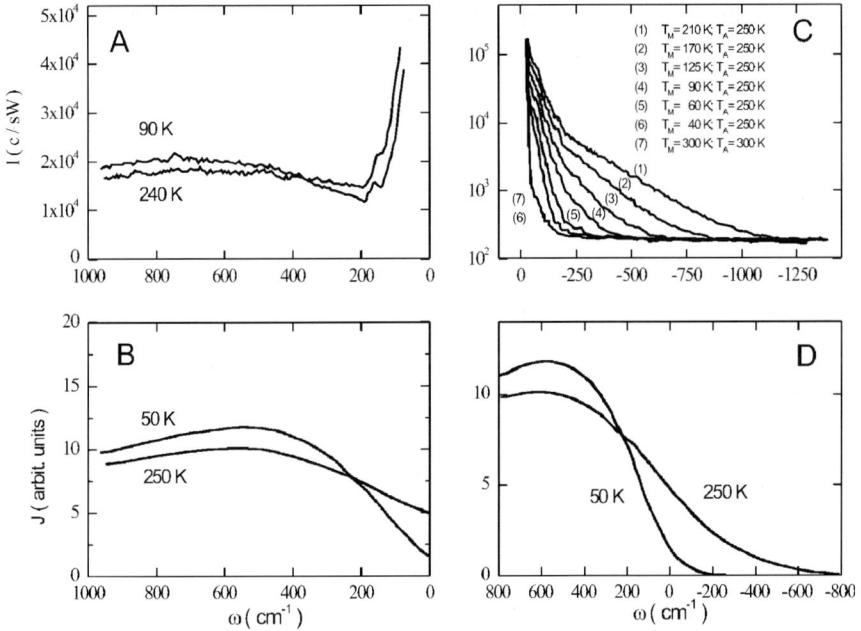

Fig. 6. (a) Background spectra of a silver film, annealed at 250 K at the recording temperatures 90 K and 240 K. (b) Computed joint density of states at 50 K and 250 K. $\Delta = 1000\,\mathrm{cm}^{-1}$. (c) Measured anti-Stokes spectra of a cold-deposited Cu film, annealed to $T_A = 250\,\mathrm{K}$, at measuring temperatures T_M from 40 K to 210 K. Spectrum 7 is from a SERS-inactive film, annealed to 300 K. Note the logarithmic intensity scale. Below $250\,\mathrm{cm}^{-1}$, phonon structure also contributes to the intensity. (d) Computed J for $\Delta = 1000\,\mathrm{cm}^{-1}$ at 50 K and 250 K on the 2 sides of the laser frequency

Within this model the spectral yield of Raman scattering by electron–hole-pair excitation in the metal is assumed to be proportional to $J(\hbar\omega)$. Equation (5) also yields the intensity at the anti-Stokes side. This corresponds to the annihilation of electron–hole pairs of energy $\hbar\omega$ as the Fermi–Dirac distribution allows for filled above empty electronic states in thermal equilibrium. The fit to the experimental data is good, given that there is only one fit parameter, see Fig. 5 and Fig. 6. More details on the inelastic background may be found in [53].

5 SERS-Active Sites

There have been many indirect indications for "SERS-active sites of atomic scale". The best indication have been achieved with ethylene C_2H_4 and parallel measurements of SERS and infrared reflection absorption spectroscopy (IRRAS) of the same copper samples [54]. Ethylene has an inversion

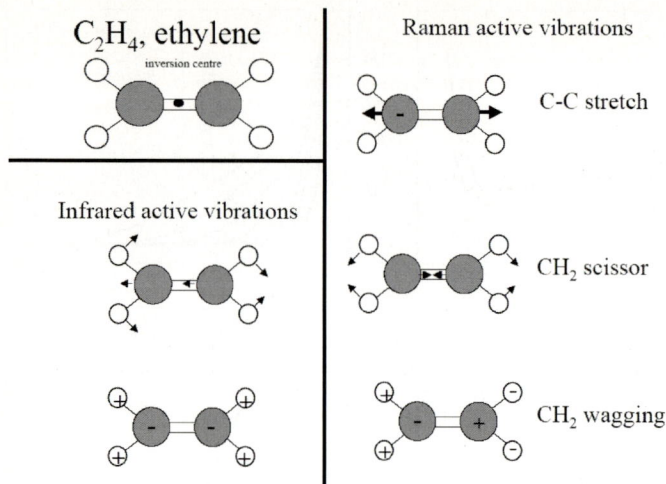

Fig. 7. Infrared- and Raman-active vibrations of ethylene (C_2H_4)

centre and therefore there is a mutual exclusion rule between infrared and Raman activity of the vibrational modes, see the top fig. 5 (the 4 C-H stretch vibrations are not displayed).

Figure 8 shows the IRRAS spectra of two Cu films [54], one prepared at room temperature (300 K), the other at 50 K. Both films have been cooled to 50 K and then exposed to a low background pressure of C_2H_4. The IRRAS spectra have been taken continuously during this time, the average achieved exposure during a run is given in the figures. As expected, for the 300 K film, only the IR modes of ethylene are observed. Closer inspection of the spectra, also in the C–H stretch region, shows that only IR-active modes with the IR dipole moment normal to the local surface are observed, the so-called surface-selection rules hold [54].

This is very different for the cold-deposited silver film. At submonolayer coverage only Raman-active modes (characterized by the letter E) are observed, infrared-active modes appear only at higher coverages. The Raman E-modes are the ones that are strongly observed in SERS from cold-deposited copper films [55]. They have been called E = extra modes, because they originate from species adsorbed at sites that are lost by annealing the sample, see Fig. 9. This is corroborated by the IRRAS results; which show no E-bands for the Cu(300 K) films (Fig. 8). The corresponding E species have a higher bonding enthalpy and a frequency lower than those adsorbed on the (111) facets (the so-called N (normal) modes) and in condensed layers [55].

The loss of E bands is also observed in Fig. 10, where a single crystal with straight steps was used. Exposing this stepped crystal to ethylene did not deliver any Raman signal. However, after cold deposition of 3 monolayers of Cu on this sample, there is a clear E signal. Also, this is lost by heating

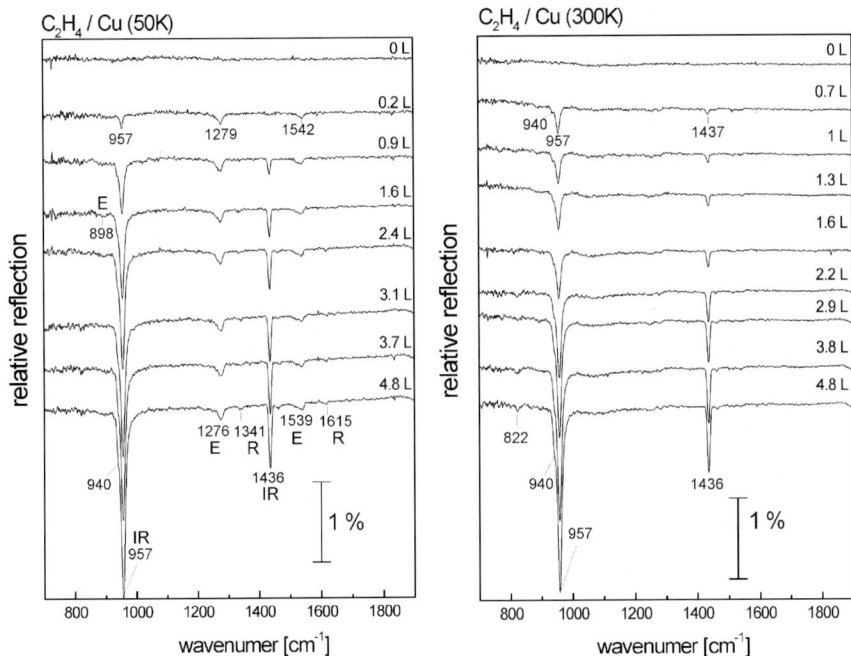

Fig. 8. Development of relative IR reflection with C_2H_4 exposure. Selection of spectra for Cu grown at 50 K (*left*) and for Cu grown at 300 K (*right*), see text

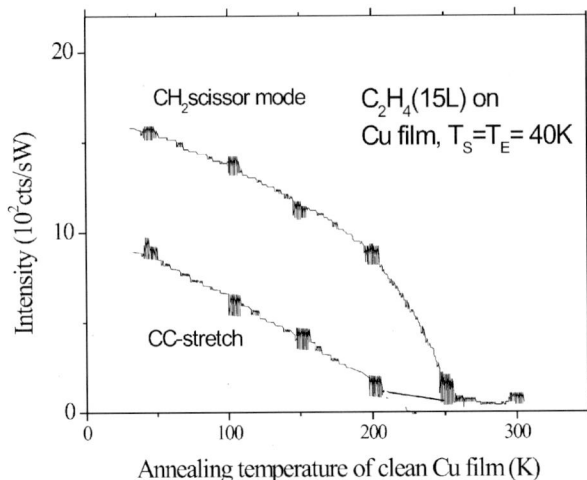

Fig. 9. The copper film was annealed from 40 K in steps of about 50 K after exposure to 15 L of C_2H_4. Spectra were taken at 40 K after recooling and replenishing the C_2H_4 desorbed in the previous annealing step. From [56]

fcc(977)=fcc(S)-[7,(111)+2,(100)]

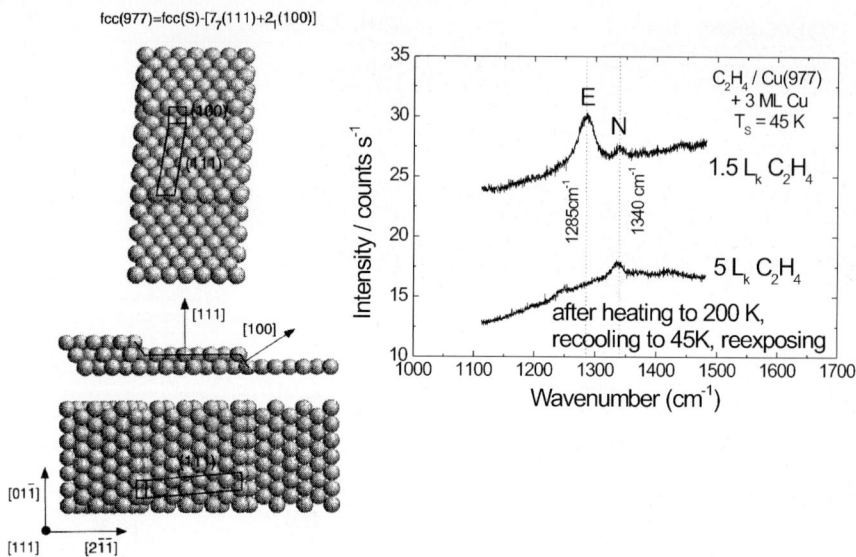

Fig. 10. *Left*: Model of the Cu(997) crystal surface. *Right, top*: Raman spectra of the Cu(977) crystal, with 3 monolayers of cold-deposited Cu, exposed to 1.5 L of ethylene in the range of the scissor mode. *Bottom*: Raman spectrum after annealing the sample to 200 K, recooling to 45 K, re-exposing to 5 L of ethylene. From [57]

to 200 K (including the desorption of C_2H_4), as seen after recooling and re-exposing to 5 L of C_2H_4. E bands originate from the species adsorbed at SERS-active sites at the surface of cold-deposited films, which start to anneal near 200 K. In the temperature range between 180 K and 250 K edge diffusion of Cu atoms becomes active and the Cu-island shape changes from fractal to compact, when small quantities of Cu have been cold deposited on Cu(111), as observed by He scattering [58]. Since the growth conditions of Pt on Pt(111) are not qualitatively different from Cu on Cu(111), scanning tunneling microscopy (STM) pictures and theoretical modeling of the first system is shown in Fig. 11 [59]. The SERS-active sites are sites within this structure where electrons are trapped for some fs, see below in section 6.

The annealing temperatures depend on the kind of Cu samples. Physical vapor deposited Cu films on a rough solid surface lose the SERS-active sites only between 300 K and 350 K, whereas they are stable on copper-island films at least up to 400 K [60]. The nonappearance of the E-bands of ethylene when annealing Cu(977) with atomic-scale roughness to temperatures of 200 K and exposing at low temperatures to ethylene, but the appearance of this band on copper-island films annealed at temperatures of 400 K, exposed at low temperatures to ethylene, indicates special adsorption sites that can be annealed on Cu(977) and on rough continuous copper films, but not on copper islands. Monocrystalline islands will have surface areas of low-index orientation and

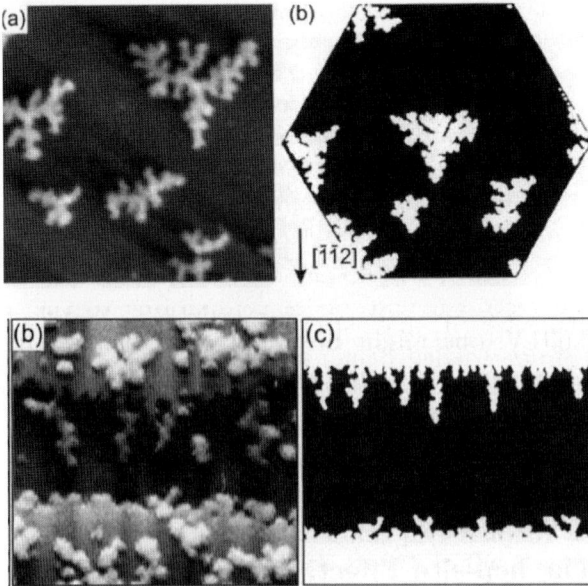

Fig. 11. *Top:* (**a**) Experimental (500 Å × 500 Å) and (**b**) simulated (482 Å × 556 Å) island shapes of Pt cold deposited at 245 K on Pt(111). *Bottom:* (**b**) Experiment and (**c**) simulation at 180 K, asperity growth is observed at the upper step edge (A step), while the lower step (B step) remains relatively smooth. For details see [59]

atomically rough, open areas, for instance areas with kinked steps. As long as there is no recrystallization to shapes with exclusively low index facets, these sites are stable. One may conjecture that the "E-sites" are among this subgroup. Multicrystalline islands will have twin and grain boundaries. We do not know whether these boundaries do provide "SERS-active" adsorption sites for molecules.

6 Theory of SERS-Active Sites

The contribution of transient charge transfer (see the Chapter by *Schatz* et al.) and of SERS-active sites to the electronic "first-layer effect" (see the Chapter by *Zou* et al.) is evident. Now it is necessary to understand, why the "first-layer effect" is not automatically active at an atomically smooth surface. Figure 12 is the summary of a more extended discussion in [12].

The left-hand side in Fig. 12 is a schematic complex of one silver atom and one ethylene molecule, characterized by the lowest unoccupied molecular orbital (LUMO) π^*. If the incoming laser field and the outgoing Stokes field have the electric vector in the direction connecting the centers of the atom

electron transfer
to LUMO (5fs)

Zone of unscreened
interaction of Laser- and
Stokes- light with
electrons

electron transfer
to LUMO (5fs)

electron density

electric field

π^*

Ag
atom

LUMO is CC-antibonding

e below
v_{Fermi}

centre of
gravity of the
screening
charge

h below
v_{Fermi}

e above v_{Fermi}

During 5 fs residence time of the electron in the
LUMO, hole „has run away" about 5nm or about
20 silver atoms. An electron hole pair remains.

Raman active CC stretch
vibration is excited

intensity

$\hbar\omega_{vibration}$

e-h excitation
continuum

$\hbar\omega_{Laser} - \hbar\omega_{emitted}$

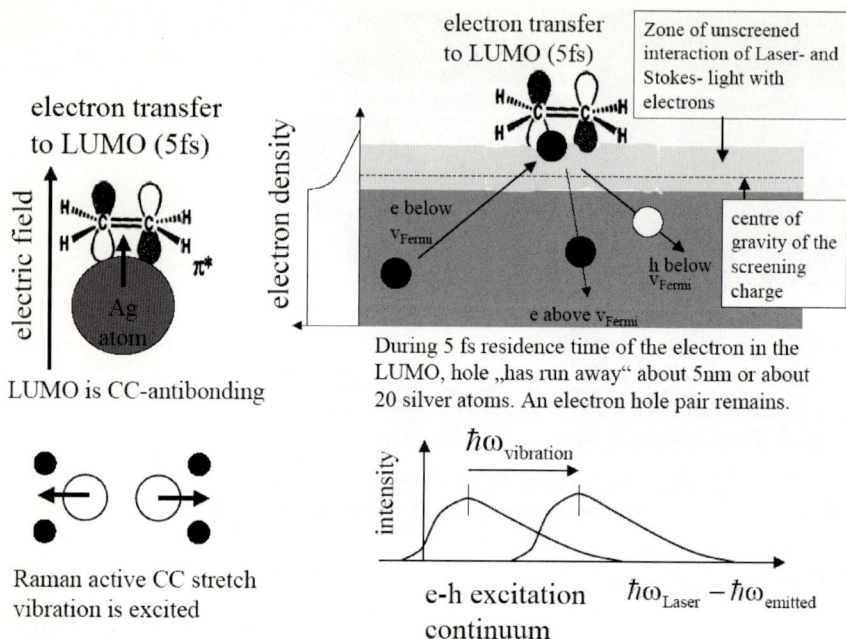

Fig. 12. Raman scattering by a silver–ethylene complex (*left*); electron–hole-pair
creation by charge transfer to ethylene at smooth metallic surface, *right*. See text

and the molecule then Raman scattering by transient CT can be described
by

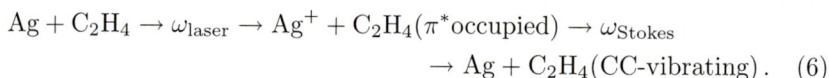

$$Ag + C_2H_4 \rightarrow \omega_{laser} \rightarrow Ag^+ + C_2H_4(\pi^* \text{occupied}) \rightarrow \omega_{Stokes}$$
$$\rightarrow Ag + C_2H_4(\text{CC-vibrating}). \quad (6)$$

The CC stretch vibration is excited, because the π^* orbital is CC antibonding.
In this case it seems trivial that the Ag atom is not excited, because that
would need a rather high excitation energy.

This is different at a smooth metal surface. In the reaction above, Ag
would now mean the extended metal, with a continuum of electron–hole-pair
excitations starting at excitation energy zero or even negative energies (see
section 4). When a metal electron in an occupied state (at 0 K below the
Fermi level) enters the LUMO of the adsorbate, a hole proceeds within the
state, the original electron would have been in after specular reflection at the
adsorbate-free surface. After the residence time of about 5 fs, (corresponding
to about 0.25 eV width of the π^* resonance) the electron leaves the surface,
in most cases in a nonspecular state. This diffuse scattering is important
in the adsorbate-induced surface resistance. During the 5 fs delay time of
the electron in the LUMO, the hole has proceeded with about the Fermi

velocity about 5 nm, equivalent to about 20 silver atom diameters. The lasting electron–hole-pair excitations must be accounted for, as given by

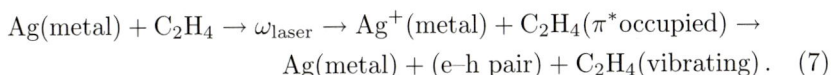

$$Ag(metal) + C_2H_4 \rightarrow \omega_{laser} \rightarrow Ag^+(metal) + C_2H_4(\pi^* occupied) \rightarrow$$
$$Ag(metal) + (e\text{–}h \text{ pair}) + C_2H_4(vibrating). \quad (7)$$

Even if the coherence of the overall process is not broken, the emission is not Raman scattering with a discrete CC stretch band, but an e–h pair continuum shifted by the energy of the vibrational energy quantum. This does not yield a discrete structure, see Fig. 12.

In the em models of SERS, the boundary between metal and vacuum (or air or aqueous electrolyte) is infinitely sharp, thus neglecting completely the results of surface science.

For metals with s-electrons at the Fermi energy (Au, Ag, Cu, alkalis, Al, Mg) there is a spill out of electrons extending about 2 Å. This is characterized in Fig. 12 by the light gray scale zone at the surface. The centroid of the charge screening the inside of the metal from the strong fields on the vacuum side is called the image plane. The zone in front of the image plane is the region, where the laser field and the Stokes field interact strongly with the electrons in the tail of the electron density. This yields, for instance, surface photoemission. (One should recall that free electrons in the bulk of a free-electron gas metal cannot absorb light, because both energy and momentum conservation cannot be fulfilled.) A weakly adsorbing molecule, like all the molecules presented in Fig. 4, is just barely embedded in the tail of the electron distribution, it pushes the tail electrons away by Pauli repulsion with the electrons in the filled molecular orbital. But the unoccupied LUMO extends further out and has an overlap with the electron tail. The electrons coming from the interior are excited by light nearly exclusively in the narrow tail zone, and in the direct "contact zone" with a LUMO the light does increase the probability of an electron entering the LUMO. This is the meaning of transient charge transfer. This process is spatially as closely confined as in the Ag atom–molecule complex with the only change that bulk electrons proceed to this place and away from this place.

To avoid the hole leaving the interaction site within 5 fs, it should be captured there for this time. In this case electron and hole would reunite (the e–h pair is "annihilated") and the discrete line of the molecular vibration excitation would prevail rather than the continuum of the e–h pair excitations. These "hole-retention" sites (local surface resonances in the language of surface science [61]) are the SERS-active sites (SAS). In other words SERS-active sites are acting for some fs like isolated metal molecule complexes acting in the following way:

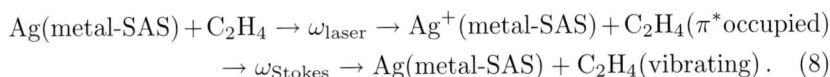

$$Ag(metal\text{-}SAS) + C_2H_4 \rightarrow \omega_{laser} \rightarrow Ag^+(metal\text{-}SAS) + C_2H_4(\pi^* occupied)$$
$$\rightarrow \omega_{Stokes} \rightarrow Ag(metal\text{-}SAS) + C_2H_4(vibrating). \quad (8)$$

The considerations about SERS-active sites will only be important for molecules with π^* orbitals. For hydrogen-saturated molecules and for C–H stretch vibrations of all molecules the excursion times to the σ^* orbitals are at least 10 times shorter and the chances for annihilation of the electron–hole-pair are good also at non-SERS-active sites [12], as already mentioned in the Chapter by *Schatz*.

The contribution of transient electron transfer at cobalt electrodes has been reported in [62]. At transition metal surfaces with "slowly moving" d-electrons, a hole moves probably more slowly than in coinage metals. The residence time in the π^*-LUMOs will be shorter due to the stronger molecular bonding to the transition metal. Both effects will render electron–hole-pair creation after transient CT unlikely. In this sense, at transition metal surfaces, there should exist a genuine first-layer effect of SERS, albeit weaker than at the coinage metals, but not restricted to SERS-active sites.

7 The Story of "Missing NO"

Figure 13 shows the SERS spectra of 5 different Cu films, deposited at 30 K, annealed at the indicated temperatures, recooled to 30 K and exposed to 5 L NO (nitric oxide) [63]. The 1284 cm^{-1} and 2228 cm^{-1} bands are assigned to the N–O and N–N stretch frequency of N_2O (nitrous oxide) (this molecule has the configuration NNO); the 585 cm^{-1} structure on the broad oxygen band may be the N–N–O bending mode of N_2O. This assignment is based on the SERS measurements of N_2O on Cu with bands at 569 cm^{-1}, 1278/1296 cm^{-1} and 2226 cm^{-1} and the gas-phase values of 589 cm^{-1}, 1285 cm^{-1} and 2224 cm^{-1} [64]. The bands at 2283 cm^{-1} and 2109 cm^{-1} can be safely assigned to adsorbed neutral and partially negatively charged N_2, because these bands were observed after exposing a cold-deposited Cu film to N_2 [65]. Surprisingly, there is no sign of the NO stretch band, the frequency of the free NO molecule is at 1876 cm^{-1}. This result seems to indicate that all NO adsorbed on cold-deposited Cu is dissociated, leading to adsorbed atomic oxygen, two species of dinitrogen and N_2O. Signs of atomic nitrogen are missing.

However, it seems that the larger part of NO is not dissociated, because a good IRRAS spectrum of dimerized NO was obtained [63], see Fig. 14. The small signal of N_2O in the IRRAS spectra occurs only after the formation of the second layer of NO [63]. Thermal desorption of NO and photoreaction or photodesorption in the laser focus can be eliminated as reasons of the missing NO signal in SERS [63].

Analogous differences between SERS and infrared spectroscopy were also observed for exposures of N_2O and CO_2. In these cases, IRRAS detects neither [66] the dinitrogen species as decay products of N_2O nor the activated anionic CO_2^- species [65]. The observations that IRRAS does not detect the

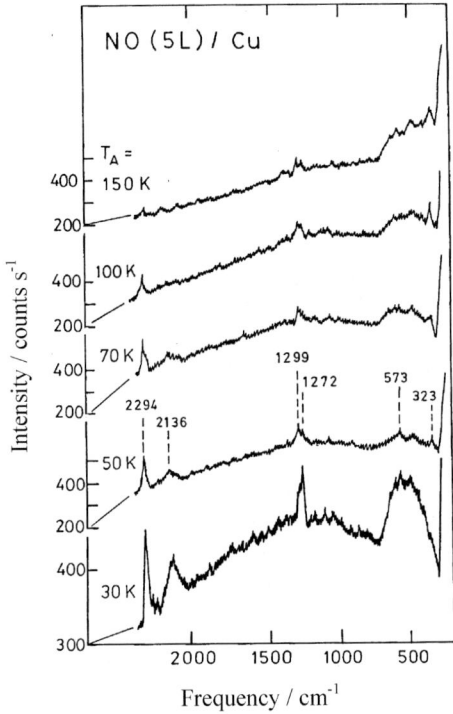

Fig. 13. SERS spectra of five different copper films that were deposited at 30 K, annealed at the indicated temperatures T_A, re-cooled to 30 K, and exposed to 5 L NO. For the assignment of bands see text

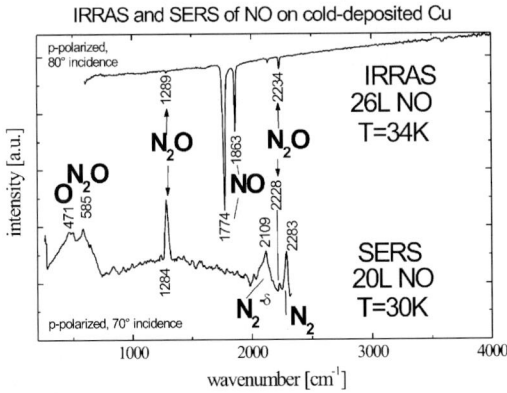

Fig. 14. Comparison of examples of IRRAS and SERS spectra of cold-deposited Cu films exposed to NO

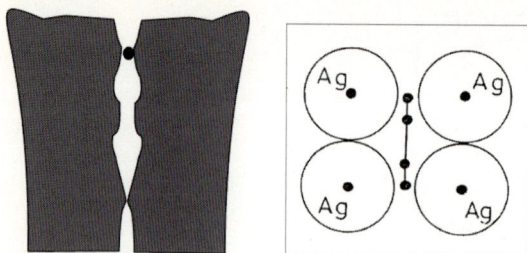

Fig. 15. *Left:* The proposal of a "tunneling site" within a "pore" of a cold-deposited film. *Right:* The narrowest gap that can accommodate an ethylene molecule

dinitrogen species from the NO dissociative reactions (see Fig. 14), but delivers a clear and pronounced spectrum, when a cold-deposited Cu film is exposed directly to molecular nitrogen [66] gives an important hint for the further explanations.

All observations can be explained by the assumption that the proto-catalytic reactions take place at special adsorption sites that cannot be monitored by IRRAS. On the other hand, the IRRAS signal is from the majority NO species adsorbed at the non-"catalytic-active sites", for instance the atomic smooth facets. That implies that the enhancement at non-"catalytic-active sites" is too weak to deliver a Raman signal of NO.

A tentative explanation as to why IR spectroscopy is "blind" to "catalytic-active sites" is based on the ratios of electric-field strength outside and inside a metal in the free-electron Drude model with the dielectric constant $\varepsilon(\omega) \approx 1 - \left(\frac{\omega}{\omega_p}\right)^2$.

Setting the normal field component equal to 1, the tangential component scales like ω/ω_p and the normal component inside the metal like $(\omega/\omega_p)^2$, according to [67]. Therefore, the electric-field strength in the metal is much weaker in the infrared than in the visible spectral range. If the "catalytic-active sites" are within narrow pores (cold-deposited films are porous [68]) maybe they are screened from the IR radiation. Of course there are also "SERS-active sites" accessible to IRRAS, otherwise one could not explain the IRRAS data of C_2H_4 adsorbed on cold-deposited Cu films, see Fig. 8.

Maybe the "catalytic-active sites" are identical to the hypothetical "tunneling sites" (see Fig. 15) that have been proposed to explain the decrease (increase) of the electrical DC resistance of cold-deposited silver film exposed to molecules with (without) π^* orbitals [69]. The decrease, respectively increase, was assigned to resonant tunneling through the π^* orbitals between the metal grains at both sides or, respectively, the hindrance of tunneling. The narrowest gap that can accommodate an ethylene molecule is given on the right side of Fig. 15. The distances from the C–C plane to the next copper atoms were taken from [70] and also applied to silver.

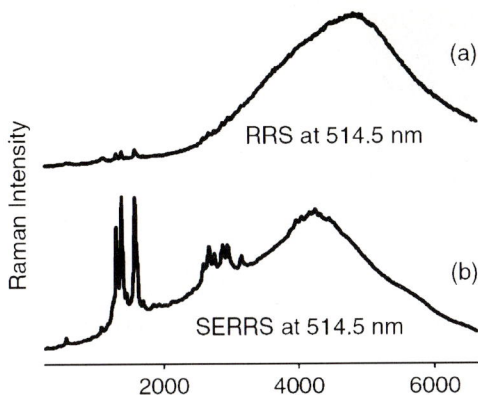

Fig. 16. (**a**) Resonance Raman spectrum of a single 10 : 1 AA-salPTCD monolayer deposited on glass. (**b**) Surface-enhanced resonant Raman scattering spectrum of a single 10 : 1 AA-salPTCD monolayer deposited on evaporated 6 nm Ag nanoparticle film, from [71]

8 EM Enhancement in Single-Molecule SERS of Dyes in Langmuir–Blodgett Films

Aroca and collaborators [71] mapped the spatial distribution of surface-enhanced resonance Raman scattering (SERRS) intensities of mixed dye–fatty acid Langmuir–Blodgett (LB) monolayers deposited on nanostructured Ag-island films under resonant Raman and resonant plasmon conditions. The ensemble-averaged spectra of a single LB layer consisting of arachidic acid (AA) and n-pentyl-5-salicyl-imido-perylene (salPTCD) with a ratio 10 : 1 on glass and on a silver-island film is given in Fig. 16

At this concentration level, the average enhancement factor for resonance Raman scattering is $\sim 10^3$ to 10^4. By high dilutions of the dye the transition of ensemble averaging to single-dye-molecule SERS could be observed. At the average distribution of one dye molecule per $1\,\mu m^2$ the chance to observe a single dye molecule, see Fig. 17, was about 0.3 %, with a rather large variance (R. Aroca, private communication). In the sense of the discussion above this number is the probability of finding a hot spot of Raman intensity among all available spots. One should note that this ratio does not depend on the area of the "hot spots". The enhancement can be obtained in the following simple way [20]:

1. One arachidic acid molecules occupies about 25 Å squared within the monolayer. In Fig. 17, each micrometer squared should contain approximately 4×10^6 molecules. Of these, about 3.64×10^6 will be arachidic acid molecules and about 3.64×10^5 will be salPTCD molecules (R. Aroca, private communication).

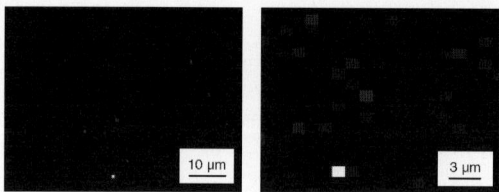

Fig. 17. Surface-enhanced resonant Raman scattering mapping images recorded from mixed Langmuir–Blodgett monolayers deposited on 6 nm Ag nanoparticle films. The concentration of the monolayers are *left* 10^5 : 1 AA-salPTCD and *right* 1 salPTCD molecule per μm^2. From [71]

2. Hence 3.64×10^5 salPTCD molecules deliver the resonant Raman signal in Fig. 16.
3. A Raman spectra from single molecule of salPTCD was observed, see Fig. 17.
4. Intensity and exposure time when monitoring the spectra in Fig. 16 were 10 times more than in the case of Fig. 17 [71].
5. Assuming that the Raman intensity is proportional to the enhancement times the number of molecules one needs an enhancement of 3.64×10^7 to observe a signal from a single molecule with a signal strength like that in Fig. 16.

Aroca has remarked: "We have not actually tested to see what the limit of resonance Raman (without surface-enhancement) detection is. They are certainly less than 3.64×10^5 molecules for this perylene system, however. This would bring the necessary hot-spot enhancement below 3.64×10^7". However, there are still problems: The same density of hot spots is observed under and outside of the resonant Raman conditions (R. Aroca, private communication).

Aroca does not believe that PTCD molecules have much mobility in the two-dimensional film. Very probably they do not have the electronic contact to the silver islands necessary for the electronic mechanisms discussed above. Therefore, the enhancement of the order of 10^7 is exclusively of em origin. Indeed, this enhancement can be quantitatively obtained by using the theory of *Shalaev* and collaborators of fluctuating intensity in semicontinuous silver films [39] in the right way [20].

Concerning the big difference in enhancement values derived here and quoted, for instance, in the pioneering works cited in the introduction, we regret to have no sensible comment.

One should note that the enhancement of the fluorescence in Fig. 16 is only about 15, much less than the enhancement of the resonance Raman scattering. As the PTCD molecules have no electronic contact to the surface, the fluorescence can only be weakened by energy transfer to the metal and nonradiative energy dissipation in the metal. Based on the electromagnetic interactions (energy transfer), *Weitz* et al. [72] found a hierarchy of enhancements, falling from normal nonresonant Raman scattering to resonance Raman and to fluorescence. This is supported by the local optics, dielectric model of *Wang* and *Kerker* [73]. In this way the relatively small enhancement of the fluorescence can be rationalized. The possibility of quenching excitations by charge transfer, following the line of *Persson* and *Avouris* [74, 75] needs no consideration at this point, but see the section 9 below.

According to [72], the normal Raman enhancement is about 2 orders of magnitude stronger than the enhancement of resonance Raman scattering, and "Shalaev's hot spots" become "hotter" on increasing the laser wavelength towards 1 μm. Both points may explain why Aroca and collaborators observe Raman scattering from single PTCD molecule also below the range of optical absorption PTCD (R. Aroca, private communication).

9 Conjectures on SM SERS in Junction Sites

According to *Brus* and collaborators [10, 76, 77] Rayleigh scattering and direct atomic force microscopic (AFM) examination shows that huge cross sections for single molecules occur in compact, nonfractal aggregates of several individual ∼ 50 nm silver nanocrystals within very narrow gaps between colloidal particles, so-called junction sites. The local field-intensity enhancement at the junction site or at the narrow gap between spherical silver particles has been calculated in [10, 78, 79, 80], following older theoretical work [81, 82, 83, 84, 85]. The finite difference time-domain (FDTD) method, which allows calculation of the local field enhancement in irregular structures, has also been applied to junction sites [86, 87]. As a rule of thumb, the enhancement of the field intensity has been squared to obtain the field enhancement of Raman scattering by molecules at the center position of the gap or junction, see one example in Fig. 18. The result of the calculation can be well approximated by the simple equation

$$\left| \frac{E^{\mathrm{loc}}}{E^0} \right| = \frac{D+d}{d} \,, \tag{9}$$

with obvious notation. This corresponds to the compression of the difference of the electric potential between the centers of the spheres (in the absence of the spheres) into the gap of the junction. The volume of the hot spot is the product of the width d of the gap between the spheres of radius r with the area of the lateral extension of the hot spot rd [79].

In this local field-enhancement model, only those molecular vibrations, that have zz-components of the Raman tensor in the Euclidian coordinate frame defined by the 2 spheres (line from center to center defines the z-direction) will contribute to the spectrum. Irrespective of the orientation of the molecule, vibrational modes that have no diagonal components of the Raman tensor in the molecular axial system would not show up. Variations

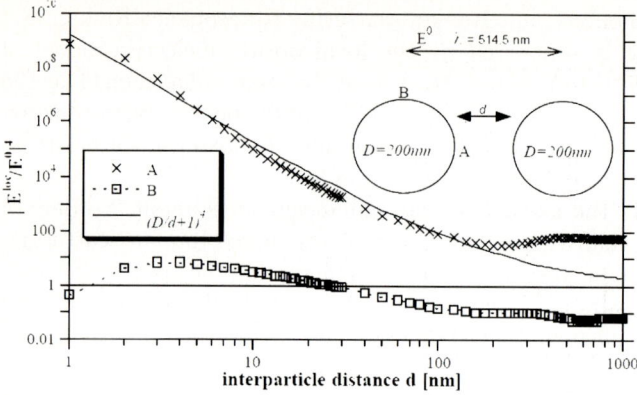

Fig. 18. Calculated SERS enhancement factor for a Ag-dimer system in the parallel polarization configuration as a function of interparticle separation d. The calculations have been performed for two positions, A and B, located 0.5 nm from the surface of one of the spheres and at a laser wavelength of 514.5 nm. The electrostatic model of interparticle coupling can quantitatively explain the dramatic increase in enhancement that occurs when $D/d > 1$. Note that the enhancement at the offaxis position B is almost always below unity (i.e., no enhancement). From [80]

of the orientation and diffusion in and out of the hot spot might explain the blinking.

Without exception, all calculations of the field enhancement have been performed with the model of local optics, in which a surface of a metal and a dielectric is described by an infinitively sharp boundary, separating the metal with the complex dielectric constant $\varepsilon(\omega)$ and, for instance, vacuum with the dielectric constant $\varepsilon_{vac} \equiv 1$. In the following, this will be called the infinite-barrier model, because electrons are not allowed to leak out towards the metal side, as they do in reality, see Fig. 19.

When using the infinite-barrier model at the distances given in Fig. 18 one can consider the gap between the two spheres as a capacitor. In this case the optical current density $J_{optical,metal}$ in the spheres, given in local optics by

$$\underline{j}_{optical,metal}(\boldsymbol{\omega}) = \boldsymbol{\sigma}_{optical}(\boldsymbol{\omega})\underline{E}_{metal}(\boldsymbol{\omega}), \tag{10}$$

with

$$\boldsymbol{\sigma}_{optical}(\boldsymbol{\omega}) = -i\varepsilon_0\boldsymbol{\omega}[\varepsilon(\boldsymbol{\omega}) - 1], \tag{11}$$

and $\underline{E}_{optical,metal}$ being the electrical-field strength in the metal spheres is going over continuously into the displacement current within the gap between the two spheres

$$\underline{j}_{optical,gap}(\omega) = -i\omega\varepsilon_0\underline{E}_{optical,gap}, \tag{12}$$

z (atomic units)

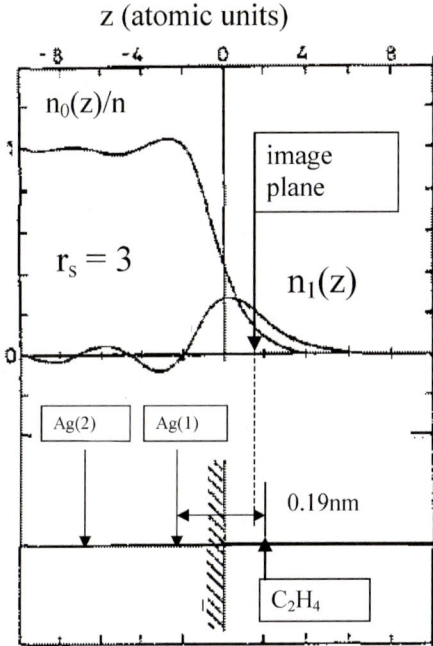

Fig. 19. *Upper part:* Ground-state electron density distribution $n_0(z)/n$ and static screening charge distribution $n_1(z)$ (in arbitrary units) with centroid called the image plane at the edge of a jellium of $r_s = (4\pi n/3)^{-1} = 3$. n = average bulk electron density, 1 au = 0.53 Å (after [88]). *Lower part:* Ag(111) surface, first and second layer of the centers of the silver atoms are indicated, the jellium edge is set equal to the geometrical surface. The distance of the center of C_2H_4 from the first layer of Cu atoms is indicated [70]

as in a capacitor. Giving up the infinite-barrier model, the density $n(\varepsilon_F, d/2)$ of electrons at the Fermi energy, propagating in the spheres in the direction of the axis between the spheres, in the middle of the empty gap of width d between the spheres can be easily approximated assuming an abrupt barrier of the height of the work function Φ:

$$\frac{n(\varepsilon_F, d/2)}{n(\varepsilon_F)} = 4e^{-\frac{\sqrt{2m\Phi}}{\hbar}d}. \tag{13}$$

For silver, with $\Phi = 4.5\,\mathrm{eV}$ and $d = 10\,\text{Å}$, one obtains for this ratio a value of about 7.7×10^{-5}.

At $d = 5\,\text{Å}$, the ratio is already 0.0157 and at a distance of the diameter of one silver atom (2.88 Å), where an ethylene molecule would still fit into the gap between the spheres [69] it is already 17.5 %. Therefore, on decreasing the gap the displacement current is more and more replaced by the electron tunneling current and the high enhancement values proportional to the fourth power of the local field will come down. Therefore the very high calculated enhancements below a gap size of 5 Å to 10 Å are unphysical. A realistic theory would bring the field enhancement of SERS to values much lower than 10^{10} at a gap of 2.28 Å.

Using the hypothesis that the "tunneling sites" discussed in Sect. 7 are related with the hot spots between the two antenna spheres discussed in this section, and considering the electronic enhancement in cold-deposited films, an electronic contribution to the overall enhancement in junction sites at tunneling distance of 10^3 has been estimated [89].

The experimental information on the atomic structure of "junction sites" is not yet complete. *Khan* et al. [90] studied a number of silver particles and aggregates of particles using surface-enhanced resonance Raman spectroscopy (SERRS), high-resolution transmission electron microscopy (HRTEM) and electron energy-loss spectroscopy (EELS). The SERRS mapping/TEM collage method allowed each SERRS-active or -inactive species to be reliably identified and analyzed by each of the techniques in three different instruments. The aim was to correlate SERRS activity, particle microstructure, chemical composition and electronic properties of each species to gain an insight into the enhancement mechanism. As yet, the findings did not reveal any clear link between particle microstructure and SERRS activity [90].

10 Special Examples of SERS at Low Coverage of Small Silver Aggregates

In the early days of SERS from adsorbates at electrode surfaces, the compelling argument for some kind of chemical effect was the very low or absent signal of the electrolytes, for instance water [91].

Single-molecule SERS was, in the beginning, mostly restricted to strongly luminescent dyes like rhodamine 6G (Rh6G), crystal violet (CV) and cyanine dye. A likely interpretation was the strong local-field enhancement and the resonance Raman effect were acting together, no "chemical effect" was present.

In this sense it is a surprise that spectra with good signal-to-noise ratios have been observed for smaller molecules without optical absorption in the visible spectral range and hence without an internal resonant Raman effect. The examples from the literature should, however, be chosen with caution. If the blinking shows lines far away from the wave numbers of the clean substance, these lines cannot be taken as vibrational bands from the applied molecule. This has been discussed for the example of the molecule tyrosine in [92]. A good test is to add up all the consecutive spectra in blinking experiments, for instance in the case of hemoglobin [78] and "spurious SERS" [93]. More information than this integration is obtained by 2D correlation analysis of *Rowlen* and collaborators [94]. Whereas for Rh6G applied, the covariance

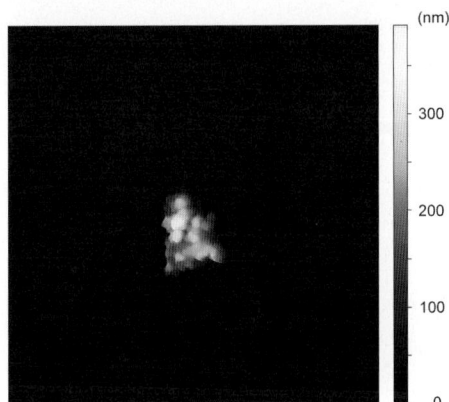

Fig. 20. AFM image of hot aggregate

slices for 3 bands of Rh6G deliver the expected Raman spectra of Rh6G, the covariance analysis of citrate-prepared Ag colloids with no added adsorbate yielded vibrational bands that remained unassigned [94].

It seems to us that the single-molecule SERS spectra of adenine, first reported by *Kneipp* et al. [95] originate from adenine. Surprisingly, the single-molecule SERS spectra of adenosine 5'-phosphate did not show a signal of the ribose and phosphate moiety. This effect cannot be easily assigned to pure em enhancement [92].

The Futamata group studied many Ag-particle aggregations prepared with the Lee–Meisel recipe [96] without any intentional addition of analyte. Neither the Raman image method nor spectral measurements showed any background emission from the "bare" particles. This is in contrast to background Raman spectra of unexposed cold-deposited films discussed in section 4 above. A possible explanation for this difference may be that narrow "open junctions" as in Fig. 18 are a very unlikely configuration, given Brownian motion of the colloidal particles and van der Waals interaction between them. Either the particles touch and eventually form a small "sinter neck" without the background of electron–hole pairs excited by the optical currents through the neck, or an analyte molecule "sitting tight" within the junction prevents this sintering, but strengthens the attraction between the colloidal particles.

The preparation of adenine samples has been described in [97]. Almost all adenine is in neutral form in aqueous solution at pH 7. NaCl was added to the solution to remove residual citrate from the Lee–Meisel process in order to allow some adenine molecules to adsorb on the Ag surface of silver aggregates. An atomic force microscopic picture is given in Fig. 20.

At concentrations lower than 10^{-6} M adenine unidentified peaks were observed and the signal-to-noise ratio of the regular adenine bands was reduced with decreasing adenine concentration. A concentration of 10^{-6} M adenine

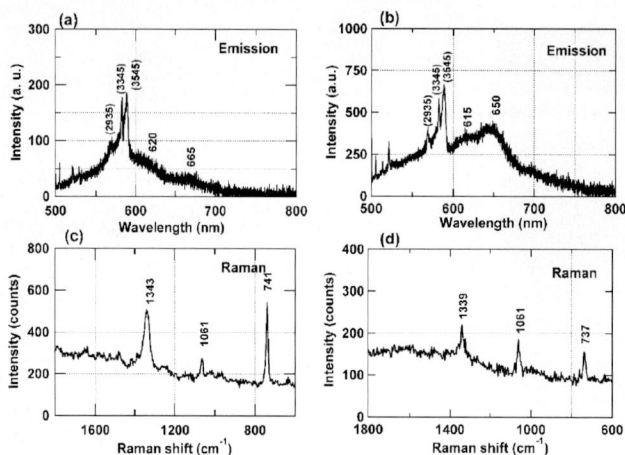

Fig. 21. Experimental SERS spectra of adenine adsorbed on hot silver aggregates. *Upper spectra* are plotted against the wavelength, *lower spectra* in a smaller spectral range against Raman shift. *Left spectra* (**a, c**) are from one aggregate, *right spectra* (**b, d**) from a second aggregate. Numbers in brackets in the spectra above are wave numbers, without brackets wavelengths. Laser wavelength was 488 nm

correspond to about 3000 adenine molecules per silver particle, but the number of adsorbed adenine molecules per silver particle is unknown. Clear blinking was observed [97], which usually is taken as a sign of single-molecule spectroscopy, which, however, does not mean that only a single molecule contributes to the light emission spectra in Fig. 21. The adenine Raman bands below 2000 cm^{-1} are probably from the vibrations of the purine double ring of adenine and the bands above 2000 cm^{-1} are perhaps stretching modes of the outer NH, NH$_2$ and CH groups.

The bands are above a background increasing with wavelength, culminating below the highest Raman bands, above which it drops. The light emission at higher wavelengths is luminescence, which does not shift with the laser wavelength and peaks between 620 nm and 650 nm, see Fig. 22.

In contrast, the background below the high-frequency modes is pure Raman scattering. Probably the luminescence is from silver, as luminescence from silver can be observed as a broad band centered at 580 nm, when a bulk silver sample is mechanically scratched, see Fig. 2 in [50]. The high-frequency Raman modes in Fig. 21 have a rather asymmetric lineshape, which may indicate a strong interaction of these modes with the metal electrons [98].

Figure 23 collects intensity versus wavelength emission spectra from rhodamine 6G (R6G), crystal violet (CV) and malachit green (MG), excited with laser radiation at 488 nm, together with the optical absorption and fluorescence spectra. Though the intensity scales are arbitrary, the signal-to-noise ratios are about the same. In this sense it does not seem to matter much

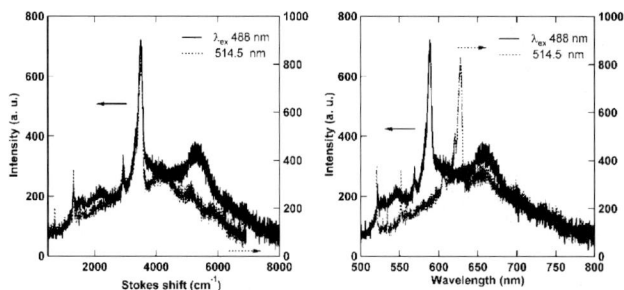

Fig. 22. Light emission spectra of adenine on silver aggregates, with laser excitation at 488 nm and 514.5 nm, plotted against Stokes shift (*left*) and emission wavelength (*right*)

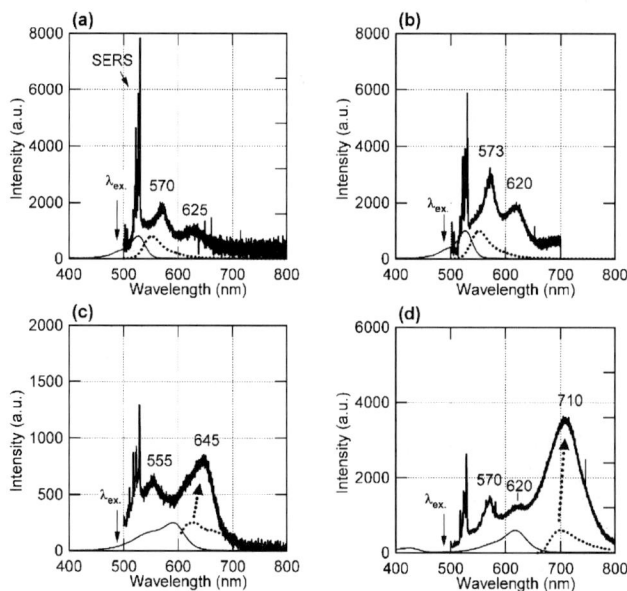

Fig. 23. Emission spectra for various dye molecules adsorbed on Ag nanoparticles: (**a**) R6G (3 molecules per particle), (**b**) R6G (30 molecules per particle), (**c**) CV (30 molecules per particle), (**d**) MG (300 molecules per particle). Marked differences were observed for the emission bands at 600 nm to 750 nm in contrast to similar features at 550 nm to 580 nm. Each spectrum was obtained by the accumulation of 20 sequential measurements. The *full* and *dotted lines* denote the absorption and fluorescence spectrum of each dye molecule in solution, respectively

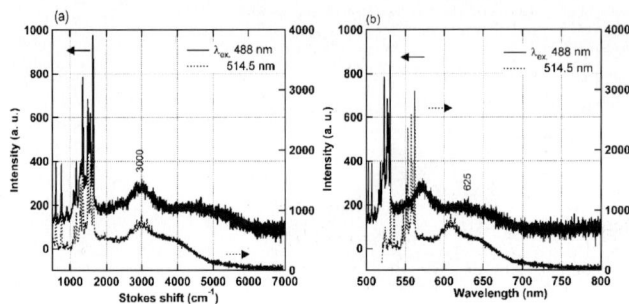

Fig. 24. Emission spectra for R6G molecules (30 molecules per particle) adsorbed on Ag nanoparticles using different excitation wavelengths: (**a**) in Stokes shift and (**b**) in wavelength scales. This figure was obtained for the same Ag particles but with different excitation wavelengths. Two distinct peaks were obtained at the same Stokes shift ($3000\,\mathrm{cm^{-1}}$) and at the same wavelength ($625\,\mathrm{nm}$)

whether the optical absorption of the free dye at 488 nm is nearly zero for MG or is in resonance for R6G.

As for adenine (Fig. 22), the Raman part of the emission of R6G extends to a Stokes shift of about $3500\,\mathrm{cm^{-1}}$, luminescence peaks at about 620 nm, see Fig. 24. Given the big optical differences between free adenine and R6G, the single-molecule spectra in Fig. 22 and Fig. 24 are not so different.

Figure 25 shows for the example of MG, that the $\sim 650\,\mathrm{nm}$ luminescence from silver (compare to Fig. 22) and the fluorescence of MG (compare to the luminescence spectra of free MG in Fig. 23d) can grow with respect to the Raman part of the emission spectra.

We propose the following explanation:

Electrons in excited states of adsorbates on metallic surfaces have a lifetime of the order of fs ($10^{-15}\,\mathrm{s}$). This can be measured with widely different methods. Direct observation of electron dynamics in the attosecond ($10^{-18}\,\mathrm{s}$) domain is possible with so-called core hole clock spectroscopy [99]. An electron in the $3\mathrm{p_z}$ level of sulfur on Ru, lying ca 1.7 eV above the Fermi level of Ru is released below one fs into Ru [99].

An electron in the π^* level of CO on Cu(111) populated in STM as intermediate state exhibits an ultrashort lifetime of 0.8 fs to 5 fs [100]. Corresponding values are observed in the frequency domain via the half-width of intermediate states in 2-photon photoemission, for example CO on Cu(111) [101]. The photoinduced transfer of an electron from Chlorophyll a to gold nanoparticles was proved by electrochemical charging of the gold particles [102]. If a molecule sits tight between two metal particles, one can expect even shorter electronic transfer times than when it is bonded to one metal side only.

In most cases, when the upper state of an optical transition is above the Fermi level, the excited electron leaves and the ensuing hole in the molecule is replenished from the pool of electrons from the metal [74, 103]. In this way

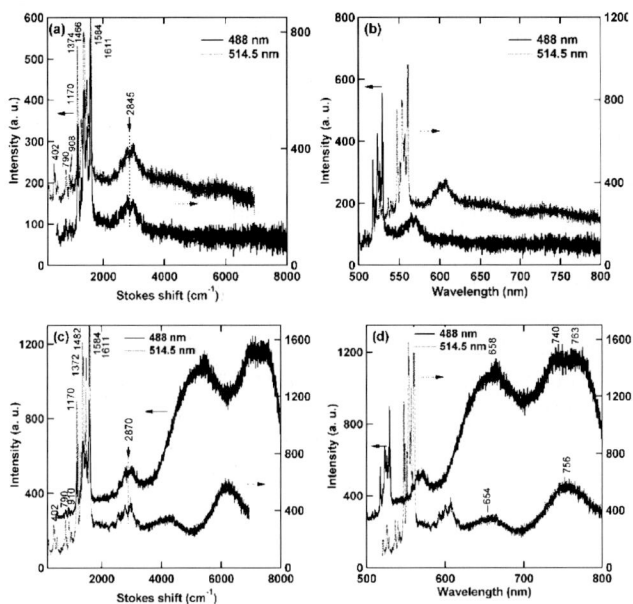

Fig. 25. Raman scattering (up to about $3000\,\mathrm{cm}^{-1}$) and luminescence (strong above 620 nm) of malachit green (3–30 molecules/Ag particle) on 2 different Ag samples (*top* and *bottom*) excited at different laser wavelengths

the optical excitation of the molecule is quenched in ultrashort time. Even if the upper state is below the Fermi level, the optical de-excitation by so-called Auger de-excitation takes place on the same order of time. Within the time of a fs, neither resonance Raman nor fluorescence is possible. The resonance Raman effect is caused by a comparatively long time in the molecular excited state, and dye molecules even need time on the order of ns to achieve the necessary electronic and nuclear reordering for fluorescence. In a way, electronic differences between molecules in this situation become less important, and no dye molecule can fluoresce in this situation.

Therefore we assign the observed dye fluorescence, which seem to be absent under real single-molecule SERS, to dye molecules not in direct contact with the metal. It is an open question why there is no pure fluorescence of dyes observed from the many "nonhot" aggregates. In view of the poor characterization of the aggregates, we give no proposal.

Molecules that are much more strongly bonded to silver than water may impede sintering and therefore are not "squeezed out" from the junction sites. Our more important proposal is that the Raman signal strengths from all these molecules do not to differ by orders of magnitude – explaining the results from adenine and the 3 dyes R6G, CV and MG in the examples given.

Finally, we come to the question of how the molecules "in" the "active sites" are excited, by the electric field or the metal electrons. As the field drives the electrons over the "gap", synonymous with driving them "through" that molecule, this question may become meaningless.

However, not all experimental results in the field of single-molecule SERS agree with those shown in this section. We regret having neither the time nor space to review the literature within this section. Most theoretical explanations are still based exclusively on local field effects (We think that this is correct for single dye molecules in Langmuir–Blodgett films, see Sect. 8, with one very noteworthy exception [104]. In the present Chapter we have reviewed the impact of surface-science concepts of bulk metals, for instance transitory charge transfer and electron–hole-pair excitations on SERS. However, based on recent work of *Dickson* and his group the discrete electronic structure of metal-atom clusters may become important in single-molecule SERS. Based on extensive work on strong visible fluorescence from several Ag_n nanoclusters [105] single-molecule Stokes and anti-Stokes spectra could be observed from dentrimer- and peptide-encapsulated Ag nanoclusters in the absence of plasmons supporting silver nanoparticles [106]. In our opinion, the field of single-molecule SERS at junction sites is still wide open.

Acknowledgements

We would like to thank R. Aroca, U. Kreibig and S. F. Alvarado for discussions.

References

[1] C. F. Bohren, D. R. Huffman: *Absorption and Scattering of Light by Small Particles* (Wiley, New York 1983)

[2] A. Otto: Adv. Solid State Phys. **XIV**, 1 (1974)

[3] L. Yin, V. K. Vlasko-Vlasov, A. Rydh, J. Pearson, U. Welp, S. H. Chang, S. K. Gray, G. C. Schatz, D. B. Brown, C. W. Kimball: Appl. Phys. Lett. **85**, 467 (2004)

[4] C. L. Haynes, A. D. McFarland, L. L. Zhao, R. P. Van Duyne, G. C. Schatz, L. Gunnarsson, J. Prikulis, B. Kasemo, M. Kall: J. Phys. Chem. B **107**, 7337 (2003)

[5] C. L. Haynes, C. R. Yonzon, X. Y. Zhang, R. P. Van Duyne: J. Raman Spectrosc. **36**, 471 (2005)

[6] L. J. Sherry, S. H. Chang, G. C. Schatz, R. P. Van Duyne, B. J. Wiley, Y. N. Xia: Nano Lett. **5**, 2034 (2005)

[7] I. Pockrand: Chem. Phys. Lett. **85**, 37 (1982)

[8] M. Moskovits: J. Raman Spectrosc. **36**, 485 (2005)

[9] L. Cui, Z. Liu, S. Duan, D. Y. Wu, B. Ren, Z. Q. Tian, S. Z. Zou: J. Phys. Chem. B **109**, 17597 (2005)

[10] A. M. Michaels, J. Jiang, L. Brus: J. Phys. Chem. B **104**, 11965 (2000)

[11] A. Campion, P. Kambhampati: Chem. Soc. Rev. **27**, 241 (1998)

[12] A. Otto: J. Raman Spectrosc. **36**, 497 (2005)

[13] B. Ren, G. Picardi, B. Pettinger, R. Schuster, G. Ertl: Angew. Chemie – Int. Edition **44**, 139 (2005)
[14] A. Bruckbauer, A. Otto: J. Raman Spectrosc. **29**, 665 (1998)
[15] E. J. Bjerneld, Z. Foldes-Papp, M. Kall, R. Rigler: J. Phys. Chem. B **106**, 1213 (2002)
[16] I. Mrozek, A. Otto: J. Electron. Spectrosc. Relat. Phenom. **54**, 895 (1990)
[17] K. Kneipp, A. S. Haka, H. Kneipp, K. Badizadegan, N. Yoshizawa, C. Boone, K. E. Shafer-Peltier, J. T. Motz, R. R. Dasari, M. S. Feld: Appl. Spectrosc. **56**, 150 (2002)
[18] E. Koglin, J. M. Sequaris: Top. Curr. Chem. **134**, 1 (1986)
[19] V. M. Shalaev: in M. Berlotti, C. Sibilia (Eds.): *Nanoscale Linear and Nonlinear Optics* (AIP 2001) p. 237
[20] A. Otto: J. Raman Spectrosc., in press, published online in Wiley InterScience (www.interscience.wiley.com). DOI: 10.1002/jrs.1524
[21] K. Kneipp, Y. Wang, R. Dasary, M. Feld: Appl. Spectrosc. **49**, 780 (1995)
[22] K. Kneipp, Y. Wang, H. Kneipp, L. T. Perelman, I. Itzkan, R. Dasari, M. S. Feld: Phys. Rev. Lett. **78**, 1667 (1997)
[23] S. M. Nie, S. R. Emory: Science **275**, 1102 (1997)
[24] P. Hildebrandt, M. Stockburger: J. Phys. Chem. **88**, 5935 (1984)
[25] B. Pettinger, A. Geralymatou: Surf. Sci. **156**, 859 (1984)
[26] B. Pettinger, A. Geralymatou: Ber. Bunsenges. Phys. Chem. **88**, 359 (1984)
[27] B. Pettinger, K. Krischer: J. Electron Spectrosc. Relat. Phenom. **45**, 133 (1987)
[28] D. A. Zwemer, C. V. Shank, J. E. Rowe: Chem. Phys. Lett. **73**, 201 (1980)
[29] C. A. Murray, D. L. Allara, M. Rhinewine: Phys. Rev. Lett. **46**, 57 (1981)
[30] R. R. Smardzewski, R. J. Colton, J. S. Murday: Chem. Phys. Lett. **68**, 53 (1979)
[31] I. Mrozek, A. Otto: Europhys. Lett. **11**, 243 (1990)
[32] I. Mrozek, A. Otto: Appl. Phys. A: Mater. Sci. Process. **49**, 389 (1989)
[33] F. Varnier, N. Mayani, G. Rasigni: Appl. Opt. **28**, 127 (1989)
[34] S. L. McCall, P. M. Platzmann, P. A. Wolf: Phys. Lett. A **77**, 381 (1980)
[35] M. Hänisch, A. Otto: J. Phys.: Condens. Matter **6**, 9659 (1994)
[36] A. Otto, D. Diesing, S. Schatteburg, H. Janssen: Phys. Stat. Sol. A: Appl. Res. **175**, 297 (1999)
[37] M. Pohl, A. Otto: Surf. Sci. **406**, 125 (1998)
[38] C. Pettenkofer, I. Mrozek, T. Bornemann, A. Otto: Surf. Sci. **188**, 519 (1987)
[39] K. Seal, M. A. Nelson, Z. C. Ying, D. A. Genov, A. K. Sarychev, V. M. Shalaev: Phys. Rev. B **67**, 035318 (2003)
[40] B. P. Johnson, R. W. Christy: Phys. Rev. B **6**, 4370 (1972)
[41] U. Kreibig, M. Vollmer: *Optical Properties of Metal Clusters* (Springer, Berlin, Heidelberg 1995)
[42] A. Otto, I. Pockrand, J. Billman, C. Pettenkofer: in R. K. Chang, T. E. Furtak (Eds.): *Surface Enhanced Raman Scattering* (Plenum, New York, London 1982) pp. 147–172
[43] J. Billmann, A. Otto: Solid State Comm. **44**, 105 (1982)
[44] A. Otto: J. Electron. Spectrosc. Relat. Phenom. **29**, 329 (1983)
[45] A. Otto, I. Mrozek, H. Grabhorn, W. Akemann: J. Pys. Condens. Matter **4**, 1143 (1992)
[46] H. Grabhorn: Dissertation (1991), Düsseldorf

[47] H. Grabhorn, A. Otto: Vacuum **41**, 473 (1990)
[48] A. Otto: Surf. Sci. **75**, L392 (1978)
[49] I. Pockrand: Springer Tracts Mod. Phys. **104**, 1 (1984)
[50] W. Akemann, A. Otto: Philos. Mag. B: Phys. Condens. Matter Statist. Mech.
      Electron. Opt. Magn. Properties **70**, 747 (1994)
[51] W. Akemann, A. Otto: Surf. Sci. **309**, 1071 (1994)
[52] W. Akemann, A. Otto, H. R. Schober: Phys. Rev. Lett. **79**, 5050 (1997)
[53] A. Otto, W. Akemann, A. Pucci: Israel J. Chemistry, accepted (2005)
[54] A. Priebe, A. Pucci, A. Otto: J. Phys. Chem. B **110**, 1673 (2006)
[55] J. Grewe, U. Ertürk, A. Otto: Langmuir **14**, 696 (1998)
[56] J. Grewe: Dissertation (1996), Düsseldorf
[57] A. Bruckbauer: Dissertation (2000), Düsseldorf
[58] W. Wulfhekel, N. N. Lipkin, J. Kliewer, G. Rosenfeld, L. C. Jorritsma,
      B. Poelsema, G. Comsa: Surf. Sci. **348**, 227 (1996)
[59] T. Michely, J. Krug: *Islands, Mounds and Atoms* (Springer, Berlin, Heidelberg
      2004)
[60] C. Siemes, A. Bruckbauer, A. Goussev, A. Otto, M. Sinther, A. Pucci: J.
      Raman Spectrosc. **32**, 231 (2001)
[61] A. Zangwill: *Physics at Surfaces* (Cambridge University Press, Cambridge
      1988)
[62] Y. Xie, D. Y. Wu, G. K. Liu, Z. F. Huang, B. Ren, J. W. Yan, Z. L. Yang,
      Z. Q. Tian: J. Electroanalyt. Chem. **554**, 417 (2003)
[63] M. Lust, A. Pucci, A. Otto: J. Raman Spectrosc. **37**, 166–174 (2006)
[64] J. Laane, J. R. Ohlsen: Progr. Inorg. Chem. **27**, 465 (1980)
[65] W. Akemann, A. Otto: Surf. Sci. **272**, 211 (1992)
[66] M. Lust, A. Pucci, A. Otto: in preparation
[67] B. N. J. Persson: Surf. Sci. **270**, 103 (1992)
[68] J. Eickmans, A. Otto, A. Goldmann: Surf. Sci. **171**, 415 (1986)
[69] C. Holzapfel, F. Stubenrauch, D. Schumacher, A. Otto: Thin Solid Films **188**,
      7 (1990)
[70] D. Arvanitis, K. Baberschke, L. Wenzel, U. Dobler: Phys. Rev. Lett. **57**, 3175
      (1986)
[71] P. J. G. Goulet, N. P. W. Pieczonka, R. F. Aroca: J. Raman Spectrosc. **36**,
      574 (2005)
[72] D. A. Weitz, S. Garoff, J. I. Gersten, A. Nitzan: J. Raman Spectrosc. Relat.
      Phenom. **29**, 363 (1983)
[73] D. S. Wang, M. Kerker: Phys. Rev. B **25**, 2433 (1982)
[74] B. N. J. Persson, P. Avouris: J. Chem. Phys. **79**, 5156 (1983)
[75] P. Avouris, B. N. J. Persson: J. Phys. Chem. **88**, 837 (1984)
[76] A. M. Michaels, M. Nirmal, L. E. Brus: J. Am. Chem. Soc. **121**, 9932 (1999)
[77] K. A. Bosnick, J. Jiang, L. E. Brus: J. Phys. Chem. B **106**, 8096 (2002)
[78] H. X. Xu, E. J. Bjerneld, M. Kall, L. Borjesson: Phys. Rev. Lett. **83**, 4357
      (1999)
[79] H. X. Xu, J. Aizpurua, M. Kall, P. Apell: Phys. Rev. E **62**, 4318 (2000)
[80] H. Xu, E. J. Bjerneld, J. Aizpurua, P. Apell, L. Gunnarsson, S. Petronis,
      B. Kasemo, P. Larsson, F. Höök, M. Käll: Proc. SPIE **4258**, 35 (2001)
[81] R. W. Rendell, D. J. Scalapino, B. Muhlschlegel: Phys. Rev. Lett. **41**, 1746
      (1978)
[82] P. K. Aravind, A. Nitzan, H. Metiu: Surf. Sci. **110**, 189 (1981)

[83] P. K. Aravind, H. Metiu: J. Phys. Chem. **86**, 5076 (1982)

[84] P. K. Aravind, R. W. Rendell, H. Metiu: Chem. Phys. Lett. **85**, 396 (1982)

[85] J. I. Gersten, A. Nitzan: Surf. Sci. **158**, 165 (1985)

[86] M. Futamata, Y. Maruyama, M. Ishikawa: Vibrat. Spectrosc. **30**, 17 (2002)

[87] M. Futamata, Y. Maruyama, M. Ishikawa: J. Phys. Chem. B **107**, 7607 (2003)

[88] M. Weber, A. Liebsch: Phys. Rev. B **35**, 7411 (1987)

[89] A. Otto: Indian J. Phys. Proc. Indian Assoc. Cultivat. Sci.: Part B **77B**, 63 (2003)

[90] I. Khan, D. Cunningham, S. Lazar, D. Graham, E. W. Smith, D. McComba: Faraday Discuss. **132**, 171 (2006)

[91] D. L. Jeanmaire, R. P. Van Duyne: J. Electroanal. Chem. **84**, 1 (1977)

[92] A. Otto: J. Raman Spectrosc. **33**, 593 (2002)

[93] P. C. Andersen, M. L. Jacobson, K. L. Rowlen: J. Phys. Chem. B **108**, 2148 (2004)

[94] A. A. Moore, M. L. Jacobson, N. Belabas, K. L. Rowlen, D. M. Jonas: J. Am. Chem. Soc. **127**, 7292 (2005)

[95] K. Kneipp, H. Kneipp, V. B. Kartha, R. Manoharan, G. Deinum, I. Itzkan, R. R. Dasari, M. S. Feld: Phys. Rev. E **57**, R6281 (1998)

[96] P. C. Lee, J. Meisel: J. Phys. Chem. **86**, 3391 (1982)

[97] Y. Maruyama, M. Ishikawa, M. Futamata: Chem. Lett. **30**, 834 (2001)

[98] J. W. Gadzuk, M. Plihal: Faraday Discuss. **117**, 1 (2000)

[99] A. Fohlisch, P. Feulner, F. Hennies, A. Fink, D. Menzel, D. Sanchez-Portal, P. M. Echenique, W. Wurth: Nature **436**, 373 (2005)

[100] L. Bartels, G. Meyer, K.-H. Rieder, D. Velic, E. Knoesel, A. Hotzel, M. Wolf, G. Ertl: Phys. Rev. Lett. **80**, 2004 (1998)

[101] M. Wolf, A. Hotzel, E. Knoesel, D. Velic: Phys. Rev. B **59**, 5926 (1999)

[102] S. Barazzouk, P. V. Kamat, S. Hotchandani: J. Phys. Chem. B. **109**, 716 (2005)

[103] W. Gebauer, A. Langner, M. Schneider, M. Sokolowski, E. Umbach: Phys. Rev. B **69**, 155431 (2004)

[104] J. Jiang, K. Bosnick, M. Maillard, L. Brus: J. Phys. Chem. B **107**, 9964 (2003)

[105] T. H. Lee, J. I. Gonzalez, J. Zheng, R. M. Dickson: Accounts Chem. Res. **38**, 534 (2005)

[106] L. P. Capadona, J. Zheng, J. I. Gonzalez, T. H. Lee, S. A. Patel, R. M. Dickson: Phys. Rev. Lett. **94**, 058301 (2005)

Index

Two-Photon Excited Surface-Enhanced Raman Scattering

Katrin Kneipp[1,2,3] and Harald Kneipp[1]

[1] Harvard-MIT Division of Health Science and Technology, Cambridge, MA 02139, USA
[2] Wellman Center for Photomedicine, Harvard University, Medical School, Boston, MA 02114, USA
[3] Quantum Protein Center, Technical University of Denmark, DK-2800, Lyngby, Denmark

1 Introduction

Surface-enhanced Raman scattering is generally considered to be "normal" linear Raman scattering performed in enhanced local optical fields. According to the electromagnetic-field enhancement model, nonlinear effects should derive a greater benefit from local optical fields and should be surface enhanced to an even greater extent as they depend on enhanced local optical fields by the power of two or higher [1]. Here we will discuss anti-Stokes SERS scattering originating from pumped vibrational levels and hyper-Raman scattering as two-photon excited surface-enhanced Raman processes.

The interest in nonlinear or two-photon excited surface-enhanced Raman scattering is twofold: First, two-photon excited processes can provide new insight into enhancement mechanisms, predominantly about electromagnetic contributions. It is useful to compare enhancement factors achieved in one-photon and two-photon excited Raman processes. Second, two-photon spectroscopy has several advantages, such as a decrease in the probed volume and the use of longer excitation wavelengths. These advantages have been demonstrated for two-photon fluorescence [2, 3]. It is a challenge to extend the concept of two-photon spectroscopy also to Raman scattering [1, 4].

In this Chapter we discuss nonlinear Raman processes in local optical fields with a focus on incoherent or spontaneous Raman scattering. In these effects, two photons of one excitation laser are involved in generating an incoherent Raman-scattering signal. This situation is different from coherent anti-Stokes Raman scattering (CARS), where two incident laser fields separated in frequency by the vibrational energy result in a coherent Raman signal at the anti-Stokes side of the higher-frequency laser [5]. Whereas in CARS experiments, a chemical compound is detected and imaged based on one or a few selected specific Raman frequencies, the incoherent nonlinear Raman scattering effects discussed in this Chapter display the entire Raman spectrum and therefore provide a structural fingerprint. Moreover, a surface-enhanced hyper-Raman spectrum can provide additional vibrational

K. Kneipp, M. Moskovits, H. Kneipp (Eds.): Surface-Enhanced Raman Scattering – Physics and Applications, Topics Appl. Phys. **103**, 183–196 (2006)
© Springer-Verlag Berlin Heidelberg 2006

information due to different selection rules for Raman scattering and hyper-Raman scattering.

In Sect. 2, we discuss surface-enhanced anti-Stokes Raman scattering under the condition of vibrational pumping. In these experiments, a very strong Stokes process at an effective cross section on the order of 10^{-16} cm^2 corresponding to SERS enhancement factors of $\sim 10^{14}$ measurably populates the first excited vibrational levels in excess of the thermal population. Anti-Stokes signals originating from these pumped vibrational levels show a quadratic dependence on the excitation intensity. Surface-enhanced Raman scattering at the anti-Stokes side became the focus of interest after unexpectedly high anti-Stokes SERS signals were reported for the first time in 1996, together with a quadratic dependence of the anti-Stokes signal on the excitation intensity [6]. The first report on the effect of vibrational pumping generated some controversial opinions [7, 8, 9, 10]. Here, we will revisit this field and discuss experimental observations that give compelling evidence of the effect [11].

As a second two-photon Raman process, we elucidate surface-enhanced hyper-Raman scattering (SEHRS). Hyper-Raman scattering (HRS) is a spontaneous nonlinear Raman process resulting in an incoherent Raman signal shifted relative to the second harmonic of the excitation laser. HRS is an extremely weak effect [12]. Cross sections on the order of 10^{-65} cm$^4 \cdot$ s prevent practical use of the effect. However, a strong electromagnetic-field enhancement in surface-enhanced hyper-Raman scattering can compensate for the extremely small intrinsic cross section [13]. In this Chapter we discuss the application of SEHRS for trace detection. Comparison between SERS and SEHRS will be used to extract information on the size of the electromagnetic enhancement factor.

2 Surface-Enhanced Anti-Stokes Raman Scattering

2.1 SERS Vibrational Pumping

Figure 1 illustrates Stokes and anti-Stokes scattering in a molecular-level scheme. Population and depopulation of the first excited vibrational state by Raman Stokes and anti-Stokes transitions can be described by the rate equation [6, 14]

$$\frac{\mathrm{d}N_1}{\mathrm{d}t} = (N_0 - N_1)\sigma^{\mathrm{SERS}} n_{\mathrm{L}} - \frac{N_1}{\tau_1}, \tag{1}$$

with N_0 and N_1 the number of molecules in the vibrational ground and first excited state, reapectively, $\sigma_{\mathrm{S}}^{\mathrm{SERS}}$ describes an effective cross section of the surface-enhanced Raman process, and n_{L} is the photon flux density of the excitation laser.

Simple equations for the anti-Stokes (2a) and Stokes signals (2b) can be derived from (1) assuming steady state and weak saturation ($\exp(-h\nu_{\mathrm{M}}/kT)$

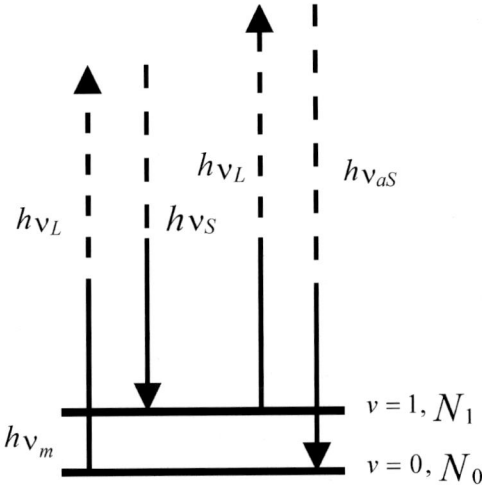

Fig. 1. Stokes and anti-Stokes scattering in a molecular-level scheme

$\leq \sigma_{\mathrm{S}}^{\mathrm{SERS}} \tau_1 n_{\mathrm{L}} \ll 1$). Under weak saturation conditions a cw-laser-excited Raman process populates the first excited vibrational state comparably or higher than the Boltzmann population, but still far away from approaching the equilibrium population between N_0 and N_1.

$$P_{\mathrm{aS}}^{\mathrm{SERS}} = \left(N_0 e^{-\frac{h\nu_{\mathrm{M}}}{kT}} + N_0 \sigma^{\mathrm{SERS}} \tau_1 n_{\mathrm{L}} \right) \sigma^{\mathrm{SERS}} n_{\mathrm{L}} \tag{2a}$$

$$P_{\mathrm{S}}^{\mathrm{SERS}} = N_0 \sigma^{\mathrm{SERS}} n_{\mathrm{L}}, \tag{2b}$$

where τ_1 is the lifetime of the first excited vibrational state, T is the sample temperature, h and k are the Planck and Boltzmann constants, respectively.

The first term in (2a) describes the anti-Stokes signal due to the thermal population of the first excited vibrational state. The second term occurs due to vibrational pumping by the strong Stokes process. In normal Raman scattering this term can be neglected compared to the thermal population since for normal Raman scattering the product $\sigma^{\mathrm{RS}}(\nu_{\mathrm{m}}) \cdot \tau_1(\nu_{\mathrm{m}})$ is on the order of $10^{-40}\,\mathrm{cm}^2 \cdot \mathrm{s}$. This means that about 10^{38} laser photons/cm$^2 \cdot$s (ca. $10^{20}\,\mathrm{W/cm}^2$ excitation intensity) are required to make the Raman pumping comparable to the thermal population of the first vibrational level.

Figure 2 compares SERS Stokes and anti-Stokes experiments performed on isolated gold nanoparticles and on small aggregates formed by these spheres. In agreement with theory [16] enhancement factors for isolated gold spheres have been estimated to be 10^3–10^4 [17]. These SERS enhancement factors are too small to measurably populate the first excited vibrational state and the anti-Stokes spectrum appears at the expected relatively low signal level (Fig. 2b). In particular, the high-frequency modes are not seen on the anti-Stokes side due to their low thermal population. This situation changes

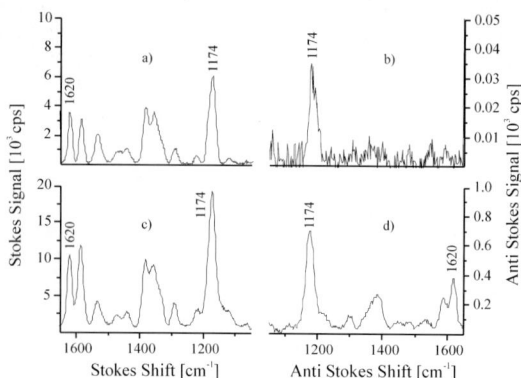

Fig. 2. SERS Stokes and anti-Stokes experiments on isolated gold nanoparticles (**a** and **b**) and on aggregates formed by these spheres (**c** and **d**). Reprinted with permission from [15], Copyright 1998 Society for Applied Spectroscopy

for SERS performed on cluster structures at 830 nm excitation (Fig. 2d). Now a strong anti-Stokes signals appears, in particular also of higher-frequency Raman modes. The ratio between anti-Stokes and Stokes signals increases linearly with the excitation intensity. Anti-Stokes to Stokes ratios for different Raman lines result in different "temperatures" [6, 18].

In order to account for the experimental observations, the product of cross section and vibrational lifetime in (2a) must be on the order of 10^{-27} cm$^2 \cdot$ s. Assuming vibrational lifetimes on the order of 10 ps, the surface-enhanced Raman cross section is then estimated to be at least $\sim 10^{-16}$ cm^2, corresponding to SERS enhancement factors on the order of 10^{14} [6]. Anti-Stokes to Stokes signal ratios measured on different clusters from 150 nm size to larger fractal silver structures show that enhancement factors on the order of 10^{14} in nonresonant SERS are independent of the size and shape of the cluster [15,17,18]. In order to make the large cross section consistent with the observed Stokes signal, we must draw the conclusion that the number of molecules involved in the scattering process must be extremely small [6, 11, 18].

These nonresonant SERS enhancement factors are confirmed in single-molecule Raman experiments [17, 18]. Figure 3 shows Raman spectra measured from an aqueous solution of silver nanoaggregates containing 10^{-14} M pseudoisocyanine (PIC) and 5 M methanol, which yields one PIC molecule and $\sim 10^{14}$ methanol molecules in the 30 pl probed volume. The experiment shows that the SERS signals of a single PIC molecule appears at the same signal level as the nonenhanced Raman signal of the 10^{14} methanol molecules. Raman lines assigned to PIC appear at varying signal levels due to Brownian motion of the silver colloidal cluster carrying single PIC molecules into and out of the probed volume [18].

Other evidence for vibrational pumping comes from the observations of so-called hot vibrational transitions [6]. This effect is illustrated in the energy

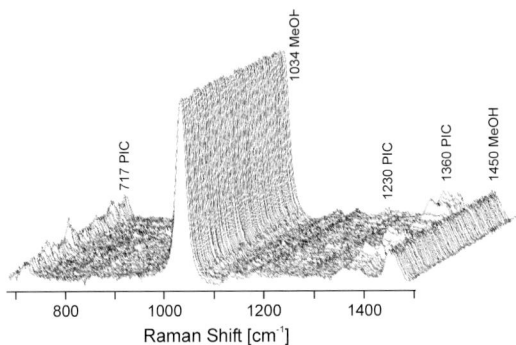

Fig. 3. Raman spectra measured from aqueous solution of silver nanoaggregates containing one PIC molecule and $\sim 10^{14}$ methanol molecules in the 30 pl probed volume. Reprinted with permission from [18], Copyright 1998 Society for Applied Spectroscopy

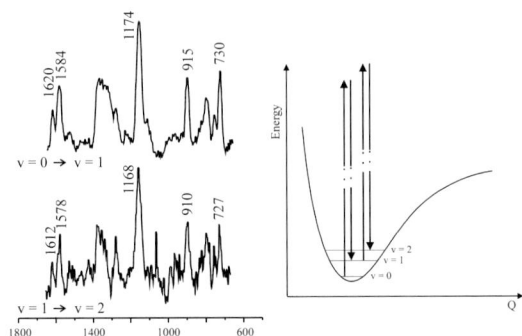

Fig. 4. Raman spectrum of the $v = 0$ to $v = 1$ transitions in comparison to a Raman spectrum of $v = 1$ to $v = 2$ transitions extracted from SERS Stokes spectra measured at excitation intensities high enough to give clearly increased anti-Stokes to Stokes ratios. Reprinted with permission from [11], Copyright 2006 Royal Society of Chemistry

level scheme in Fig. 4. Appreciably populating the first excited vibrational state permits the observation of $\{v = 1$ to $v = 2\}$ Stokes scattering in addition to $\{v = 0$ to $v = 1\}$ transitions usually observed in Raman scattering. The frequency shifts observed between these vibrational transitions are proportional to the absolute Raman shifts of the vibrational modes, and they provide direct information about the anharmonicity of the electronic ground-state potentials.

Anomalies in anti-Stokes to Stokes SERS signal ratios could be also explained by molecular or charge-transfer resonance effects, which can be selectively efficient for Stokes and anti-Stokes scattering [8, 10]. However, these

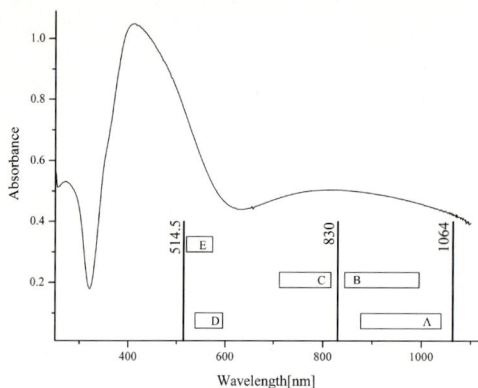

Fig. 5. Plasmon resonance curve of silver nanoclusters displaying also different excitation lasers and the wavelength ranges covered by different linear and non-linear Raman signals: A is the anti-Stokes region for 1064 nm excitation, B and C are Stokes and anti-Stokes regions, respectively, for 830 nm excitation, D is the hyper-Raman scattering region for 1064 nm excitation, and E is the Stokes region for 514 nm excitation. Reprinted with permission from [11], Copyright 2006 Royal Society of Chemistry

abnormal anti-Stokes to Stokes ratios are independent of the excitation intensity and, consequently, no quadratic dependence of the anti-Stokes signal can be observed.

In particular, different plasmon-resonance conditions for the Stokes and anti-Stokes frequency range can also result in different effective cross sections for Stokes and anti-Stokes SERS. In our experiments we avoided this effect by selecting a favorable excitation wavelength.

Figure 5 displays the extinction curve of a suspension of colloidal silver clusters together with the frequencies of excitation lasers and the frequency ranges, where the scattering signals appear for different Raman processes. By applying near-infrared excitation at 830 nm, Stokes and anti-Stokes light should experience the same electromagnetic enhancement level. This figure also demonstrates that experiments performed at excitation wavelengths below 700 nm will experience a high asymmetry in electromagnetic enhancement factors between Stokes and anti-Stokes SERS.

2.2 Pumped Anti-Stokes SERS – A Two-Photon Raman Effect

As (2a) shows, anti-Stokes Raman scattering starting from pumped vibrational levels is a two-photon Raman process, one photon populates the excited vibrational state, a second photon generates the anti-Stokes scattering. The anti-Stokes signal P_{aS} depends quadratically on the excitation laser intensity whereas the Stokes signal P_S remains linearly dependent [6].

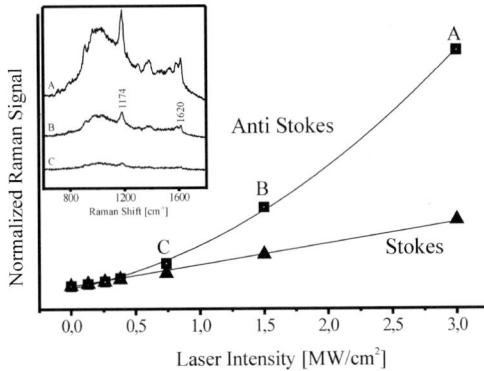

Fig. 6. Surface-enhanced Stokes and anti-Stokes Raman scattering vs. excitation intensity. The insert shows surface-enhanced anti-Stokes Raman scattering of crystal violet on colloidal gold clusters at $3\,\mathrm{MW/cm^2}$ (A), $1.4\,\mathrm{MW/cm^2}$ (B), and $0.7\,\mathrm{MW/cm^2}$ (C) excitation intensity. Reprinted with permission from [19], Copyright 2000 SPIE Publishing Services

Figure 6 shows plots of SERS anti-Stokes and Stokes signal powers of crystal violet vs. excitation intensity. The lines indicate quadratic and linear fits to the experimental data, displaying the predicted quadratic and linear dependence. The insert shows that the anti-Stokes SERS signal is increased compared to the Rayleigh background for increasing laser intensities due to the nonlinear dependence of the anti-Stokes signal compared to the linear fluorescence background.

It should be noted that the experimental verification of the quadratic dependence requires a weak saturation condition [6, 11].

In general, a nonlinear dependence of the anti-Stokes signal on the excitation intensity could also be caused by an increase of temperature due to laser heating [9, 20]. In order to exclude this effect and to provide evidence that the quadratic dependence is due to vibrational pumping, SERS studies have been performed on single-wall carbon nanotubes (SWNTs) on silver-aggregate clusters [20, 21]. SWNTs show a strong dependence of their Raman frequencies on temperature. This allows monitoring of the temperature changes during the measurement. The observed very small shifts in frequencies along with a quadratic dependence of the anti-Stokes signal show that temperature changes in the sample are too small to explain the observed quadratic dependence of the anti-Stokes signal on the excitation intensity.

Considering the two-photon process, we can write the second term in (2a), using an effective two-photon cross section $\sigma_{\mathrm{aS},nl}^{\mathrm{SERS}}$

$$P_{\mathrm{aS},nl}^{\mathrm{SERS}} = N_0 \sigma_{\mathrm{aS},nl}^{\mathrm{SERS}} n_{\mathrm{L}}^2 \,, \tag{3a}$$

with

$$\sigma_{\mathrm{aS},nl}^{\mathrm{SERS}} = \sigma_{\mathrm{S}}^{\mathrm{SERS}} \sigma_{\mathrm{aS}}^{\mathrm{SERS}} \tau_1 \,. \tag{3b}$$

Assuming a SERS cross section of approximately $10^{-16}\,\mathrm{cm}^2$ and a vibrational lifetime on the order of $10\,\mathrm{ps}$, effective two-photon cross sections can be inferred to be about $10^{-43}\,\mathrm{cm}^4\cdot\mathrm{s}$. This provides a two-photon excited Raman probe at a cross section more than 7 orders of magnitude larger than typical cross sections for two-photon excited fluorescence [22, 23].

Because of its very high cross section, this two-photon Raman effect can be observed at relatively low excitation intensities using cw lasers. As a further advantage for spectroscopic applications, anti-Stokes spectra are measured at the high-energy side of the excitation laser, which is free from fluorescence. Two-photon excited fluorescence appears at much higher frequencies.

3 Surface-Enhanced Hyper-Raman Scattering (SEHRS)

3.1 SEHRS Enhancement Factors

Hyper-Raman scattering (HRS) is a spontaneous nonlinear process, i.e., different molecules independently scatter light due to their hyperpolarizability and generate an incoherent Raman signal shifted relative to the second harmonic of the excitation laser. The number of surface-enhanced hyper-Raman Stokes photons P^{SEHRS} can be written as [1]

$$P^{\mathrm{SEHRS}} = N_0 \sigma^{\mathrm{SEHRS}} n_{\mathrm{L}}^2 \,, \tag{4}$$

where σ^{SEHRS} is the effective cross section for the surface-enhanced hyper-Raman process.

The "normal" hyper-Raman cross section σ^{HRS} is extremely small, on the order of $10^{-65}\,\mathrm{cm}^4\cdot\mathrm{s}$. HRS can be enhanced in an analogous fashion to normal Raman scattering. Then the surface-enhanced hyper-Raman cross section is

$$\sigma^{\mathrm{SEHRS}} = \sigma_{\mathrm{ads}}^{\mathrm{HRS}} |A(\nu_{\mathrm{L}})|^4 |A(\nu_{\mathrm{HS}})|^2 \,, \tag{5}$$

where $\sigma_{\mathrm{ads}}^{\mathrm{HRS}}$ describes an enhanced hyper-Raman cross section compared to that of a "free" molecule $A(\nu)$ describes the enhancement of the optical fields. We can write an enhancement factor for SEHRS as

$$G_{\mathrm{SEHRS}} = \frac{\sigma_{\mathrm{ads}}^{\mathrm{HRS}}}{\sigma_{\mathrm{free}}^{\mathrm{HRS}}} |A(\nu_{\mathrm{L}})|^4 |A(\nu_{\mathrm{HS}})|^2 \tag{6a}$$

compared to

$$G_{\mathrm{SEHRS}} = \frac{\sigma_{\mathrm{ads}}^{\mathrm{RS}}}{\sigma_{\mathrm{free}}^{\mathrm{RS}}} |A(\nu_{\mathrm{L}})|^2 |A(\nu_{\mathrm{S}})|^2 \tag{6b}$$

for SERS. Strong surface-enhancement factors can overcome the inherently weak nature of hyper-Raman scattering and SERS and SEHRS can appear at comparable signal levels [4, 13].

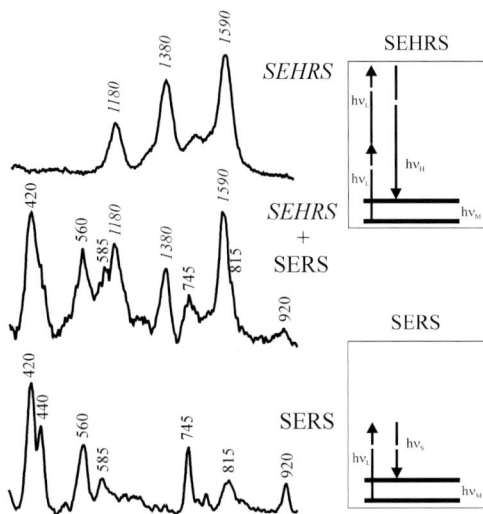

Fig. 7. SEHRS and SERS signals of crystal violet on silver colloidal clusters measured in the same spectrum using 10^7 W/cm^2 NIR excitation (*middle trace*, see also Fig. 1). In the *upper* and the *lower trace*, SEHRS and SERS spectra are differentiated by placing a NIR absorbing filter in front of the spectrograph or by switching off the mode-locked regime of the Ti:sapphire laser, respectively. Reprinted with permission from [1], Copyright 2002 Springer Verlag

This is demonstrated in the middle spectrum in Fig. 7, which displays surface-enhanced hyper-Raman and Raman signals of crystal violet on colloidal silver clusters measured in the same spectrum by using 830 nm excitation [4, 13].

Taking into account the different sensitivity of the Raman system in the near-infrared and blue region, the SEHRS signal is weaker by a factor of 100 than the SERS signal. The experimental ratios between SERS and SEHRS intensities can be combined with the corresponding estimated intensity ratio between RS and HRS, which is on the order of 10^8 for the applied 10^7 W/cm^2 excitation intensity [24]. This means that the ratio between surface enhancement factors of hyper-Raman scattering and Raman scattering must be on the order of 10^6. Combining this ratio with NIR-SERS enhancement factors of crystal violet on colloidal silver clusters on the order of 10^{14} total surface enhancement factors of hyper-Raman scattering on crystal violet adsorbed on colloidal silver clusters can be inferred to be on the order of 10^{20} [1, 4, 13].

3.2 Resonant and Nonresonant Surface-Enhanced Hyper-Raman Scattering

Surface-enhanced hyper-Raman scattering (SEHRS) can also benefit from an additional molecular resonance enhancement when the excitation laser is

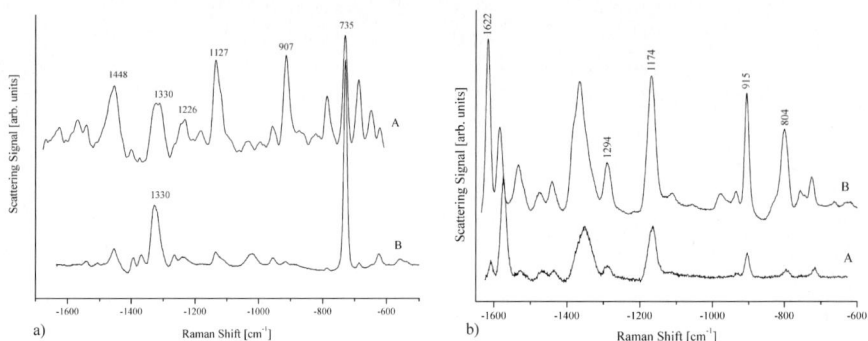

Fig. 8. (a) Surface-enhanced hyper-Raman (A) – and surface-enhanced Raman (B) spectra of 10^{-4} mol/l adenine in silver colloidal solution. Excitation (A): 2.4 W 1064 nm mode-locked laser (7 ps, 74 MHz); collection time 10 s and (B): 1 W 514.5 nm cw. laser; collection time 0.1 s. (b) Surface-enhanced hyper-Raman (A) – and surface-enhanced Raman (B) spectra of crystal violet(CV) in silver colloidal solution. (A): 10^{-8} mol/l CV, excitation: 2.4 W, 1064 nm, 7 ps, 74 MHz; collection time 10 s, and (B): 10^{-6} mol/l CV, excitation: 1.4 W, 514.5 nm, cw; collection time 0.01 s. Reprinted with permission from [27], Copyright 2005 John Wiley & Sons, Ltd

either one-photon or two-photon resonant with electronic transitions in the molecule (SEHRRS) [25, 26].

We want to compare surface-enhanced hyper-Raman scattering under nonresonant and resonant conditions [27]. Adenine and crystal violet in silver colloidal solution are test molecules for SEHRS and SEHRRS, respectively. For hyper-Raman studies the sample solutions were excited by a mode-locked Nd laser at 1064 nm excitation. For comparison, SERS spectra were collected using a cw argon ion laser line at 514.5 nm for excitation.

Figure 8a shows SERS and SEHRS spectra of adenine in silver colloidal solution [27]. Additionally to strong bands at 735 cm^{-1} and 1330 cm^{-1}, which also occur in the SERS spectrum, the SEHRS spectrum shows new lines. These lines are not so strong or do not even exist in the SERS spectrum, but they also appear at 938, 1123, 1247, and 1480 cm^{-1} in the infrared absorption spectra of adenine [28]. In general, SERS and SEHRS lines are slightly shifted relative to the Raman frequencies measured in normal Raman scattering and infrared absorption, respectively. SEHRS measurements at 1064 nm excitation as well as SERS experiments at 514.5 nm excitation are absolutely nonresonant experiments for adenine. Even potential multiphoton excitations of the 1064 nm pulsed excitation light would not be able to reach the absorption levels of adenine below 250 nm. Nevertheless, Fig. 8a shows very strong SEHRS spectra measured from 10^{-4} M adenine in 10 s collection time. Increasing collection times allow detection of lower concentrations of adenine.

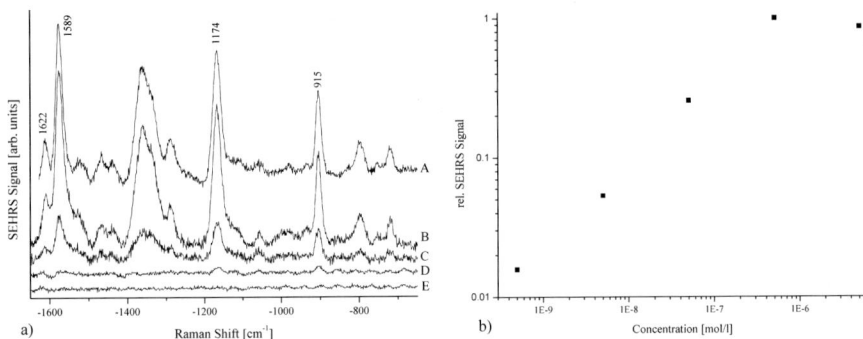

Fig. 9. (a) Surface-enhanced hyper-Raman spectra of crystal violet at different concentrations in silver colloidal solution (A) 5×10^{-6} mol/l, (B) 5×10^{-7} mol/l, (C) 5×10^{-8} mol/l, (D) 5×10^{-9} mol/l, (E) 5×10^{-10} mol/l, excitation: 2.4 W, 1064 nm, 7 ps, 74 MHz; collection times 1 s (spectra A–C), 10 s (spectra D, E); all spectra shown are normalized to 1 s collection time. (b) Normalized relative SEHRS signal of the 1174 cm^{-1} line vs. concentration of crystal violet. Reprinted with permission from [27], Copyright 2005 John Wiley & Sons, Ltd

Figure 8b displays SEHRS and SERS spectra of crystal violet. Excitation at 514.5 nm for SERS as well as the 1064 nm excitation for SEHRS are one- and two-photon resonant, respectively, with electronic transitions in this molecule. In contrast to adenine, SERRS and SEHRRS spectra of crystal violet are very similar, showing basically the same lines, sometimes with altered relative intensities [27].

Figure 9a shows the SEHRRS spectra of crystal violet at different concentrations between 5×10^{-6} mol/l and 5×10^{-10} mol/l. Figure 9b plots the scattering signal of the line at 1170 cm^{-1} vs. the concentration of crystal violet. In the measured concentration range between 10^{-10} and 10^{-7} mol/l the SEHRS signal follows the concentration of the target molecule. Saturation occurs relatively suddenly at about 10^{-6} mol/l, where obviously the first monolayer of crystal violet has been completed. Our experiments demonstrate the quantitative detection of molecules based on their SEHRS signature.

4 Summary and Conclusion

Strongly enhanced local optical fields provide particularly favorable conditions for nonlinear Raman scattering. We have discussed anti-Stoke SERS signals originating from pumped vibrational levels as well as surface-enhanced hyper-Raman signals as incoherent two-photon excited Raman effects.

We have demonstrated and discussed experiments that provide clear evidence for SERS vibrational pumping. This includes: 1. anti-Stokes to Stokes signal ratios that cannot be explained in terms of a Boltzmann distribution, and that are determined by the product of the SERS cross section and vibrational lifetime, and no longer by temperature [6, 17, 18, 20], 2. a quadratic dependence of the anti-Stokes signal on the excitation intensity [6, 20, 21], and 3. an additional component of $\{v = 1$ to $v = 2\}$ Stokes transitions [6]. On the other hand, the fact that vibrational pumping exists raises questions about the mechanisms behind this high effective cross section operative in pumping [29].

Pumped anti-Stokes Raman scattering provides a two-photon excited Raman probe at cross sections orders of magnitude larger than observed in two-photon absorption. The large effective cross section can be explained by the nature of the process, which is a two-photon process exploiting the vibrational level as an actual intermediate state.

In surface-enhanced hyper-Raman scattering, strongly enhanced local optical fields compensate for the extremely small cross section of HRS. Experiments show that SEHRS must be enhanced 10^6 times more than SERS. This factor can be understood when considering the quadratic dependence of the effect on the enhanced local optical fields. Enhancement of field strengths for silver and gold nanoclusters on the order of 10^3 ($A(\nu_{L,S})$ in (6a) and (6b)) yield electromagnetic enhancement factors 10^6 times higher for SEHRS than for SERS.

Experiments performed with crystal violet on silver nanoclusters demonstrate sensitive and reproducible surface-enhanced hyper-Raman measurements down to nanomolar concentration ranges.

Acknowledgements

This work is supported in part by DOD grant # AFOSR FA9550-04-1-0079 and by the generous gift of Dr. and Mrs. J.S. Chen to the optical diagnostics program of the Massachusetts General Hospital Wellman Center for Photomedicine.

References

[1] K. Kneipp, H. Kneipp, I. Itzkan, et al.: Nonlinear Raman probe of single molecules attached to colloidal silver and gold clusters, in V. M. Shalaev (Ed.): *Optical Properties of Nanostructured Random Media* (Springer, Berlin, Heidelberg 2002) pp. 227–247

[2] W. Denk, J. H. Strickler, W. W. Webb: Science **248**, 73 (1990)

[3] P. T. C. So, C. Y. Dong, B. R. Masters, et al.: Ann. Rev. Biomed. Eng. **2**, 399 (2000)

[4] K. Kneipp, H. Kneipp, I. Itzkan, et al.: Chem. Phys. **247**, 155 (1999)

[5] A. Volkmer, J. X. Cheng, L. D. Book, et al.: Biophys. J. **80**, 656 (2001)

[6] K. Kneipp, Y. Wang, H. Kneipp, et al.: Phys. Rev. Lett. **76**, 2444 (1996)

[7] T. L. Haslett, L. Tay, M. Moskovits: Chem. Phys. **113**, 1641 (2000)

[8] A. G. Brolo, A. C. Sanderson, A. P. Smith: Phys. Rev. B **69**, 045424 (2004)

[9] R. C. Maher, L. F. Cohen, P. Etchegoin, et al.: J. Chem. Phys. **120**, 11746 (2004)

[10] R. C. Maher, J. Hou, L. F. Cohen, et al.: J. Chem. Phys. **123**, 084702 (2005)

[11] K. Kneipp, H. Kneipp: Faraday Discuss. **132**, 27 (2006)

[12] V. N. Denisov, B. N. Mavrin, V. B. Podobedov: Phys. Rep. **151**, 1 (1987)

[13] H. Kneipp, K. Kneipp, F. Seifert: Chem. Phys. Lett. **212**, 374 (1993)

[14] K. Kneipp, H. Kneipp, I. Itzkan, et al.: Chem. Rev. **99**, 2957 (1999)

[15] K. Kneipp, H. Kneipp, R. Manoharan, et al.: Appl. Spectrosc. **52**, 1493 (1998)

[16] D. S. Wang, H. Chew, M. Kerker: Appl. Opt. **19**, 2256 (1980)

[17] K. Kneipp, H. Kneipp, V. B. Kartha, et al.: Phys. Rev. E **57**, R6281 (1998)

[18] K. Kneipp, H. Kneipp, G. Deinum, et al.: Appl. Spectrosc. **52**, 175 (1998)

[19] K. Kneipp, H. Kneipp, I. Itzkan, et al.: Near-infrared surface-enhanced Raman spectroscopy of biomedically relevant single molecules on colloidal silver and gold clusters, in S. Nie, E. Tamiya, E. S. Yeung (Eds.): *Photonics West – BIOS 2000* (SPIE, SanJose, CA 2000) p. 49

[20] K. Kneipp, H. Kneipp, P. Corio, et al.: Phys. Rev. Lett. **84**, 3470 (2000)

[21] P. V. Teredesai, A. K. Sood, A. Govindaraj, et al.: Appl. Surf. Sci. **182**, 196 (2001)

[22] W. Denk, J. H. Strickler, W. W. Webb: Science **248**, 73 (1990)

[23] D. R. Larson, W. R. Zipfel, R. M. Williams, et al.: Science **300**, 1434 (2003)

[24] L. D. Ziegler: J. Raman Spectrosc. **21**, 769 (1990)

[25] J. T. Golab, J. R. Sprague, K. T. Carron, et al.: J. Chem. Phys. **88**, 7942 (1988)

[26] X. Y. Yu, Q. J. Huang, Q. Luo, et al.: J. Infrared Millimeter Waves **22**, 51 (2003)

[27] H. Kneipp, K. Kneipp: J. Raman Spectrosc. **36**, 551 (2005)

[28] B. Schrader: *Raman/Infrared Atlas of Organic Compounds* (VCH Publishers, Weinheim 1989)

[29] M. Moskovits: J. Raman Spectrosc. **36**, 485 (2005)

Index

Applications of the Enhancement of Resonance Raman Scattering and Fluorescence by Strongly Coupled Metallic Nanostructures

Nicholas P. W. Pieczonka, Paul J. G. Goulet, and Ricardo F. Aroca

Materials and Surface Science Group, Department of Chemistry and Biochemistry, University of Windsor, Windsor, ON, N9B 3P4, Canada
raroca1@cogeco.ca

1 Introduction

Resonance Raman scattering (RRS), absorption, and luminescence occur when a molecule is irradiated with exciting radiation in resonance with one of its electronic transitions. These processes can be enhanced (sometimes quite significantly) when the molecule is placed at or near the surface of appropriate aggregates of surface-plasmon resonance-supporting metal nanoparticles, leading to, among others, the phenomena of surface-enhanced resonance Raman scattering (SERRS), surface-enhanced absorption, and surface-enhanced fluorescence (SEF). Examining this concept more deeply, we find that the interaction of light with molecules is most commonly discussed in terms of the semiclassical picture of harmonic or anharmonic oscillators. For example, for electronic absorption and emission, the Frank–Condon idea of rapidly moving electrons and vertical transitions is broadly used. Practical quantum computations of the absorption and emission spectra of large polyatomic molecules are still challenging. However, the approach, whereby electronically resonant spectroscopies are expressed in terms of overlaps of nuclear wavepackets moving on Born–Oppenheimer potential energy surfaces, has become popular, particularly for calculating resonance Raman intensities [1]. The challenge is even greater for molecules adsorbed onto metal nanoparticles where there is a need to include the electronic resonances of the nanoparticle, the effect of the interaction on the electronic density of the molecule, and the polarization of the nanoparticle by the molecule. Consequently, the discussion of the experimental data obtained in surface-enhanced absorption, surface-enhanced fluorescence (SEF), and surface-enhanced resonance Raman scattering (SERRS) has been mainly in terms of classical electrodynamics [2, 3].

As an introduction to the practical considerations of these effects, let us consider an example. SEF and SERRS are discussed here for a Langmuir–Blodgett (LB) monolayer of bis (benzimidazo) thioperylene (Monothio BZP) [4] deposited on glass and on silver-island films. First, the results for excitation with a laser at 633 nm are shown in Fig. 1.

The LB monolayer of Monothio BZP on glass gives fluorescence due to excimer formation in the film, but there is little evidence of Raman vibra-

K. Kneipp, M. Moskovits, H. Kneipp (Eds.): Surface-Enhanced Raman Scattering – Physics and Applications, Topics Appl. Phys. **103**, 197–216 (2006)
© Springer-Verlag Berlin Heidelberg 2006

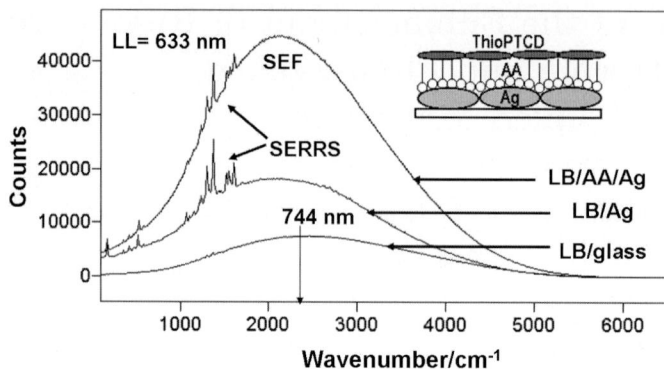

Fig. 1. 633 nm excited SERRS and SEF of thioPTCD monolayer deposited onto glass, Ag-island film, and Ag-island film with intervening spacer layer of arachidic acid

tional bands. The same LB film transferred to silver islands shows both strong SERRS and enhanced excimer fluorescence. However, the enhancement factor for SERRS is orders of magnitude higher than that of SEF. Notably, exciting with the 633 nm laser leads to the Raman bands overlapping with the fluorescence. Also, the SEF can be further enhanced by separating the metal and the dye monolayer by depositing a spacer LB layer between them, thus confirming the long-range action of the electromagnetic (em) mechanism. At the same time, the SERRS signal is slightly reduced, since SERRS is maximized for the first layer. Since the fluorescence is fixed in the electromagnetic spectrum and the inelastic scattering follows the excitation laser line, the overlapping SEF and SERRS can be separated by changing the excitation laser line. Emission spectra of an LB monolayer of Monothio BZP on glass and on a silver-island film, excited with a 514.5 nm laser, are presented in Fig. 2.

A marker for the excimer fluorescence at 744 nm has been added and the spectrum of the LB film on glass has been multiplied by 10 to facilitate visual comparison. It can be seen that the overtones and combinations, lost in the fluorescence when excitation is at 633 nm, are clearly observed in the SERRS spectra obtained with the 514.5 nm laser. For molecules adsorbed on metallic nanoparticles, according to the electromagnetic mechanism, the polarization of the substrate by the excited molecule is the cause for the enhancement of luminescence, which is particularly strong if the emission frequency matches that of a radiative surface resonance. The expressions for computation of properties such as absorption, lineshape, lifetime, and quantum yield, for the emission of molecules adsorbed on metal particles, were developed early on by *Gersten* and *Nitzan* [2].

As illustrated by the spectra shown in Fig. 1 and Fig. 2, there are important differences between the two enhancements, the most evident of which is

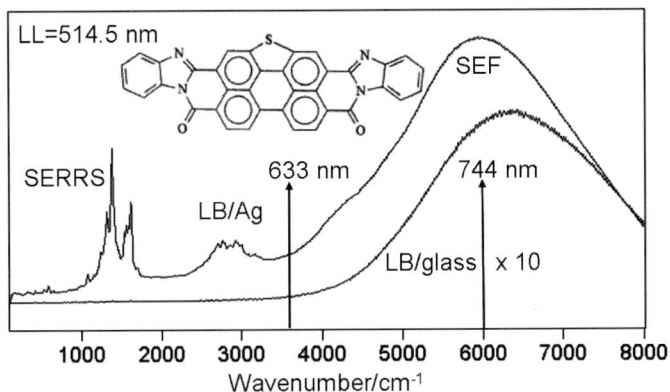

Fig. 2. 514.5 nm excited SERRS and SEF of thioPTCD monolayer deposited onto glass and Ag-island film

their distance dependence. SERRS intensities have been shown to decrease as the distance between adsorbates and enhancing metal nanostructures is increased. In other words, SERRS is an effect that is strengthened when the adsorbate is placed directly on the metal nanostructure. By contrast, in SEF, molecules are understood as free-running oscillators, and intensity is expected to exhibit a maximum enhancement at a certain distance of separation between the molecule and the surface. In SEF, energy transfer to the metal can take place during the finite lifetime of the excitation. It is thus common to observe a reduction in luminescence when a fluorophore is placed directly on the surface of a metal (i.e., enhancement factor EF < 1). The latter is explained by a direct energy transfer from the dipole to the planar surface, as elegantly shown in the classical work of *Chance* et al. [5]. Förster-type dipolar energy transfer is proportional to d^{-3} (where d is the dipole–metal surface separation), and the energy is dissipated into the phonon bath. The distance dependence of em enhancement indicates that it decreases appreciably when d approaches the dimension of the metal particle. SEF is thus a balance between the two opposing processes of surface enhancement, and nonradiative energy loss to the metal surface. It is predicted by the em mechanism that for a spherical metal particle of radius a, separated from a molecule by a distance d, enhancement will decay according to $(a/a+d)^{12}$. For a monolayer at a distance d, enhancement is expected to decrease as $(r/r + d)^{10}$. The electromagnetic enhancement is a much slower decaying function of molecule–surface separation than that of surface quenching, and as a result of this, maximum fluorescence will occur at distances that are intermediate between the maximum quenching ($d = 0$) and "no-effect" conditions. Experimentally, the highest enhancement factors reported for molecular fluorophores are in the hundreds, but there are reports of much greater enhancement of the photoluminescence of the enhancing metallic nanoparticles themselves.

Photoluminescence (PL) is a broad term used to describe any luminescence arising from the photoexcitation of molecules or solids; thereby it includes fluorescence, excimer-fluorescence, phosphorescence, and the emission
observed in solid-state materials. Before the SERS (surface-enhanced Raman
scattering) revolution, the photoluminescence of gold and copper films was
demonstrated using several excitation frequencies [6]. The observed PL from
these samples is due to electron–hole recombination that occurs between the
excited electrons near the Fermi surface and holes in the d band. The peak
energy of the emitted photons is thus closely connected to the energy separation between the Fermi surface and the holes in the d band. In both metals,
the peak emission was consistent with the optically observed energy gap between the upper d band and the Fermi energy (2.0 eV for Cu and 2.2 eV
for Au). Recently, *Imura* et al. [7] observed the emission peaks for single gold
nanorods, near 550 nm and 650 nm, in a two-photon PL spectrum.

After the discovery of SERS, *Boyd* et al. [8] measured the single-photon-
induced luminescence from clean rough surfaces (10^{-9} Torr) of Cu, Au, and
Ag, and compared these experimental values with luminescence measured
from smooth Cu, Au, and Ag surfaces. Modest enhancements of up to one
order of magnitude were reported for the rough surface. The enhanced luminescence was attributed to the effect of localized plasmon resonances, in
agreement with an emission enhancement that begins below the interband
absorption edge of the metal. Since then, it has become clear that surface-
enhanced phenomena are strongly correlated with nanometer-sized particles,
and nanostructures have received much attention in recent years due to the
tunability of their optical properties with particle size [9, 10].

Mohamed et al. [11] reported fluorescence enhancement of over a million
times for gold nanorods. They described their findings as follows: "We observed that while nanodots of 35 nm diameter did not emit, nanorods are
found to have fluorescence of quantum efficiency, which is 6–7 orders of magnitude higher than that of the metal. Furthermore, its wavelength maximum
and quantum efficiency are found to increase linearly on increasing the rod
length and the square of the length, respectively." [11] The same theoretical
model used by *Boyd* et al. [8] in 1986, was called upon in this work to explain
fluorescence enhancement as coupling of the molecular resonance with the local plasmon resonance of the nanoparticle. The theoretical model applied to
gold nanorods is based on the same mechanism used to explain particle-as-
sisted enhancement of molecules interacting with spheres and hemispheroids
sustaining surface-plasmon resonance, where the optical signals are amplified
by the local field created around the metal particle by the plasmon resonance [3]. What sets the El-Sayed report apart from the most commonly observed SEF enhancement, is the large enhancement factor (EF) of 10^6. This
is the highest enhancement factor that has ever been reported for enhanced
luminescence. The authors explain their observation by arguing that longitudinal plasmon resonances absorbing at longer wavelengths (prominent in
nanorods) are more effective in amplifying the fluorescence of gold nanopar-

ticles than the surface-plasmon resonance of spheres because the longitudinal plasmon resonance is less damped. Therefore, proximity to the radiative resonance, and low damping at longer wavelengths, can be taken as the primary reasons for the large EF observed for luminescence in Au nanorods.

2 Applications

2.1 SERRS of Thin Solid Films and Langmuir–Blodgett Monolayers

Understanding the basic properties of thin films is of great interest. However, the analysis of monolayers and submonolayer coatings can be quite challenging. Raman spectroscopy, a nondestructive vibrational technique with a high degree of spatial resolution and molecular specificity, is an extremely powerful analytical technique that can be used toward this analysis throughout a wide range of the electromagnetic spectrum (from the ultraviolet to near-infrared). However, the extensive utilization of this technique has always been held back by the intrinsically low cross section of Raman scattering (RS). Under electronic resonance conditions, though, where the energy of excitation matches the energy of one of the electronic transitions of the molecular system, the signal intensity will be further enhanced due to the effect of resonance Raman enhancement [12]. This signal can be further strengthened through surface enhancement, as in SERS and SERRS, if thin films or monolayers of interest are deposited onto specially prepared substrates [3, 13, 14, 15]. These phenomena can provide extremely large enhancement factors (often upward of 10^6), making it possible for them to compete with fluorescence as ultrasensitive spectroscopic techniques capable of single-molecule detection [16, 17]. This potential for observation of vibrational spectra from monolayers and surface-adsorbed species, with single-molecular sensitivity, thus makes SERS and SERRS two of the most powerful tools available for chemical analysis.

In this work, several properties and applications of SERRS are discussed using examples of studies carried out using two classes of dye molecules that have been the focus of many fruitful studies. These are derivatives of the phthalocyanine (Pc) and perylenetetracarboxylic diimide (PTCD) chromophores. These aromatic systems have large Raman cross sections and have been shown to have many unique and interesting properties. Of particular interest is their use in alternating-layer films, codeposited films, and mixed monomolecular layers. As an example, exciplex emission from mixed films has been clearly identified where the intensity of emitted near-IR light depends on the degree of mixing, molecular orientation, and film structure of coevaporated mixed thin solid films of vanadyl-phthalocyanine (VOPc) and N,N′-bis(neopentyl)-3,4,9,10-perylenebis (dicarboximide) (BNPTCD) [18].

In many of our studies, films are deposited using two primary methods, depending on the desired type of film. Films greater then a few nanometers

(a)

(b)   **(c)**

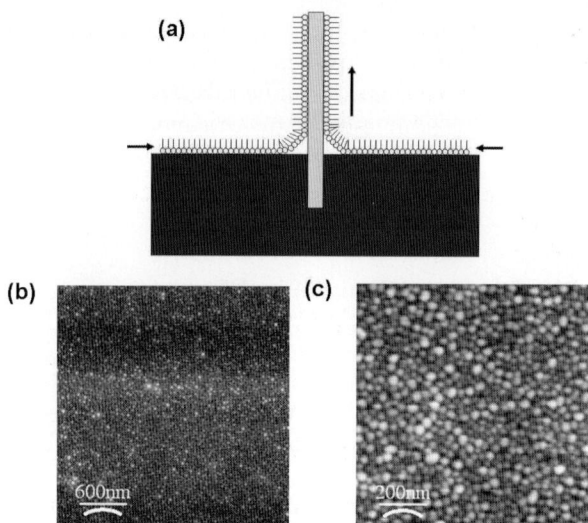

600nm   200nm

Fig. 3. (a) Schematic of z-deposition for Langmuir–Blodgett monolayer transfer to a solid substrate. (b) $3\,\mu m \times 3\,\mu m$ noncontact tapping mode AFM image of a $6\,nm$ mass thickness silver-island film. (c) $1\,\mu m \times 1\,\mu m$ noncontact tapping mode AFM image of a $6\,nm$ mass thickness silver-island film

mass thickness are deposited by vapor deposition in a high-vacuum system where species of interest are thermally evaporated under high-vacuum conditions. For films composed of a singular molecular layer, on the other hand, the Langmuir–Blodgett technique is utilized. This method depends on the fact that certain classes of molecules organize themselves on the surface of liquid subphases (typically water) in single-molecule thick layers. After monitored compression, these layers can then be carefully transferred to solid substrates through controlled deposition, as depicted in Fig. 3. The primary advantage of the LB technique is that it allows for a great degree of control over film architecture and composition.

The enhancing substrates used in all of the SERRS work to be discussed in this section are metal-island films of Ag, Au, or a Ag/Au bimetallic mix, fabricated by metal vapor deposition. These films are deposited to a mass thickness of between $4\,nm$ and $6\,nm$ on glass substrates. This creates surfaces composed of nanostructures with diameters ranging between ca. $20\,nm$ and $100\,nm$, as shown in the AFM (atomic force microscopy) images in Fig. 3. Films created in this fashion are very uniform and give relatively reproducible enhancement factors (for ensemble measurements) on the length scales involved in these experiments (typically $1\,\mu m^2$ to $40\,\mu m^2$). By acquiring single-point spectra, 2D maps, and global images of SERS/SERRS, one can examine various properties of thin films that otherwise are very difficult to attain. These include the degree of miscibility; surface coverage; homogeneity; phase separation;

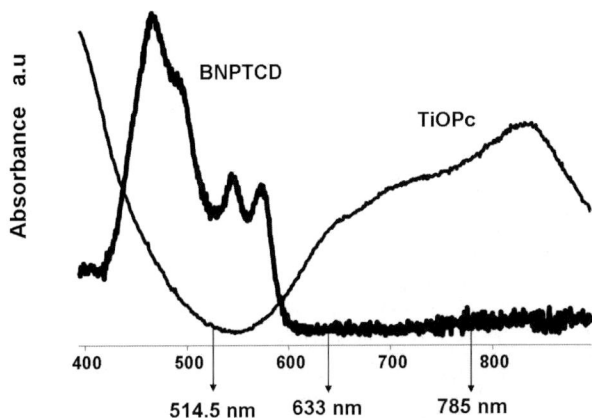

Fig. 4. Absorbance of single LB monolayers of BNPTCD and TiOPc. Also indicated are some commonly used laser excitation lines

molecular orientation; surface chemistry; and nature of the adsorption. Well-characterized films can also be used as sensors. For example, the response of thin films to N_xO_x gas exposure has been measured [19], and temperature-dependent SERS/SERRS measurements have been employed toward the observation of phase changes [20, 21]. Here, we present several examples of how SERS/SERRS can be used for thin-film characterization, and also how controlled architectures have been employed to shed light on the nature of SERRS.

2.1.1 Spatial Spectroscopic Tuning

Spatial spectroscopic tuning is an application of SERRS that exploits both the enhanced fields of SERS and the molecular specificity of the enhancement of RRS [22]. By tuning into the molecular absorptions of different resonant species one can selectively isolate the response from each desired component. This can be applied to stacked films where the layers are constructed from differing moieties, and also to dyes in non-resonant media. SERRS generally shows at least 2 orders of magnitude enhancement over SERS. This means that even for absorbing layers away from the surface, strong SERRS should be possible since it may be stronger than the SERS directly on the surface. This was demonstrated with a sample composed of an enhancing substrate (Ag-covered Sn spheres) coated with an LB monolayer of a PTCD compound followed by an LB monolayer of a Pc molecule [22]. (The absorption spectra of single monolayers of similar PTCD and Pc derivatives on glass are shown in Fig. 4.)

When the excitation line was in resonance with the PTCD derivative ($\lambda = 514.5\,\text{nm}$) then the spectrum contained the fingerprint bands of PTCD. However, when the excitation was in resonance with the Pc absorption ($\lambda =$

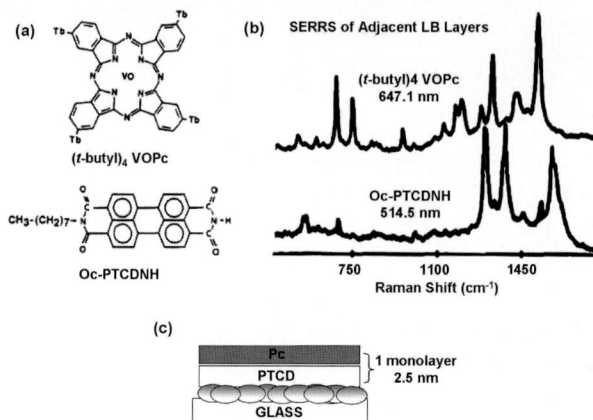

Fig. 5. (a) Structures of (t-butyl)$_4$VOPc and Oc-PTCDNH. (b) SERRS of a Ag-island film coated with a monolayer of Oc-PTCDNH followed by a monolayer of (t-butyl)$_4$VOPc recorded with 647.1 nm and 514.5 nm excitation. (c) Depiction of the layered architecture used. (Reproduced with permission from [22])

647.1 nm), the SERRS spectrum corresponding to the Pc derivative was seen (Fig. 5). This allows a spectroscopic tuning that can be used to study a variety of samples including doped monolayers and surfaces.

2.1.2 SERRS Mapping and Imaging

When studying the SERRS of thin films (monolayers in particular) it is often desirable to obtain information about the two-dimensional distributions of signal on the surface [23, 24, 25, 26]. Investigations of this type can provide essential information regarding the distribution of analyte molecules, metallic nanoparticles, or hot spots of em enhancement, and can be achieved using two different techniques: point-by-point mapping, and global imaging. The first makes use of a mechanical stage that is raster scanned while Raman spectra are taken at all selected positions. This method allows one to obtain complete spectra at every measured spot, with the lateral spatial resolution being determined by the beam waist of the excitation source. Intensity maps can be generated from this data cube to illustrate various relations, such as variance in peak parameters, relative intensities, and peak width. Also, if the system is composed of multiple components, then particular bands, or the ratio of bands from different systems, can be mapped. This often allows for one to make conclusions regarding mixing and distribution on a surface. Global imaging, on the other hand, differs in that only photons of a very narrow range of frequencies are collected. Filters are used to block all light except that of the frequency of the desired band. In effect, a picture is taken of the spatial distribution of intensity of a specific fingerprint band.

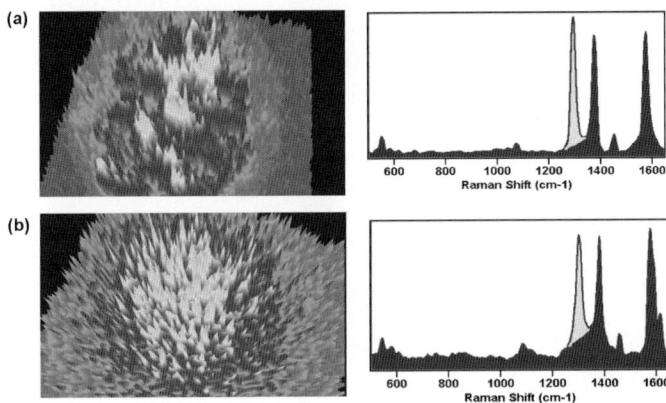

Fig. 6. Global Raman imaging of mixed evaporated films on Ag islands. Images are generated by the narrow band filtering of photons corresponding to the highlighted band in each spectrum. Dimensions $= 40 \times 40\,\mu$m, with excitation 514.5 nm defocused through a 50× objective. (**a**) Sample is CuPc/Bis-PTCD mixed film on Ag-island film. (**b**) Sample is CoPc/Bis-PTDA mixed film on Ag-island film. (Adapted from [21])

These methods have been successfully applied to a number of monomolecular films where they have aided in determining homogeneity and quality of the films [21, 27, 28, 29, 30]. By mapping or imaging the fingerprint bands of the component species, the miscibility of different pairs of molecular moieties can be determined within the resolution of the experiment [21, 31]. As an example, we can look at a typical study where films composed of Pc and PTCD pairs were examined using SERRS [21]. These films consisted of mixed thin solid films of either CoPc or CuPc with bis(n-propylimodo)perylene (Bis-PTCD) coevaporated onto silver-island films. Each molecule has a distinct fingerprint spectrum with little overlap in bands. This is especially true if the excitation source is in resonance with an absorption band as discussed earlier. Global Raman images were attained by collecting light for the fundamental vibrational band of Bis-PTCD, centered at 1297 cm^{-1}, for both CoPc/Bis-PTCD and CuPc/Bis-PTCD mixed films, as shown in Fig. 6. Through an examination of the images, one can see the general homogeneity and quality of each of the films. In this particular case, it was found that the CoPc shows a greater tendency for phase separation than the CuPc, which mixes much more uniformly with the Bis-PTCD.

Often it is beneficial to probe molecular interactions more directly through the study of films that are only one molecule thick. These are fabricated using the LB technique, as discussed above. In a recent work, SERRS mapping results for single mixed PTCD/Pc monolayers on a mixed Ag/Au evaporated island film were presented [32]. The molecular mixing in the monolayer film was studied down to areas of $\sim 1\,\mu$m^2. The Ag/Au metal-island film that was used had broad surface-plasmon absorption, allowing the easy observation of SERRS for both molecules. The SERRS mapping results obtained with excitation at 633 nm showed that, whether the map was plotted for the fingerprint band of the Pc ($1512\,cm^{-1}$), or for the characteristic perylene vibration ($1378\,cm^{-1}$), the intensity profile was fairly uniform across the experimental region ($40\,\mu m^2$). This result implied that, down to the micrometer scale, these molecules are very miscible with one another.

2.1.3 Single-Molecule SERRS from Langmuir–Blodgett Monolayers

The nanoscale architectural control afforded by the Langmuir–Blodgett technique has also been utilized in our laboratory to test the limits of trace detection using SERRS. Our group has pioneered a technique that involves the dilution of a probe molecule in a matrix of arachidic acid with the eventual limit of a single target dye molecule within a $1\,\mu m^2$ probing area. This monolayer is then deposited onto metal-island films prepared by vacuum evaporation. Figure 7 depicts the method that is employed in these experiments. Our group has used this novel method to successfully detect the single-molecule (SM) SERRS spectra of several molecules [33,34,35,36,37,38,39]. Notably, we have also been able to demonstrate some interesting aspects of SM SERRS. For instance, we have observed the overtone and combination bands of several perylene molecules down to the single-molecule level. Utilizing 2D SERRS mapping we were able to observe the transition from ensemble to single-molecule SERRS where only molecules sitting in em hot spots are observed. These measurements provided an insight into the highly localized nature of em hot spots on island-film substrates and also into the rarity of coincidence between single isolated target molecules and hot spots. Figure 8 shows an example of a typical SM SERRS map and typical SM SERRS spectrum.

2.2 SERRS from Colloids and Nanocomposite Films

The most widely employed enhancing substrates for SERRS measurements today are colloidal solutions of Ag and Au nanoparticles (*diameter* $\sim 10\,nm$ to $100\,nm$), or nanoscale assemblies incorporating them. Metallic nanoparticle solutions are most commonly prepared by chemical reduction of metal salts in the presence of one or more stabilizing agents in aqueous media. These preparations tend to be extremely simple, requiring no specialized equipment or techniques, and so colloidal metal solutions are a viable and easy alternative for nearly all laboratories wishing to carry out enhanced spectroscopic measurements.

Recent developments in the field of nanoparticle synthesis have been numerous, and nanoparticles can now be readily engineered with an amazing degree of control over a wide variety of properties. A broad range of different

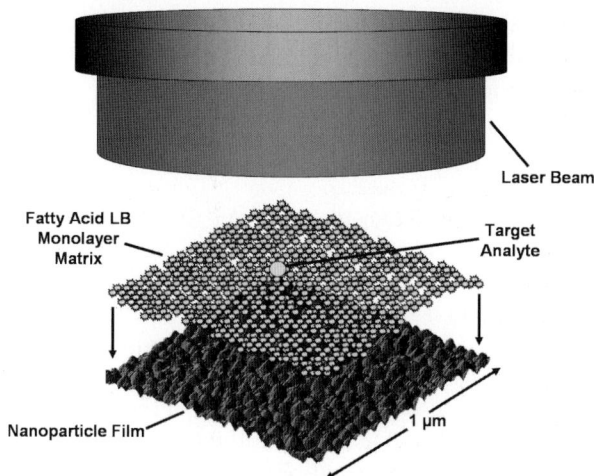

Fig. 7. Depiction of the Langmuir–Blodgett method for single-molecule SERRS

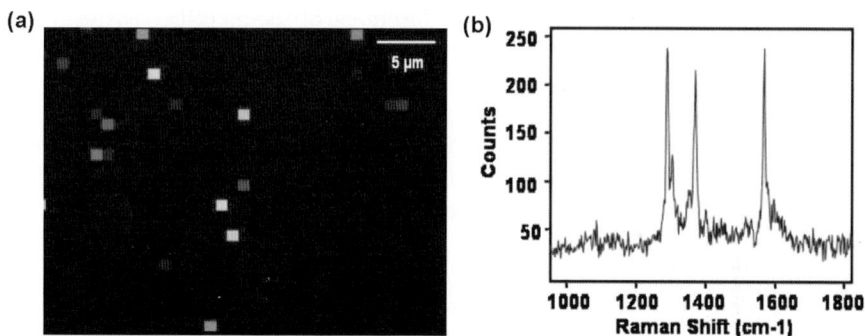

Fig. 8. (a) 2D SERRS intensity map of mixed monolayer containing 1 PTCD molecule/μm^2 on a Ag-island film (514.5 nm excitation). (b) Single-molecule PTCD spectrum derived from mapping experiment

particle shapes, sizes, configurations, and compositions have been prepared through careful manipulation of experimental conditions such as concentration, temperature, pH, ionic strength, stabilizing agent, and more. The versatility of Ag and Au nanoparticles for incorporation into complex architectures with increased functionality, along with their powerful optical properties, bestow upon them a huge potential for inclusion into cutting-edge materials now being developed. It is interesting to note that much of the progress that has been seen in this field has been driven by the search for improved and specialized substrates for surface enhancement of optical signals.

In this section, we discuss recent work in our laboratory where we have employed Ag and Au nanoparticles, or nanocomposite films of these, as substrates for SERRS measurements. In particular, our focus will be on the colloidal SERRS of PTCD derivatives down to the SM level and SERRS on layer-by-layer (LbL) films of Ag nanowires and dendrimer.

2.2.1 SERRS from Ag and Au Metal Colloids

The ease of production, and strong enhancement, of Ag and Au colloidal solutions has long made them the most widely employed substrates for SERS/SERRS [40, 41, 42]. The first reports of single-molecule SERS [17] and SERRS [16] in 1997, both from colloidal Ag nanoparticles, have served only to further promote this trend. In fact, nearly all single-molecule Raman reports since have focused on the use of these substrates. The vast majority of these studies have employed dye molecules that have resonances with the excitation laser used. The double-resonance condition achieved in these studies, by tuning into the molecular electronic transition and also into the surface-plasmon resonance of the metallic nanoparticle, takes advantage of two separate enhancement mechanisms; the molecular resonance enhancement of RRS and the surface enhancement of SERS, and is termed SERRS. Together, these mechanisms provide the necessary enhancement of the inelastic scattering to make SM detection a common analytical technique.

In the work discussed here, we have employed Ag and Au colloidal solutions as substrates for the collection of ensemble and SM SERRS spectra of perylene dye derivatives [36, 43]. Silver colloids were prepared according to the citrate method reported by *Lee* and *Meisel* [44], and had a turbid gray appearance. The Au colloids, on the other hand, were prepared by the citrate method described by *García-Ramos* and coworkers, and were violet in color [45].

The surface-plasmon absorption spectrum for each of these colloidal solutions is shown in Fig. 9a with maxima at 408 nm and 528 nm, respectively. Figure 9b shows a representative SEM (scanning electron microscopy) image of these Ag citrate colloids, showing a relatively broad distribution of particle shapes and sizes. Finally, Fig. 9c shows the structure of one of the amine-terminated PTCD derivatives employed for this work, n-(n-butyl)-n'-(4-aminobutyl) perylene-3,4,9,10-tetracarboxylic acid diimide, or nBu-PTCD-$(CH_2)_4$–NH_2 for short. This molecule has a strong absorption in the visible, and the $\pi \rightarrow \pi^*$ transition of the PTCD chromophore, with its associated vibronic structure (0–0 at 528 nm, 0–1 at 493 nm, and 0–2 at 464 nm), is clearly in resonance with the 514.5 nm laser line used for excitation.

SERRS experiments on Ag colloids were carried out by transferring 50 μl of solutions of nBu-PTCD-$(CH_2)_4$–NH_2, with concentrations ranging between 10^{-3} and 10^{-12} M, into 3 ml of the sol. Spectra were then collected from solution directly using an oil immersion objective (NA 0.80), that focuses the laser to a scattering volume of about 1 pl. The results of these experiments are shown in Fig. 10a.

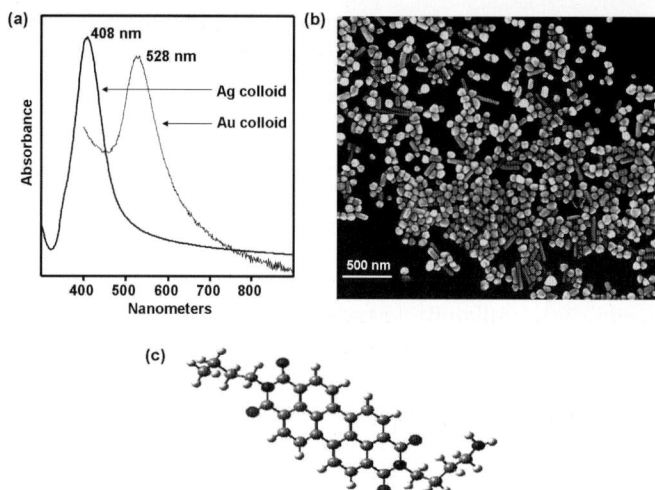

Fig. 9. (a) Surface-plasmon extinction spectra for Ag and Au citrate colloidal solutions. (b) SEM image of cast Ag citrate colloid. (c) Structure of nBu-PTCD-$(CH_2)_4$–NH_2

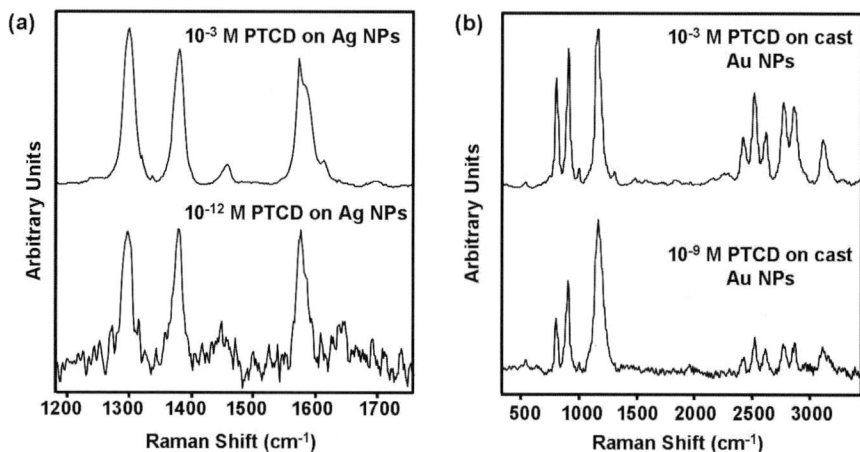

Fig. 10. (a) SERRS spectra collected from Ag citrate colloidal solution after addition of 10^{-3} M (*top*) and 10^{-12} M (*bottom*) nBu-PTCD-$(CH_2)_4$–NH_2 solution in CH_2Cl_2. (b) SERRS spectra collected from cast Au citrate colloidal solution incubated with 10^{-3} M (*top*) and 10^{-9} M (*bottom*) nBu-PTCD-$(CH_2)_4$–NH_2 solution in CH_2Cl_2. All spectra recorded using 514.5 nm laser excitation

A second set of experiments was carried out using Au colloids where the Au nanoparticle–analyte complex was cast onto glass slides for measurement. The sample was prepared by adding 40 μl of the analyte (10^{-3} M to 10^{-9} M in concentration) to 3 ml of gold colloidal solution. 20 μl of the solution was then cast onto a glass slide and left to dry before measurement. The scattering area of the 50× objective used is ca. 1 μm^2 and the number of molecules in this scattering area is estimated to be ca. 1 for the lowest concentration tested. The SERRS spectra on cast Au colloids are shown in Fig. 10b, where the fundamental ring-stretching vibrational frequencies of the PTCD moiety, together with overtone and combinations, are clearly seen in the average SERRS spectra down to the single-molecule spectrum.

2.2.2 SERRS from Silver Nanowire Layer-by-Layer Film Substrates

The development of improved substrates for enhancement of Raman signals remains one of the most critical challenges facing SERS/SERRS research today. Also, SERS/SERRS are increasingly being recognized as tools for the characterization of nanomaterials [46]. In this work, we fabricated unique, highly stable dendrimer/silver nanowire layer-by-layer (LbL) films for use as substrates for SERS/SERRS [47]. The layer-by-layer(LbL) technique, as a method for nanodimension film fabrication, was pioneered by *Decher* [48] and involves the electrostatic assembly of alternating positively and negatively charged layers. It allows for the controlled production of layered nanoarchitectures and represents a great potential for broadening the possibilities of surface-enhanced spectroscopy. In fact, several groups have now begun to incorporate metallic nanoparticles into LbL thin films [49, 50] and to examine the use of these films as substrates for SERS measurements [47, 51, 52]. Also, nanowires of Ag are particularly interesting for use in applications such as this due to the unique optical properties they possess as a result of their high aspect ratios. Some of the key results of this work are shown in Fig. 11.

The solution-phase synthesis of uniform nanowires of silver was carried out using the polyol method employed in the earlier report of *Sun* et al. [53]. The synthesized nanowires of diameters of ca. 100 nm and lengths of several micrometers, were characterized by their surface-plasmon absorption in the visible region of the electromagnetic spectrum, and the results are shown in Fig. 11a. The spectrum of the solution clearly shows a shoulder at 353 nm and a maximum at 383 nm, with the former being assigned to the transverse plasmon mode of the particles. To prepare the LbL films, clean glass substrates were immersed into a 0.1 g/l aqueous solution of DAB-Am 64-polypropylenimine tetrahexacontaamine dendrimer, generation 5, for 10 min, followed by rinsing with deionized water, drying with nitrogen gas, immersion in a Ag nanowire solution for 3 h, and finally drying with nitrogen gas. The surface-plasmon absorption spectrum for this film is also shown in Fig. 11a with two well-resolved maxima at 358 nm and 396 nm. These absorptions are

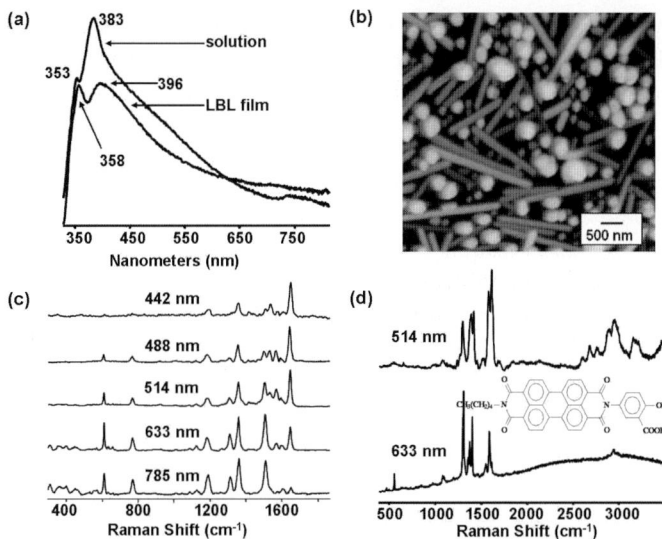

Fig. 11. (**a**) Surface-plasmon extinction spectra for Ag nanowire solution and Ag nanowire/G5 DAB-Am dendrimer LBL film. (**b**) AFM image of Ag nanowire/G5 DAB-Am dendrimer LBL film. (**c**) SERS/SERRS spectra of R6G on LBL substrate recorded using laser excitation at 442 nm, 488 nm, 514 nm, 633 nm, and 785 nm. (**d**) SERS/SERRS spectra of 1 : 10 mixed sal-PTCD/arachidic acid Langmuir–Blodgett monolayer on LBL substrate recorded using laser excitation at 514 nm and 633 nm. Also shown is the molecular structure of sal-PTCD. (Reproduced with permission from [47])

noticeably redshifted and broadened relative to those for the Ag nanowire solution.

The morphology, as revealed by AFM imaging, of the Ag nanowires self-assembled into LbL films is shown in Fig. 11b. Higher magnification images (not shown here) reveal lengthwise aggregation and also show the structures to be many-sided rather than smooth cylinders. This confirms previous results that demonstrated pentagonal cross sections for these wires [54]. The AFM images of the water-insoluble LbL film also show an abundance of metal nanostructures on its surface, making it an ideal candidate for SERS/SERRS measurements.

In this work, the ability of Ag nanowire/G5 DAB-Am dendrimer LbL substrates to provide SERS/SERRS was tested in two ways. First, a known volume (10 μl) of dilute rhodamine 6G (R6G) solution containing the dye molecule in picomole quantities was cast onto the LbL film and a section of the surface covered by the solution ($\sim 1\,\mu m^2$) was excited with several laser wavelengths to gather the inelastic scattering in a backscattering collection geometry. Second, a concentrated mixed monolayer of n-pentyl-5-salicylimidoperylene (sal-PTCD) and arachidic acid was transferred onto the water-insoluble LbL substrate using the Langmuir–Blodgett technique and probed similarly. This perylenetetracarboxylic diimide (PTCD) derivative is of the same series of molecules that we have been testing for single-molecule detection using SERRS in our laboratory [37, 38]. Both R6G and sal-PTCD have strong absorption profiles in the visible, and were selected to investigate the SERS/SERRS enhancement of the substrates. At excitation wavelengths of 442 nm, 488 nm, and 514 nm the laser is in resonance with the $\pi \rightarrow \pi^*$ electronic transition of the molecular systems and SERRS is observed. On the other hand, when excitation at 633 nm or 785 nm is used there is no resonance with the molecular system, and the resultant enhanced spectra are classified as SERS.

The SERS/SERRS spectra of R6G and salPTCD are shown in Fig. 11c and Fig. 11d, respectively. Notably, all laser excitation lines from the visible, starting at 442 nm, to the near-infrared, at 785 nm, provide strong SERS/SERRS with these substrates, as shown in Fig. 11c for the R6G molecule. The relative intensities of the peaks in the SERS/SERRS spectra vary as the laser excitation is brought in and out of resonance with the molecular electronic transition. The SERRS spectra presented in Fig. 11d are from a single monolayer, and represent the detection of less than 1 attomol of sal-PTCD molecules. The change in the relative intensities of the vibrational bands with different excitation energies is once more the result of molecular resonance with the dye molecules at 514.5 nm and not at 633 nm. The clearest spectral difference observed between the results from these two excitation lasers, however, is that the SERRS spectrum recorded at 514 nm shows strong overtone and combination bands above $2500 \, \text{cm}^{-1}$ while the nonresonant SERS spectrum does not.

2.3 SEF of LB

Finally, we provide an illustrative example of how SEF can be used as a tool to further understand the impact that molecular organization has on the optical properties of thin films. In this example, the film-packing properties in LB films, of a new class of PTCD derivatives, were investigated [55]. How these molecules organize themselves on the nanoscale is of vast importance as it has a direct impact on the film's optical properties, in particular its electronic absorption and emission. In this set of molecules, the side chains of the main PTCD chromophore were modified with alkyl chains of increasing length C_n ($n = 5, 10, 12, 15$). Langmuir–Blodgett films for each molecule were fabricated on glass substrates, with, and without, a silver-island film. By depositing these monolayers onto an enhancing surface, it is possible to observe SEF, and the spectral properties of the resulting SEF spectrum can provide insights into how the molecules orient themselves with respect to the metal. Some key results obtained in this work are shown in Fig. 12.

It was observed that larger SEF enhancement was correlated with increasing alkyl chain length, with $n = 15$ displaying a 2-fold increase over $n = 5$.

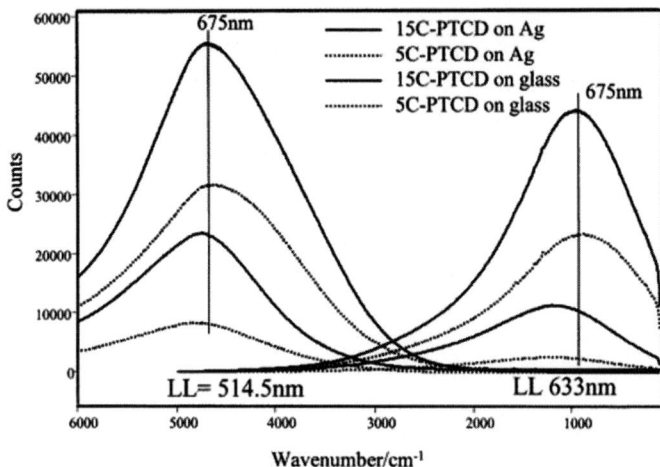

Fig. 12. Fluorescence spectra for monolayer LB films of 5C-PTCD and 15C-PTCD on silver-island films and glass using 514.5 nm and 633 nm laser lines. (Reproduced with permission from [55])

Both the evidence of SEF and this trend provide insight into how the molecule must be oriented relative to the substrate's surface and therefore how it must be packed in the LB monolayer. The fact that the SEF enhancement factor is greater than unity indicates that the molecule must not be lying directly on the surface. Since the excimer emission is observable, the molecule must be positioned in such a way so as to reduce contact with the metal surface. This can be rationalized by the molecule being orientated "head on" to the surface with the long axis of the chromophore perpendicular to the metal. This is further supported by the apparent trend in increasing SEF enhancement with increasing alkyl chain length. By lengthening the alkyl chain the main chromophore is, in effect, being distanced from the metal surface.

3 Conclusion

In this Chapter, we have described several applications of the enhancement of resonance Raman scattering and molecular fluorescence toward the characterization of single molecules, monolayers, and thin solid films on metal-island films, colloidal metal nanoparticles, and nanocomposite films. While this review is by no means extensive, we believe it does serve to shed light on some of the more critical considerations of this work, and also to highlight some of the very powerful potential uses of resonant surface-enhanced spectroscopies.

Acknowledgements

Financial assistance from the Natural Science and Engineering Research Council of Canada (NSERC) is gratefully acknowledged.

References

[1] A. B. Myers: J. Raman Spectrosc. **28**, 389 (1997)
[2] J. Gersten, A. Nitzan: J. Chem. Phys. **75**, 1139 (1981)
[3] M. Moskovits: Rev. Mod. Phys. **57**, 783 (1985)
[4] C. Constantino, J. Duff, R. Aroca: Spectrochim. Acta, Part A **57A**, 1249 (2001)
[5] R. R. Chance, A. Prock, R. Silbey: Adv. Chem. Phys. **37**, 1 (1978)
[6] A. Mooradian: Phys. Rev. Lett. **22**, 185 (1969)
[7] K. Imura, T. Nagahara, H. Okamoto: J. Am. Chem. Soc. **126**, 12730 (2004)
[8] G. T. Boyd, Z. H. Yu, Y. R. Shen: Phys. Rev. B **33**, 7923 (1986)
[9] K. L. Kelly, E. Coronado, L. L. Zhao, et al.: J. Phys. Chem. B **107**, 668 (2003)
[10] S. Link, M. A. El-Sayed: Annu. Rev. Phys. Chem. **54**, 331 (2003)
[11] M. B. Mohamed, V. Volkov, S. Link, et al.: Chem. Phys. Lett. **317**, 517 (2000)
[12] P. P. Shorygin, L. L. Krushinskij: J. Raman Spectrosc. **28**, 383 (1997)
[13] M. G. Albrecht, J. A. Crieghton: J. Am. Chem. Soc. **99**, 5215 (1977)
[14] D. L. Jeanmaire, R. P. Van Duyne: Electroanal. Chem. **84**, 1 (1977)
[15] K. Kneipp, H. Kneipp, I. Itzkan, et al.: Chem. Rev. **99**, 2957 (1999)
[16] S. Nie, S. R. Emory: Science **275**, 1102 (1997)
[17] K. Kneipp, Y. Wang, H. Kneipp, et al.: Phys. Rev. Lett. **78**, 1667 (1997)
[18] R. Aroca, T. Del Cano, J. A. de Saja: Chem. Mater. **15**, 38 (2003)
[19] R. E. Clavijo, D. Battisti, R. Aroca, et al.: Langmuir **8**, 113 (1992)
[20] E. Johnson, R. Aroca, J. Pahapill: J. Mol. Struct. **293**, 331 (1993)
[21] R. Aroca, N. Pieczonka, A. P. Kam: J. Porphyrins Phthalocyanines **5**, 25 (2001)
[22] R. Aroca, U. Guhathakurta-Ghosh: J. Am. Chem. Soc. **111**, 7681 (1989)
[23] U. Guhathakurta-Ghosh, R. Aroca, R. O. Loutfy, et al.: J. Raman Spectrosc. **20**, 795 (1989)
[24] R. Aroca, C. Jennings, G. J. Kovac, et al.: J. Phys. Chem. **89**, 4051 (1985)
[25] C. Jennings, R. Aroca, A. M. Hor, et al.: Anal. Chem. **56**, 2033 (1984)
[26] R. Aroca, R. O. Loutfy: J. Raman Spectrosc. **12**, 262 (1982)
[27] R. F. Aroca, C. J. L. Constantino: Langmuir **16**, 5425 (2000)
[28] R. F. Aroca, C. J. L. Constantino, J. Duff: Appl. Spectrosc. **54**, 1120 (2000)
[29] C. J. L. Constantino, R. F. Aroca: J. Raman Spectrosc. **31**, 887 (2000)
[30] C. J. L. Constantino, R. F. Aroca, C. R. Mendonca, et al.: Adv. Funct. Mater. **11**, 65 (2001)
[31] A. P. Kam, R. Aroca, J. Duff: Chem. Mater. **13**, 4463 (2001)
[32] T. del Cano, P. J. G. Goulet, N. P. W. Pieczonka, et al.: Synth. Met. **148**, 31 (2005)
[33] C. J. L. Constantino, T. Lemma, P. A. Antunes, et al.: Spectrochim. Acta, Part A **58A**, 403 (2002)
[34] C. J. L. Constantino, T. Lemma, P. A. Antunes, et al.: Anal. Chem. **73**, 3674 (2001)

[35] C. J. L. Constantino, T. Lemma, P. A. Antunes, et al.: Appl. Spectrosc. **57**, 649 (2003)
[36] T. Lemma, R. F. Aroca: J. Raman Spectrosc. **33**, 197 (2002)
[37] P. J. G. Goulet, N. P. W. Pieczonka, R. F. Aroca: Anal. Chem. **75**, 1918 (2003)
[38] P. J. G. Goulet, N. P. W. Pieczonka, R. F. Aroca: J. Raman Spectrosc. **36**, 574 (2005)
[39] P. Goulet, N. Pieczonka, R. Aroca: Can. J. Anal. Sci. Spectrosc. **48**, 146 (2003)
[40] X.-M. Dou, Y. Ozaki: Rev. Anal. Chem. **18**, 285 (1999)
[41] P. Mulvaney: Langmuir **12**, 788 (1996)
[42] R. F. Aroca, R. A. Alvarez-Puebla, N. Pieczonka, et al.: Adv. Colloid Interf. Sci. **116**, 45 (3)
[43] B. Tolaieb, C. J. L. Constantino, R. F. Aroca: Analyst (Cambridge, UK) **129**, 337 (2004)
[44] M. Lee, D. Meisel: J. Phys. Chem. **86**, 3391 (1982)
[45] L. E. Camafeita, S. Sánchez-Cortés, J. V. García-Ramos: J. Ram. Spectrosc. **26**, 149 (3)
[46] D. Roy, J. Fendler: Adv. Mater. **16**, 479 (2004)
[47] R. F. Aroca, P. J. G. Goulet, D. S. Dos Santos, Jr., et al.: Anal. Chem. **77**, 378 (2005)
[48] G. Decher: Science **277**, 1232 (1997)
[49] K. Esumi, T. Hosoya, A. Suzuki, et al.: J. Colloid Interface Sci. **226**, 346 (2000)
[50] V. Salgueirino-Maceira, F. Caruso, L. M. Liz-Marzan: J. Phys. Chem. B **107**, 10990 (3)
[51] X. Li, W. Xu, J. Zhang, et al.: Langmuir **20**, 1298 (2004)
[52] P. J. G. Goulet, D. S. Dos Santos, Jr., R. A. Alvarez-Puebla, et al.: Langmuir **21**, 5576 (3)
[53] Y. G. Sun, B. Gates, B. Mayers, et al.: Nano Lett. **2**, 165 (2002)
[54] A. Tao, F. Kim, C. Hess, et al.: Nano Lett. **3**, 1229 (2003)
[55] P. A. Antunes, C. J. L. Constantino, R. F. Aroca, et al.: Langmuir **17**, 2958 (2001)

Index

Tip-Enhanced Raman Spectroscopy (TERS)

Bruno Pettinger

Fritz Haber Institute of the Max Planck Society, Faradayweg 4–6, D-14195, Berlin, Germany
pettinger@fhi-berlin.mpg.de

1 Introduction

The understanding and knowledge of interfaces is of vital importance for science and technology. Yet, their investigation represents a great challenge still today. It becomes more and more important to achieve information at the level of individual structural blocks of the interfacial region. For the full understanding of interfacial processes, one needs both structural and chemical information at the atomic level. One example for developments in this direction is the advent of the scanning probe microscopies (SPM) [1, 2, 3, 4]. They permit monitoring of surface structures and even their changes in time with atomic resolution. While initially the SPM techniques could not reveal the chemical nature of atoms seen in SPM images, recent development, leading to ISTS (inelastic scanning tunneling spectroscopy), opens the view to the chemical nature of surface atoms and molecules [5, 6]. A major limitation in these experiments is the need for UHV conditions and cryogenic temperatures for the necessary stability of the sample–probe arrangement [6].

IST spectroscopy belongs to the class of vibronic spectroscopies, as well as Raman or infrared spectroscopy. The latter two are optical spectroscopies and provide information on the chemical nature of the investigated species. However, the information (spectrum) is usually averaged over a large ensemble and can, therefore, contain only a certain amount of structural information. While initially Raman spectroscopy was not considered to be sensitive enough to become a suitable tool for interfacial studies (in contrast to infrared spectroscopy), detections and developments in the last decades and even more in the last several years has turned this picture upside down: Due to the availability of huge surface enhancement(s), Raman spectroscopy belongs to the most sensitive vibronic spectroscopies today[1] [7, 8, 9, 10, 11].

The field of surface-enhanced Raman spectroscopy (SERS) started in 1973 with the famous papers of *Fleischmann* et al. [12] and *Jeanmarie* and *Van Duyne* [13], giving rise to more than 4000 papers in the following decades. Most recent key words for intriguing approaches in the SERS field are "hot

---

[1] Certainly, this will hold also for the next several years. The sensitivity of STS is similar; however, its spectral resolution is comparatively poor.

K. Kneipp, M. Moskovits, H. Kneipp (Eds.): Surface-Enhanced Raman Scattering – Physics and Applications, Topics Appl. Phys. **103**, 217–240 (2006)
© Springer-Verlag Berlin Heidelberg 2006

spots" and "single-molecule SERS" (this book will contain a number of contributions for both of these hot fields).

It is not necessary to present the case of surface-enhanced Raman spectroscopy in detail in this Chapter. What we would like to address here is the general problem of SERS: (Large) enhancements for adsorbates are only found for the coinage metals (copper, silver and gold) and only if these metals have nonplanar surfaces[2]. There is a commonly accepted explanation for these observations: One of the enhancement mechanisms (denoted as *electromagnetic enhancement*) is correlated with the excitation of surface plasmons within small metal structures [14]. Excited surface plasmons are associated with collective electron oscillations in the surface region; they create a localized electromagnetic field with a field strength often much higher than that of the incident wave. In a sense, the nonplanar surface structures act as an antenna for the amplification of the incident *and* radiated electromagnetic fields. The second enhancement mechanism is denoted as the *chemical enhancement* mechanism and is often considered as a kind of resonance Raman enhancement for adsorbate–substrate complexes that provide new or modified orbitals for (pre-)resonant excitation of the Raman processes [15].

The general obstacle for SERS is that adsorption or chemisorption, electromagnetic and chemical enhancement are strongly interconnected. Adsorption and chemisorption can change the (local) surface structure, with impacts on the chemical, but also on the electromagnetic enhancement (particularly if very small structures in the nanometer regime are responsible for huge enhancements). In recent years, a powerful concept was proposed: The concept of a few "hot spots" making up most of the SER signal while the rest of the surface is comparatively SERS inactive [16, 17, 18]. Consensus has been reached that nanostructures or nanoparticles approaching each other very closely provide the scenario for such hot spots: The interstitial region between the particles provide new, very localized modes of surface plasmons with extraordinary field strengths. Hence, these regions produce the strongest enhancements. The field of *single-molecule SERS* is based on this fact (see other Chapters in this book) [14]. In a sense, the lightning-rod effect [19] can be considered as a precursor of the concept of hot spots [20], as both concepts describe a sort of focusing of field energy into narrow spatial regions.

As a consequence of the hot-spot concept, one could have considered to create a single hot spot, for example, by moving a sharp protrusion or a tip made of silver or gold towards the sample under investigation. Illuminating the whole unit should also produce enhanced Raman scattering. In contrast, the development of SERS towards tip-enhanced Raman spectroscopy (TERS) went along a different route (see the next Chapter). In fact, the use of the hot spot in terms of an external device was developed in the SNOM field (scanning

---

[2] Periodic and nonperiodic structures support SERS; to the latter belong roughened surfaces and fractal structures. SERS is also observed for small single spheres as well as for aggregated nanoparticles.

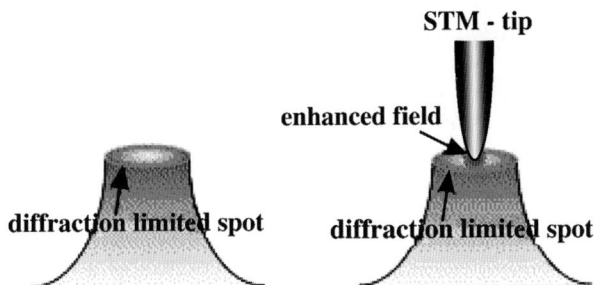

Fig. 1. Conceptual figure of (**a**) normal micro-Raman detection (far-field detection) as compared to (**b**) near-field Raman detection by a metallic probe tip. (Reprinted with permission from [21])

near-field optical microscopy) and was first denoted as apertureless SNOM (a-SNOM).

Nevertheless, once the TERS approach had been born, it was quickly understood that TERS belonged to one of the most intriguing developments in optical spectroscopy [21, 22, 23, 24, 25, 26, 27, 28, 29, 30, 31, 32, 33, 34]. It uses an illuminated scanning probe tip of a scanning probe device to greatly enhance the Raman scattering from the sample underneath the tip. In this way, Raman spectroscopy and scanning probe microscopy are married, forming a new spectroscopic–microscopic tool with hitherto unprecedented spectroscopic and spatial sensitivity. Suitable tips are either silverized AFM tips or STM tips made of a thin wire of silver or gold. As illustrated in Fig. 1, upon illumination of the tip with a sufficiently focused laser beam, localized surface plasmons are excited mainly near the tip apex leading to an enhanced electromagnetic (em) field in a very small region near or underneath the tip apex. It has a radius r_{tip} that is usually much smaller than the radius of the focal region r_{focus}. In TERS experiments, one can measure both the Raman scattering intensities in the absence ($I_{\mathrm{RS,\ tip\ retracted}}$) and in the presence of the tip ($I_{\mathrm{RS,\ tip\ down}}$). The ratio of the two intensities is often $\gg 1$, but in many cases < 100. Thus, the TERS enhancement may reach values between two and four orders of magnitude, which is sufficient to monitor TERS for strong Raman scatterers such as dyes. Compared to theoretically predicted enhancement factors of more than 10^{10} and in view of giant enhancements reported for single-molecule SERS ($\geq 10^{14}$), TERS shows a rather moderate enhancement range. Obviously, one can expect much higher TERS signals if the experimental conditions are improved with respect to more efficient excitation and radiation of localized surface plasmons.

In this contribution, we would like to sketch the developments in the field of TERS from 2000 to 2005. During these years, a number of advances have taken place. A striking one is an enhancement three or four orders of magnitude higher than in the initial reports that allows monitoring of even small molecules adsorbed at single-crystalline samples.

The present chapter is organized as follows: After this introduction, also briefly sketching SERS and the principles of TERS (Sect. 1), Sect. 2 will report a number of TERS studies, including our own recent investigations with single-crystalline substrates. Section 3 gives a short outlook.

2 TERS Results

Before beginning with the reports on tip-enhanced Raman scattering, we would like to comment on surface-enhanced resonance Raman scattering (SERRS), relevant for both the field of single-molecule SERS and TERS, when using strong Raman scatterers such as dyes.

2.1 Remarks on SERRS

Particularly huge surface enhancements are observed for dye molecules that show a strong absorption in the visible frequency regime [35, 36, 37], among them rhodamine 6g or crystal violet. Adsorbed, for example, at aggregated Ag colloids, they show extremely large total enhancements [38, 39, 40, 41]. In such systems, the differential cross sections for SERRS can easily be measured to be around $(\mathrm{d}\sigma/\mathrm{d}\Omega)_{\mathrm{SERRS}} \simeq 10^{-18}\,\mathrm{cm}^2\cdot\mathrm{sr}^{-1}$ for individual Raman lines [39, 40, 41]. The total Raman cross section including an intense SERS background amounts to about $\sigma \simeq 10^{-16}\,\mathrm{cm}^2$. It is substantially more difficult, however, to determine the actual enhancement factors directly, because the intense fluorescence obstructs any resonance Raman measurement of the free rhodamine 6g molecules excited around 514.5 nm. Fortunately, the differential cross section for normal and resonance Raman scattering (RRS) of these molecules can be roughly estimated. As is well known, in the absence of optical resonance, the polarizability and, hence, the differential cross section scale linearly with the molecular volume or the number of electrons in a molecule. Thus, based on the low cross section of simple gases [42], the dye's cross section for normal Raman scattering (NRS) may be estimated as $(\mathrm{d}\sigma/\mathrm{d}\Omega)_{\mathrm{NRS}} \simeq 10^{-28}\,\mathrm{cm}^2\cdot\mathrm{sr}^{-1}$. Obviously, the total enhancement factor (surface × resonance Raman enhancement) is $F_{\mathrm{SERRS}} \approx 10^{10}$. For the $1184\,\mathrm{cm}^{-1}$ vibrational mode of rhodamine 6G adsorbed at glassy carbon (for this system no electromagnetic (em) enhancement is expected [43]), *Kagan* and *McCreery* reported a differential cross section of $(\mathrm{d}\sigma/\mathrm{d}\Omega)_{\mathrm{RRS}} \simeq 10^{-25}\,\mathrm{cm}^2\cdot\mathrm{sr}^{-1}$. This indicates the presence of a classical RRS enhancement of 10^3 in accordance with the usually expected 10^2- to 10^5-fold resonance Raman enhancement for free dye molecules [44]. Consequently, for the colloidal solutions considered here, one estimates the (average) em enhancement factor as about 10^7. If one takes into account that only a small fraction of dye molecules ($\sim 1\%$) is adsorbed within the interstitial volume of two particles that presumably is the location of giant field enhancement, the em enhancement factor is "only" around 10^9. Although the cross sections presented above are comparable to these recently reported in the field of single-molecule SERS, the estimated em enhancement ($\sim 10^9$) is rather conservative compared to values reportedly one to four orders of magnitude higher [8, 14, 45]. Please note that in the above estimate, the contribution of resonance Raman processes to the total enhancement is most likely underestimated. Yet, the remaining electromagnetic enhancement of around 10^9 remains huge. Important in this context is that localized surface-plasmon (LSP) resonances and a few (or even single), light-absorbing dye molecules adsorbed in the high-field zone must show mutual interactions: For dyes showing bleaching behavior such as rhodamine 6g excited at 488 nm or 514.5 nm, the enhanced field provided by the LSP should result in an enhanced rapid bleaching rate. In turn, the higher the field enhancement due to LSP resonances, the more sensitive these resonances must respond to light-absorbing species. These effects seem to be (nearly) absent in the field of single-molecule SERS, while a few reports (in somewhat different fields) show a decrease of the enhancement with increasing dye coverage for colloidal systems and a rapid dye bleaching in the case of TERS (this effect will be addressed below).

2.2 Initial Steps in TERS

The Zenobi group put substantial effort into improving the SNOM approach and adapting Raman spectroscopy to it. One key problem has been, and still is, the insensitivity of the tapered fiber tip that reduces the transmission of both the incident and scattered light. Although the choice of a suitable tip shape and its metallization have substantially improved the throughput of light [46], there still is an unavoidable limitation to this technique: The higher the required spatial resolution of this technique, the narrower the aperture of the tip has to be; this, however, reduces the field strength and the area from which the signal is collected. In other words, the SNOM resolution cannot be much better than 50 nm. To overcome this limitation, *Stöckle* et al. [22] designed an apertureless SNOM (a-SNOM) by using a silverized AFM tip with a granular silver film of about 20 nm thickness. Under illumination of the AFM tip, one of the particles at the tip apex acts as an antenna for the incident light and permits the excitation of surface plasmons. This leads to an increased local electromagnetic field. If the illuminated AFM tip is in (close) contact with the sample, an enhanced light scattering is observed. In fact, *Stöckle* et al. [22] reported a large gain in the Raman scattering intensity[3] for a dye film (brilliant cresyl blue) deposited on a glass substrate if a metalized AFM tip is set to close contact with this sample. Figure 2 reproduces the upper part of Fig. 1 in [22]. The two spectra, denoted as curves *a* and *b*, refer

---

[3] As indicated by the insert in Fig. 2, an inverted microscope objective has been used to focus the laser beam through the glass slide covered by a dye film onto the AFM tip.

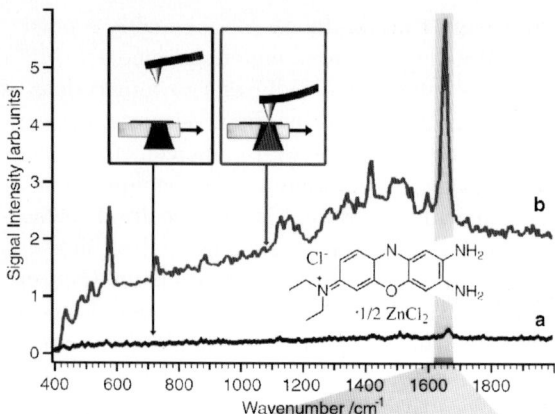

Fig. 2. Tip-enhanced Raman spectra of brilliant cresyl blue (BCB) dispersed on a glass support and measured with a silver-coated AFM probe. The two Raman spectra were measured with the tip retracted from the sample (**a**) and with the tip in contact with the sample (**b**), respectively. (Reprinted with permission from [22])

to the tip retracted from and in contact mode with the sample, respectively. The authors report a more than 30-fold net increase[4] of the Raman signal by the tip. Taking into account the small area of the enhanced field underneath the tip of less than 50 nm in diameter, a 2000-fold enhancement of Raman scattering in the vicinity of the tip has been deduced. Similar effects were reported for C_{60} molecules and an electrochemically etched Au tip with an apex of 20 nm in diameter (this tip is mounted in a shear-force setup). In addition, the authors also illustrated that the spatial resolution of this approach was about 55 nm [22].

A similar approach was presented by *Nieman* et al. [26]. They used an illuminated AFM tip covered with a 100 nm gold film to enhance Raman scattering from samples of polydiacetylene para-toluene sulfonate and two-photon-induced fluorescence from crystallites of coumarin I dye. An example of the latter effect is shown in Fig. 3. In this experiment, coumarin I was dried on a coverslip. A mode-locked Ti:Sa laser served as the excitation source with a 200 fs pulse width, an excitation line around 790 nm and a power of less than 100 μW. In order to ensure that a tip enhancement occurs, the fluorescence signals (recorded in the wavelength region of 535 nm to 565 nm) were recorded for polarizations with major components perpendicular or parallel to the tip axis.

According to Fig. 3, the fluorescence signal for parallel polarization is about twice as large as for the perpendicular polarization. This suggests

---

[4] By net increase we mean the n-fold ($f_{n\mathrm{gain}}$) increase of the overall Raman signal in the presence of the tip compared with the Raman intensity observed in the absence or retraction position of the tip.

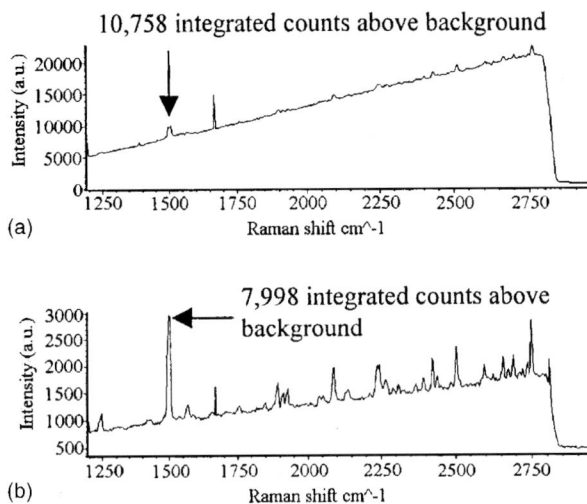

Fig. 3. Fluorescence spectra taken with a LN–CCD camera in the reflection path with an exposure time of 2 min. During acquisition, the gold-coated tip was imaging in noncontact mode with a scan rate of 0.3 Hz over a 5 μm × 5 μm area. (**a**) Incident beam polarization perpendicular to the plane of the tip. (**b**) Incident beam polarization parallel to the plane of the tip. The integrated peak intensity from polarization parallel to the plane of the tip is a factor 2.4 larger than from polarization perpendicular to the plane of the tip. (Reprinted with permission from [26])

a strong enhancement of the two-photon-induced fluorescence by the gold-covered AFM tip, because there is no influence of the polarization on the fluorescence signal in the absence of the tip. The authors observed images produced by the near-field of the illuminated tip with a resolution of about 100 nm.

Hayazawa compared far-field- and near-field-excited Raman scattering for rhodamine 6g on a silver-island film. A silverized cantilever tip was used with a coating thickness of 40 nm. The Ag-island film was evaporated on a thin glass slide that was brought into optical contact with the objective (NA = 1.4) using an immersion oil. Both the excitation of the Raman scattering and the recording of the emitted radiation were performed with the same objective. Figure 4 shows the near-field spectrum of rhodamine 6g, denoted as spectrum (a), obtained as the difference between spectra recorded in the presence and absence of the AFM tip. For comparison, a spectrum is shown where a pure cantilever is used.

This shows that only the thin silver film, i.e., only one of the silver particles at the tip apex, generates the enhancement again by the excitation of localized surface plasmons that produce an enhanced field near the tip apex. Hayazawa et al. estimate a 40-fold enhancement for Raman scattering. It is noteworthy that the dye on the silver-island film already showed a surface

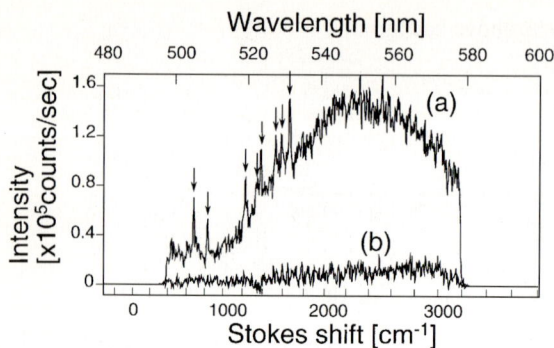

Fig. 4. (a) Near-field spectrum corresponding to the difference of spectra with and without a cantilever at a tip–sample distance of 0 nm. **(b)** The same as **(a)**, but with a silicon cantilever. The laser power is 0.5 mW and the exposure time is 5 s. (Reprinted with permission from [24])

enhancement; in the presence of the tip, a further 40-fold enhancement could be achieved. In a later paper, Hayazawa et al. presented an interesting experiment: Two different dyes, rhodamine 6g and crystal violet, were deposited on a substrate and near-field Raman images and topographic AFM images were recorded. These images are reproduced in Fig. 5, each showing an area of $1\,\mu m \times 1\,\mu m$. To identify the two dye molecules and to differentiate between them, two characteristic Raman bands were used: 1. The $607\,cm^{-1}$ line of the C–C–C in-plane bending mode of rhodamine 6g for Fig. 5a, and 2. the $908\,cm^{-1}$ line of the C–H out-of-plane bending mode of crystal violet for Fig. 5b. A third line, the C–H inplane bending mode of crystal violet at $1172\,cm^{-1}$, is used to record another near-field Raman image (Fig. 5c) of the crystal violet distribution on the substrate. The fourth figure (Fig. 5d) shows a topologic AFM image that was simultaneously recorded with the image of Fig. 5a.

By using the different Stokes-shifted Raman lines, the distribution of the two molecules within the imaged area can be monitored, which is impossible with the topological image [21]. Although the resolution and the differentiation between the two molecules are limited, these results show the high potential of TERS for providing chemical information about adsorbed species with a spatial resolution in the nanometer region.

As shown above, in the first TERS studies, strong Raman scatterers were studied such as brilliant cresyl blue (BCB), fullerenes, sulfur layers or rhodamine 6g and crystal violet. In these cases, the high Raman cross sections allowed measurements of both the Raman scattering in the absence and in the presence of the tip. From the ratio of these signals, one can evaluate the net gain observed in these experiments. In the publications on TERS between 2000 and 2003, the reported net gains range between 1.5 and 40, which is associated with a local enhancement of Raman scattering by about

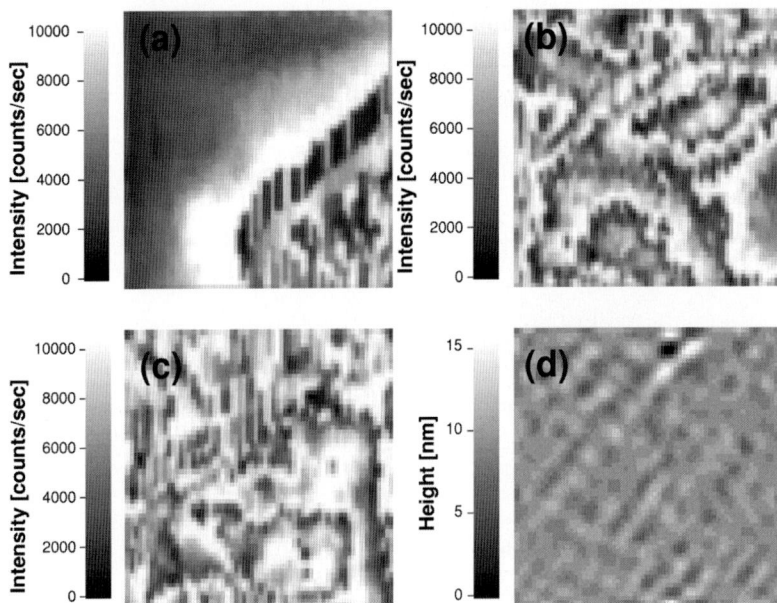

Fig. 5. Near-field Raman images obtained at (**a**) 607 cm^{-1} for the C–C–C inplane bending mode of rhodamine 6g, (**b**) 908 cm^{-1} for the C–H out-of-plane bending mode of crystal violet, and (**c**) 1172 cm^{-1} for the C–H inplane bending mode of crystal violet. (**d**) is the corresponding topologic image. Recording time: 10 min for one image consisting of 64 by 64 pixels covering a 1 μm × 1 μm scanning area. (Reprinted with permission from [21])

two to four orders of magnitude. In most cases, AFM tips covered with Ag or Au films were used for the TERS experiments. A wide variety of samples indicate the general applicability of this approach. Let me just list a few examples: Thin BCB films on glass slides [22], or C$_{60}$ molecules on glass or quartz slides [22, 23], thin sulfur films on quartz [23], rhodamine 6g on a thin Ag film [21, 24], tip-induced SERS of a polydiacetylene (PDA–PTS) [26], diamond particles on glass [47], single-walled carbon nanotubes on glass [28].

STM devices implemented for TERS are less common, probably for one reason: One needs a conductive surface, and, hence, TERS using an STM device seems to be a less-general approach. However, this approach has its own merits: It provides an excellent distance control, and, if a suitable metal substrate is used, it has a higher spatial resolution according to theory. An additional advantage is a very efficient fluorescence quenching for fluorescent species adsorbed at a metal substrate. This latter property permits us to work even in spectral regions that otherwise would be inaccessible due to an overwhelming fluorescence. *Pettinger* et al. [25, 27] reported on TERS for BCB adsorbed on a thin, smooth gold film and exciting the Raman process

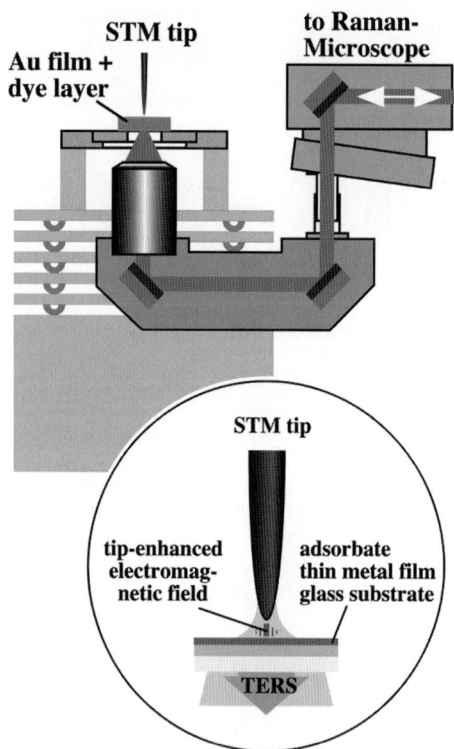

Fig. 6. Experimental scheme of the TERS experiment. Please note that the laser beam is passed through the thin metal film to excite the optical modes of the tip–metal junction. Light scattered by the adsorbate is collected in the backscattering mode and also passed through the thin metal film. (Reproduced with permission from [27])

with a He–Ne laser line at 632.8 nm. Figure 6 shows the TERS setup using the inverted microscope approach.

The Raman system is a LabRam1000 from Dilor that consists of a microscope stage attached to a Raman spectrograph. A U-turn inverts the light beam that is then focused on the sample through a glass slide and through a thin metal film by an inverted microscope objective of 50× or 100× magnification. The scattered light is collected in the backscattering mode. It is then passed through a cinematic notch filter system to the spectrograph and dispersed on a liquid-nitrogen-cooled CCD camera. An in-house-built STM is mounted on an x–y-stage driven by piezoelements in order to bring the tunneling tip into the illuminated focal spot. During the experiments, the bias voltage is kept at 50 mV with a constant tunneling current of 1 nA. In these TERS experiments, silver tips were made from 0.25 mm silver wire (99.998 % purity, Alfa Aesar) by electrochemical etching in perchloric acid, ethanol and

200 μm

Etched Ag tip (2)

~200 nm

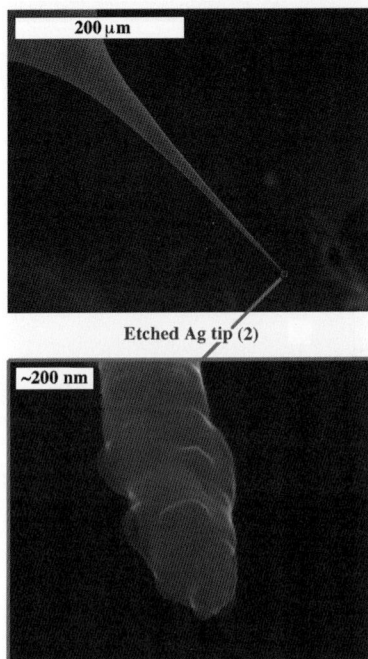

Fig. 7. Images of an electrochemically etched silver tip. The etching procedure is described in the text. (Reprinted with permission from [27])

water (Streuers) using a gold ring as a counter electrode that results in tip radii of about 100 nm (see Fig. 7) [27].

As seen from the fluorescence spectrum in Fig. 8 measured for a BCB layer deposited on glass, the laser line lies within the fluorescence region of this dye (its maximum is around 660 nm). If the dye is adsorbed on a thin Au film (12 nm evaporated on a glass slide) in monolayer or submonolayer quantities, no fluorescence is observable, but Raman scattering is seen (see curve 1 in Fig. 9). This Raman scattering is rather weak for two reasons: The coverage is a monolayer or less, and the gain in signal provided by the resonance Raman effect is about three orders of magnitude. Therefore, only the strongest peak (here at about 540 cm^{-1} due to the excitation in the red at 633 nm) is discernible while the other BCB bands are more or less hidden in the noise. If, however, the STM tip is moved into the laser focus and into a tunneling position, a significant rise in the overall Raman signal is observed with a net increase of about $f_{n\text{gain}} \sim 16$ for all Raman lines (curve 2 in Fig. 9).

Based on a tip radius of about 80 nm and a focal radius of 500 nm (100× magnification), the TERS enhancement in these experiments amounts to about three orders of magnitude ($f_{\text{TERS}} = f_{n\text{gain}} * (r_{\text{focus}}/r_{\text{tip}})^2 \sim 630$).

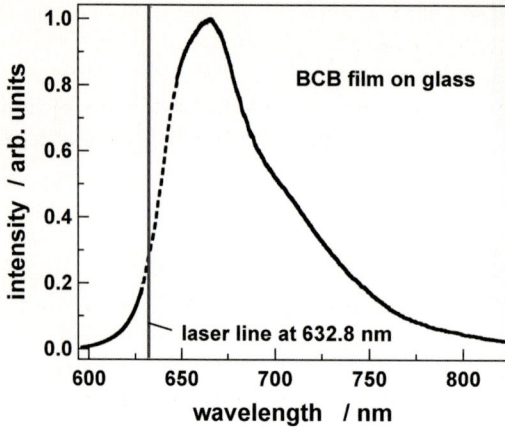

Fig. 8. Fluorescence spectrum of brilliant cresyl blue, recorded in backscattering mode. (Reprinted with permission from [27])

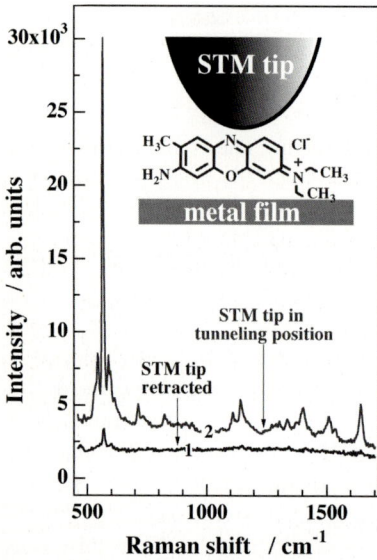

Fig. 9. Resonance Raman scattering and TERS for brilliant cresyl blue (BCB) on a smooth gold film (thickness $d = 12$ nm). (Reprinted with permission from [27])

The weak points of these initial TERS studies are: 1. The polarization of the laser light perpendicular to the tip axis (which is less efficient for surface-plasmon excitations)[5], 2. both the incident and scattered light having to pass the metal film, 3. the adsorbates are possibly not in a favorable adsorption geometry, and 4. tips made of silver, because they are not inert enough, as long as one works in air.[6]

2.3 Towards Higher Efficiencies in TERS

The above-discussed point forced us to also test tips made of gold, a material well known for its inertness.[7] In our experiments, Au tips yielded similar enhancements as Ag tips. However, SEM images showed that electrochemically etched tips were often very rough. The experience with SERS tells us, however, that roughened tip surfaces provide a number of locations for local surface-plasmon (LSP) excitations with varying enhancements, but without any guarantee that the one particle at the end of tip apex is the most suitable one for TERS. In contrast to that, the presumably optimal and single location of LSP excitation would be the small tip apex of a smooth tip with a sufficiently narrow tip cone. Indeed, theoretical studies indicate huge enhancements for rather narrow tips, with field enhancements of 300 to more than 1000. According to the famous g^4 law, this should lead to about a 10^{10}- to more than 10^{12}-fold TERS enhancements [48, 49]. Obviously, there is a large gap between the theoretically predicted and the (so far) experimentally reported TERS enhancements of 10^4. To close this gap, tips with improved performance are needed. In other words, we were searching for a tip preparation procedure that leads to sharp tips with a smooth surface. In a detailed study, *Ren* et a. [34] developed an electrochemical etching procedure that meets the requirement just mentioned. It is a rather simple procedure that requires only a beaker filled with a 1 : 1 mixture of fuming HCl and ethanol as shown in Fig. 10. The Au wire is dipped into the etching solution with an immersion length of the wire of 2 mm to 3 mm. The etching is performed by applying a potential of 2.4 V between the Au wire and a gold ring counter electrode [34].

It is important that the mixture of fuming HCl and ethanol leads to a reduction of bubbling in the solution, and the high electrode potential to a fast and more homogeneous removal of material. For both parameters, the mixing and the electrode potential, optimal conditions will lead to sharp and smooth tips. In Fig. 11, a series of SEM images is shown that illustrates

---

[5] This point holds also for AFM-based experiments if the illumination occurs along the tip axis.

[6] Silver exposed to ambient conditions will quickly be contaminated by impurities. In addition, it is continuously oxidized even down to bulk silver layers.

[7] From the ambient this is still a problem, though gold is much more inert than silver; therefore, we seldom use tips for a time longer than one day.

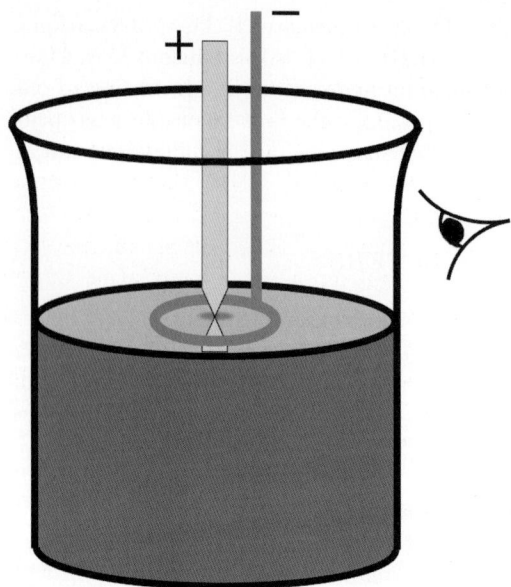

Fig. 10. Scheme of electrochemical etching of gold tips. (Reprinted with permission from [34])

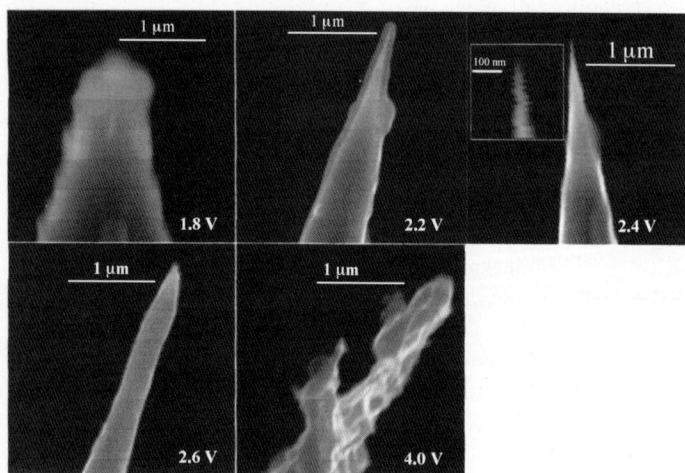

Fig. 11. Images of gold tips for different electrochemical etching potential. (Reprinted with permission from [34])

remarkable changes in surface structures with a varying electrode potential. Optimal smoothness is found for a potential of 2.4 V.

For the desired etching, it is also advantageous that a meniscus forms between the Au wire and electrolyte that leads to the conical form of the etched tips. If the wire is getting too thin to carry its lower part, it will break and the rest of the wire will drop. At this stage the electric circuit was cut off manually in earlier experiments. Now, an electronic device cuts off the electric circuit, when a preset current threshold is reached. Under these conditions, Au tips with apex radii of 20 nm to 30 nm can be produced routinely.

In our view, another important improvement was the use of side illumination of the tip. In this way, the disadvantageous polarization of the inverse-microscope approach, where the laser beam and the backscattered light propagate along the tip axis, can be avoided. Because the main polarization is oriented perpendicularly to the tip axis, the LSP modes will be less efficiently excited. In addition, the recording of the scattered light is inefficient, as its emission cone is not centered in the backscattering direction [49, 50].

As depicted in Fig. 12, the side-illumination approach means the redirection of the incident laser beam to an angle of 60° relative to the surface normal and the focusing onto the sample by a correspondingly oriented objective. The scattered light is collected by the same objective in the backscattering mode; it sends the scattered radiation backwards to the spectrograph. In this configuration, a strong polarization component exists parallel to the tip axis; also, the radiation of the scattered light is centered along the backscattering direction, and, last but not least, massive, opaque samples can be used. Together with suitable tips, this experimental configuration supports both an efficient excitation of localized surface plasmons (LSP) and an optimal recording of inelastically scattered light.

Malachite green isothiocyanate (MGITC) chemisorbed at a smooth Au(111) surface was chosen as a test system together with a Au tip. The adsorbate under investigation is located in the cavity formed between the tip and the Au(111) surface and is exposed to the field enhancement upon illumination. Before the experiment, the Au(111) sample was flame-annealed, which leads to a well-defined surface structure and removes contaminations [51]. In order to form a submonolayer of dye molecules, the crystal was wetted with a droplet of a 10^{-6} M ethanolic MGITC solution and dried. During the drying procedure, the dye molecules that came into contact with the metal surface were chemisorbed irreversibly through their sulfur atoms. In this way, a self-assembled (sub-)monolayer was formed. Subsequently, the sample was washed with copious amounts of ethanol to remove all molecules not directly bound to the gold substrate.

TERS experiments with MGITC on Au(111) generally showed a bleaching behavior. This means that the TER signals decay under illumination at a rate that depends on the incident laser intensity. Figure 13 shows a 3D plot of the TER intensity vs. Raman shift and vs. total time. To avoid too fast a bleaching, the laser power was reduced to 0.5 mW. Yet, the TER intensities

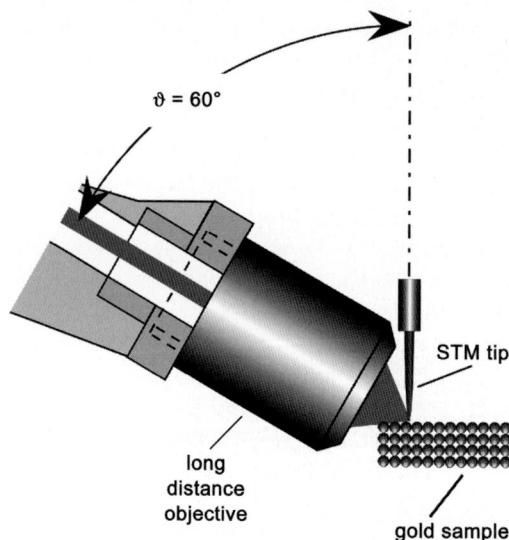

Fig. 12. 60° set up for TERS. Olympus long-distance objective: 50× magnification, NA 0.5. He–Ne laser: 5 mW at the sample, $\lambda_{ex} = 632.8$ nm. (Reprinted with permission from [33])

in Fig. 13 drop quickly within a few seconds and then level off at a slower rate.

For a deeper understanding of the bleaching let us investigate the time dependence of the $1618\,\mathrm{cm}^{-1}$ band. It shows the fastest intensity decay: After $70\,\mathrm{s}$ of illumination, it has only $\frac{1}{20}$ of its initial value, whereas the band at $1584\,\mathrm{cm}^{-1}$ still exhibits $25\,\%$ of its initial level. Obviously, the $1618\,\mathrm{cm}^{-1}$ band is more affected by the photofragmentation and growing disorder during the photobleaching than the $1584\,\mathrm{cm}^{-1}$ band [52]. Moreover, because of the sensitivity of the bleaching to the (local) intensity, differences in the bleaching rates for the cases "tip retracted" and "tip tunneling" can be related directly to the corresponding different (local) intensities. In other words, the ratio of the bleaching time constants represents a measure of the local field enhancement.

Figure 14 shows the time dependence of the resonance Raman intensity of the $1618\,\mathrm{cm}^{-1}$ band (no tip, hence, no field enhancement is present). The intensity is rather weak, initially around $15\,\mathrm{cps}$ to $20\,\mathrm{cps}$. The semi-logarithmic plot exhibits a time dependence that can not be described by first-order rate law. The (small) deviation from an exponential decay rate can be attributed to radially varying bleaching rates caused by a spatially varying intensity in the laser focus.

In the presence of the tip, the bleaching time constant drops dramatically, as is evident from Fig. 15 that plots the TER intensity of the $1618\,\mathrm{cm}^{-1}$ band vs. time on a semilogarithmic scale. Here again, we observe a deviation of the

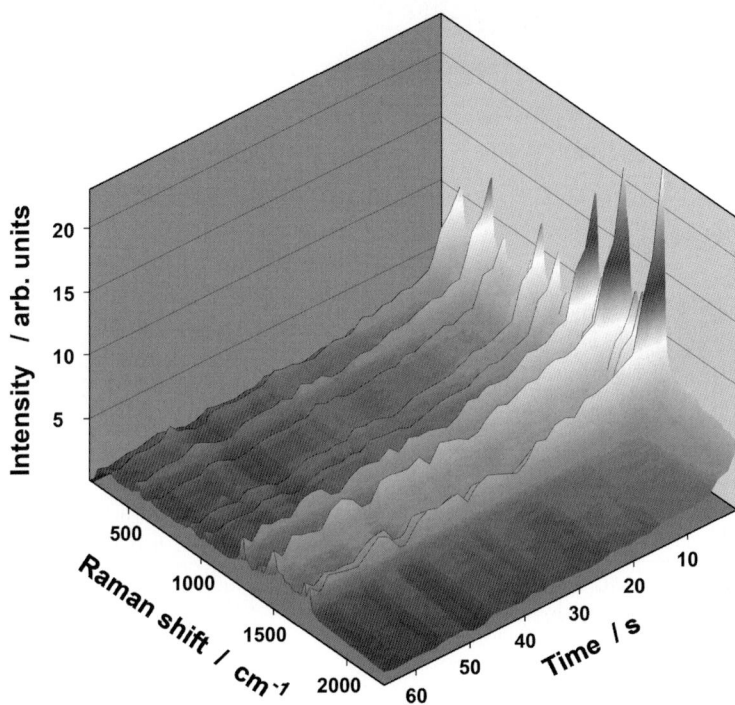

Fig. 13. Three-dimensional plot of the TER intensity of MGITC vs. wave number and time. Laser power: 0.5 mW; acquisition time: 1 s; delay between spectra: 1 s. (Reprinted with permission from [52])

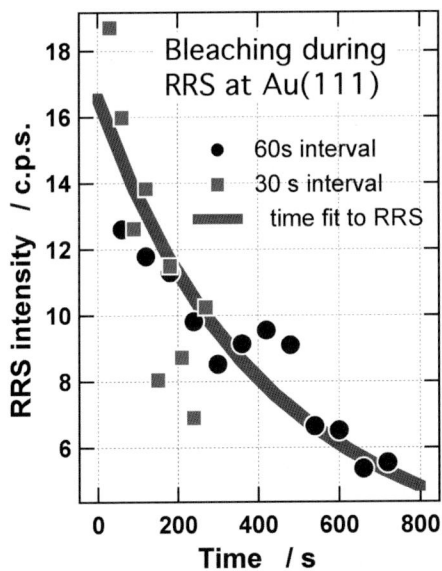

Fig. 14. Time dependence of the RRS signal of MGITC (the 1616 cm^{-1} band) adsorbed on an Au(111) surface. *Filled squares* and circles represent experimental data; the *solid line* is a fit curve using (1). No tip present. Laser power: 5 mW (Reprinted with permission from [52])

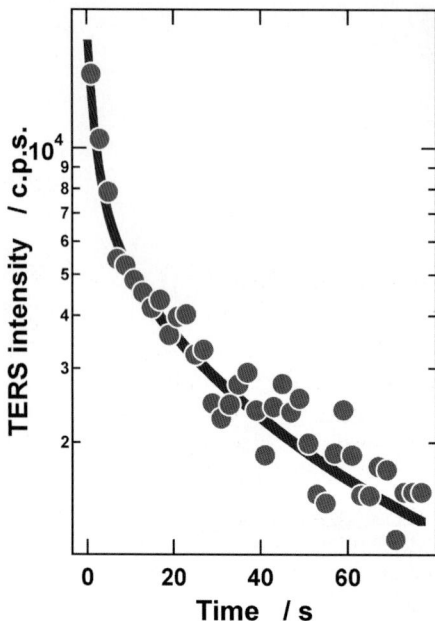

Fig. 15. Time dependence of the TER intensity of the $1618\,\mathrm{cm}^{-1}$ band of MGITC adsorbed on an Au(111) surface. *Filled circles* represent experimental data, the *solid line* is a fit curve using (2). Tip in tunneling position. Laser power: $0.5\,\mathrm{mW}$. (Reprinted with permission from [52])

curve from a first-order rate law. For a complete description of the bleaching behavior, two different bleaching processes have to be taken into account, but each of them deviates from a first-order rate law (i.e. following a non-exponential decay). The two bleaching processes are associated with different bleaching constants and amplitudes (initial intensities). Obviously, the bleaching occurs differently for MGITC located in an intact environment (fast) and for MGITC in a disordered/bleached environment (slow). Important here is the initial bleaching process with its short time constant.

The ratio of the bleaching time constants can be related to the square of the field enhancement g^2 [52]. Assuming Gaussian distributions for the incident and enhanced light intensities and different radii R_f (RRS) and R_T (TERS) $(I(r) = I_L \exp(-r^2/R_f^2)$ and $g^2(r)I_L = g_0 I_L \exp(-r^2/R_T^2)$, respectively)[8], the bleaching behavior can be given in analytical form:

$$
\begin{aligned}
I_{\mathrm{RRS}} &\propto N_0 \left(\frac{\mathrm{d}\sigma}{\mathrm{d}\Omega}\right)_{\mathrm{RRS}} \frac{I_L}{2} \left(4\pi \int_0^\infty e^{-e^{-r^2/R_f^2} I_L \gamma t - r^2/R_f^2} r \, \mathrm{d}r\right) \\
&= N_0 \left(\frac{\mathrm{d}\sigma}{\mathrm{d}\Omega}\right)_{\mathrm{RRS}} \frac{I_L}{2} R_f^2 \pi \left[\frac{1}{I_L \gamma t}\left(1 - e^{-I_L \gamma t}\right)\right],
\end{aligned} \tag{1}
$$

---

[8] This formulation of the bleaching behavior is more stringent than the one given in [52], because the quantities R_f and R_T are here defined as the radii of the focus and of the enhanced intensity underneath the tip that are, in principle, measurable.

$$I_{\text{TERS}} \propto N_0 \left(\frac{d\sigma}{d\Omega}\right)_{\text{RRS}} g_0^4 \frac{I_\text{L}}{2} \left(4\pi \int_0^\infty e^{-e^{-r^2/R_\text{T}^2 g_0^2 I_\text{L}\gamma t - 2r^2/R_\text{T}^2}} r \, dr\right)$$

$$= N_0 \left(\frac{d\sigma}{d\Omega}\right)_{\text{RRS}} g_0^4 \frac{I_\text{L}}{2} 2R_\text{T}^2 \pi$$

$$\times \left\{ \frac{2}{(g_0^2 I_\text{L}\gamma t)^2} \left[1 - e^{-g_0^2 I_\text{L}\gamma t} \left(1 + g_0^2 I_\text{L}\gamma t\right)\right]\right\}, \tag{2}$$

where N_0 is the number of the adsorbates per unit area, $\left(\frac{d\sigma}{d\Omega}\right)_{\text{RRS}}$ is the differential (resonance) Raman cross section of the adsorbate, I_L is the incident intensity in the center of the laser focus (i.e., twice the average intensity), γ is the bleaching constant, r is the radial coordinate underneath the tip and along the substrate surface, and t is time. Note that the large brackets on the right sight of (1) and (2) represent the time dependences, which are normalized to one for $t \to 0$ or $\gamma \to 0$. The bleaching time constants can be associated with $\tau_{\text{RRS}} = 1/(I_\text{L}\gamma)$ and $\tau_{\text{TERS}} = 1/\left[g_0^2(I_\text{L}/10)\gamma\right]$, where the divisor takes into account the laser intensity reduced by a factor of 10.[9] Building the ratio of the bleaching time constants, one can express the field enhancement g_0 as

$$g_0 = \sqrt{\frac{\tau_{\text{RRS}}}{\tau_{\text{TERS}}/10}}.$$

Using (1) and (2) for a fit to the experimental data (gray curves in Fig. 14 and Fig. 15), we obtain: $\tau_{\text{RRS}} = 244\,\text{s}$ and $\tau_{\text{TERS}} = 1.78\,\text{s}$.[10] This leads to $g_0 \approx 37$; thus, the TER enhancement is about $F_{\text{TERS}} \approx 2 \times 10^6$.

Next, taking the ratio of (2) and (1) at $t \to 0$ and equalizing it to the experimentally determined ratio $I_{\text{TERS}}/I_{\text{RRS}} = q$, one gets

$$R_\text{T} = \sqrt{\frac{q}{2} \frac{R_\text{f}}{g_0^2}}.$$

Since $q \approx 9200$ [11] and $R_\text{f} \approx 1000\,\text{nm}$, we obtain for the radius of the enhanced field $R_\text{T} \approx 50\,\text{nm}$. This means that the TERS radius is $R_{\text{TERS}} = \frac{1}{2}R_\text{T} \approx 25\,\text{nm}$, which agrees with the experimental findings of $R_{\text{TERS}} \approx R_{\text{tip}}$, i.e., the spatial resolution of TERS lies in the order of the size of the tip.

For a more than million-fold enhancement of Raman scattering by the local increase of the electromagnetic field, vibrational spectroscopy of adsorbates that belong to the class of normal Raman scatterers should be achiev-

---

[9] In (2), a reduced laser intensity is not considered.

[10] Please note that (1) and (2) describe nonexponential curves. Hence, the time constants must be evaluated from the fits or taken from the tangential lines at the points for $t \to 0$.

[11] For the nonreduced laser power, the TER intensity of the 1618 cm^{-1} line is about 150 000 cps, i.e., ten times higher than shown in Fig. 15.

able. A simple estimate yields the intensities to be expected: For the Raman intensity one can write

$$I_{\mathrm{NRS}} = N_0 \left(\frac{\mathrm{d}\sigma}{\mathrm{d}\Omega} \right)_{\mathrm{NRS}} \frac{I_{\mathrm{L}}}{2} \eta \,.$$

For an adsorbate density $N_0 = 8 \times 10^{14}\,\mathrm{cm}^{-2}$ (a monolayer of small molecules), a typical differential Raman cross section of small molecules $(\mathrm{d}\sigma/\mathrm{d}\Omega)_{\mathrm{NRS}} = 4 \times 10^{-30}\,\mathrm{cm}^2$, an incident power of 0.5 mW at 633 nm at the sample, i.e., of 1.6×10^{15} photons per second, and an overall spectrometer sensitivity of $\eta = 0.0005$, one obtains: $I_{\mathrm{NRS}} \approx 0.025$ cps. In other words, for a monolayer of weak Raman scatterers adsorbed at a smooth surface, the expected count rate is much below 1 cps; such low signals are not detectable. The situation is different in the case of TERS for sufficiently large enhancements. Then, $I_{\mathrm{TERS}} = I_{\mathrm{NRS}}\, g_0^4 R_{\mathrm{T}}^2 / R_f^2$. This means the expected TERS intensity is about $I_{\mathrm{TERS}} = 125$ cps; these intensity levels are easy to measure. In fact, TERS is achievable for a number of small molecules adsorbed at smooth Au(111) or other single crystals, as listed in Table 1, together with a rough indication of the height of the TER intensity (very strong, strong, middle,weak: vs, s, m, w) [53].

Table 1. TERS for weak Raman scatterers

Adsorbate	Substrate	TERS intensity level	Refs.
CN⁻	Au(111)	w	[33]
Benzenethiol	Au(110)	m	[54]
Benzenethiol	Pt(110)	w	[54]
Mercaptopyridine	Au(111)	m	[54]
ClO₄⁻	Au(111)	m	[53]
Adenine	Au(111)	w	[53]
Guanine	Au(111)	w	[53]
Cytosine	Au(111)	w	[53]
Thymine	Au(111)	w	[53]

All listed species, exhibit normal Raman scattering, but adsorbed on a gold(111) or on other single crystals they also exhibit TER scattering under proper experimental conditions. Table 1 indicates that TERS is operative (virtually) for all adsorbates. Besides, the kind of substrate metal does not pose severe restrictions on TERS. All experimental evidence shows that TERS is a more general spectroscopical tool than surface-enhanced Raman spectroscopy (SERS) that has a number of severe limitations concerning the kind of adsorbate, the substrates that support SERS, and the surface quality (roughness is required). Moreover, the TER signals originate from a subwavelength region.

The last few years also saw other very interesting and promising approaches that should be mentioned here briefly. *Ichimura* et al. [55] reported on tip-enhanced coherent anti-Stokes Raman scattering (tip-enhanced CARS or TE-CARS) of a nanometric DNA network structure. The authors used the $1337\,cm^{-1}$ diazole ring-breathing mode of adenine molecules to record images of DNA nanocrystals at a glass substrate with resolution much beyond the Abbe's diffraction limit [55]. *Keilmann* and *Hillenbrand* [56] developed a microscopy with ultraresolution that is far beyond the limitations of optical microscopy ($< \lambda/10$). Using a tip with radius $< 20\,nm$, a resolution of $\lambda/60$ was found in the visible and of $\lambda/500$ in the midinfrared wavelength region at $\lambda = 10\,\mu m$. *Hartschuh* et al. [57] used a metal tip to enhance optical spectroscopy such as two-photon fluorescence or Raman imaging with a resolution better than $20\,nm$. *Bulgarevich* and *Futamata* [58] built a microscope using confocal epi-illumination/collection optics and an AFM operated in semicontact mode. The authors observed a 1000-fold enhancement of the Raman scattering. *Methani* et al. [59] developed a TERS setup using side illumination optics with the goal of achieving larger enhancement factors and, more importantly, larger imaging contrast. Various molecular, polymeric and semiconducting materials and carbon nanotubes were investigated. *Raschke* et al. [60] used an infrared scattering-type near-field microscope to probe thin films of polymers at a wavelength of $3.39\,\mu m$ ($2950\,cm^{-1}$) and with a spatial resolution better than $10\,nm$. *Billot* et al. [61] focused on an optimized manufacturing of gold tips for TER microscopy and spectroscopy. The authors report on a set of parameters that enable a reproducible production of gold tips with $20\,nm$ radius of curvature. An analogous effort was made by *Saito* et al. [62] for the manufacturing of Ag-coated tips by using a chemical mirror reaction. This leads to a reproducible formation of particles of $40\,nm$ diameter at the tip apex.

Recently, an increased activity in theoretical studies of the optical properties of tip–substrate configurations was noticed. Quite realistic 2D and 3D finite element models were built that include the description of typical tip-sample configurations and geometries in TERS experiments. Huge enhancements for Raman scattering were predicted together with a high spatial resolution [31, 63, 64, 65]. A fruitful, mutual stimulation of experimental and theoretical studies in this rather young scientific field of TERS can be foreseen.

3 Conclusion/Outlook

Tip-enhanced Raman spectroscopy is a vibrational spectroscopy with hitherto unprecedented sensitivity and spatial resolution. Since the enhancement is mainly provided by the near-field excited at the apex of a suitable tip,

TERS appears to be a widely applicable spectroscopy *and* microscopy tool, in contrast to its parents, surface-enhanced Raman spectroscopy (SERS) and scanning near-field optical microscopy (SNOM). TER scattering has been observed for a number of molecules adsorbed at various substrates, including single-crystalline metal surfaces, showing thereby a more than million-fold enhancement of the Raman scattering. It is important to note that the field-enhancement provides, beyond TERS, promising avenues for applications to other optical techniques, such as tip-enhanced CARS, two-photon fluorescence and infrared scattering-type near-field microscopy.

Common to all these approaches is the high spatial resolution that is by far better than Abbe's diffraction limit of $\lambda/2$. The lateral resolution achieved today is in the range of 10 nm to 20 nm. Optical microscopy with such an excellent resolution has a very promising future.

The keys for further advances in the application of enhanced near-fields to scientific and technological (analytical) tasks include the optimization of tips, excitation and collection optics as well as of imaging techniques. Last but not least, it is necessary to achieve a deeper theoretical understanding of the optical properties of the cavity formed between tip and substrate as well as of the influence of (light-absorbing) adsorbates on the optical resonances of this cavity.

Acknowledgements

The author thanks K. F. Domke for carefully reading the manuscript.

References

[1] G. Binning, H. Rohrer, C. Gerber, et al.: Phys. Rev. Lett. **49**, 57 (1982)
[2] P. K. Hansma (Ed.): *Tunneling Spectroscopy: Capabilities, Applications, and New Technologies* (Plenum, New York 1982)
[3] G. Binning, N. Garcia, H. Rohrer: Phys. Rev. B **32**, 1336 (1985)
[4] H. K. Wickramasinghe: Acta Mater **48**, 347 (2000)
[5] B. C. Stipe, M. A. Rezai, W. Ho: Science **280**, 1732 (1998)
[6] J. I. Pascual, J. Gómez-Herrero, D. Sánchez-Portal, et al.: J. Chem. Phys. **117**, 9531 (2002)
[7] K. Kneipp, Y. Wang, H. Kneipp, et al.: Phys. Rev. Lett. **78**, 1667 (1997)
[8] S. M. Nie, S. R. Emory: Science **275**, 1102 (1997)
[9] A. Campion, P. Kambhampati: Chem. Soc. Rev. **27**, 241 (1998)
[10] K. Kneipp, H. Kneipp, I. Itzkan, et al.: Chem. Rev. **99**, 2957 (1999)
[11] A. M. Michaels, J. Jiang, L. Brus: J. Phys. Chem. B **104**, 11965 (2000)
[12] M. Fleischmann, P. J. Hendra, A. J. Mcquillan: Chem. Phys. Lett. **26**, 163 (1974)
[13] D. L. Jeanmaire, R. P. Van Duyne: J. Electroanal. Chem. **84**, 1 (1977)
[14] M. Moskovits, L. L. Tay, J. Yang, T. Haslett: SERS and the single molecule, in V. M. Shalaev (Ed.): *Optical Properties of Nanostructured Random Media* (Springer, Berlin, Heidelberg 2002) pp. 215–226

[15] A. Otto, I. Mrozek, H. Grabhorn, et al.: J. Phys. Condens. Matter **4**, 1143 (1992)

[16] V. M. Shalaev: Optical nonlinearities of fractal composites, in V. M. Shalaev (Ed.): *Optical Properties of Nanostructured Random Media* (Springer, Berlin, Heidelberg 2002) pp. 93–112

[17] V. A. Markel, V. M. Shalaev, P. Zhang, et al.: Phys. Rev. B **59**, 10903 (1999)

[18] M. Moskovits, D. H. Jeong: Chem. Phys. Lett. **397**, 91 (2004)

[19] J. I. Gersten, A. Nitzan: Electromagnetic theory: A spheroidal model, in R. K. Chang, T. E. Furtak (Eds.): *Surface Enhanced Raman Scattering* (Plenum, New York 1982) p. 89

[20] V. M. Shalaev, A. K. Sarychev: Phys. Rev. B **57**, 13265 (1998)

[21] N. Hayazawa, Y. Inouye, Z. Sekhat, et al.: J. Chem. Phys. **117**, 1296 (2002)

[22] R. M. Stöckle, Y. D. Suh, V. Deckert, R. Zenobi: Chem. Phys. Lett. **318**, 131 (2000)

[23] M. S. Anderson: Appl. Phys. Lett. **76**, 3130 (2000)

[24] N. Hayazawa, Y. Inouye, Z. Sekkat, et al.: Opt. Commun. **183**, 333 (2000)

[25] B. Pettinger, G. Picardi, R. Schuster, et al.: Electrochem. Jpn. **68**, 942 (2000)

[26] L. T. Nieman, G. M. Krampert, R. E. Martinez: Rev. Sci. Instrum. **72**, 1691 (2001)

[27] B. Pettinger, G. Picardi, R. Schuster, et al.: Single Molec. **3**, 285 (2002)

[28] A. Hartschuh, E. J. Sanchez, X. S. Xie, et al.: Phys. Rev. Lett. **90**, 95503 (2003)

[29] N. Hayazawa, T. Yano, H. Watanabe, et al.: Chem. Phys. Lett. **376**, 174 (2003)

[30] M. Micic, N. Klymyshyn, Y. D. Suh, et al.: J. Phys. Chem. B **107**, 1574 (2003)

[31] D. Hu, M. Micic, N. Klymyshyn, et al.: Rev. Sci. Instrum. **74**, 3347 (2003)

[32] B. Pettinger, G. Picardi, R. Schuster, et al.: J. Electroanal. Chem. **554**, 293 (2003)

[33] B. Pettinger, B. Ren, G. Picardi, R. Schuster, et al.: Phys. Rev. Lett. **92**, 96101 (2004)

[34] B. Ren, G. Picardi, B. Pettinger: Rev. Sci. Instrum. **75**, 837 (2004)

[35] T. Watanabe, B. Pettinger: Chem. Phys. Lett. **89**, 501 (1982)

[36] K. Kneipp, G. Hinzmann, D. Fassler: Chem. Phys. Lett. **99**, 503 (1983)

[37] K. Kneipp, D. Fassler: Chem. Phys. Lett. **106**, 498 (1984)

[38] B. Pettinger, A. Gerolymatou: Ber. Buns.-Gesellsch. Physik. Chem. **88**, 359 (1984)

[39] B. Pettinger, K. Krischer: J. Electr. Spectrosc. Relat. Phenom. **45**, 133 (1987)

[40] B. Pettinger, K. Krischer, G. Ertl: Chem. Phys. Lett. **151**, 151 (1988)

[41] K. Kneipp, Y. Wang, R. R. Dasari, et al.: Appl. Spectrosc. **49**, 780 (1995)

[42] H. W. Schrötter, H. W. Klöckner: *Raman Spectroscopy of Gases and Liquids* (Springer, Berlin, Heidelberg 1979) p. 123

[43] M. R. Kagan, R. L. McCreery: Langmuir **11**, 4041 (1995)

[44] M. D. Morris, D. J. Wallan: Anal. Chem. **51**, 182 A (1979)

[45] H. X. Xu, J. Aizpurua, M. Kall, et al.: Phys. Rev. E **62**, 4318 (2000)

[46] R. Stöckle, C. Fokas, V. Deckert, et al.: Appl. Phys. Lett. **75**, 160 (1999)

[47] M. S. Anderson, W. T. Pike: Rev. Sci. Instrum. **73**, 1198 (2002)

[48] F. Demming, J. Jersch, K. Dickmann, et al.: Appl. Phys. B **66**, 593 (1998)

[49] D. L. Mills: Phys. Rev. B **65**, 125419 (2002)

[50] R. W. Rendell, D. J. Scalapino, B. Mühlschlegel: Phys. Rev. Lett. **25**, 1746 (1978)

[51] J. Clavilier, R. Faure, G. Guinet, et al.: J. Electroanal. Chem. **107**, 205 (1980)
[52] B. Pettinger, B. Ren, G. Picardi, et al.: J. Raman Spectrosc. **36**, 541 (2005)
[53] D. Zhang, K. F. Domke, B. Pettinger: Chem. Phys. Lett. (to be publ. 2006)
[54] B. Ren, G. Picardi, B. Pettinger, et al.: Angew. Chem. Int. Ed. **44**, 139 (2005)
[55] T. Ichimura, N. Hayazawa, M. Hashimoto, et al.: Phys. Rev. Lett. **92**, 220801 (2004)
[56] F. Keilmann, R. Hillenbrand: Philos. Trans. Roy. Soc. Lond. Ser. A Math. Phys. Eng. Sci. **362**, 787 (2004)
[57] A. Hartschuh, M. R. Beversluis, A. Bouhelier, et al.: Philos. Trans. Roy. Soc. Lond. Ser. A Math. Phys. Eng. Sci. **362**, 807 (2004)
[58] D. S. Bulgarevich, M. Futamata: Appl. Spectrosc. **58**, 757 (2004)
[59] D. Mehtani, N. Lee, R. D. Hartschuh, et al.: J. Raman Spectrosc. **36**, 1068 (2005)
[60] M. B. Raschke, L. Molina, T. Elsaesser, et al.: Chem. Phys. Chem. **6**, 2197 (2005)
[61] L. Billot, L. Berguiga, M. L. De La Chapelle, et al.: Eur. Phys. J. Appl. Phys. **31**, 139 (2005)
[62] Y. Saito, T. Murakami, Y. Inouye, et al.: Chem. Lett. **34**, 920 (2005)
[63] F. Festy, A. Demming, D. Richards: Ultramicroscopy **100**, 437 (2004)
[64] A. L. Demming, F. Festy, D. Richards: J. Chem. Phys. **122**, 184716 (2005)
[65] I. Notingher, A. Elfick: J. Phys. Chem. B **109**, 15699 (2005)
[66] S. Wu, D. L. Mills: Phys. Rev. B **65**, 205420 (2002)

Index

Tip-Enhanced Near-Field Raman Scattering: Fundamentals and New Aspects for Molecular Nanoanalysis/Identification

Prabhat Verma[1], Yasushi Inouye[2,3], and Satoshi Kawata[1,3]

[1] Department of Applied Physics, Osaka University, Suita, Osaka 565-0871, Japan
[2] Graduate School of Frontier Biosciences, Osaka University, Suita, Osaka 565-0871, Japan
[3] RIKEN, Wako, Saitama, 351-0198, Japan
   verma@ap.eng.osaka-u.ac.jp

1 Introduction

Raman scattering from molecules adsorbed on metallic nanostructures is strongly enhanced due to excitation of local surface-plasmon polaritons (SPPs). This gives rise to the well-known surface-enhanced Raman scattering (SERS) effect when the sample is dispersed on a rough metallic surface. Even a single metallic nanostructure, e.g., a metallic nanosized needle tip, can induce SERS at the tip apex. Raman scattering enhanced by such a tip is called tip-enhanced Raman scattering (TERS). The essential feature of TERS utilizing an apertureless metallic tip is the use of the incident electric-field-enhancement effect at the proximity of the metallic tip apex. The most important feature that makes TERS so attractive is its capability of optical sensing with high spatial resolution beyond the diffraction limits of light. We have demonstrated enhancement with high spatial resolution in TERS measurements for several nanomaterials [1, 2, 3, 4, 5], using a metallic probe tip.

In this Chapter, we discuss some TERS results from different kinds of samples and show the nanoanalysis capabilities of this technique. In Sect. 2, we discuss briefly the localized SPPs at a metallic tip apex, followed in Sect. 3 by an introduction to the instrumentation utilized in the experiments. The TERS measurements on carbon nanotubes and rhodamine R6 are discussed in Sect. 4 and Sect. 5, respectively. Section 6 discusses the TERS results for a DNA-based adenine molecules, including a discussion on the specific Raman band shifts that occurred by pressurizing adenine molecules by the tip. Finally, the tip-enhanced coherent anti-Stokes Raman scattering is discussed in Sect. 7, where we show an extremely high spatial resolution of 15 nm.

2 Localized Surface-Plasmon Polaritons at the Tip

Plasmons are defined as electromagnetic excitations coupled to the free charges of a conductive medium, and the surface-plasmon polaritons (SPPs)

K. Kneipp, M. Moskovits, H. Kneipp (Eds.): Surface-Enhanced Raman Scattering – Physics and Applications, Topics Appl. Phys. **103**, 241–260 (2006)
© Springer-Verlag Berlin Heidelberg 2006

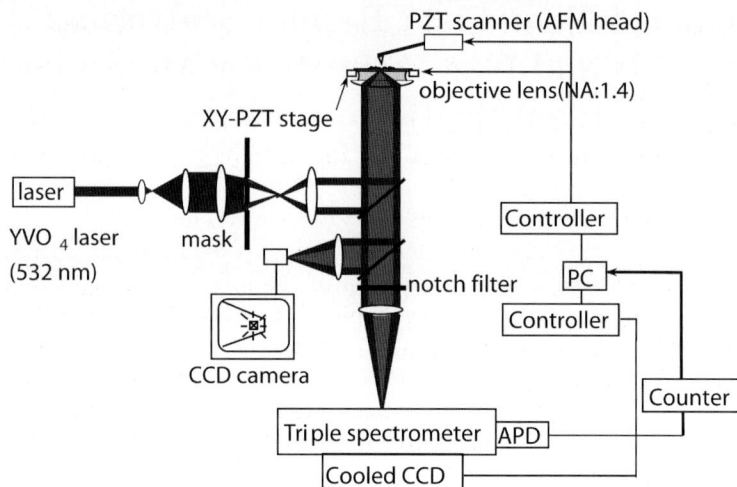

Fig. 1. Experimental setup for tip-enhanced near-field Raman spectroscopy

are such plasmons bound to an interface with a dielectric medium. Physically, one can understand it as photons coupled to collective excitations of conduction electrons on a metal surface. The speed of these SPPs is lower than the speed of light in the medium adjacent to the metal surface, so the electromagnetic field is evanescent or, in other words, these photons are non-propagative. Principally, SPPs can be excited in any kind of metal structure, but excitations in structures much smaller than the wavelength of the exciting light are distinguished by a pronounced resonance character. Such SPPs in the bounded geometry of a small structure are called the localized SPPs. These localized SPPs can be easily excited on small metallic structures and the incident light field is enhanced strongly. In the tip-enhanced near-field Raman scattering, a sharp apertureless metallic tip (e.g., a silver-coated AFM-cantilever tip) with a tip apex of about a few tens of nanometer is utilized as the small metallic structure to generate the localized SPPs. The electric field that is coupled to the localized SPPs at the apex of this metallic tip comprises an evanescent field. Since the evanescent field is localized around the tip apex, a super-resolution capability is attained by detecting this field. Therefore, an enhancement in the scattered field along with a super-resolution can be achieved by using a sharp metallic tip in TERS measurements.

3 Instrumentation

The basic difference in the instrumentation of micro-Raman spectroscopy (far-field detection) and the near-field TERS is related to the fact that in the former, Raman scattering is collected from all the sample molecules present in the focal spot of the probing light, whereas in the latter, the Raman scattering cross sections of the sample molecules just below the enhanced electric field near the metallic tip apex are selectively enhanced and detected to provide a sufficiently high spatial resolution depending on the size of the enhanced electric field that corresponds to the diameter of the metallic tip, e.g., less than 30 nm. The experimental setup for the near-field TERS is shown in Fig. 1. An expanded and collimated light beam from a frequency-doubled YVO_4 laser (wavelength: 532 nm, power: 50 mW) enters into the epi-illumination optics. A circular mask is inserted in the beam path of the illumination light and is located at the conjugate plane of the pupil of the objective lens with a numerical aperture (NA) equal to 1.4. This mask rejects part of the beam corresponding to focusing angles that are less than NA = 1.0, and the transmitted light is focused to produce an evanescent field on the sample surface [6]. As the metallic tip is moved into the focused spot, a highly p-polarized evanescent field excites surface-plasmon polaritons at the tip apex, and an enhanced electric field that is localized at the tip is generated [7, 8]. The localized enhanced electric field at the tip is scattered inelastically by the Raman-active molecules, which constitutes the near-field Raman scattering. The Raman scattering is collected by the same objective lens, and is directed to the triple-grating spectrophotometer operated in subtractive dispersion that is equipped with a liquid-nitrogen-cooled CCD camera for Raman spectra measurement and with an avalanche photodiode (APD) for Raman imaging. The APD is located after the exit slit of the spectrophotometer so that a specific Stokes-shifted line can be detected. The metallic tip is a commercial AFM silicon cantilever that is coated with a silver layer by a thermal evaporation process. The evaporation rate was adjusted to a relatively slow rate of 0.3 Å per second to avoid undesirable bending of the silicon tip. The silver-coated tip diameter was around 30 nm. The distance between the sample and the cantilever is regulated by contact-mode AFM operation, and the sample is scanned with piezoelectric transducers (PZT) in the X–Y plane. Scanning the XY-PZT sample stage while simultaneously detecting the Raman signal with the avalanche photodiode can perform near-field Raman imaging at the specific Stokes-shifted line.

Some additional features of the experimental setup utilized for CARS experiments are discussed further in Sect. 7.

4 TERS from Single-Wall Carbon Nanotube

Far-field Raman spectroscopy has recently been recognized as a very powerful tool to evaluate single-wall carbon nanotubes (SWNT) and it achieved an individual SWNT detection [9, 10] because much structural information including chirality and diameter could be deduced from the vibration modes [11]. To further detail the specific structure-dependent single-molecule property, a single-molecule addressing capability at the nanometer scale is needed,

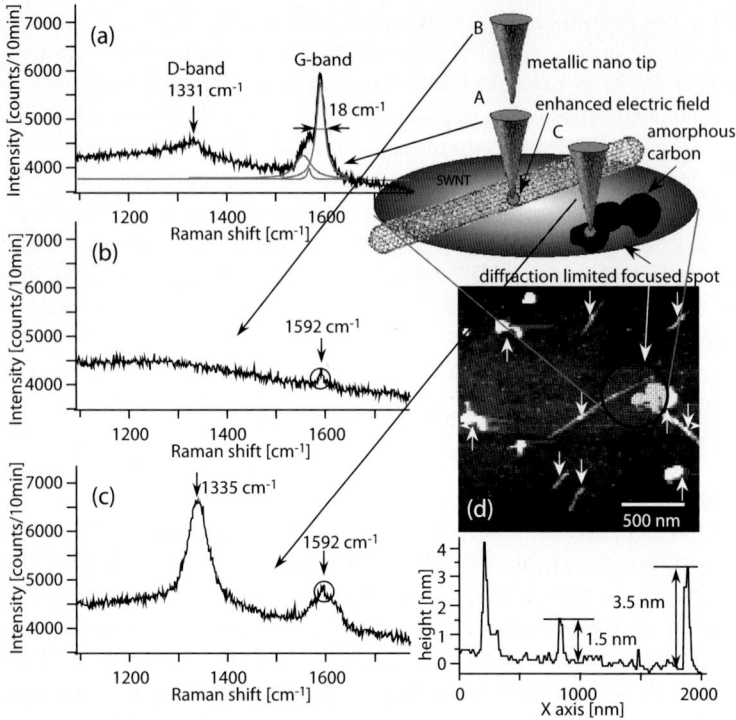

Fig. 2. (**a**) TERS spectra of an individual SWNT, which is obtained at the schematically shown tip position of "A". Remarkable tangential G-band peaks of SWNT and a small D-band peak at 1331 cm^{-1} are observed. The G-band peaks are well fitted by a Lorentzian lineshape at the peak positions of 1593 cm^{-1}, 1570 cm^{-1}, and 1558 cm^{-1} (*gray lines*). (**b**) Far-field Raman spectrum of SWNT with the tip being 500 nm away from the sample surface shown in "B". (**c**) TERS spectrum of an amorphous carbon, which is obtained at schematically shown tip position of "C". (**d**) Topographic AFM image and the cross section (*white line in the image*) of the isolated SWNTs dispersed on a coverslip. Some isolated SWNTs indicated by *downwards arrows* and amorphous carbon particles indicated by *upwards arrows* can be seen in the image. The dimension of the image is 2 μm × 2 μm consisting of 200 × 200 pixels

which can be provided by the TERS technique. For these measurements [3], we sparsely spread out SWNTs on a coverslip, and focus a laser beam at a tube (Fig. 2d). Generally, our sample consisted of bundles of SWNTs, however, after spincoating, some individual SWNTs could be separated out from aggregated bundles of SWNTs. Figure 2a shows a TERS spectrum of an individual SWNT, obtained at the schematically shown tip position "A". Remarkable tangential G-band peaks, representing a graphite mode, of a SWNT that is split into three peaks and fitted well to the Lorentzian lineshape at 1593 cm^{-1}, 1570 cm^{-1}, and 1558 cm^{-1} were observed (gray lines

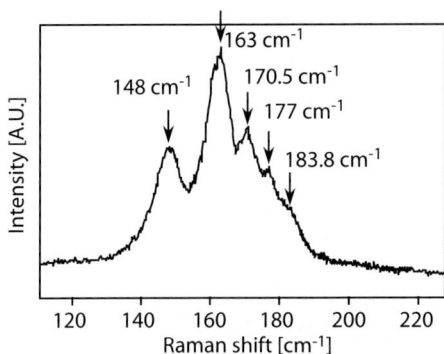

Fig. 3. High-resolution far-field Raman spectrum of radial breathing mode from aggregated bulk of SWNTs detected with a double-grating spectrometer

in Fig. 2a). These Lorentzian lineshapes feature semiconducting SWNTs, excluding the Breit–Wigner–Fano lineshape [12] that is observed in the resonant Raman spectra of metallic SWNTs. The fitted linewidth at $1593\,\mathrm{cm}^{-1}$ is $18\,\mathrm{cm}^{-1}$. Taking into account that the FWHM of the excitation laser detected by our single spectrometer with an entrance slit width of $250\,\mu\mathrm{m}$ has a linewidth of $10\,\mathrm{cm}^{-1}$, the deduced linewidth at $1593\,\mathrm{cm}^{-1}$ is $8\,\mathrm{cm}^{-1}$. This narrow linewidth suggests that the observed isolated SWNTs consist of only one or a few SWNTs [13]. A small D-band peak, representing a defect mode, at $1331\,\mathrm{cm}^{-1}$ was also detected.

Due to the strong coupling between electrons and phonons in the resonant Raman effect, the unique one-dimensional electronic density of states of SWNTs plays an important role in the exceptionally strong resonant Raman spectra of SWNTs associated with the interband optical transitions between two van Hove singularities [14]. According to [15], while the 532 nm wavelength of our light source, carrying a photon energy of 2.33 eV, can resonantly excite electrons contained in semiconducting SWNTs whose diameter is from 1.35 nm to 1.55 nm, the optical absorption spectrum does not show the corresponding strong absorption [15]. However, the tip-enhancement effect compensates for this less efficient resonant Raman effect. Furthermore, the high-resolution far-field Raman spectrum detected with a double-grating spectrometer at the aggregated bulk of SWNTs included in the sample, shown in Fig. 3, can give us the diameter distribution of the SWNTs [16]. Five apparent peaks are detected as radial breathing modes. By using a simple approximation between mode frequency W_{RBM} (cm^{-1}) and the diameter d (nm), i.e., $W_{\mathrm{RBM}} = 224/d$ [11], the peaks of the radial breathing modes at $148\,\mathrm{cm}^{-1}$, $163\,\mathrm{cm}^{-1}$, $170.5\,\mathrm{cm}^{-1}$, $177\,\mathrm{cm}^{-1}$, and $183.8\,\mathrm{cm}^{-1}$ correspond to the diameters of 1.51 nm, 1.37 nm, 1.31 nm, 1.27 nm, and 1.22 nm, respectively. This diameter feature is also supported by the TERS spectrum of the radial breathing mode shown in Fig. 4. The spectra a–c in Fig. 4 correspond to

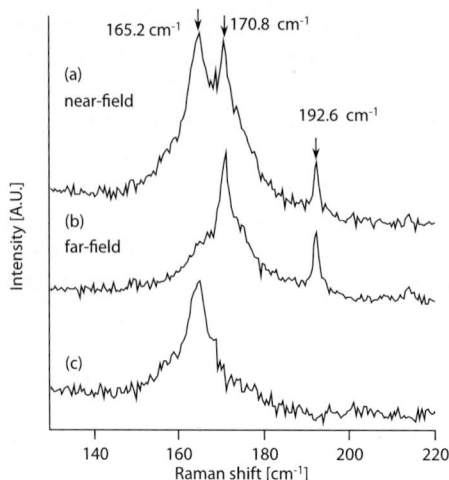

Fig. 4. (**a**) TERS and (**b**) far-field Raman spectra of radial breathing mode of isolated SWNTs. (**c**) The subtracted Raman spectra between far-field and TERS. The Raman peak at $165.2\,\mathrm{cm}^{-1}$ corresponding to the diameter of $1.50\,\mathrm{nm}$ is selectively enhanced by the metallic probe tip

tip-enhanced near-field, far-field, and the subtracted Raman spectra, respectively. The metallic probe tip selectively enhanced the SWNT just below the tip so that the enhanced Raman peak at $165.2\,\mathrm{cm}^{-1}$ is clearly observed, while other peaks such as $170.8\,\mathrm{cm}^{-1}$ and $192.6\,\mathrm{cm}^{-1}$, which are generated from the far-field focused spot are not enhanced. By using the modified approximation for individual SWNTs, $W_{\mathrm{RBM}} = 248/d$ [9], the Raman peak at $165.2\,\mathrm{cm}^{-1}$ corresponds to the diameter of $1.50\,\mathrm{nm}$. The topographic image of the sample (Fig. 2d) confirms the height of the SWNT ($\sim 1.5\,\mathrm{nm}$) and clearly shows the distribution of SWNTs and amorphous carbon. This image was obtained in AFM operation by the tip-enhanced near-field microscope. The white lines in the figure indicated by downwards arrows are isolated SWNTs. The height of the observed isolated SWNT is $\sim 1.5\,\mathrm{nm}$ corresponding to the diameter of an individual SWNT, while the other one has the height of $3.5\,\mathrm{nm}$ consisting of several SWNTs. White spots with a diameter of $\sim 100\,\mathrm{nm}$ indicated by upwards arrows are amorphous carbon inherently included in our sample as the impurities of the arc-discharge process of SWNTs.

Accordingly, the most possible chirality for the semiconducting SWNT observed in the TERS spectrum is (17.4) $\sim 1.513\,\mathrm{nm}$. Note that the peak at $192.6\,\mathrm{cm}^{-1}$ in Fig. 4 is not observable in the high-resolution far-field Raman spectra from the aggregated bulk of SWNTs in Fig. 3. This can be attributed to the fact that the number of SWNTs in Fig. 4 is extremely small compared to that of Fig. 3. The averaging effect overwhelmed the characteristic peak at $192.6\,\mathrm{cm}^{-1}$ in Fig. 3.

Figure 2b was obtained when the metallic probe tip was 500 nm away from the sample surface (the tip position of "B"). This spectrum represents a far-field Raman spectrum, where the featured lines are gone due to the absence of a field-enhancing metallic probe tip in the near-field except for a small protrusion at $1592\,cm^{-1}$. Since the sample contains not only SWNTs but also impurities (amorphous carbon), this small peak represents an average within a focused spot of the laser beam, however, the spectrum dominantly reflects the non-enhanced G-band feature of SWNTs. Note that although isolated SWNTs are randomly dispersed and aligned parallel to the substrate, we can observe the G-band of isolated SWNTs only at several far-field focused positions by an objective lens, and the observed far-field Raman intensity at each position is different. Because of the polarization effect of both illumination light field and SWNTs and the severe selectivity of the resonant Raman effect [15], the number of observable isolated SWNTs is quite limited. These facts suggest that the observed isolated SWNT consists of only one or a few SWNTs. Figure 2c shows another TERS spectrum from the same sample when the probe enhances the field at a different location. A broad D-band representing the defect mode of amorphous carbon is dominantly detected around $1335\,cm^{-1}$ [17]. The G-band is also seen but it is wider than that seen in Fig. 2a. This can be attributed to the tip position schematically shown by "C". A small G-band peak shown in Fig. 2b (circle in the figure) is recognizable on the top of the broad G-band peak in Fig. 2c (circle in the figure). The broad G-band peak represents the tip-enhanced near-field Raman scattering of amorphous carbon due to the tip, while the small G-band peak is the far-field background signal coming from an isolated SWNT. The sharp tangential G-band feature of an individual semiconducting SWNT (Fig. 2a) and the broad D-band and G-band feature of amorphous carbon (Fig. 2c) are explained on the basis that the enhanced electric field at the metallic probe tip selectively enhances the near-field Raman scattering of the SWNT and amorphous carbon inside of the tightly focused spot by the objective lens.

5 TERS Measurements on Rhodamine 6G

Rhodamine 6G (R6G) molecules dispersed on a silver-island film were studied by TERS and the corresponding spectra are shown in Fig. 5 [2, 18]. The TERS spectra a–d, h, and i together with spectrum j in Fig. 5 are quite similar to the far-field surface-enhanced resonant Raman scattering (SERRS) spectrum [18]. On the other hand, some of the spectra, such as the spectra e, f, and g in Fig. 5 exhibit anomalous spectral patterns that are different from those in the far-field SERRS [18]. Such anomalous spectral patterns, in detail, seem to have an overlap of several new peaks, in addition to the TERS spectral pattern analogous to the far-field SERRS. The frequencies of most TERS bands are found to be consistent with those of the SERRS peaks. In addition, the intensities of some TERS bands such as those at $702\,cm^{-1}$, $1027\,cm^{-1}$,

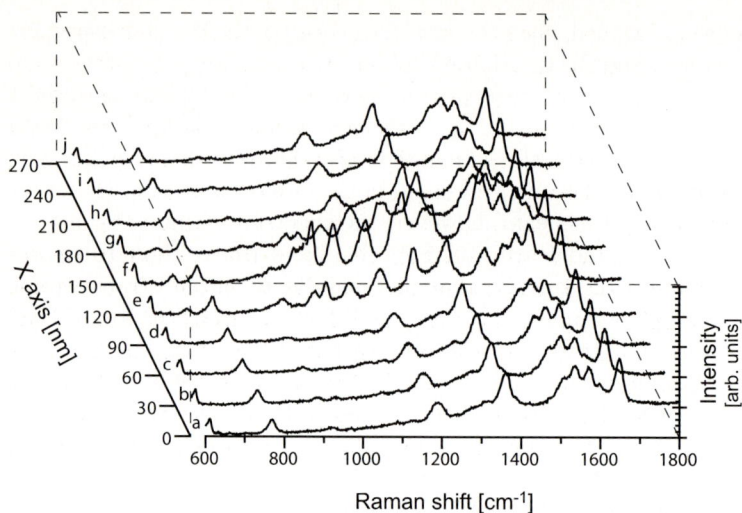

Fig. 5. TERS spectral mapping of R6G adsorbed on silver film at several positions. Labels a–j correspond to various tip positions where the TERS spectra were observed

$1061\,\mathrm{cm}^{-1}$, $1120\,\mathrm{cm}^{-1}$, $1269\,\mathrm{cm}^{-1}$, and $1457\,\mathrm{cm}^{-1}$ are quite inconsistent with those of the SERRS being strongly enhanced.

In the TERS spectra, such as spectra a–d, h, and i in Fig. 5, an averaged enhancement factor by the metallic tip is estimated from the TERS/SERRS signal ratio at $1647\,\mathrm{cm}^{-1}$ [4]. The averaged enhancement factor by the tip reaches approximately 1×10^2, which is extrapolated for the spot area (400 nm ϕ) from the metallic tip diameter (40 nm ϕ). Meanwhile, a specific enhancement factor is estimated from the signal ratio of the most strongly enhanced peak at $1054\,\mathrm{cm}^{-1}$ in spectrum f. The specific enhancement factor reaches approximately 2×10^4. In TERS spectroscopy, enhancement mechanisms caused by the metallic tip are thought to be the same as the SERS, the electromagnetic (em), and the charge transfer (CT). In addition to these SERS-like enhancement mechanisms, the third mechanism is proposed as a mechanical pressure mechanism due to molecular deformation by the tip [5]. According to the former two enhancement mechanisms, a ratio of the specific/averaged enhancement factors at least correspond to the contribution of the CT mechanism to the enhancement. The ratio obtained in the TERS experiment is less than or equal to 10^2; this magnitude is as much as a generally accepted degree of the enhancement effect caused by the CT mechanism [19, 20, 21].

The vibrating regions of all the peaks with anomalous TERS signal are located on the opposite side of the amino group that is bonded to the silver surface. The CT mechanism can be explained by the resonant Raman mechanism in which charge-transfer excitations (either from the metal to the adsorbed molecule, or vice versa) occur at the energy of incident laser frequency [21]. In the case of the TERS, the electronic wavefunctions overlap between the silver atoms of the metallic tip apex and molecular orbitals of the adsorbed species is a requirement for the charge-transfer excitations. With an excitation frequency at 488 nm (2.54 eV), adsorbates such as R6G having a narrow HOMO–LUMO bandgap are electronically excited from the lower occupied orbitals to a Fermi level of silver surfaces, or from the Fermi level to higher virtual orbitals. In the R6G molecule, occupied orbitals (ϕ_{116} and ϕ_{117}) and a virtual orbital (ϕ_{120}) are the candidates for the CT electronic excitations in comparison with their energy differences from the Fermi level and the excitation frequency. An electron density distribution of the occupied orbitals (ϕ_{116} and ϕ_{117}) is localized in the xanthene ring, and that of the virtual orbital (ϕ_{120}) is centralized in the phenyl ring. In our Raman NSOM setup, the silver metallic tip approaches from the opposite direction toward the adsorbed R6G molecules on the silver-island film and touches the molecules by the contact mode of the AFM standard operation. As the silver atoms of the tip at the apex interact with the virtual orbital (ϕ_{120}) at the phenyl ring of the R6G molecules, the bands due to symmetric deforming motions of the phenyl ring could be highly enhanced in their TERS intensities. Hence, the specific intensity enhancement of some of these TERS bands could be understood with the CT mechanism caused by the silver tip together with the approaching direction of the tip toward the molecule. From the viewpoint of the molecular orientations, it is strongly suggested that R6G molecules in the anomalous regions turn the opposite part of the molecules against the adsorption site to the direction of the metallic tip. In other words, the R6G molecules are in a state of higher order such as assembly or aggregation in the anomalous regions.

Comparing the enhancement factors of SERRS/RRS and TERS/SERRS, the averaged enhancement factor of TERS/RRS and the factor at the specific band are estimated to be approximately 1×10^6 and 2×10^8, respectively. In particular, the enhancement factor at the specific band is beyond the values that are usually obtained in the conventional SERS measurement. Recent progress in single-molecule detection of SERS reveals that a gigantic local plasmonic field is generated at a junction between two connecting metallic nanoparticles [22, 23]. In our TERS system, R6G molecules are tightly sandwiched between the silver-island film (thickness = 8 nm) and the silver tip (40 nm ϕ) by the contact mode of the AFM. These contacting areas could be the junctions (hot spots). Supposing that the hot spot is an area where all particles (40 nm ϕ) are contacting within a gap of 1 nm, the diameter of the hot-spot area would be less than 1 nm on the basis of the finite difference time-domain (FDTD) method [23, 24]. Accordingly, the enhancement factor at the specific band would reach the 10^{11} to 10^{12} level, which is estimated for the conventional TERS observable area by the area of the hot spot. This means a metallic tip surface doubly enhanced Raman scattering has a possibility to detect the molecular vibration of subnanometer size at the level of

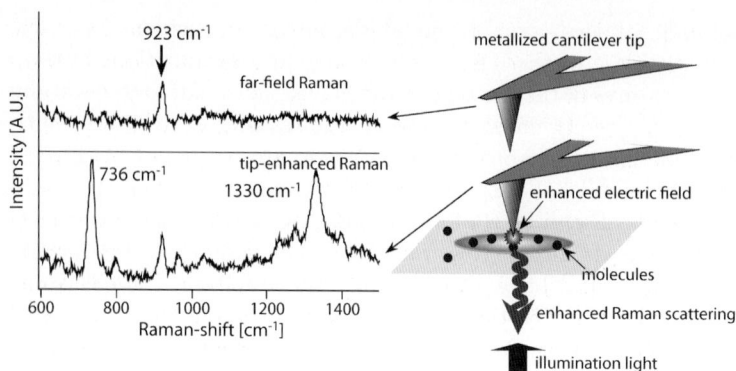

Fig. 6. Tip-enhanced near-field Raman (*lower spectrum*) and far-field Raman (*upper spectrum*) spectra of adenine nanocrystals. The peak at $923\,\mathrm{cm}^{-1}$ is from the glass substrate, and shows the same intensity both in TERS and the far-field Raman spectra. A schematic of the tip position in the two cases is also shown

a single molecule. Under such a strong electric field, two enhancement mechanisms would be tentatively considered. One is an extended CT mechanism where the bands of SERRS-inactive vibrational modes could be enhanced either with a vibronic coupling (the Albrecht B term [25]) or with electronic excitations that are not excited by the classical, linear CT mechanism. Another is a polarized em mechanism where Raman bands could be enhanced by a highly polarized electric field coupled with their Raman polarizability tensors. Intensity fluctuation of specific vibrational modes of R6G was reported in the single-molecule Raman detection [5, 26], and the calculated polarizability tensors of the specific Raman bands of the R6G molecule are parallel to the approaching direction of the tip.

6 Tip Force on DNA-Based Adenine Molecules

In this section, we will first discuss the normal TERS results from a DNA-based adenine molecule and demonstrate how the metallic tip enhances the weak Raman signal from a nanometer region, which cannot be detected by a conventional micro-Raman configuration. In the next part, we will show the effect of pressure that a tip can apply on the sample during the TERS experiments.

6.1 TERS from Adenine

Adenine nanocrystals are spread out on a coverslip by casting adenine molecules dissolved in ethanol [3]. The crystal size is laterally $7\,\mathrm{nm} \times 20\,\mathrm{nm}$ wide and $15\,\mathrm{nm}$ high. Figure 6 shows a TERS spectrum of a single nanocrystal of adenine molecules. For a comparison, a far-field Raman spectrum of the same sample without a metallic tip is also shown. Figure 6 also schematically shows the tip position in the two cases. In the TERS spectrum, several characteristic Raman peaks of adenine molecules are enhanced and become visible, such as peaks at $736\,cm^{-1}$ and $1330\,cm^{-1}$, while no Raman peak at these positions is observed when the tip is withdrawn from the near-field region (far-field Raman spectrum). The experimental condition was exactly the same for both experiments shown in Fig. 6 except for the tip–sample distance. The strong Raman peaks at $736\,cm^{-1}$ and $1330\,cm^{-1}$ are assigned to ring–breathing mode of a whole molecule and the ring-breathing mode of a diazole, respectively. Other Raman peaks are also assigned to the normal modes of adenine molecules [27, 28, 29]. The observed TERS peaks are shifted slightly from those obtained in the far-field. For example, the ring-breathing mode at $720\,cm^{-1}$ in the far-field Raman spectrum of the bulk sample is shifted to $736\,cm^{-1}$ in the TERS spectrum, while the ring-breathing mode of a diazole is not shifted. These phenomena in the spectral shifts are in good agreement with the SERS spectra of adenine molecules [30], and ensure that the metallic probe tip works as a surface enhancer for the SERS effect.

Assuming that the enhanced electric field is $30\,nm$ ϕ corresponding to the tip diameter and the focused light spot is $400\,nm$ ϕ, the enhancement factor for the ring-breathing mode of a whole molecule is 2700. Furthermore, the factor for the ring-breathing mode of a diazole is uncountable because the far-field Raman signal without a metallic probe tip is too weak to be detected.

6.2 TERS from Tip-Pressurized Adenine Molecules

One of the major differences in the experimental conditions between the TERS from those of the SERS is the interaction mechanism of adenine with silver surfaces. In SERS, adenine is adsorbed in equilibrium onto silver surfaces and it interacts with the silver atoms electromagnetically and chemically. However, in TERS, apart from these two interactions, the tip and the sample can also interact mechanically, if the metal molecules of the tip push against the sample molecules. This effect can be observed if the sample is pressurized, intentionally or unintentionally, by the tip [5]. Unlike the isotropic hydraulic pressure usually studied in high-pressure Raman scattering [31, 32, 33], the pressure applied by a tip is unidirectional, and hence it can change the bond lengths uniaxially, resulting in modifications of the molecular vibrations. The Raman modes tend to shift and get broadened. Assuming that the atomic force is applied only to the contraction of the bond between the silver atom of the tip and the adjacent nitrogen of the adenine molecule, the bond distance would be expected to shrink and the vibrational frequencies may then shift upwards.

In our experiment, we used a cantilever with a spring constant of $0.03\,N/m$ and a silver-coated tip apex diameter of this cantilever was $5\,nm$ to $10\,nm$.

The atomic force was kept constant at 0.3 nN by the feedback loop. As deduced from the unit-cell parameters of single-crystal 9-methyladenine [34], a couple of adenine molecules exist in a rectangle approximately 0.77 nm long by 0.85 nm wide. If we assume that the force is equally applied to all the molecules that are adjacent to the tip apex, the adenine molecules are subjected to a pressure of ~ 1 to 5 pN per molecule by the silver atom attached on the surface of the tip. As the metallic tip approaches the surface of the adenine nanocrystal, the tip is at first subject to a van der Waals attractive force and after passing through the equilibrium point, the tip receives a repulsive force. When the bond distance of the Ag–N linkage is reduced by 10 %, the repulsive force of 7 pN per molecule is derived from a harmonic oscillation of the displacement by 0.025 nm and the energy difference by 1.7 kcal/mol in the case of the Ad–N3. The repulsive force coincides with the atomic force obtained with our system.

For the further understanding of the tip-enhanced near-field Raman active species of adenine molecules, we theoretically investigated (using the density functional theory calculations) the transition states of both the Ad–N3 and the Ad–N7 isomers by changing the bond distance (in the model) between the nitrogen of the adenine molecule and the silver atom. The bond distances for the calculations were 2.502 (equilibrium), 2.75 (10 % elongation), 2.25 (10 % contraction), and 2.0 Å (20 % contraction). The binding energies, the vibrational frequencies of free adenine molecules, and the two complex isomers of three states are calculated. The partially optimized geometries exhibit imaginary frequencies, indicating transition states for the Ad–N3 and the Ad–N7. The ring–breathing mode of the Ad–N3 shows a significant shift towards a higher frequency as a function of the contracted bond distance between the silver atom and the N3 nitrogen of adenine. The frequency of this mode is shifted upwards by 5 cm^{-1} when the bond distance is reduced by 10 %. The frequency is shifted upwards by 17 cm^{-1} when the bond distance is reduced by 20 %. The frequency shift of this band agrees well with those obtained in TERS experiments. On the other hand, the Ad–N7 shows only small frequency shifts. This difference in frequency shift suggests that interaction between an adenine molecule and the silver tip (e.g., the Ad–N3) may be one of the possible reasons that we see large enhancement effects in TERS.

In addition to the higher-frequency shift, Raman-band broadening is also observed in TERS. The line broadening as well as the Raman frequency shift have also been reported in far-field high-pressure Raman scattering studies [31, 32, 33]. The line broadening occurs not only as a result of surface interaction, but also as a result of pressures caused by the silver tip.

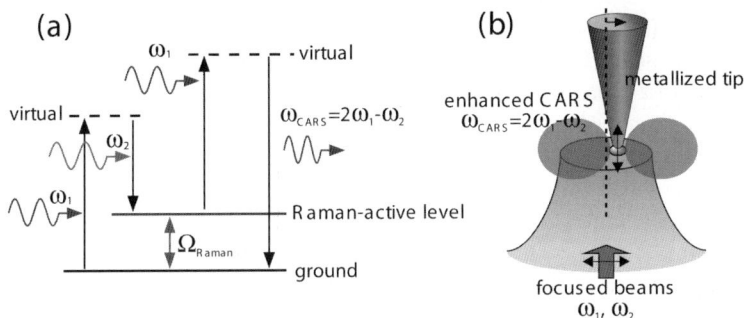

Fig. 7. (a) Energy diagram of coherent anti-Stokes Raman scattering process.
(b) Tip enhancement of CARS polarization of molecules near the metallic tip in a
tightly focused spot

7 Tip-Enhanced Coherent Anti-Stokes Raman Scattering

Coherent anti-Stokes Raman scattering (CARS) is one of the most powerful
four-wave mixing spectroscopic methods, which includes a pump field (ω_1),
a Stokes field (ω_2; $\omega_2 < \omega_1$), and a probe field (ω_1), and induces a nonlinear
polarization at the frequency of CARS [35], as shown in Fig. 7a. When the
frequency difference of ω_1 and ω_2 ($\omega_1 - \omega_2$) coincides with one of the specific molecular vibrational frequencies (Raman) of a given sample, the anti-
Stokes Raman signal is resonantly generated. CARS spectroscopy is sensitive
to molecular species and molecular conformation, resulting in a method of
molecule identification without staining. As a CARS spectrum is obtained
by scanning the frequency difference between two lasers, a high-resolution
spectrometer is not required. Biological samples often have strong autoflu-
orescence that overlaps that of Stokes-shifted Raman scattering. Since the
frequency of CARS is higher than those of CARS excitation lasers, CARS
emission is separable from fluorescence. Although CARS spectroscopy has the
above-mentioned advantages, it has not been utilized for microscopy since the
report of CARS microscopy by *Duncan* et al. in 1982 [36]. Phase matching
is required for CARS, and consequently it was thought to be difficult that
high spatial resolution consists with the phase-matching condition. Recently,
CARS microscopy has been realized to have a three-dimensional resolving
power by collinear illumination optics [37, 38], using a high-NA objective
lens. In this case, the phase-matching condition is relaxed at the tightly focused light spot. CARS is generated only at the focal point in the sample
medium, similar to two-photon fluorescence microscopy [39].

In order to circumvent the problems of extremely small signals in nano-
scaled spectroscopic sensing, we have proposed a combination of a third-order
nonlinear optical effect with the field-enhancement effect of a metallic tip,
that is, a technique for vibrational nanoimaging with tip-enhanced coher-

ent anti-Stokes Raman scattering (TE-CARS), one of third-order coherent nonlinear Raman scattering [35]. The alternative type of NSOM using an aperture-type probe was previously combined with CARS spectroscopy by *Schaller* et al. [40], where they employed a fiber probe for the signal collection of CARS excited by an external illumination and demonstrated chemical selective imaging of a biological specimen. The use of the tip-enhancement effect is, however, more advantageous with respect to spatial resolution, and is indispensable for observation of small numbers of molecules. The excitation of CARS polarization can be further confined spatially and highly enhanced at the very end of the probe tip owing to its third-order nonlinearity, providing higher spatial resolution than tip-enhanced spontaneous Raman scattering. We realized TE-CARS imaging of a specific vibrational mode of DNA molecules in the fingerprint region with a spatial resolution beyond the diffraction limit of light.

Recently, several scientists have reported that tight focusing of the excitation fields with a high-NA objective lens can achieve CARS microscopy with three-dimensional imaging capability at a submicrometer scale [37, 38]. The phase-matching condition can be satisfied automatically in the focused fields of multiple angles [41]. In other words, the phase-matching condition is not necessary to be considered when the CARS polarizations are generated only in a volume smaller than the propagation wavelength of CARS light [42, 43]. In our previous work, CARS was strongly amplified by isolated gold nanoparticles, which verified the possibility of the local enhancement of CARS by a metallic nanostructure [44]. Based on the concept mentioned above, one can observe CARS signals generated by the enhanced electric field at a metallic tip end of nanometric scale [45]. Figure 7b shows a schematic illustration of CARS generation by a metallic probe tip. Both the incident fields (ω_1 and ω_2) are strongly amplified by the metallic tip in the tightly focused spot, and induce CARS polarization with $\omega_{CARS} = 2\omega_1 - \omega_2$ in the molecules located near the tip. As the z-polarized component of the electric field along the tip axis is dominant in the tip-enhanced field [46], the CARS polarizations are induced along the z-direction. For effective coupling of incident fields and tip-enhanced fields, the tip has to be in a position where the incident electric field in the z-direction is strong. Since we use linearly polarized beams, the peaks of the z-component are found at 200 nm from the center of the focused spot in the direction parallel to the polarization of the incident fields. Thus, the tip is displaced to one of the peaks, as shown in Fig. 7b. The CARS polarization of molecules is locally generated within the very small volume near the tip so that the ensemble of the induced polarizations behaves as a dipole oscillating in the z-direction, as seen in Fig. 7b. The backscattered component of the radiation can be efficiently collected with the high-NA focusing lens. Scanning the sample stage, while keeping the tip at the focused spot, can acquire two-dimensional TE-CARS images of a specific vibrational mode with a high spatial resolution that is determined by the size of the tip end rather than the diffraction-limited focused spot.

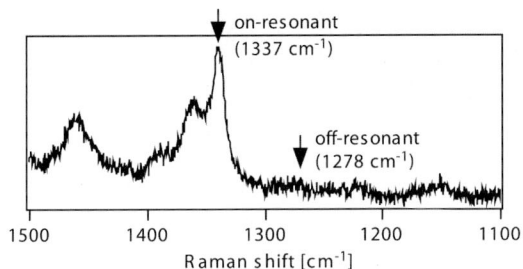

Fig. 8. A spontaneous Raman spectrum of the DNA of poly(dA-dT)-poly(dA-dT). The two frequencies adopted for our TE-CARS imaging are indicated by the *downward arrows*. The on-resonant frequency at $1337\,\text{cm}^{-1}$ can be assigned to the ring-breathing mode of diazole adenine molecule in the DNA

The experimental setup mainly consists of two mode-locked Ti:sapphire lasers (pulse duration 5 ps, spectral band width $\sim 4\,\text{cm}^{-1}$, repetition rate: 80 MHz), an inverted optical microscope, and an AFM using a silicon cantilever tip coated with a 20 nm thick silver film [1, 2, 3]. The ω_1 and ω_2 beams are collinearly overlapped in time and space, and introduced into the microscope with an oil-immersion objective lens (NA = 1.4) focused onto the sample surface. The AFM-controlled tip contacts the sample surface with a constant force and is illuminated by the focused spot. The repetition rate of the excitation lasers is controlled by an electro-optically (EO) modulated pulse picker. The backscattered CARS emission is collected with the same objective lens and detected with an APD based photon-counting module through an excitation-cut filter and a monochromator. The observing spectral width through the detection system is $\sim 12\,\text{cm}^{-1}$. The pulse signals from the APD are counted by a gated photon counter synchronously triggered with the pulse picker. The dark counts are effectively reduced to ~ 0 counts/s with the gate width of 5 ns. The repetition rate was reduced to 800 kHz to avoid the thermal damage on both the sample and the silver tip while keeping the peak power high.

We used DNA molecules of poly(dA-dT) aggregated into clusters for TE-CARS imaging. The poly(dA-dT) solution in water (250 μg/ml) was cast and dried on a coverslip at room temperature with the fixation time of ~ 24 h. The dimensions of the clusters were typically ~ 20 nm in height and ~ 100 nm in width. The frequency difference of the two excitation lasers for TE-CARS was set to be $1337\,\text{cm}^{-1}$, corresponding to a Raman mode of adenine (ring-breathing mode of diazole) by tuning the excitation frequencies ω_1 and ω_2 to be $12\,710\,\text{cm}^{-1}$ (λ_1: 786.77 nm) and $11\,373\,\text{cm}^{-1}$ (λ_2: 879.25 nm), respectively. After the "on-resonant" imaging, the frequency of ω_2 was changed such that the frequency difference corresponds to none of the Raman-active vibrations ("off-resonant"). Figure 8 shows a spontaneous Stokes Raman spectrum of the DNA in a part of the fingerprint region. The solid arrows on the spec-

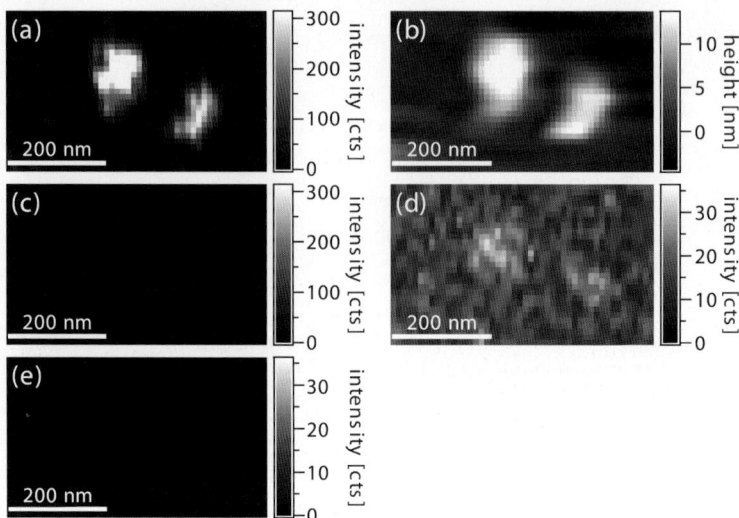

Fig. 9. TE-CARS images of the DNA clusters. (**a**) TE-CARS image at on-resonant frequency (1337 cm^{-1}), and (**b**) the simultaneously obtained topographic image. (**c**) TE-CARS image at the off-resonant frequency (1278 cm^{-1}). (**d**) The same image as (**c**) shown with a different gray scale. (**e**) Far-field CARS image of the corresponding area obtained without the silver tip. The scanned area is 500 nm by 300 nm. The number of photons counted in 100 ms was recorded for one pixel. The acquisition time was ∼ 3 min for the image. The average powers of the 1 and 2 beams were 30 μW and 15 μW at the 800 kHz repetition rate

trum denote the frequencies adopted for the "on-resonant" and "off-resonant" conditions in TE-CARS imaging.

Figure 9 shows the TE-CARS images of the DNA clusters obtained by the TE-CARS microscope. Figures 9a and b are the TE-CARS image at the on-resonant frequency (1337 cm^{-1}) and the simultaneously acquired topographic AFM image. The DNA clusters of ∼ 100 nm diameter are visualized in Fig. 9a. The two DNA clusters with a distance of ∼ 160 nm are obviously distinguished by the TE-CARS imaging. This indicates that the TE-CARS imaging successfully achieved a super-resolving capability beyond the diffraction limit of light. At the off-resonant frequency (1278 cm^{-1}), the CARS signals mostly vanished in Fig. 9c. Figures 9a and c verify that vibrationally resonant CARS is emitted from the DNA molecules at the specific frequency. However, there remains some slight signal increase at the clusters at the off-resonant frequency, as seen in Fig. 9d, which is the same as Fig. 9c but is shown with a different gray scale. This can be caused by both the frequency-invariant (nonresonant) component of the nonlinear susceptibility of DNA [35] and the topographic artifact [47]. Figure 9e is the far-field CARS image at the on-resonant frequency, which was obtained after removing the tip from the sample. The CARS signal was not detected in the far-field CARS image, which confirms that the CARS polarization is effectively induced by the tip-enhanced field. It can be noted that there exists background light in the presence of the tip, as is obvious from Fig. 9d. This background light is emitted from the silver-coated tip. The tip emits light at the same frequency as the CARS ($2\omega_1 - \omega_2$) by the third-order nonlinear susceptibility of silver, which is attributed to local four-wave mixing. In addition, noble metals such as gold and silver generate a white-light continuum, which is induced by multiphoton excited photoluminescence due to recombination radiation between electrons near the Fermi level and photoexcited holes in the d band [48, 49]. These two components become background light and compete with the CARS process. In our experiments the dominant background source is the FWM emission as the monochromator was utilized to selectively detect the signal at $2\omega_1 - \omega_2$. The background light can be seen at both the on-resonant and off-resonant frequencies, as they are independent of the molecular vibrations of the sample. Such light emission from a metallic tip degrades the image contrast and signal-to-noise ratio, and subsequently limits the smallest number of molecules that can be observed. In this experiment, however, the TE-CARS signal intensity largely surpasses the background because the number of molecules in the excited volume is enough to induce the signal.

In order to assess the capability of the sensitivity of the TE-CARS microscopy, we prepared a DNA network of poly(dA-dT)-poly(dA-dT) [50]. DNA (poly(dA-dT)-poly(dA-dT)) dissolved in water ($250 \mu g/ml$) was mixed with $MgCl_2$ ($0.5 mM$) solution, then the DNA solution was cast on a coverslip and blow-dried after the fixation time of $\sim 2 h$. Mg^{2+} has a role for the linkage between DNA and oxygen atoms of the glass surface. Figure 10a shows a typical topographic image of the DNA network sample. The DNA network consists of bundles of DNA double-helix filaments aligned parallel on the glass substrate. Since the diameter of single DNA double-helix filaments is $\sim 2.5 nm$, the height of the bundle structures is $\sim 2.5 nm$, and the width is from 2.5 nm (for single filaments) to a few tens of nanometer (for ca. 10 filaments). The TE-CARS images at the on- and off-resonant frequencies are shown in Figs. 10b,c. The DNA bundles are observed at the resonant frequency in Fig. 10b, while they cannot be visualized at the off-resonant frequency in Fig. 10c. This indicates that the observed contrast is dominated by the vibrationally resonant CARS signals. Figure 10d shows one-dimensional line profiles at $y = 270 nm$, which were acquired with a step of $\sim 5 nm$. The line profile of far-field CARS acquired without the silver tip is also added for comparison. Only the TE-CARS in the on-resonant condition has peaks at the positions of $x \sim 370 nm$ and $x \sim 700 nm$ where adenine molecules exist in the DNA double helix, while the other line profiles do not sense the existence of the molecules. The full width at half maximum of the peak at $x \sim 700 nm$ is 15 nm, which confirms that with the existing TE-CARS setup, we are able to obtain a spatial resolution down to 15 nm. This extremely high resolution is attributed to the combination of the near-field effects of TERS and the nonlinearities of CARS. The intensity enhancement factor for

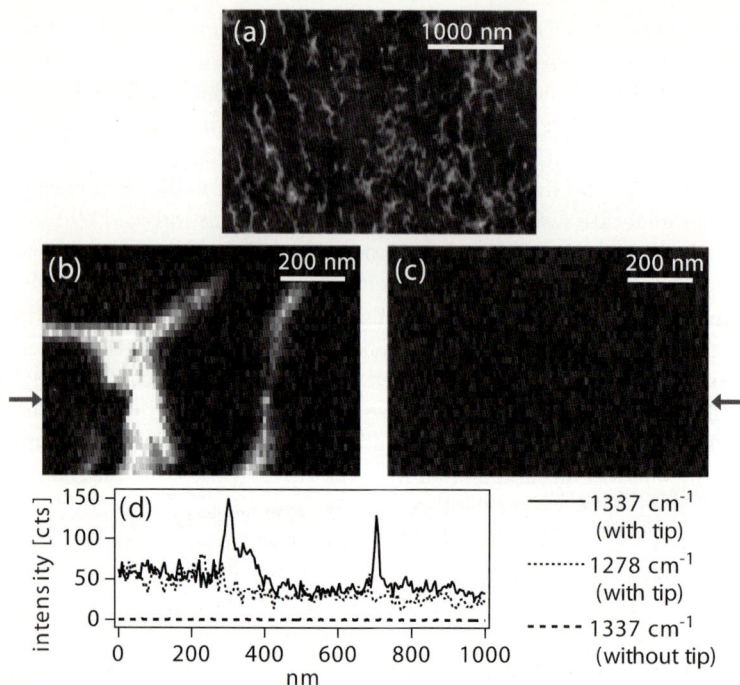

Fig. 10. TE-CARS images of the DNA network. (**a**) Topographic image of the DNA network. (**b**) TE-CARS image at on-resonant frequency ($1337\,\mathrm{cm}^{-1}$). (**c**) TE-CARS image at the off-resonant frequency ($1278\,\mathrm{cm}^{-1}$). (**d**) Cross-sectional line profiles of $y = 270\,\mathrm{nm}$ (indicated by the *solid arrows*). The scanned area is 1000 nm by 800 nm. The number of photons counted in 100 ms was recorded for one pixel. The acquisition time was $\sim 12\,\mathrm{min}$ for the image. The average powers of beams 1 and beam 2 were $45\,\mu\mathrm{W}$ and $23\,\mu\mathrm{W}$ at the 800 kHz repetition rate

each electric field is estimated to be ~ 100-fold. The estimated value of the enhancement factor (~ 100) is quite realistic and reasonable, as compared to previous numerical results [8, 51], although this estimation is very much subject to the changes in each parameter with high-order dependency. We also estimated the size of the locally excited volume of the DNA structure to be ~ 1 zeptoliter. The smallest detectable volume of DNA under the current experimental condition is estimated as $\sim \frac{1}{4}$ zeptoliter, which is derived from the signal-to-noise ratio of $\sim 15 : 1$ in Fig. 10d and the quadratic dependence of the CARS intensity on interaction volume. This indicates that our TE-CARS microscope is capable of sensing a vibrational-spectroscopic signal from an extremely small subzeptoliter volume.

References

[1] N. Hayazawa, Y. Inouye, Z. Sekkat, et al.: Opt. Commun. **183**, 333 (2000)
[2] N. Hayazawa, Y. Inouye, Z. Sekkat, et al.: Chem. Phys. Lett. **335**, 369 (2001)
[3] N. Hayazawa, T. Yano, H. Watanabe, et al.: Chem. Phys. Lett. **376**, 174 (2003)
[4] N. Hayazawa, Y. Inouye, Z. Sekkat, et al.: J. Chem. Phys. **117**, 1296 (2002)
[5] H. Watanabe, Y. Ishida, N. Hayazawa, et al.: Phys. Rev. B **69**, 155418 (2004)
[6] N. Hayazawa, Y. Inouye, S. Kawata: J. Microsc. **194**, 472 (1999)
[7] S. Kawata: *Near-Field Optics and Surface Plasmon Polaritons* (Springer Berlin 2001)
[8] H. Furukawa, S. Kawata: Opt. Commun. **148**, 221 (1998)
[9] A. Jorio, R. Saito, J. H. Hafner, et al.: Phys. Rev. Lett. **86**, 1118 (2001)
[10] G. S. Duesberg, I. Loa, M. Burghard, et al.: Phys. Rev. Lett. **85**, 5436 (2000)
[11] M. S. Dresselhaus, P. C. Eklund: Adv. Phys. **49**, 705 (2000)
[12] S. D. M. Brown, A. Jorio, P. Corio, et al.: Phys. Rev. B **63**, 155414 (2001)
[13] K. Kneipp, H. Kneipp, P. Corio, et al.: Phys. Rev. Lett. **84**, 3470 (2000)
[14] R. Saito, M. Fujita, G. Dresselhaus, et al.: Appl. Phys. Lett. **60**, 2204 (1992)
[15] H. Kataura, Y. Kumazawa, Y. Maniwa, et al.: Synth. Met. **103**, 2555 (1999)
[16] A. M. Rao, E. Richter, S. Bandow, et al.: Science **275**, 187 (1997)
[17] F. Tuinstra, J. L. Koenig: J. Chem. Phys. **53**, 1126 (1970)
[18] H. Watanabe, N. Hayazawa, Y. Inouye, et al.: J. Phys. Chem. B **109**, 5012 (2005)
[19] M. Moskovits: Rev. Mod. Phys. **57**, 783 (1985)
[20] A. Otto, I. Mrozek, H. Grabhorn, et al.: J. Phys.: Condens. Matter **4**, 1143 (1992)
[21] A. Campion, P. Kambhampati: Chem. Soc. Rev. **27**, 241 (1998)
[22] A. M. Michaels, M. Nirman, L. E. Brus: J. Am. Chem. Soc. **121**, 9932 (1999)
[23] M. Futamata, Y. Maruyama, M. Ishikawa: J. Phys. Chem. B **107**, 7607 (2003)
[24] Y. Inouye, S. Kawata: Opt. Lett. **19**, 159 (1994)
[25] A. C. Albrecht: J. Chem. Phys. **34**, 1476 (1961)
[26] U. C. Fischer, D. W. Pohl: Phys. Rev. Lett. **62**, 458 (1989)
[27] A. D. Becke: J. Chem. Phys. **98**, 5648 (1993)
[28] C. Lee, W. Yang, R. G. Parr: Phys. Rev. B **37**, 785 (1988)
[29] M. J. Frisch, G. W. Trucks, H. B. Schlegel, et al.: computer code GAUSSIAN98 Revision A.9 (1998)
[30] C. Otto, T. J. J. van den Tweel, F. F. M. de Mul, et al.: J. Raman Spectrosc. **17**, 289 (1986)
[31] J. R. Ferraro: *Vibrational Spectroscopy at High External Pressures* (Academic, New York 1984) p. 2
[32] R. A. Crowell, E. L. Chronister: Chem. Phys. Lett. **195**, 602 (1992)
[33] S. A. Hambir, J. Franken, D. E. Hare, et al.: J. Appl. Phys. **81**, 2157 (1997)
[34] K. Hoogsten: Acta Crystallogr. **12**, 822 (1959)
[35] Y. R. Shen: *Principle of Nonlinear Optics* (Wiley, New York 1984)
[36] M. D. Duncan, J. Reintjes, T. J. Manuccia: Opt. Lett. **7**, 350 (1982)
[37] A. Zumbusch, G. R. Holtom, X. S. Xie: Phys. Rev. Lett. **82**, 4142 (1999)
[38] M. Hashimoto, T. Araki, S. Kawata: Opt. Lett. **25**, 1768 (2000)
[39] W. Denk, J. H. Strickler, W. W. Webb: Science **248**, 73 (1990)
[40] R. D. Schaller, J. Ziegelbauer, L. F. Lee, et al.: J. Phys. Chem. B **106**, 8489 (2002)

[41] M. Hashimoto, T. Araki: J. Opt. Soc. Am. A **18**, 771 (2001)
[42] J.-X. Cheng, A. Volkmer, X. S. Xie: J. Opt. Soc. Am. B **19**, 1363 (2002)
[43] Z. Xiaolin, R. Kopelman: Ultramicrosc. **61**, 69 (1995)
[44] T. Ichimura, N. Hayazawa, M. Hashimoto, et al.: J. Raman Spectrosc. **34**, 651 (2003)
[45] T. Ichimura, N. Hayazawa, M. Hashimoto, et al.: Phys. Rev. Lett. **92**, 220801 (2004)
[46] A. Bouhelier, M. Beversluis, A. Hartschuh, et al.: Phys. Rev. Lett. **90**, 013903 (2003)
[47] B. Hecht, H. Bielefeldt, Y. Inouye, et al.: J. Appl. Phys. **81**, 2492 (1997)
[48] G. T. Boyd, Z. H. Yu, Y. R. Shen: Phys. Rev. B **33**, 7923 (1986)
[49] J. P. Wilcoxon, J. E. Martin: J. Chem. Phys. **108**, 9137 (1998)
[50] S. Tanaka, L. T. Cai, H. Tabata, et al.: Jpn. J. Appl. Phys. **116**, L407 (2002)
[51] J. T. Krug II, E. J. Sánchez, X. S. Xie: J. Chem. Phys. **116**, 10895 (2002)

Index

Single-Molecule SERS Spectroscopy

Katrin Kneipp[1,2,3], Harald Kneipp[2], and Henrik G. Bohr[3]

[1] Harvard-MIT Division of Health Sciences and Technology, Cambridge, MA
02139, USA
[2] Wellman Center for Photomedicine, Harvard University, Medical School,
Boston, MA 02114, USA
[3] Quantum Protein Center, Technical University of Denmark, DK-2800, Lyngby,
Denmark
kneipp@usa.net

1 Introduction

It has long been the dream of chemists to examine a single molecule and
monitor its structural changes. The observation of a single molecule and its
individual properties and structural transformations can provide useful insight into the nature of processes that cannot be studied in an ensemble of
molecules due to averaging. Moreover, high-throughput structurally selective detection of single molecules and quantification of matter by counting
single molecules represent the ultimate limit in chemical analysis and trace
detection. Today, single-molecule studies are a topic of rapidly growing scientific and practical interest in many fields, such as chemistry, physics, life
sciences, and nanotechnology [1, 2, 3]. Photons have been identified as nearly
perfect tools for detecting and studying single molecules. For example, the
first single-molecule detection under ambient conditions was achieved using
laser-induced fluorescence [4]. However, particularly under ambient conditions, there is a limited amount of molecular information that can be obtained
from a fluorescence signal.

Raman spectroscopy as a vibrational technique provides a high degree
of structural information but the extremely small cross section of the effect,
typically $\sim 10^{-30}\,\mathrm{cm}^2$ to $10^{-25}\,\mathrm{cm}^2$, with the larger values occurring only
under favorable resonance Raman conditions, precludes the use of Raman
spectroscopy as a method for single-molecule detection. For example, the
number of Stokes photons can be estimated as the product of Raman cross
section and excitation intensity: Assuming a Raman line with a scattering
cross section of $10^{-29}\,\mathrm{cm}^2$ and $100\,\mathrm{mW}$ excitation light focused to $1\,\mathrm{\mu m}^2$, a
single molecule scatters $\sim 10^{-4}$ photons per second, which means that one
must wait more than an hour for a Stokes photon from a single molecule. In
comparison, fluorescence cross sections are $\sim 10^{-16}\,\mathrm{cm}^2$.

Estimated enhancement factors for the Raman signals in SERS started
from modest factors of 10^3 to 10^5 in the initial SERS experiments [5, 6].
For excitation laser wavelengths in resonance with the absorption band of
the target molecule, surface-enhanced resonance Raman scattering (SERRS)

K. Kneipp, M. Moskovits, H. Kneipp (Eds.): Surface-Enhanced Raman Scattering – Physics and
Applications, Topics Appl. Phys. **103**, 261–278 (2006)
© Springer-Verlag Berlin Heidelberg 2006

benefits from a superposition of surface enhancement and resonance Raman effect and can result in higher total effective Raman cross sections. Enhancement factors on the order of about 10^{10} to 10^{11} for rhodamine 6G and other dyes adsorbed on colloidal silver and excited under molecular resonance conditions have been reported by several authors in independent experiments [7, 8, 9, 10, 11, 12, 13, 14], based on a comparison of SERRS signals with fluorescence or with normal Raman signals of nonenhanced molecules such as methanol. In 1988, we measured SERRS spectra from about 100 rhodamine 6G molecules in a solution of small silver aggregates and concluded that in principle, even lower detection limits in SERS may be possible than were obtained with fluorescence [12, 15]. These results were confirmed in 1995 when we collected SERRS signals from 60 rhodamine 6G molecules and found that single-molecule Raman spectroscopy could be possible by further improving the experimental conditions [16].

However, SERS enhancement factors were underestimated in all these experiments for many years as they were inferred from a comparison between surface-enhanced Raman signals and normal Raman scattering or fluorescence. The main reason for this underestimation is that we do not know how many molecules are involved in the SERS process. The applied assumption that all molecules in a SERS experiment contribute to the observed SERS signal always results in estimating too small enhancement factors. In order to avoid this problem, we invented a different approach in which both surface-enhanced Stokes *and* anti-Stokes Raman scattering were used to extract information on the effective SERS cross section [17]. The idea behind this experiment was that a very strong Raman process can measurably populate the first vibrational level in addition to the thermal population. The effective Raman cross section, operative in this vibrational pumping process, can be inferred from the deviation of the anti-Stokes to Stokes signal ratio from the expected Boltzmann population. We discuss this effect in more detail in another Chapter in this book. Using this approach, we inferred effective SERS cross sections in nonresonant experiments that exceeded all previous assumptions and result in SERS enhancement factors of at least 10^{14}, amounting to nearly ten orders of magnitude higher than assumed in the first SERS experiments, but also at least 2 to 3 orders of magnitude higher than concluded from former SERRS studies. Effective SERS cross sections inferred from vibrational pumping were on the same order of magnitude as good fluorescence cross sections, making SERS spectroscopy of single molecules a realistic goal.

Two options have been used to achieve single-molecule sensitivity in a SERS experiment. One is based on the extremely large SERS enhancement factors obtained on silver and gold colloidal clusters at near-infrared excitation [18, 19, 20]. No resonance Raman effect for the target molecule is required in these experiments. A second approach exploits a favorable superposition of surface enhancement and resonance Raman enhancement of the analyte molecule, and is performed using excitation wavelengths within or close to the absorption band of the analyte molecule [21, 22, 23].

In this Chapter we will focus on single-molecule surface-enhanced Raman scattering experiments using near-infrared excitation where no resonance Raman effect contributes to the observed total enhancement. This is particularly true for adenine, which absorbs in the UV at 270 nm and is excited at 830 nm. Two types of SERS-active nanostructures will be used, colloidal nanoclusters in solution and fractal cluster structures on surfaces. These experiments will confirm many of the theoretical results discussed in the previous chapters, but will also show that there are still unanswered questions related to single-molecule Raman spectroscopy. In Sect. 3, we will consider some aspects of SERS as a single-molecule tool, and compare it with single-molecule fluorescence.

2 Single-Molecule SERS Experiments

Electromagnetic and chemical effects can account for the highest observed SERS enhancement on the order of 10^{14}. Theoretical estimates show that in principle, two kinds of composites of nanoparticles can account for the hot spots providing giant electromagnetic SERS enhancement: 1. Dimers and small aggregates formed from silver and gold nanoparticles [24, 25, 26, 27], where strong field enhancement exists particularly at the intersection between two nanoparticles, and 2. fractal types of nanostructures, which are comprised of aggregates and clusters formed by the self-assembling of nanoparticles [28, 29, 30, 31].

Electromagnetic effects in fractal structures can result in enhancement factors up to 10^{12} and even higher. Recent studies show that combining plasmon resonances and photonic resonances can give rise to electromagnetic enhancement factors sufficient for single-molecule Raman detection without chemical enhancement [32]. The extent of electronic enhancement to single-molecule SERS remains a subject of discussion [22, 33, 34], but there is compelling evidence that single-molecule Raman spectroscopy is primarily a phenomenon associated with the enhanced local optical fields in the vicinity of metal nanostructures. The high local optical fields in the hot spots of silver and gold cluster structures can be considered the key effect in SERS at an extremely high enhancement level [18, 35, 36, 37, 38]. This is also supported by the scaling of the enhancement factors for SERS and SEHRS as we have discussed previously in another Chapter. Large effective SERS cross sections allow the detection of single molecules based on pumped anti-Stokes SERS spectra [19, 39].

Figure 1 shows gold nanoclusters in various size ranges providing a high enough enhancement level for single-molecule Raman experiments. Larger aggregates show a fractal nature [40, 41].

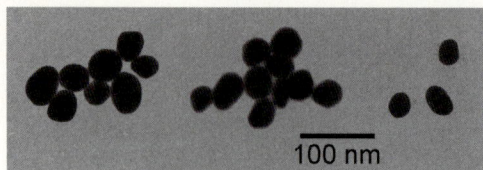

Fig. 1. Isolated gold nanoparticles and gold aggregates

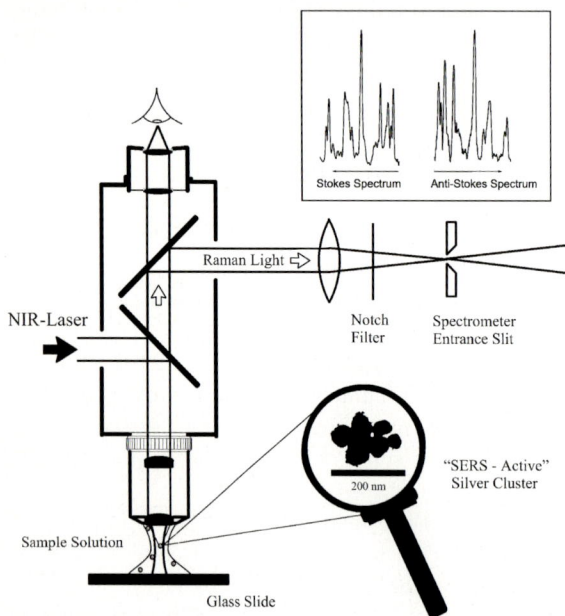

Fig. 2. Schematic of a typical single-molecule SERS experiment. Reprinted with permission from [37], Copyright 2002 Institute of Physics Publishing, Ltd

2.1 Single-Molecule SERS on Silver and Gold Nanoclusters in Solution

Figure 2 shows a schematic of a typical single-molecule SERS experiment performed in silver or gold colloidal solution [18, 19, 20]. Spectra are excited by a cw Ti:sapphire laser operating at 830 nm. A microscope attachment is used for laser excitation and collection of the Raman scattered light. The analyte is added to the solution of small silver or gold colloidal clusters (for example, see Fig. 1).

Concentration ratios of nanoclusters and target molecules of at least 10 make it unlikely that more than one analyte molecule will attach to the same colloidal cluster, avoiding formation of aggregates of the target molecules on the surface. Application of the analyte at such low concentration does not induce coagulation of the colloidal particles/colloidal clusters, and avoids for-

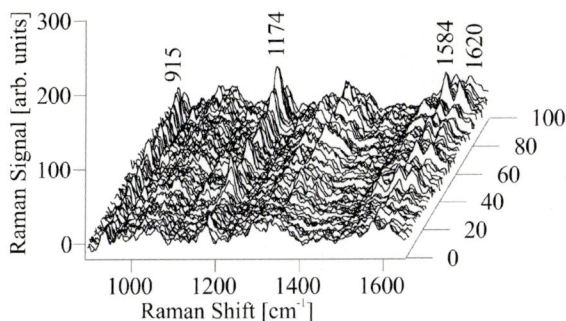

Fig. 3. 100 single-molecule SERS spectra of crystal violet on silver colloidal clusters. Collection time was one second for one spectrum, 830 nm excitation. Reprinted with permission from [18], Copyright 1997 American Physical Society

mation of larger clusters. The described procedure results in individual single molecules that are adsorbed on silver colloidal clusters. Analyte concentrations on the order of 10^{-12} M to 10^{-14} M and probed volumes whose sizes are on the order of femtolitre to picolitre result in average numbers of one or fewer target molecules in the focus volume.

The Brownian motion of single analyte molecule-loaded silver or gold clusters into and out of the probed volume results in strong statistical changes in the height of Raman signals measured from such a sample in time sequence. This is demonstrated in Fig. 3 that shows typical unprocessed SERS spectra measured in a time sequence from a sample with an average of 0.6 crystal violet molecules in the probed 30 pl volume [18]. Figure 4 displays the peak heights of the 1174 cm^{-1} crystal violet Raman line for the 100 SERS spectra (top), the background level of the colloidal solution with no analyte present (middle), and 100 measurements of the 1030 cm^{-1} Raman line of 3 M methanol in colloidal silver solution with about 10^{14} molecules of methanol in the scattering volume (bottom).

The SERS signals appear at different power intervals, which can be assigned "0"-, "1"-, "2"-, and "3"-molecule events. The normal Raman signal of the 10^{14} methanol molecules appears at about the same level as the SERS signal of a single-crystal violet molecule, confirming an enhancement factor on the order of 10^{14}. The inserts in Fig. 4 show the statistical distribution of the Raman and SERS signals. As expected, the methanol Raman signals collected in time sequence displays a Gaussian distribution. In contrast, the statistical distribution of the "0.6-molecules SERS signal" exhibits four relative maxima that are reasonably fit by the superposition of four Gaussian curves. The gradation of the areas of the four statistical peaks are roughly consistent with a Poisson distribution for an average number of 0.5 molecules. This reflects the probability of finding 0, 1, 2 or 3 molecules in the scattering volume during the actual measurement. Comparing the measured Poisson distribution with the 0.6-molecule concentration/volume estimate,

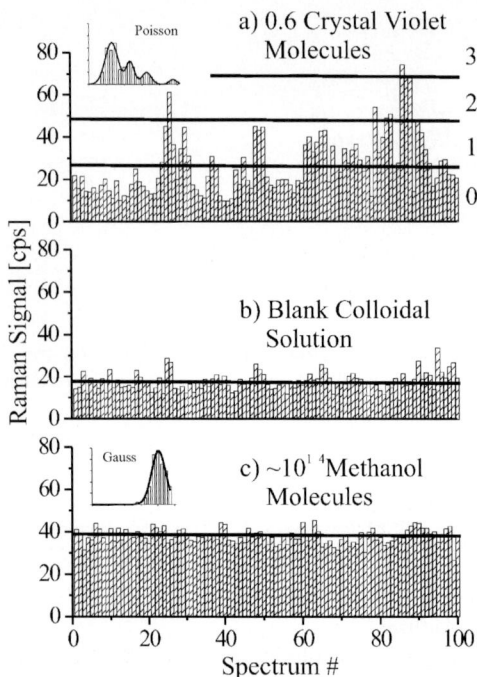

Fig. 4. Peak heights of the 1174 cm^{-1} SERS line from an average of 0.6 crystal violet molecules (*top*), background signal (*middle*), peak heights of a Raman line measured from 10^{14} methanol molecules (*below*). The *horizontal lines* display the thresholds for one-, two- and three-molecule signals (*top*), the average background (*middle*), and the average 10^{14}-molecule signal (*below*). The *inserts* show the Poisson and Gaussian statistics of single-molecule and many-molecule Raman signals, respectively. Reprinted with permission from [18], Copyright 1997 American Physical Society

we conclude that about 80 % of molecules are detected by SERS. This statistical behavior of single-molecule SERS signals has been corroborated in several other studies [19, 20, 42]. The change in the statistical distribution of the Raman signal from Gaussian to Poisson when the average number of target molecules in the scattering volume is one or less is evidence for single-molecule detection by SERS. Fluctuations in the SERS signal of the target molecule are described by the Poisson statistics, and therefore reflect only Brownian motion of the silver or gold nanoclusters.

A look at Fig. 3 also shows fluctuations and changes in the SERS spectra collected in time sequence. Sometimes, new Raman lines appear in a spectrum, which do not exist in the following trace. The "new" spectral features do not correlate with the spectral signal strengths of the target molecule. We ascribe these Raman lines to the surface-enhanced spectra of impurities at the surface of the colloidal particles, which are probably introduced in the chemical preparation process. The impurity spectra change due to different colloidal clusters, which move into the focal volume loaded with different impurities.

The SERS signal of a target molecule at an extremely low concentration can vanish in a background SERS signal of impurities on the colloidal particles. This observation was discussed in our first studies of SERS spectroscopy

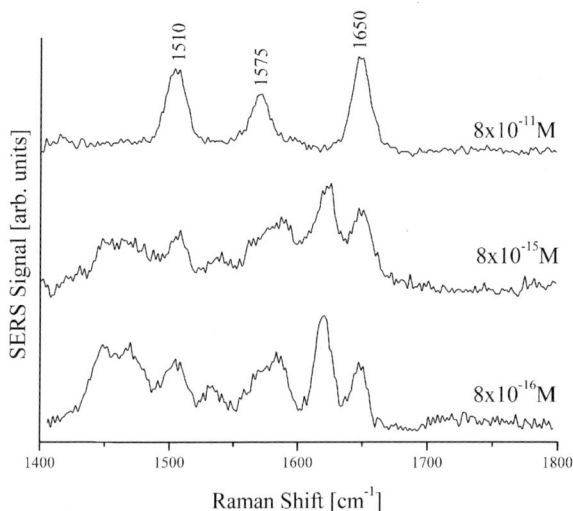

Fig. 5. SERS spectra of rhodamine 6G in silver colloidal solution in concentrations between 10^{-11} M and 10^{-16} M along with SERS spectra of impurities

at the single-molecule level [12, 16]. Figure 5 demonstrates this effect for the SERS spectrum of rhodamine 6G in a silver colloidal solution.

2.2 Single-Molecule SERS on Fixed Fractal Silver and Gold Cluster Structures

As discussed above, areas of particularly high local optical fields can also exist on silver and gold fractal structures formed, for example, by aggregation of colloidal particles of these metals. Near-field images taken on top of silver-cluster structures confirm the existence of optical hot spots and "cold zones" [43]. The dimensions of these hot spots can be orders of magnitude smaller than the wavelengths of light [44]. Since SERS takes place in the local fields of metallic nanostructures, the probed volume is determined by the confinement of the local fields, which can be two orders of magnitude smaller than the diffraction limit. The strongly confined hot spots provide the opportunity to spectroscopically select single nano-objects or molecules within a larger population. This was demonstrated in SERS experiments performed on bundles of single-wall carbon nanotubes (SWNTs) on fractal silver cluster structures [45, 46]. Usually a bundle of nanotubes gives rise to inhomogeneously broadened Raman lines due to tubes of different diameters present in the bundle. By scanning over nanotube bundles in contact with a fractal silver surface, one can sometimes measure SWNT spectra showing a very narrow linewidth without the sign of inhomogeneous broadening, indicating that this spectrum must originate from a single SWNT, or at most, from

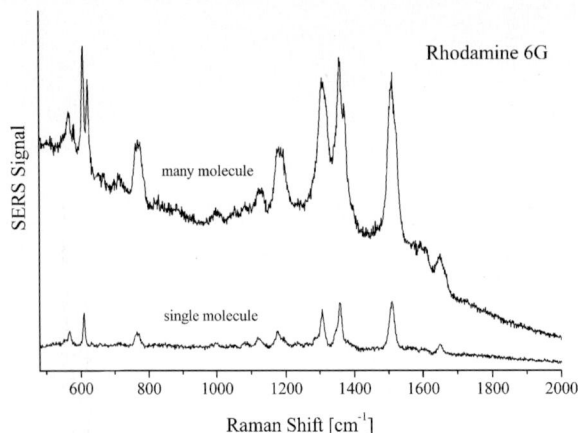

Fig. 6. Many-molecule (*top*) and single-molecule (*bottom*) SERS spectra of rhodamine 6G measured at different places on a fractal silver surface

very few tubes. The spectroscopic selection of single SWNTs within a bundle shows that SERS signals are collected from dimensions smaller than 5 nm.

It is interesting to compare SERS on the hot spots of nanoclusters with SERS exploiting local optical fields of metal tips [47, 48, 49]. In spite of the high field confinement for these tips, enhancement factors are reported to be orders of magnitude below those required for nonresonant single-molecule Raman spectroscopy. However, tip-enhanced Raman spectroscopy might approach the single-molecule level in combination with a very strong resonance Raman effect. For example, the technique can be used for studying single SWNTs because these tubes show an extremely strong resonance Raman effect based on resonances of the excitation and/or scattered photons with the van Hove singularities in the electronic energy levels of these one-dimensional carbon structures [50, 51].

The strong confinement of local optical fields is also demonstrated in Fig. 6, which shows two typical SERS spectra collected with a $\sim 1\,\mu m$ laser spot size from two different places at a fractal silver structure with an average of about 100 rhodamine 6G molecules in the $1\,\mu m$ spot. The upper spectrum reveals a many-molecule spectrum showing typical inhomogeneous broadening of the Raman lines due to the different environment of the rhodamine 6G molecules contributing to the spectrum on the silver surface. The bottom spectrum in Fig. 6 shows the Raman signal of a single rhodamine 6G molecule measured in a situation where a hot spot in the probed focal area selects a single molecule from a larger population of rhodamine 6G molecules present in the $1\,\mu m$ spot [52]. The single-molecule spectrum shows a clear decrease in linewidth compared to many-molecule SERS spectra as well as other changes due to different adsorption forms [53].

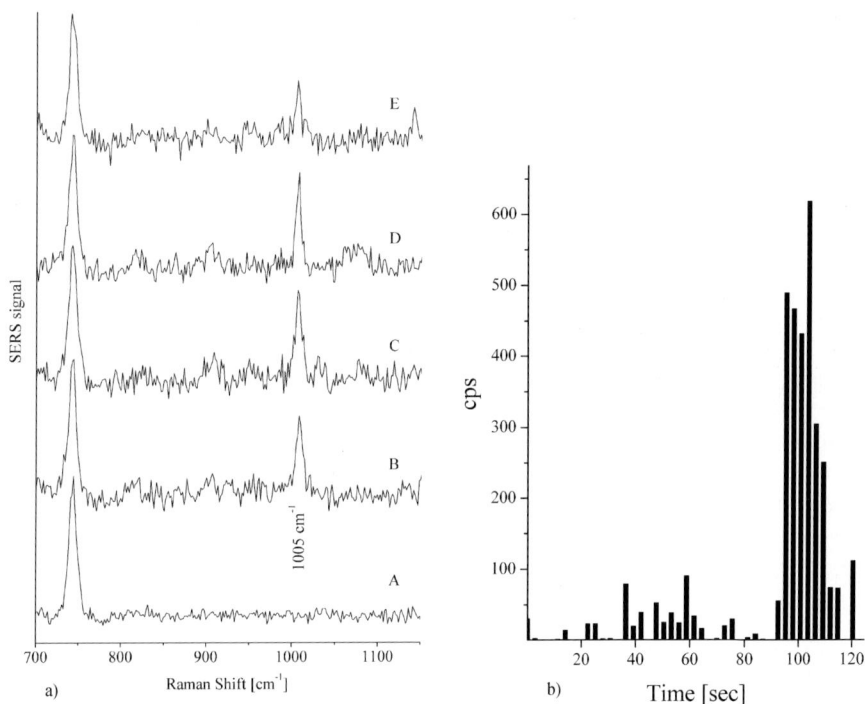

Fig. 7. (**a**) Selected single-molecule SERS spectra of enkephalin from one fixed spot on a fractal silver surface showing the 1000 cm^{-1} SERS line of phenylalanine and a line at 750 cm^{-1}, which can be ascribed to citrate on the silver colloids. (**b**) Time series of the 1000 cm^{-1} phenylalanine signal measured from one fixed spot on a sample with an average of one enkephalin molecule in the laser spot. Spectra were observed over a time interval of 120 s, 1 s collection time each. The signal level of 100 cps represents approximately the background level. Reprinted with permission from [54], Copyright 2004 IOS Press

We demonstrate the capability of fractal silver surfaces for single-molecule detection in Fig. 7, using the small protein enkephalin as a target molecule [54]. Enkephalin is a mixture of two pentapeptides, [Leu]enkephalin and [Met]enkephalin. Here we detect enkephalin at the single-molecule level based on the strongly enhanced ring-breathing mode of phenylalanine ($\sim$ 1000 cm^{-1}), which is a building block of both pentapeptides. Figure 7a shows selected single-molecule SERS spectra measured in a time interval in a spectral window, which displays the 1000 cm^{-1} SERS line of phenylalanine and a line at about 750 cm^{-1}, which can be ascribed to a citrate impurity on the silver colloids. Figure 7b displays the Raman signal measured at 1000 cm^{-1} shift from the 830 nm excitation from the same $\sim$ 1 μm spot in a time sequence (1 s collection time each). Over the first 95 s, no SERS signal was measured. Then, a SERS signal appears relatively abruptly, stays for about 20 s at the same level, and vanishes again. Such behavior, namely, the appearance of the $1000 \, cm^{-1}$ SERS signal within a 10 s to 30 s time window, has been observed in many measurements on single enkephalin molecules. If the enkephalin signal appears, it always appears at the same level. Spectra 7aB) to 7aE) were measured in the time window between 95 s and 115 s, spectrum 7aA) shows a situation when the $1000 \, cm^{-1}$ SERS signal does not exceed the noise level. A possible explanation for these changes in scattering power is that a single enkephalin molecule is diffusing on the colloidal silver cluster and can only be "seen" when it enters a hot spot.

Fluctuations in scattering power and/or sudden spectral shifts and changes that appear as a "blinking" of SERS signals have been reported by several authors [21, 23, 55]. Different effects can account for these spectral changes, such as thermally and nonthermally activated diffusion of molecules, as well as real transformations, such as protonation and deprotonation [56, 57]. Although blinking had been claimed as a hallmark of single-molecule detection, it has become evident that this behavior is not necessarily connected to single-molecule SERS. The effect has also been observed in lower-concentration "many-molecule" SERS spectra [58]. Fluctuations and changes in SERS spectra are discussed in more detail in several places in this book.

3 SERS as a Single-Molecule Analytical Tool – Comparison between SERS and Fluorescence

Ideally, one would like to have a spectroscopic tool that detects a single molecule and simultaneously identifies its chemical structure. A SERS spectrum, containing different vibrational modes, provides high structural information content on a molecule that is the equivalent of its fingerprint. At present, SERS is the only way to detect a single molecule and simultaneously identify its chemical structure. But, as we have demonstrated for enkephalin, also measuring only one typical SERS line of the target molecule and using this Raman line as an intrinsic spectroscopic signature for the specific molecule is a useful tool for detecting and tracking a known molecule without the use of a fluorescence tag.

If labeling of the target molecule should be necessary for any reason, SERS labels can provide several advantages compared to fluorescence tags. Under biologically relevant conditions, such as in solutions and at room temperature, fluorescence signals are very broad spectral signatures, typically $400 \, cm^{-1}$ to $800 \, cm^{-1}$. Compared to a fluorescence band, SERS results in spectrally narrow lines typically $10 \, cm^{-1}$ to $20 \, cm^{-1}$. Because of the high specificity of a Raman spectrum, spectral overlap between different labels is very unlikely, even when their molecular structures are very similar [36]. This results in a huge arsenal of potential SERS labels making possible multiplex detection [59]. Moreover, detecting a label based on a signature that comprises several sharp lines allows for the application of spectral correlation techniques, and therefore improved contrast [60].

A principal advantage of SERS over fluorescence for single-molecule detection involves the total number of photons that can be emitted by a molecule. This is determined by the maximum number of excitation–emission cycles a molecule survives. In fluorescence experiments, this number is limited by the rate of photobleaching of the molecule. No photodecomposition has been observed in nonresonant SERS, and also in SERRS experiment the effect of bleaching is strongly reduced.

Another number that is of particular interest for the rapid detection and screening of single molecules is the maximum number of photons that can be emitted by a molecule per unit time. Under saturation conditions, this number is inversely proportional to the lifetime of the excited molecular states involved in the optical detection process. Due to the shorter vibrational relaxation times compared to the electronic relaxation times, a molecule can go through more Raman cycles than fluorescence cycles per time interval. Therefore, the number of Raman photons per unit time that can be emitted by a molecule under saturation conditions can be higher than the number of fluorescence photons by a factor of 10^2 to 10^3 [12].

In single-molecule SERS experiments in solution, the target molecule has to be attached to SERS-active nanoclusters (compare Sect. 2.1). This attachment of the target molecule to a much bigger particle can be an advantage for single-molecule detection as increased diffusion times minimize the possibility of the target molecule escaping out of the focal volume of the laser too rapidly.

For practical applications, it can be of interest to detect single molecules in a fluorescent environment. In this case, the background problem can be rigorously treated by using anti-Stokes SERS signals for single-molecule detection at the high-energy side of the excitation laser [19, 20].

3.1 Potential Applications of Single-Molecule Raman Spectroscopy

Single-molecule capabilities of SERS open up exciting perspectives and new aspects in the field of biophysics and biochemistry as compared to fluorescence, which is widely used as a single-molecule spectroscopic tool in this field [37]. For example, SERS opens up opportunities for monitoring protein at the single-molecule level in their natural aqueous environment.

One of the most spectacular potential applications of single-mo'
SERS is in the field of rapid DNA sequencing at the single-molecule le'
These applications range from the use of Raman spectroscopic c'
zation of specific DNA fragments down to structurally sensiti'
of single bases without the use of fluorescent or radioactive t
the intrinsic surface-enhanced Raman scattering of the bas

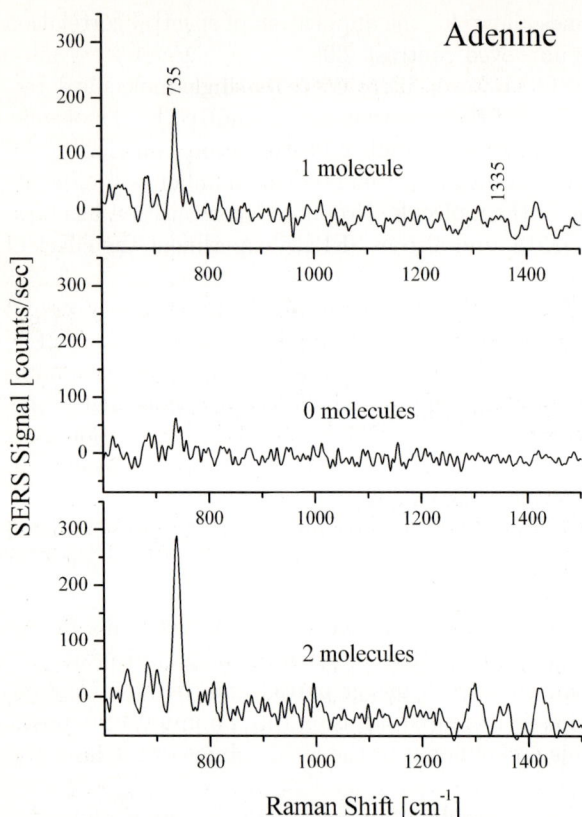

Fig. 8. SERS spectra of single adenine molecules in silver colloidal solution measured in one second collection time using 830 nm cw excitation. Reprinted with permission from [20], Copyright 1998 American Physical Society

SERS cross sections of the order of 10^{-16} cm^2 can be inferred for adenosine monophosphate (AMP) and for adenine on colloidal silver clusters. NIR-SERS spectra of adenine and adenosine monophosphate (AMP) are identical, indicating sugar and phosphate bonds do not interfere with the strong SERS effect of adenine.

Figure 8 displays situations in aqueous solution where "1", "0", or "2" adenine molecules attached to silver nanoclusters were present in the probed volume during the 1 s collection time. Due to the electromagnetic origin of the enhancement, it should be possible to achieve SERS cross sections for other bases in the same order of magnitude as adenine when they are attached to colloidal silver or gold clusters. The nucleotide bases show well-distinguished surface-enhanced Raman spectra, also shown in Fig. 9. Thus, after cleaving single native nucleotides from the DNA or RNA strand into a medium containing colloidal silver or gold clusters, direct detection and identification of

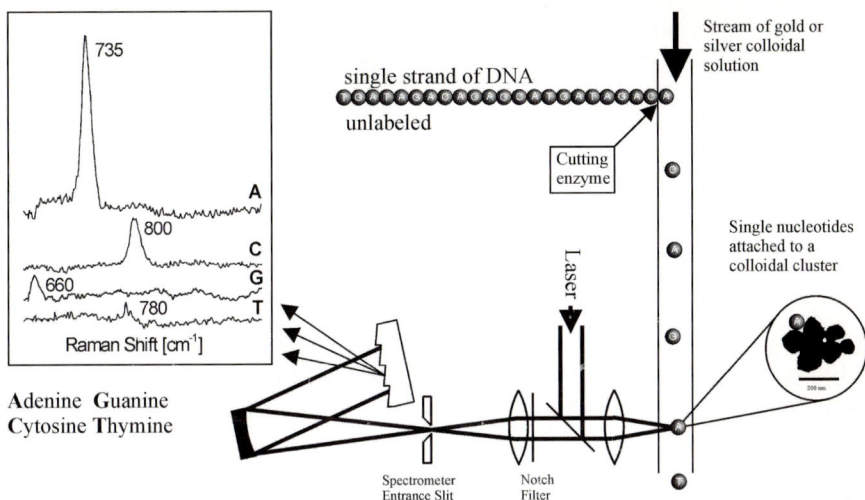

Fig. 9. Schematic of DNA sequencing based on the intrinsic Raman spectra of the four bases. Reprinted with permission from [37], Copyright 2002 Institute of Physics Publishing, Ltd

single native nucleotides should be possible, due to the unique SERS spectra of their bases. SERS active silver or gold nanoclusters can be provided in a flowing stream of colloidal solution or onto a moving surface with silver or gold cluster structures.

It is interesting to estimate the detection rate of single nucleotides in such an experiment. Single-molecule adenine spectra can be measured at good signal-to-noise ratios of 10 in a 1 s collection time at 3×10^5 W/cm^2 excitation. Assuming a SERS cross section on the order of 10^{-17} to 10^{-16} cm^2 and a vibrational lifetime on the order of 10 ps, saturation of SERS will be achieved at 10^8 to 10^9 W/cm^2 excitation intensity. Extrapolation to saturation conditions shows that single-molecule SERS spectra over the complete fingerprint region (ca. 700 cm^{-1} to 1700 cm^{-1}) should be measurable in ms or at kHz rates.

4 Conclusion

Numerous experiments conducted during the last decade have demonst‐
that SERS enables the measurement of Raman spectra of single mc'
one at a time. Presently, single-molecule experiments without the '
Raman contribution of the target molecule require SERS enhancer
of 10^{13} to 10^{14}. In principle, electromagnetic and/or chemic'
account for this enhancement level. Theoretical estimates for
and gold nanostructures from dimers to self-similar structur

optical fields in hot spots can give rise to SERS enhancement factors of up to 10^{12} and even higher. However, in many experiments, electronic effects might contribute in order to match the experimentally observed effective cross sections. The extent of electronic enhancement remains a subject of discussion. There is evidence that SERS is primarily a phenomenon associated with the enhanced local optical fields in the vicinity of metal nanostructures. Enhanced and strongly confined fields are always associated with high field gradients. These large field gradients may direct molecules to the hot spot and hold them there for detection.

As a single-molecule tool, SERS opens up some exciting potential capabilities as compared with fluorescence, another widely used single-molecule technique. To detect and identify single molecules by fluorescence, in many cases the molecules must be labeled to achieve large enough quantum yields and distinguishable spectral properties. However, in each case, the structural information on the target molecule is very limited, particularly under ambient conditions. SERS is a method for detecting *and* identifying a molecule without any labeling because it is based on the intrinsic Raman signature of the molecule. Moreover, Raman scattering is a very general property of a molecule, and almost every molecule has Raman-active molecular vibrations that can be "seen" by the Raman effect at a cross section of at least 10^{-30} cm^2, whereas far fewer molecules show fluorescence. For the detection of a molecule by SERS, it has to be attached to SERS-active substrates in order to increase its effective Raman cross section by 12–14 orders of magnitude. Due to the primarily electromagnetic origin of this enhancement, it should be possible to achieve an equally strong SERS effect for each molecule, thus making SERS a broadly applied tool for single-molecule detection.

Acknowledgements

This work is supported in part by DOD grant # AFOSR FA9550-04-1-0079 and by the generous gift of Dr. and Mrs. J.S. Chen to the optical diagnostics program of the Massachusetts General Hospital, Wellman Center for Photomedicine.

One of us (KK) would like to thank Michael S. Feld and Ramachandra R. Dasari for kind support and collaboration when performing the first single-molecule Raman experiments as Heisenberg Fellow at the G. R Harrison Spectroscopy Laboratory at MIT.

References

[1] M. Eigen, R. Rigler: Proc. Nat. Acad. Sci. USA **91**, 5740 (1994)
[2] X. S. Xie, J. K. Trautman: Ann. Rev. Phys. Chem. **49**, 441 (1998)
[3] F. Kulzer, M. Orrit: Ann. Rev. Phys. Chem. **55**, 585 (2004)

[4] R. A. Keller, W. P. Ambrose, P. M. Goodwin, et al.: Appl. Spectrosc. **50**, A12 (1996)

[5] D. L. Jeanmaire, R. P. Van Duyne: J. Electroanal. Chem. **84**, 1 (1977)

[6] M. G. Albrecht, J. A. Creighton: J. Am. Chem. Soc. **99**, 5215 (1977)

[7] A. Bachackashvilli, S. Efrima, B. Katz, et al.: Chem. Phys. Lett. **94**, 571 (1983)

[8] K. Kneipp, G. Hinzmann, D. Fassler: Chem. Phys. Lett. **99**, 5 (1983)

[9] K. Kneipp, D. Fassler: Chem. Phys. Lett. **106**, 498 (1984)

[10] P. Hildebrandt, M. Stockburger: J. Phys. Chem. **88**, 5935 (1984)

[11] K. Kneipp, H. Kneipp, M. Rentsch: J. Molec. Struct. **156**, 3 (1987)

[12] K. Kneipp: Exp. Techn. Phys. **36**, 161 (1988)

[13] B. Pettinger, K. Krischer: J. Electron. Spectrosc. Relat. Phenom. **45**, 133 (1987)

[14] B. Pettinger, K. Krischer, G. Ertl: Chem. Phys. Lett. **151**, 151 (1988)

[15] D. C. Nguyen, R. A. Keller, M. Trkula: J. Opt. Soc. Am. B: Opt. Phys. **4**, 138 (1987)

[16] K. Kneipp, Y. Wang, R. R. Dasari, et al.: Appl. Spectrosc. **49**, 780 (1995)

[17] K. Kneipp, Y. Wang, H. Kneipp, et al.: Phys. Rev. Lett. **76**, 2444 (1996)

[18] K. Kneipp, Y. Wang, H. Kneipp, et al.: Phys. Rev. Lett. **78**, 1667 (1997)

[19] K. Kneipp, H. Kneipp, G. Deinum, et al.: Appl. Spectrosc. **52**, 175 (1998)

[20] K. Kneipp, H. Kneipp, V. B. Kartha, et al.: Phys. Rev. E **57**, R6281 (1998)

[21] S. Nie, S. R. Emory: Science **275**, 1102 (1997)

[22] A. M. Michaels, M. Nirmal, L. E. Brus: J. Am. Chem. Soc. **121**, 9932 (1999)

[23] H. X. Xu, E. J. Bjerneld, M. Kall, et al.: Phys. Rev. Lett. **83**, 4357 (1999)

[24] H. Metiu: Possible causes for surface enhanced raman scattering, in W. F. Murphy (Ed.): *Proc. VIIth Int. Conf. Raman Spectrosc. Linear and Non Linear Processes* (North-Holland, Amsterdam 1980) pp. 382–384

[25] M. Inoue, K. Ohtaka: J. Phys. Soc. **52**, 3853 (1983)

[26] H. Xu, J. Aizpurua, M. Kaell, et al.: Phys. Rev. E **62**, 4318 (2000)

[27] L. Gunnarsson, T. Rindzevicius, J. Prikulis, et al.: J. Phys. Chem. B **109**, 1079 (2005)

[28] M. I. Stockman, V. M. Shalaev, M. Moskovits, et al.: Phys. Rev. B **46**, 2821 (1992)

[29] E. Y. Poliakov, V. M. Shalaev, V. A. Markel, et al.: Opt. Lett. **21**, 1628 (1996)

[30] Y. Yamaguchi, M. K. Weldon, M. D. Morris: Appl. Spectrosc. **53**, 127 (1999)

[31] Z. J. Wang, S. L. Pan, T. D. Krauss, et al.: Proc. Nat. Acad. Sci. USA **100**, 8638 (2003)

[32] S. L. Zou, G. C. Schatz: Chem. Phys. Lett. **403**, 62 (2005)

[33] A. Otto: J. Raman Spectrosc. **36**, 497 (2005)

[34] L. P. Capadona, J. Zheng, J. I. Gonzalez, et al.: Phys. Rev. Lett. **94**, 058301 (2005)

[35] K. Kneipp, H. Kneipp, R. Manoharan, et al.: Appl. Spectrosc. **52**, 1493 (1998)

[36] K. Kneipp, H. Kneipp, I. Itzkan, et al.: Chem. Rev. **99**, 2957 (1999)

[37] K. Kneipp, H. Kneipp, I. Itzkan, et al.: J. Phys.: Condens. Matter **14**, R597 (2002)

[38] M. Moskovits: J. Raman Spectrosc. **36**, 485 (2005)

[39] K. Kneipp, H. Kneipp, I. Itzkan, et al.: Chem. Phys. **247**, 155 (1999)

[40] D. A. Weitz, M. Oliveria: Phys. Rev. Lett. **52**, 1433 (1984)

[41] K. Güldner, R. Liedtke, K. Kneipp, et al.: Morphological study of colloidal silver clusters employed in surface-enhanced Raman spectroscopy (SERS), in H. D. Kronfeldt (Ed.): *29th EGAS Berlin* (European Physical Society Berlin 1997) p. 128

[42] P. Etchegoin, R. C. Maher, L. F. Cohen, et al.: Chem. Phys. Lett. **375**, 84 (2003)

[43] P. Zhang, T. L. Haslett, C. Douketis, et al.: Phys. Rev. B **57**, 15513 (1998)

[44] V. M. Shalaev: Phys. Rep. **272**, 61 (1996)

[45] K. Kneipp, H. Kneipp, P. Corio, et al.: Phys. Rev. Lett. **84**, 3470 (2000)

[46] K. Kneipp, H. Kneipp, I. Itzkan, et al.: Nonlinear Raman probe of single molecules attached to colloidal silver and gold clusters, in V. M. Shalaev (Ed.): *Optical Properties of Nanostructured Random Media* (Springer, Berlin, Heidelberg, New York 2002) pp. 227–247

[47] N. Hayazawa, Y. Inouye, Z. Sekkat, et al.: Opt. Commun. **183**, 333 (2000)

[48] A. Hartschuh, M. R. Beversluis, A. Bouhelier, et al.: Philos. Trans. Roy. Soc. London: A Math. Phys. Eng. Sci. **362**, 807 (2004)

[49] B. Pettinger, B. Ren, G. Picardi, et al.: Phys. Rev. Lett. **92**, 096101 (2004)

[50] M. S. Dresselhaus, G. Dresselhaus, A. Jorio, et al.: J. Nanosci. Nanotechnol. **3**, 19 (2003)

[51] N. Anderson, A. Hartschuh, S. Cronin, et al.: J. Am. Chem. Soc. **127**, 2533 (2005)

[52] K. Kneipp, H. Kneipp: Canadian J. Anal. Sci. Spectrosc. **48**, 125 (2003)

[53] T. Vosgrone, A. J. Meixner: Chem. Phys. Chem. **6**, 154 (2005)

[54] K. Kneipp, H. Kneipp, S. Abdali, et al.: Spectrosc. Int. J. **18**, 433 (2004)

[55] M. Futamata, Y. Maruyama, M. Ishikawa: J. Molec. Struct. **735-36**, 75 (2005)

[56] A. Weiss, G. Haran: J. Phys. Chem. B **105**, 12348 (2001)

[57] Z. J. Wang, L. J. Rothberg: J. Phys. Chem. B **109**, 3387 (2005)

[58] C. J. L. Constantino, T. Lemma, P. A. Antunes, et al.: Appl. Spectrosc. **57**, 649 (2003)

[59] Y. C. Cao, R. C. Jin, J. M. Nam, et al.: J. Am. Chem. Soc. **125**, 14676 (2003)

[60] J. Kneipp, H. Kneipp, W. L. Rice, et al.: Anal. Chem. **77**, 2381 (2005)

[61] W. P. Ambrose, P. M. Goodwin, J. H. Jett, et al.: Berichte der Bunsen-Gesellschaft – Phys. Chem. Chem. Phys. **97**, 1535 (1993)

Index

Temporal Fluctuations in Single-Molecule SERS Spectra

Anna Rita Bizzarri and Salvatore Cannistraro

Biophysics and Nanoscience Centre, CNISM, Dipartimento di Scienze Ambientali, Università della Tuscia, I-01100, Viterbo, Italy
{bizzarri,cannistr}@unitus.it

1 Introduction

New advances in ultrasensitive spectroscopy, able to detect a single molecule (SM), are of utmost interest in different disciplines from both fundamental and application aspects [1, 2, 3]. Earlier SM fluorescence studies at ambient conditions, showed blinking, spectral jumps, intensity fluctuations, yielding thus new insight into phenomena usually wiped out in ensemble average [4, 5].

Although Raman spectroscopy is characterized by an extremely low cross section, SM detection can be reached when the molecules are adsorbed onto metallic surfaces with nanometer-scale roughness or onto metal nanoparticles [6, 7]. Such an approach, called surface-enhanced Raman spectroscopy (SERS), gives rise in a drastic increase of the Raman cross section (up to 10^{14}) [8, 9, 10], as due to two, likely cooperating, mechanisms: an electromagnetic (em) local field enhancement and a charge transfer (CT) between the molecule and the metal surface [6, 7]. The latter, which requires a tight, chemical interaction between the molecule and the metal [11, 12], has been recently suggested to play a dominant role in SM SERS spectra [13]. With respect to other SM spectroscopies, SERS is endowed with a high sensitivity coupled to a rewarding chemical and structural specificity [14]. Therefore, SERS is one of the most sensitive spectroscopic approaches available for analytical chemistry, nanomedicine and nanotechnology [15, 16].

SERS spectra in the SM regime exhibit drastic temporal fluctuations in both line intensity and frequency [17, 18]. The vibrational mode emission of a SM has been shown to undergo a characteristic intermittent behavior that might encode both the dynamics of the molecule and the details of its interaction with the environment [19]. Information about the structural conformations sampled by the molecule can also be obtained; this being particularly relevant for proteins, due to the complex energy landscape explored during their dynamical evolution [20].

Therefore, an investigation of the SM spectral fluctuations, based also on suitable statistical approaches, may help in addressing some fundamental issues, such as ergodicity, statistical aging, entanglement of vibrational modes, nonstationarity of the emission processes, etc. [21, 22, 23]; these being of high relevance for a detailed knowledge of the fundamental dynamical processes in

K. Kneipp, M. Moskovits, H. Kneipp (Eds.): Surface-Enhanced Raman Scattering – Physics and Applications, Topics Appl. Phys. **103**, 279–296 (2006)
© Springer-Verlag Berlin Heidelberg 2006

molecular systems [24]. Moreover, an analysis of SM SERS fluctuations could also lead to new insights into the charge or electron-transfer (ET) mechanisms regulating the molecule–metal interaction under light excitation [25]. Such an aspect may deserve some interest to unravel the role played by the CT in SERS and even in the interfacial photophysical processes occurring in optoelectronic devices [26, 27].

On such a basis, we have performed a statistical analysis on the intensity and frequency fluctuations of SERS spectra of single Myoglobin (Mb) and protoporphyrin IX (FePP) molecules adsorbed onto silver nanoparticles. Such an analysis has allowed us to investigate the role of the continuum background, underneath the SERS spectra, on the intensity fluctuations.

The spectral jumps observed in the FePP vibrational modes that are fingerprints of the iron oxidation states, evidenced a light-induced, reversible, ET occurring between the molecule and the metal surface. These SERS experiments have also been conjugated with scanning tunneling microscopy (STM) measurements to closer elucidate the ET properties of the molecules at a metal surface.

An analysis of the temporal emission of most of the molecular vibrational modes has shown a random switching from a bright (on) state to a dark (off) state under continuous laser excitation. The probability distribution associated with the time intervals between the off states follows a power law that has been traced back to a Lévy statistics. Such a behavior has been suggested to arise from an anomalous diffusion of the FePP molecule on the metal surfaces, likely in connection with a modulation of the CT between the molecule and the metal. Furthermore, the presence of the protein milieu around the heme group has been shown to induce a higher variability on the intensity fluctuations of the heme vibrational modes.

All these findings constitute a new insight into both the photophysical behavior of molecules and their interaction with the nanoenvironment.

2 Materials and Methods

Solutions of colloidal silver were prepared by standard citrate reduction of $AgNO_3$ (Sigma) by following the procedure of *Lee* and *Meisel* [28]. The concentration of the silver colloidal particles, estimated from the optical extinction spectrum, was about 10^{-11} M, corresponding to about 7×10^{12} particles per liter. As activation agent NaCl was added to reach the final concentration of 0.25 mM. FePP (Sigma) was dissolved into a KOH solution 0.5 M at pH = 12 and 10^{-6} M. Purified horse Mb (Sigma) was dissolved into a phosphate buffer solution at pH 6.8 and a concentration of 1 mM. An aliquot of successive dilutions of each of these solutions was incubated with silver colloidal suspension for 3 h at room temperature to obtain a final concentration of 10^{-11} M with an approximate ratio of 1 : 1 between molecules and colloidal particles. The samples were obtained from a 10 μl droplet of these solutions

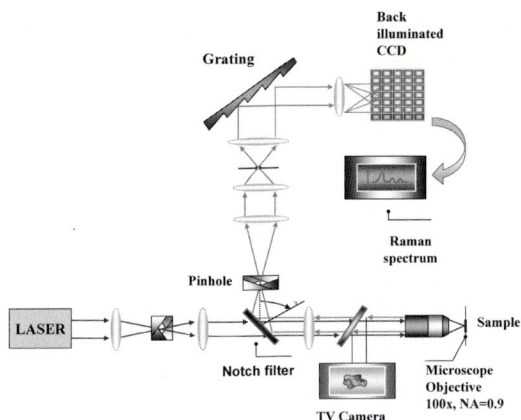

Fig. 1. A schematic representation of the Raman equipment

deposited onto a glass slide, of area $15 \times 15 \, \mathrm{mm}^2$, coated with polymerized 3-aminopropyltriethoxysilane (APES, Sigma) and, finally, dried. A characterization of immobilized silver colloidal particles, performed by atomic force microscopy (AFM) under ambient conditions, revealed single spherical and rod-shaped particles and aggregates of two, three, up to many particles; the heterogeneous particle-size distribution being characterized by an average of about 70 nm [29]. Silver STM tips, prepared by cutting a silver wire, were washed in acetone, ethanol, and finally in ultrapure water, and incubated for 1 h at room temperature with an FePP solution at 10^{-9} M.

SERS spectra were collected with a confocal SuperLabram (Jobin–Yvon) equipment using the 514.5 nm radiation line of an argon laser. The illumination and backscattering collection system consisted of a confocal microscope coupled to a single-grating spectrometer (300 mm focal length spectrograph with a 1800 grooves/mm grating) and a liquid-nitrogen-cooled backilluminated CCD detector; a schematic representation of the equipment is shown in Fig. 1.

The microscope objective was 100× with NA = 0.9, producing a laser spot size of about 1 μm in diameter. The spectral resolution is better than $5 \, \mathrm{cm}^{-1}$. The laser power was varied from 0.1 mW to 8 mW, resulting into a range of 0.01 mW to 0.8 mW for the power impinging on the sample. The spectra were collected in sequence with 1 s of integration time and 1 s from one measurement to the next by an automatic acquisition routine.

Tunneling spectroscopy was performed with a PicoSPM (Molecular Imaging). I–V curves were recorded on FePP adsorbed on silver tips in close proximity to a freshly cleaved highly oriented pyrolitic graphite (HOPG) electrode in a nitrogen atmosphere. The scan was disabled and the tip set at a tunneling resistance of $4 \times 10^{10} \, \Omega$ (200 mV, 5 pA).

The total noise N_t of a SERS signal can be determined as the root of the sum of the square noise components:

$$N_t = [N_{SN}^2 + N_R^2 + N_D^2]^{\frac{1}{2}}, \tag{1}$$

where the readout noise N_R, and the dark charge noise N_D, are specified by the chip manufacturer; N_{SN} is the shot noise associated with the signal and whose amplitude is given by the square root of the measured signal. For our CCD chip, N_R was 5 electrons rms and N_D one electron/pixel/h. The CCD chip had, at a wavelength of 514.5 nm, a quantum efficiency of about 0.92 with one count per two collected photons. In a typical SM SERS experiment, a rate of about 100–400 counts per second are usually detected; the raw data including an underlying continuum. Accordingly, a total noise, dominated by the shot noise, of about 7–18 counts is expected for measurements obtained for an integration time of 1 s [8, 30]. As long as the shot noise is dominant, a Poisson distribution with a standard deviation equal to N_t, is expected to describe the spread of the Raman intensity around the average value.

3 Results and Discussion

3.1 Fluctuations in SM SERS Spectra

Scanning the sample area under the microscope objective of the Raman equipment, generally reveals emitting sites with a different average intensity. Most of these sites are characterized by an intensity almost equivalent to that of the noise. Only a few bright emitting sites, usually called hot spots, corresponding to about 1 % of the total, are observed.

Figure 2 shows a sequence of SERS spectra with 1 s of integration time and excited at a wavelength of 514.5 nm, from a typical bright site of immobilized Ag colloids incubated with FePP at 10^{-11} M. From spectrum to spectrum, there is a great variability in both the relative peak intensity as well as in the peak frequency. Lines appear and disappear suddenly and the peak intensity, as well as the frequency of the lines, drastically change in time; similar results having been obtained for Mb [29] and cytochrome c [31], at the SM level. The emission behavior resembles that found in the SM SERS spectra of heme-proteins [31, 32, 33], of green fluorescent protein [34], and of other organic molecules [17, 18, 35].

The drastic fluctuations and blinking of modes, in connection with the extremely low concentration of molecules present in the initial solution (an average number of less than one molecule in the laser-illuminated spot is estimated [18, 31]) are indicative that SM detection has been reached in our SM SERS spectra, in agreement with what was found in the literature [18, 32].

Figure 3a and 3b show the spectra obtained by summing up a large number of successive SM SERS traces from single Mb and FePP molecules,

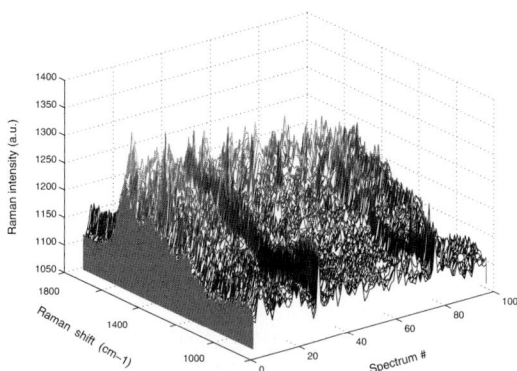

Fig. 2. A sequence of 100 SERS spectra with 1 s of integration time from a bright site of immobilized silver colloidal particles incubated with FePP at a concentration of 10^{-11} M

respectively. These spectra appear very similar to those taken from a dried droplet of bulk solutions (10^{-4} M) (see dashed lines in Fig. 3). The almost complete recovery of the bulk vibrational spectra, reached either by summing several SM spectra or, equivalently by using a long integration time, suggests a substantial ergodicity of the system, and that no significant photodegradation has occurred in our sample during the measurements. However, in a few cases, the SERS signals become dominated by two characteristic, high-intensity, peaks at about $1350\,\mathrm{cm}^{-1}$ and $1580\,\mathrm{cm}^{-1}$. These peaks, usually attributed to carbon contamination [36], might arise from a photoinduced degradation of aromatic molecules [37]; these SERS series having been discarded in the further analysis.

3.2 Statistical Analysis of Intensity Fluctuations in SM SERS Spectra

As already mentioned, intensity and spectral fluctuations in SM SERS spectra may encode information about both the dynamics of the molecule and its interactions with the environment including the metal surface. We have taken into account the intensity fluctuations of the vibrational modes of FePP and Mb.

Figure 4 shows the intensity as a function of time (a), and the related histograms (b), of two of the most intense vibrational modes of Mb and FePP [38]: the iron spin marker ν_3 (at around $1480\,\mathrm{cm}^{-1}$) and the vinyl stretching mode ν_{10} (at around $1620\,\mathrm{cm}^{-1}$). For comparison, the same analysis has been performed at $1800\,\mathrm{cm}^{-1}$, where no Raman signal is registered. Drastic temporal fluctuations for both of the molecules are observed at all three frequencies. Such a recurrent feature of the SERS spectra at the SM level [17,18] discloses processes strongly modulating the Raman scattering of

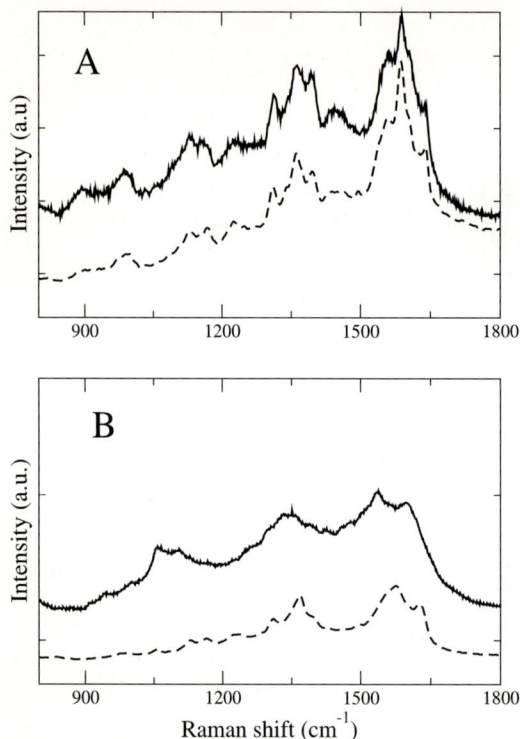

Fig. 3. (a) *Continuous line*: Collection of 600 1 s SERS spectra from immobilized silver colloids incubated with Mb at 10^{-11} M. *Dashed line*: Raman spectrum from a dry droplet of bulk Mb (10^{-4} M in phosphate buffer) with 60 s of integration time. (b) *Continuous line*: Collection of 600 1 s SERS spectra from immobilized silver colloids incubated with FePP at 10^{-11} M. *Dashed line*: Raman spectrum from a dry droplet of bulk FePP (10^{-4} M in alkaline aqueous medium) with 60 s of integration time. The SERS spectra were recorded in sequence lasting 1 s between two successive measurements. The spectral resolution is better than $5\,\mathrm{cm}^{-1}$

the molecules at the metal surface. The histograms of Fig. 4b are consistent with a single-mode distribution for all the analyzed lines of the two systems. From the analysis of the distribution spread, quantified by their standard deviations, some observations can be made: 1. The standard deviations measured at $1480\,\mathrm{cm}^{-1}$ and $1620\,\mathrm{cm}^{-1}$ are found to be higher than those at $1800\,\mathrm{cm}^{-1}$ that, in turn, significantly exceed the values expected from the CCD noise (see values in Fig. 4b). The intensity fluctuations at $1800\,\mathrm{cm}^{-1}$, where no Raman lines occur, monitor the variability of the continuum background that always underlies the SERS lines [10, 18] (see also below). On the other hand, the intensity fluctuations detected at $1480\,\mathrm{cm}^{-1}$ and $1620\,\mathrm{cm}^{-1}$ are expected to arise from the variability of both the continuum background and the specific vibrational modes; the latter encompassing the intrinsic dy-

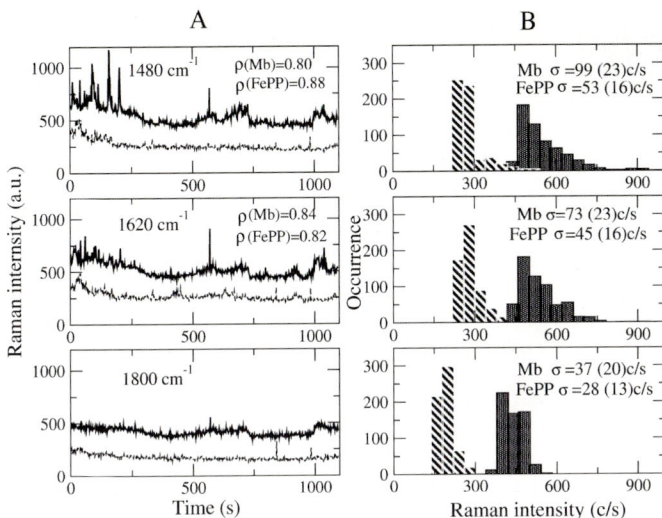

Fig. 4. (a) Intensity, as a function of time, detected at $1480\,\mathrm{cm}^{-1}$, $1620\,\mathrm{cm}^{-1}$, and $1800\,\mathrm{cm}^{-1}$ from SM SERS spectra of Mb (*black lines*) and FePP (*gray lines*). The values of the crosscorrelation ρ between the intensity at $1480\,\mathrm{cm}^{-1}$ and $1800\,\mathrm{cm}^{-1}$ and at $1620\,\mathrm{cm}^{-1}$ and $1800\,\mathrm{cm}^{-1}$ were calculated by (2). (b) Histograms of data in (a) for Mb (*dark pattern*) and for FePP (*dashed pattern*). σ indicates the standard deviations of the data; in parentheses the corresponding σ values expected from the CCD noise (see (1))

namics of the molecule, and its interaction with the environment. 2. The standard deviations related to Mb appear higher than those of FePP. Actually, the presence of the protein, which dynamically explores the complex energy landscape of its conformational substates [20], reasonably leads to a larger variability of the heme group vibration intensities [29, 31].

On visual inspection, the intensity trend of the three frequencies in Fig. 4a seems to indicate that the Raman emissions are strongly correlated in time; drastic intensity jumps simultaneously occurring at the three frequencies for Mb and FePP. The same trend in time has also been observed for the other main vibrational modes and even for the total intensity [18]. To quantify such a correlation, we have evaluated the crosscorrelation, between the intensities at the two frequencies ν_1 and ν_2, at a 0 time delay [39]:

$$\rho_{\nu_1,\nu_2} = \frac{\sigma_{\nu_1,\nu_2}}{\sigma_{\nu_1}\sigma_{\nu_2}}, \tag{2}$$

where σ_{ν_1} and σ_{ν_2} are the standard deviations of the intensity at ν_1 and ν_2, respectively, and σ_{ν_1,ν_2} is the covariance given by:

$$\sigma_{\nu_1,\nu_2} = \frac{1}{N}\sum_i [I_i(\nu_1) - \overline{I(\nu_1)}][I_i(\nu_2) - \overline{I(\nu_2)}]. \tag{3}$$

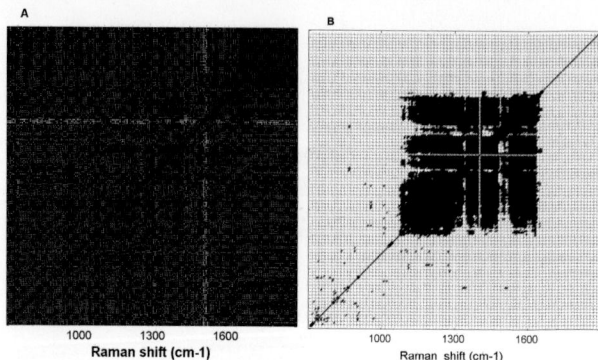

Fig. 5. Two-dimensional crosscorrelation map (see (2)) from a series of 600 SERS spectra of FePP obtained from (**a**) the raw data and (**b**) after removing the continuum background. The continuum background has been removed from each spectrum by subtracting the intensity measured at $1800\,cm^{-1}$ at all frequencies. *Dark points* mark regions, over a grating, with crosscorrelation values higher than 0.8

$I_i(\nu_j)$ $(j = 1, 2)$ is the Raman intensity, detected at the wavelength ν_j at time t_i; $\overline{I(\nu_j)}$ is the Raman intensity, at the wavelength ν_j, averaged over all the measurement time intervals. Accordingly, high ρ_{ν_1,ν_2} values indicate that the intensity at ν_1 fluctuates, around the average, in parallel with the intensity at ν_2.

The crosscorrelation, calculated by (2), between the two intensities at $1480\,cm^{-1}$ and $1620\,cm^{-1}$ is found to be higher than 0.8 for both Mb and FePP (see the legend of Fig. 4). Strikingly enough, similar high correlation values are also obtained for the crosscorrelation between the $1800\,cm^{-1}$, and the $1480\,cm^{-1}$ or $1620\,cm^{-1}$ intensities (see values in Fig. 4). Furthermore, high correlation values are observed for the intensities between each couple of frequencies of the FePP SERS spectrum. This can be visualized from the two-dimensional (2D) map, shown in Fig. 5a, where crosscorrelation values higher than 0.8 are detected throughout almost all the frequency range.

Such a high correlation finds a correspondence with that reported for rhodamine [40], for which it has been put into relationship to the presence of SERS signals due to bare colloids. These results, also in connection with a variability at $1800\,cm^{-1}$ higher than that due to the noise, suggest that the continuum significantly contributes to the intensity modulation of the vibrational-mode intensities. Provided that such a background is not due to fluorescence, commonly affecting Raman spectra and quenched when molecules are adsorbed onto metal surfaces [7, 41], several hypotheses have been put forward about its origin. Indeed, it could be suggested to arise from an electronic scattering from the metal [42], or from a byproduct of a nonradiative CT process between the surface and the molecule [12].

To put into evidence the correlation between the real FePP signals, we have to get rid of the continuum background. Figure 5b shows the map of the 2D crosscorrelation after having removed the continuum background from each spectrum. Correlation values higher than 0.8 are detected only at the crossing points between the frequencies of the main FePP vibrational modes, at variance with what is observed in Fig. 5a, in which a specific intensity correlation was observed. The evidence that a high correlation in the intensity of the FePP vibrational modes persists after removing the continuum, suggests a common origin for their intensity fluctuations. It could be hypothesized that the intensity fluctuations of the specific vibrational modes could be driven by the correspondent fluctuations of the enhancement factor. In particular, the presence of the continuum could be connected to the SERS mechanism, through, for example, a CT process between the molecule and the metal surface, in good agreement with the recent findings supporting the dominant role of the CT to obtain SM SERS signals [13].

3.3 ET in SM SERS Spectra of FePP

The capability of FePP iron to switch from Fe^{2+} to Fe^{3+} and backwards, when illuminated in the presence of a metallic surface, can have a high impact from an application standpoint. Indeed, FePP has been proposed to be a suitable, robust molecular element for assembling electronic and photonic nanodevices [43, 44]. Furthermore, the iron oxidation state of FePP, which constitutes the prosthetic group of several metalloproteins, plays a key role in many biological functions [45]. Therefore, it would be interesting to follow the temporal evolution, at the SM level, of the vibrational markers of the iron oxidation state in FePP. We have then focused our attention on the spectral region of the ν_4 band, corresponding to the pyrrole half-ring vibration, marker of the heme iron oxidation state. In bulk, this band is peaked at $1363\,cm^{-1}$ for the iron ferrous state, and at a higher frequency, $1375\,cm^{-1}$, for the ferric state [38]. The intensity detected at both these frequencies, reveals drastic jumps, as a function of time, similarly to what was previously shown for the other vibrational modes. Moreover, a temporal analysis of the peak occurrence at each of the two frequencies (see Fig. 6 and its legend), indicates that the two peaks are alternatively observed in almost all the spectra registered in a time sequence. A few cases with a simultaneous detection of both lines have also occurred (triangles in Fig. 6).

The observed bimodal behaviour in the peak appearance can be linked to a fast and reversible change in the oxidation state of the FePP iron ion during the measurement [46, 47]. The detection of both the oxidation states in a single spectrum, observed for a few cases, can be due to a switching between Fe^{2+} and Fe^{3+} during the integration time.

Further information about the ET properties of FePP at a metal interface, can be obtained by performing a SERS/STM combined experiment in which FePP is adsorbed on a silver STM tip. Therefore, it has been verified that

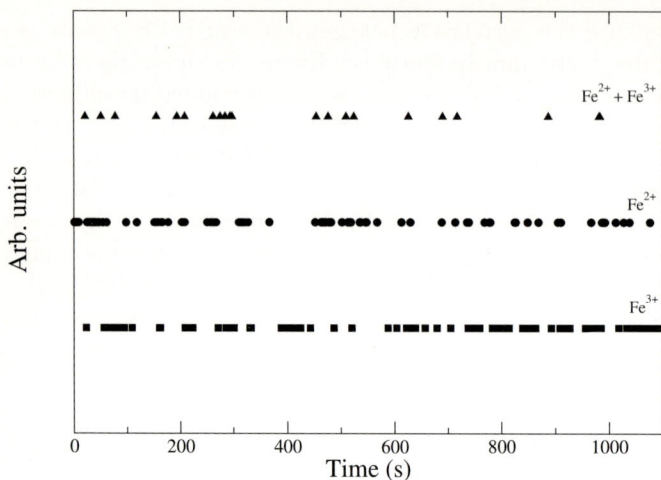

Fig. 6. Appearance of Raman peaks, over the continuum background, as a function of time, in SM SERS spectra of FePP, at $1363\,cm^{-1}$ (*squares*), at $1375\,cm^{-1}$ (*circles*) and at both $1363\,cm^{-1}$ and $1375\,cm^{-1}$ (*triangles*). A peak at a given frequency is detected when its intensity overcomes the threshold of 2σ above the noise level of the spectrum

the tip roughness is suitable to yield a huge SERS enhancement down to SM detection; the occurrence of ET, similarly to what was observed with silver colloids, having also been revealed [25]. Then, we have investigated the electron flow in the molecule–tip tunneling junction, mounted on a STM equipment.

The recorded I–V curve is shown in Fig. 7a (continuous line) together with data related to a bare tip (dashed line). The tunneling current as a function of the bias in the presence of FePP appears markedly asymmetric, increasing rapidly up to about $0.8\,nA$ for negative bias lower than $-0.7\,V$, while it remains small in reverse bias. By contrast, the tunneling curve for the bare silver tip is highly symmetric, with the absolute current intensity smaller than $0.2\,nA$. Generally, the observed trend for the I–V curve, indicative of a diode-like curve (current rectification) [48], reflects the redox properties of the molecule [49, 50]. Then, the current flowing through the tunneling junction obtained by approaching the tip to a HOPG substrate has been recorded in a sequence of $10\,s$ at a fixed bias of $-0.2\,V$, with the feedback control transiently disabled. While the current recorded on a bare tip is almost constant with very small fluctuations, on the contrary, the FePP–tip system exhibits drastic tunneling-current fluctuations (see Fig. 7b). The reported results on an FePP molecule adsorbed onto a silver tip are indicative of a significant variability in the tip–molecule substrate junction during the tunneling experiments, and are closely reminiscent of the variability observed in SM SERS spectra.

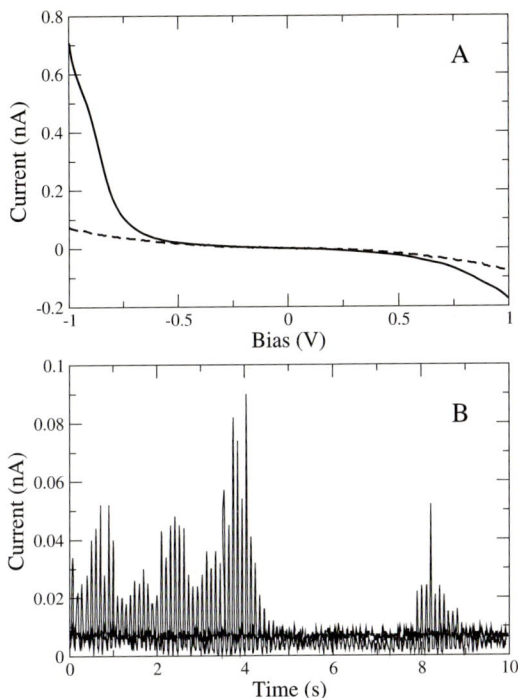

Fig. 7. (a) *I–V* characteristics as recorded by tunneling spectroscopy of FePP–tip (*solid line*) and bare tip (*dashed line*). Measurements were performed under nitrogen atmosphere at a starting tunneling resistance of $4 \times 10^{10}\,\Omega$ (bias, $0.2\,\mathrm{V}$; tunneling current, $5\,\mathrm{pA}$). (b) Tunneling current, as a function of time, at $-0.2\,\mathrm{V}$ bias for bare tip (*black curve*) and for a silver tip incubated with FePP (*gray curve*)

The switching of the iron oxidation state of FePP could be induced by visible light that triggers a nonradiative CT between the FePP molecule and the silver surface. Such a picture finds a correspondence with the Otto–Persson model for which ballistic electrons generated via excitation of the surface plasmon couple to the chemisorbed molecule [12, 42]. In other words, ballistic electrons from the silver surface could jump nonradiatively towards the FePP molecule and backwards. In this framework, the highest occupied molecular orbital (HOMO) and the lowest unoccupied molecular orbital (LUMO) of the adsorbed molecule should be symmetrically located in energy with respect to the Fermi level of the metal that acts as an initial or final state in the resonance Raman process [7, 12, 42]. Therefore, only molecules with low-lying LUMO levels can interact with ballistic electrons giving rise to high SERS signals [51]. The related positions among the HOMO, LUMO levels and the Fermi energy of the FePP could be slightly modulated by different factors, such as the presence of oxygen or some doping elements, and the arrangement of the molecule with respect to the surface [7, 46]. Furthermore, the observed fluctuations in the tunneling current, reflecting a variability in the electronic coupling between the molecule and the metal, can be traced back to migrations of the molecule and/or changes of the molecule orientations with respect to the metal surface during the measurements.

Finally, we mention that the occurrence of an ET between FePP and the metal surface could provide a support to the CT mechanism in yielding the SERS enhancement required in the SM regime [13].

3.4 Lévy Statistics in SM SERS Spectra of FePP

At first glance, the intermittent behavior of the vibrational modes in SM SERS signals appears to be essentially erratic. However, it would be interesting to assess if some law could be hidden in this phenomenon. To such a purpose, we have focused our attention on the appearance and disappearance of three main vibrational modes in the SERS spectrum of single FePP molecules.

The vertical lines in Fig. 8 mark the times, within the acquisition sequence, at which peaks at $1480\,\mathrm{cm}^{-1}$, $1570\,\mathrm{cm}^{-1}$ and $1620\,\mathrm{cm}^{-1}$ are detected. These peaks reveal a binary trace (intermittence), which is common to the other vibrational modes. This means that the molecular vibrational modes may switch from a bright (on) state to a dark (off) state under continuous laser excitation. Notably, no correlation (or anticorrelation) among the temporal emission of the three peaks is registered (see also legend of Fig. 8). Accordingly, it can be suggested that the various vibrational modes follow independent activation channels during the temporal evolution of the molecule dynamics.

To analyze the statistical properties of the on/off state occurrence for the selected vibrational emission modes, we have collected for ten spectra series in sequence, the time intervals, τ_{off}, during which no peak is revealed in the series of spectra for each analyzed frequency. The P_{off} distributions, calculated by evaluating the occurrence of the corresponding τ_{off} within the acquisition time, are well described by a power law (continuous lines in Fig. 9a):

$$P_{\mathrm{off}} \sim 1/\tau_{\mathrm{off}}^{(1+\alpha)}, \tag{4}$$

with α values ranging from 0.46 to 0.56. The assessed independence of the α exponents on the laser intensity rules out the possibility that such an effect could be due to irreversible photoinduced processes [37]. The observed nonexponential behavior of the P_{off} distribution suggests some space–temporal heterogeneity in the off (dark) state of the FePP vibrational modes [52]. Indeed, the presence of a single off state would have given rise to a simple exponential decay for P_{off} [53]. On the other hand, the rather similar temporal decay observed for the three lines suggests that, although the vibrational modes may follow a different activation channel, a common mechanism is

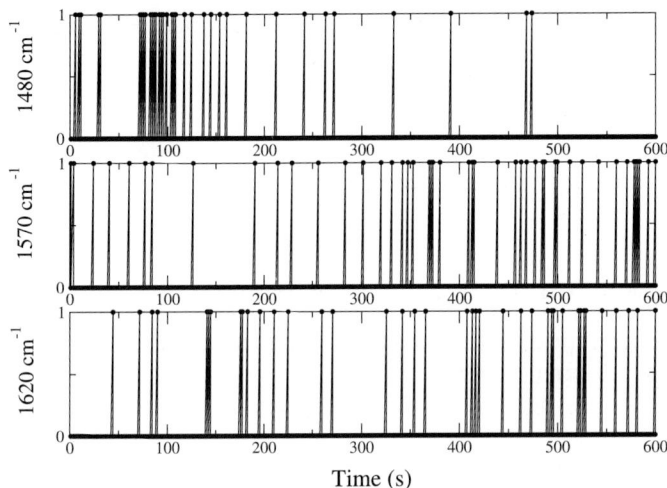

Fig. 8. On/off behavior of signals at the $1480\,\mathrm{cm}^{-1}$, $1570\,\mathrm{cm}^{-1}$, and $1620\,\mathrm{cm}^{-1}$ frequencies as a function of time in SM SERS of FePP. The *vertical lines* mark the spectrum number at which a signal overcoming a threshold of 2σ above the noise level of the spectrum is detected. The values of the crosscorrelation ρ (see (2)) calculated between the appearance of a peak at the two different frequencies are: $\rho_{1480,1570} = 0.03$, $\rho_{1480,1620} = 0.05$, $\rho_{1570,1620} = -0.05$

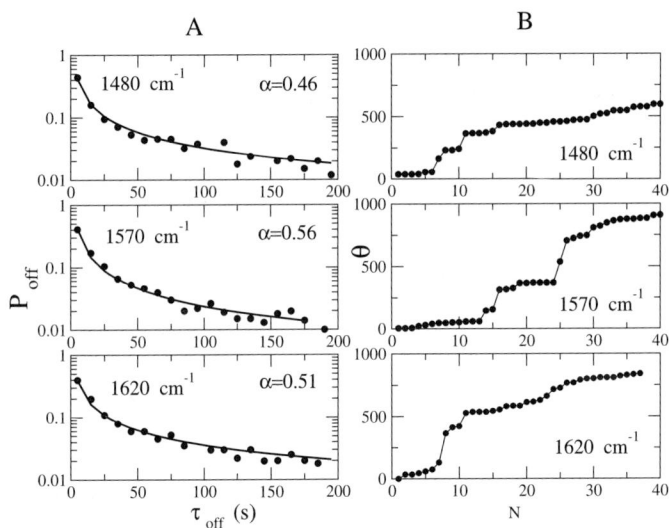

Fig. 9. (a) Distribution of the time interval τ_{off} between the appearance of two successive peaks, from a collection of 10 spectra series (each series containing 600 spectra) at $1480\,\mathrm{cm}^{-1}$, $1570\,\mathrm{cm}^{-1}$, and $1620\,\mathrm{cm}^{-1}$. *Continuous curves* give the fit by $1/\tau_{\mathrm{off}}^{1+\alpha}$. (b) Time interval spent in the off state during the first N intervals, $\theta(N)$, as a function of N at the three frequencies (see (5))

expected to regulate the sampling of the FePP vibrational modes, and then of the on–off process.

Furthermore, the nonexponential trend for the P_{off} distributions points to a progressive increase in the probability to detect longer off events [23]. To better investigate such an aspect, we have analyzed the overall temporal behavior of the off states by evaluating the quantity:

$$\theta(N) = \sum_{i=1}^{N} \tau_{off}(i), \tag{5}$$

which indicates the time spent in the off state during the first N time intervals. Figure 9b shows the trend of θ as a function of N for the three analyzed frequencies. In all cases, as time increases, longer events are observed and θ scales more rapidly than N.

From the statistical point of view, a power-law distribution for P_{off} with an exponent between 1 and 2 (corresponding to α between 0 and 1), together with the peculiar trend of $\theta(N)$, can be interpreted in the framework of the Lévy statistics [54, 55]. Lévy statistics is a natural generalization of the Gaussian distribution when analyzing sums of independent, identically distributed, random variables and is characterized by a diverging variance and broad distributions with power-law tails [56, 57]. Lévy statistics, encountered in a variety of fields (economy, physics, biology, SM fluorescence emission, etc.) has been taken into account to analyze many different phenomena (laser cooling of atoms, relaxation processes of glasses, inhomogeneous line broadening, anomalous diffusion, etc.) of long-range interaction systems [21, 23, 54, 55, 58].

From the microscopic point of view, the occurrence of Lévy statistics in SM signals has been generally linked to relaxation phenomena arising from some disorder around the molecule [21]. The disorder could arise from a static, heterogeneous distribution of traps (off states) from which the molecules should escape. Alternatively, it could be due to dynamical changes in the molecule's environment, likely connected with the presence of a random walk, with a multiple pathway of the molecule onto the surface [53, 59].

The strong fluctuations in SERS signals of FePP are expected to reflect, either the intrinsic molecule dynamics determining the gating/activation of the different vibrational modes, or its interaction with the external environment. It is well known that SERS arises from a strong interaction between a molecule and its metallic substrate. Both the em and the CT enhancement mechanisms drastically depend on the details of the molecule arrangements with respect to the metal surface (position, orientation, distance, etc). Furthermore, diffusive processes of FePP on the colloidal surface could take place during the measurements [44, 60]. Accordingly, FePP molecules may experience continuously changing interactions able to modulate the emission and the gating of the vibrational modes. Indeed, SM SERS detection is reached when molecules are located near a fractal-like metal surface [61], whose complex topology might give rise to anomalous diffusive processes, [62, 63], or, more specifically, to Lévy flights [22]. In this respect, we recall that the fluorescence intermittency of quantum dots, following power-law statistics, has been traced back to diffusion-controlled ET processes [64].

Therefore, the vibrational-mode emissions, which follow Lévy statistics, might originate from some heterogeneity in the molecule–colloids interactions and/or some anomalous diffusive processes, likely involving a CT.

4 Conclusions and Perspectives

The statistical analysis of the fluctuations appearing in SM SERS spectra has allowed us to enlighten specific phenomena usually averaged in the ensemble measurements and to disclose temporal events hidden when many different molecules are followed at the same time. In particular, our results have contributed to elucidating the role of the continuum background on the intensity fluctuations, suggesting a tight connection between these aspecific Raman signals and the SERS enhancement mechanism. The clear evidence of a switching between the two oxidation states of iron in the SM SERS spectra of FePP has led us to suggest the occurrence of a reversible ET between the molecule and the metal surface. More information about the conductive properties of a SM at the metal surface, and also their temporal variability, has been obtained by coupling SM SERS and STM measurements by adsorbing the molecule on a STM silver tip. Furthermore, it has been found that the temporal behavior of the vibrational mode emission in SM SERS spectra follows a peculiar power law. Such a trend has been traced back to a Lévy statistics whose characteristic parameters (α-exponent, etc.) may remarkably encode significant information about the underlying photophysical processes.

In the perspective to deeper investigate the phenomena occurring at the molecule/metal interface, and to fully understand the mechanisms at the basis of SERS, it appears quite promising to conjugate SERS with other experimental techniques, such as fluorescence and STM. Actually, single molecules located in close proximity to metallic nanoparticle surfaces, exhibit fluorescence quenching or enhancement, and may undergo ET processes, likely coupled to the excitation of its vibrational modes [26, 65]. Such an approach could, therefore, offer a comprehensive view of the photophysical and -conductive properties of molecules at the metal surfaces; this information may prove rewarding also in the optonanoelectronics applications.

References

[1] A. M. Kelley, X. Michalet, S. Weiss: Science **292**, 1671 (2001)
[2] T. Basche, W. E. Moerner, M. Orrit, U. P. Wild (Eds.): *Single-Molecule Optical Detection, Imaging and Spectroscopy* (VCH, Weinheim 1997)

[3] J. A. Veerman, M. F. Garcia-Parajo, L. Kuipers, N. F. van Hulst: Phys. Rev. Lett. **83**, 2155 (1999)

[4] C. Blum, F. Stracke, S. Becker, K. Muellen, A. J. Meixner: J. Phys. Chem. A **105**, 6983 (2001)

[5] X. S. Xie, R. C. Dunn: Science **265**, 361 B (1994)

[6] M. Moskovits: Rev. Mod. Phys. **57**, 783 (1985)

[7] A. Campion, P. Kambhampati: Chem. Soc. Rev. **27**, 241 (1998)

[8] K. Kneipp, Y. Wang, H. Kneipp, L. T. Perelman, I. Itzkan, R. R. Dasari, M. S. Feld: Phys. Rev. Lett. **78**, 1667 (1997)

[9] S. Nie, S. R. Emory: Science **275**, 1102 (1997)

[10] A. M. Michaels, M. Nirmal, L. E. Brus: J. Am. Chem. Soc. **121**, 9932 (1999)

[11] P. Hildebrandt, M. Stockburger: J. Phys. Chem. **88**, 5935 (1984)

[12] A. Otto, I. Mrozek, H. Grabhorn, W. Akemann: J. Phys.: Condens. Matter **4**, 1143 (1992)

[13] L. P. Capadona, J. Zheng, J. I. Gonzalez, T. H. Lee, S. A. Patel, R. M. Dickson: Phys. Rev. Lett. **94**, 058301 (2005)

[14] K. Kneipp, H. Kneipp, I. Itzkan, R. R. Dasari, M. S. Feld: J. Phys.: Condens. Matter **14**, R597 (2002)

[15] Y. C. Cao, R. Jin, J. M. Nam, C. S. Thaxton, C. A. Mirkin: J. Am. Chem. Soc. **125**, 14676 (2003)

[16] M. Culha, D. Stokes, L. R. Allain, T. Vo-Dinh: Anal. Chem. **75**, 6196 (2003)

[17] A. Weiss, G. Haran: J. Phys. Chem. B **105**, 12348 (2001)

[18] A. R. Bizzarri, S. Cannistraro: Chem. Phys. **290**, 297 (2003)

[19] A. R. Bizzarri, S. Cannistraro: Phys. Rev. Lett. **94**, 068303 (2005)

[20] H. Frauenfelder, F. Parak, R. D. Young: Ann. Rev. Biophys. Biophys. Chem. **17**, 451 (1988)

[21] E. Barkai, Y. J. Jung, R. Silbey: Ann. Rev. Phys. Chem. **55**, 457 (2004)

[22] P. Allegrini, P. Grigolini, B. J. West: Phys. Rev. E **54**, 4760 (1996)

[23] X. Brokmann, J. P. Hermier, G. Messin, P. Desbiolles, J. P. Bouchard, M. Dahan: Phys. Rev. Lett. **90**, 120601/1 (2003)

[24] D. M. Adams, L. Brus, C. E. D. Chidsey, S. Creager, C. Creutz, C. R. Kagan, P. V. Kamat, M. Lieberman, S. Lindsay, R. A. Marcus, R. M. Metzger, M. E. Michel-Beyerle, J. R. Miller, M. D. Newton, D. R. Rolison, O. Sankey, K. S. Schanze, J. Yardley, X. Zhu: J. Phys. Chem. B **107**, 6668 (2003)

[25] A. R. Bizzarri, S. Cannistraro: J. Phys. Chem. B **109**, 16571 (2005)

[26] M. W. Holman, R. Liu, D. M. Adams: J. Am. Chem. Soc. **125**, 12649 (2003)

[27] C. Joachim, J. K. Gimzewski, A. Aviram: Nature **408**, 541 (2000)

[28] P. C. Lee, D. J. Meisel: J. Phys. Chem. **86**, 3391 (1982)

[29] A. R. Bizzarri, S. Cannistraro: Appl. Spectrosc. **56**, 1531 (2002)

[30] A. R. Bizzarri, S. Cannistraro: Chem. Phys. Lett. **349**, 503 (2001)

[31] I. Delfino, A. R. Bizzarri, S. Cannistraro: Biophys. Chem. **113**, 41 (2003)

[32] H. Xu, E. J. Bjerneld, M. Kaell, L. Boerjesson: Phys. Rev. Lett. **83**, 4357 (1999)

[33] E. J. Bjerneld, Z. Foeldes-Papp, M. Kaell, R. Rigler: J. Phys. Chem. B **106**, 1213 (2002)

[34] S. Habuchi, M. Cotlet, R. Gronheid, G. Dirix, J. Michiels, J. Vanderleyden, F. C. De Schryver, J. Hofkens: J. Am. Chem. Soc. **125**, 8446 (2003)

[35] K. A. Bosnick, J. Jiang, L. E. Brus: J. Phys. Chem. B **106**, 8096 (2002)

[36] A. Otto: J. Raman Spectrosc. **33**, 593 (2002)

[37] E. J. Bjerneld, F. Svedberg, P. Johansson, M. Kaell: J. Phys. Chem. A **108**, 4187 (2004)

[38] P. Hildebrandt, M. Stockburger: Vib. Spectra. Struct. **17**, 443 (1986)

[39] J. R. Taylor: *An Introduction to Error Analysis* (University Science Books, Sausalito 1982)

[40] A. A. Moore, M. L. Jacobson, N. Belabas, K. L. Rowlen, D. M. Jonas: J. Am. Chem. Soc. **127**, 7292 (2005)

[41] J. R. Lakowicz, J. Malicka, S. D. Auria, I. Gryczynski: Anal. Biochem. **320**, 13 (2003)

[42] B. N. Persson: Chem. Phys. Lett. **82**, 561 (1981)

[43] F. Moresco, G. Meyer, K. H. Rieder, H. Tang, A. Gourdon, C. Joachim: Phys. Rev. Lett. **86**, 672 (2001)

[44] Y. He, T. He, E. Borguet: J. Am. Chem. Soc. **124**, 11964 (2002)

[45] H. M. Marques, K. L. Brown: Coord. Chem. Rev. **225**, 123 (2002)

[46] S. Rywkin, C. M. Hosten, J. R. Lombardi, R. L. Birke: Langmuir **18**, 5869 (2002)

[47] A. R. Bizzarri, S. Cannistraro: Chem. Phys. Lett. **395**, 222 (2004)

[48] I. Lee, J. W. Lee, E. Greenbaum: Phys. Rev. Lett. **79**, 3294 (1997)

[49] N. J. Tao: Phys. Rev. Lett. **76**, 4066 (1996)

[50] W. Han, E. N. Durantini, T. A. Moore, A. L. Moore, D. Gust, P. Rez, G. Leatherman, G. R. Seely, N. J. Tao, S. M. Lindsay: J. Phys. Chem. B **101**, 10719 (1997)

[51] A. M. Michaels, J. Jiang, L. Brus: J. Phys. Chem. B **104**, 11965 (2000)

[52] M. Kuno, D. P. Fromm, H. F. Hamann, A. Gallagher, D. J. Nesbitt: J. Chem. Phys. **112**, 3117 (2000)

[53] M. Nirmal, B. O. Dabbousi, M. G. Bawendi, J. J. Macklin, J. K. Trautman, T. D. Harris, L. E. Brus: Nature **383**, 802 (1996)

[54] W. Feller: *An Introduction to Probability Theory and Its Applications* (Wiley, New York 1970)

[55] M. Shlesinger, G. M. Zaslavsky, U. Frisch (Eds.): *Lévy Flights and Related Topics in Physics* (Springer, Berlin, Heidelberg 1995)

[56] A. M. Stoneham: Rev. Mod. Phys. **41**, 82 (1969)

[57] G. Wornell: *Signal Processing with Fractals* (Prentice Hall PTR, Upper Saddle River 1995)

[58] F. Bardou, J. P. Bouchaud, A. Aspect, C. Cohen-Tannoudji: *Lévy Statistics and Laser Cooling* (Cambridge Univ. Press, Cambridge 2002)

[59] K. T. Shimizu, R. G. Neuhauser, C. A. Leatherdale, S. A. Empedocles, W. K. Woo, M. G. Bawendi: Phys. Rev. B **63**, 205316 (2001)

[60] T. Komeda, Y. Kim, M. Kawai, B. N. J. Persson, H. Ueba: Science **295**, 2055 (2002)

[61] Z. Wang, S. Pan, T. D. Krauss, H. Du, L. J. Rothberg: Proc. Natl. Acad. USA **100**, 8638 (2003)

[62] J. M. Sancho, A. M. Lacasta, K. Lindenberg, I. M. Sokolov, A. H. Romero: Phys. Rev. Lett. **92**, 250601 (2004)

[63] M. Schunack, T. R. Linderoth, F. Rosei, E. Laegsgaard, I. Stensgaard, F. Besenbacher: Phys. Rev. Lett. **88**, 156102 (2002)

[64] J. Tang, R. A. Marcus: Phys. Rev. Lett. **95**, 107401 (2005)

[65] B. C. Stipe, M. A. Rezaei, W. Ho: Science **280**, 1732 (1998)

[66] P. Etchegoin, H. Liem, R. C. Maher, L. F. Cohen, R. J. C. Brown, H. Hartigan, M. J. T. Milton, J. C. Gallop: Chem. Phys. Lett. **366**, 115 (2002)

Index

Single-Molecule Surface-Enhanced Resonance Raman Spectroscopy of the Enhanced Green Fluorescent Protein EGFP

Satoshi Habuchi[1,2] and Johan Hofkens[2]

[1] Department of Chemistry, Katholieke Universiteit Leuven, Celestijnenlaan 200F, 3001 Heverlee, Belgium
Johan.Hofkens@chem.kuleuven.be

[2] Present Address: Department of Biological Chemistry and Molecular Pharmacology, Harvard Medical School, 240 Longwood Avenue, SGM209, Boston, MA 02115, USA
Satoshi_Habuchi@hms.harvard.edu

1 Introduction

One of the most intriguing findings in single-molecule spectroscopy is the observation of Raman spectra of individual molecules adsorbed on silver nanoparticles [1, 2], which can be explained by the surface-enhanced resonance Raman scattering (SERRS) effect [3, 4, 5, 6]. Single-molecule (SM) detection by SERRS is attracting increasing interest for several reasons. Firstly, by detecting SM-SERRS one hopes to get a deeper understanding of the mechanism of surface-enhanced Raman scattering [7, 8, 9, 10]. Secondly, by looking at the Raman signal of a single molecule it should be possible to detect structural dynamics and chemical reactions of those individual molecules [11]. Finally, SM-SERRS spectroscopy opens up exciting opportunities in the field of biophysical and biomedical spectroscopy where it could provide ultrasensitive detection and characterization of biophysically relevant molecules and processes. One of the first SM-SERRS experiments on proteins was conducted on the well-known heme protein [12]. However, it was suggested and debated in the literature that the noncovalently bound porphyrin group, present in heme proteins, might diffuse out of the protein to be adsorbed directly on the Ag nanoparticles [13].

To avoid this discussion, we used green fluorescent proteins (GFPs) for the SM-SERRS studies. GFPs are a class of proteins in which the chromophore is formed autocatalytically, upon expression of the protein, inside a barrel structure (length 4 nm, height 2 nm). Thus, the chromophore (constituted by three amino acid residues of the protein) is a structural part of the protein and is kept in place by a hydrogen-bonding network. GFP and its mutants are widely used as fluorescent probes for cellular and biological studies [14]. GFP-like fluorescent proteins can have a protonated and deprotonated form of the chromophore, and it is commonly accepted that conversion between the two forms takes place. However, the conversion has not been detected directly at the single-molecule level using fluorescence spectroscopy because

K. Kneipp, M. Moskovits, H. Kneipp (Eds.): Surface-Enhanced Raman Scattering – Physics and Applications, Topics Appl. Phys. **103**, 297–312 (2006)
© Springer-Verlag Berlin Heidelberg 2006

of the dimness of the protonated form. Here, we show that the conversion between the protonated and deprotonated forms of the chromophore of enhanced green fluorescent protein (EGFP) can be monitored directly at the single-molecule level using SERRS spectroscopy via the vibrational fingerprints of the protonated and deprotonated forms.

2 Experimental Section

2.1 Purification of EGFP

Standard methods were used for in-vitro DNA manipulations. The gene coding for EGFP was first removed from pEGFP (Clontech) as an 800 bp *Eco*RI-*Pst*1 fragment and subsequently inserted into pBAD/HisA (Invitrogen). The expression of the protein was induced for 12 h in *Escherichia coli* Top10 cells grown at an optical density at 600 nm of 0.5 using 0.2 % arabinose. The His-tagged protein was subsequently isolated and purified under native conditions by Ni-chelation chromatography using the Xpress Protein Purification System. The purity of the samples was confirmed by polyacrylamide gel electrophoresis. Finally, the proteins were concentrated in a Vivaspin 6 concentrator, and were stored in a phosphate buffer saline solution (PBS, pH 7.4, Sigma).

2.2 Sample Preparation for SM-SERRS Measurements

Ag sols were prepared according to the description of *Lee* and *Meisel* [15]. A 45 mg sample of silver nitrate was added to 250 mg of ultrapure water, purged with Ar, and heated to 100 °C under vigorous magnetic stirring. 5 ml of a 35 mM sodium citrate solution (1 % sodium citrate dihydrate by weight) was injected. The resulting solution was refluxed for 1 h. A 250 µl aliquot of silver colloid was incubated with 1250 µl of 1 mM sodium chloride solution containing EGFP (2×10^{-10}) for 1 h. 40 µl of the sol was dropped on a polylysine-coated coverslip, and the solvent was allowed to evaporate at room temperature. In this condition, each particle carries an average of one EGFP molecule. The colloidal particles are immobilized on the polylysine-coated coverslips due to the electrostatic interactions between the negative charges on the particles and the positive charges of the surface. After evaporation of the solvent, the coverslips were rinsed with ultrapure water to remove loosely bound colloidal particles, and dried by flowing ultrapure nitrogen gas.

2.3 Experimental Setup

SM-SERRS measurements were performed by using a confocal microscope [16] (IX70, Olympus) with an oil immersion lens (Olympus, 100×, NA 1.4). The excitation source was the 488 nm line of an Ar ion laser (Stabilite 2017, Spectra Physics). All spectra were measured with a collection time of either 5 s or 1 s at the incident laser power of 1 μW. The wavelength resolution of the system is approximately 10 cm^{-1}. The SERRS signal was collected by the same objective, passed through a dichroic mirror (Chroma Technology), a 50 μm pinhole, a notch filter (Kaiser Optical System), divided into two beams by using a beamsplitter cube (05BC17MB.1, Newport), and focused on an avalanche photodiode (APD) (SPCM AQ15, Perkin Elmer) and into a 250 mm polychromater (250IS/SM, Chromex) coupled to a cooled CCD camera (LN/CCD-512SB, Princeton Instruments). The SERRS signal detected by the APD was registered with a time-correlated single-photon counting PC card (SPC 630, Picoquant) operated in the FIFO-mode (first in, first out) [17]. The data set obtained in this way allows the reconstruction of a SERRS intensity time trace with a user-defined dwell time with a minimum value of 50 ns.

The ensemble Raman spectra of EGFP were measured by using the same setup as for SM-SERRS measurements. 50 μl of EGFP (1 mM) in phosphate buffer (100 mM, pH 7.4) or acetate buffer (100 mM, pH 5.0) was dropped in a sample chamber and sealed to avoid evaporation. As the excitation source, the 647.1 nm line of an Ar–Kr ion laser (Stabilite 2018, Spectra Physics) was used (preresonance conditions), allowing selective intensity enhancement of vibrational bands originating from the chromophore in the protein.

Atomic force microscopy (AFM) measurements were performed using a Discoverer TMX2010 AFM (ThermoMicroscopes, San Francisco, CA) operating in tapping mode and using a silicon nitride oxide sharpened cantilever (Nanosensor, Germany) with a resonance frequency of 250 kHz to 300 kHz. The image of the Ag particles (vide infra) has been second-order background corrected on the Topometrix SPMLab 5.0 software.

2.4 Autocorrelation Analysis

The autocorrelation function $G(\tau)$, used to analyze fluctuations in the SERRS and fluorescence intensity trajectories, is defined as

$$G(\tau) = \frac{\langle F(t)F(t+\tau)\rangle}{\langle F(t)\rangle^2}, \tag{1}$$

where $F(t)$ is the SERRS intensity obtained at time t, angled brackets denote time averages, and τ is the correlation time. The autocorrelation function was constructed from the data set with a homemade program. The obtained autocorrelation function was analyzed using a (multi-)exponential decay model function.

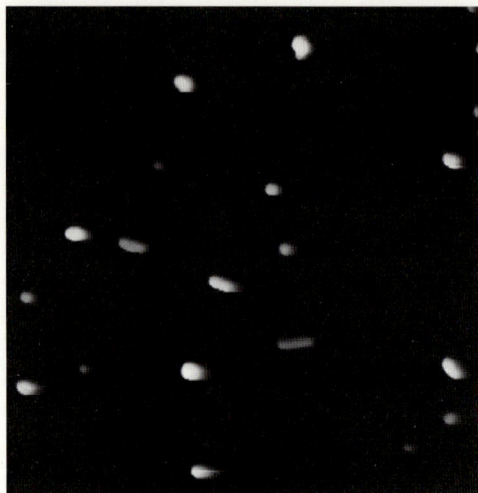

Fig. 1. An AFM image of Ag particles adsorbed on a polylysine-coated coverslip. Scale bar = 500 nm

3 Results and Discussion

3.1 Single EGFP Molecules Detected by SERRS

A major contribution for the enhancement of the Raman signal in SERRS has been described in terms of an electromagnetic (em) field enhancement at a metal-particle surface that originates from surface-plasmon resonances [3, 4, 5, 6]. It is well known that the enhancement factor strongly depends on the size and the shape of the particle. More recent reports have suggested that most of the SERS active particles are compact clusters consisting of a minimum of two individual Ag particles [7, 8, 18]. These observations have been interpreted by the fact that the em field is enhanced tremendously in a gap region between two particles that are separated in the order of a few nanometer [12, 19, 20, 21, 22]. Figure 1 displays an AFM image of the silver particles used for the SERS measurements. The majority of the particles consist of single particles that have a spherical or rod-like shape with an average particle size of 70 nm. A minor fraction of the particles is actually a tightly packed small aggregate. The ratio of aggregates versus single particles is roughly one to ten. According to the previous findings, these aggregates could have the highest SERRS activity. EGFP is a relatively small protein, having a barrel-like structure with a width of 2 nm and height of 4 nm [14]. Although the molecular size of EGFP is larger than the size of a typical organic dye molecule, the gap-distance dependence of the enhancement in this distance range (a few nanometer) is not dramatic. Consequently, for single EGFP molecules residing in gaps between particles or aggregates, detectable Raman signals can be expected.

Figure 2 shows a series of SERRS images of EGFP adsorbed on a coverslip on which Ag particles were immobilized. The ratio of EGFP molecules to Ag

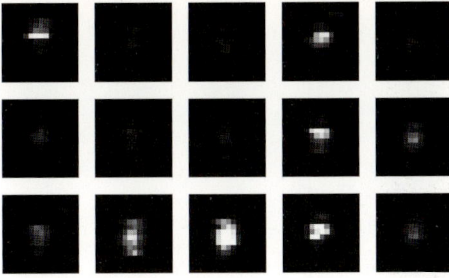

Fig. 2. The SERRS signal from an individual EGFP molecule adsorbed on an Ag particle. The different images were recorded with time lapses of one minute

particles is 1 : 1. This is a standard condition for single-molecule SERRS measurements. Some particles will carry a few molecules, and some of them will not carry any molecules at all, following a Poissonian distribution. As can be seen in Fig. 2, the SERRS images show intensity fluctuations. These fluctuations are a typical phenomena observed for all single-molecule Raman measurements at room temperature [7,11,23,24]. The observation is a strong indication that the SERRS signal from a single molecule is measured. For rhodamine 6G (R6G) deposited in the same way on a similar Ag-particle-coated coverslip, the number of observed spots is roughly ten times higher (data not shown) and in good agreement with the number of aggregates seen in AFM. This difference in SERRS-active spots may be due to a lower binding affinity to silver for EGFP, the lower SERRS cross section of EGFP compared to R6G (vide infra), the lower probability to enter the gaps due to the larger size of EGFP or a combination thereof.

3.2 SM-SERRS Spectra of EGFP

EGFP can have two different forms of the chromophore, a protonated and a deprotonated form depending on the pH of the solution [25, 26]. Ensemble Raman spectra of EGFP at neutral (pH = 7.4) and acidic pH (pH = 5.0) are depicted in Fig. 3. It should be mentioned that excitation with 752 nm results in a preresonance Raman enhancement that allows one to detect the Raman signal from the chromophoric site with minimal spectral interference from the surrounding environment [27, 28]. As expected from the identical chemical structure of the chromophore as well as the nearly similar amino acid sequences of EGFP and the Ser65Tyr mutant of GFP (S65T), the Raman spectra of EGFP are almost identical to those that were published for the S65T mutant [27,29]. Most of the bands have been assigned in the literature to specific vibrational modes of the chromophore on the basis of a model compound of the chromophore [30]. In the ensemble Raman spectra, the most important peaks are those located at $1536\,\mathrm{cm}^{-1}$ and $1556\,\mathrm{cm}^{-1}$. Both bands can be attributed to the delocalized imidazolinone/exocyclic C=C mode of

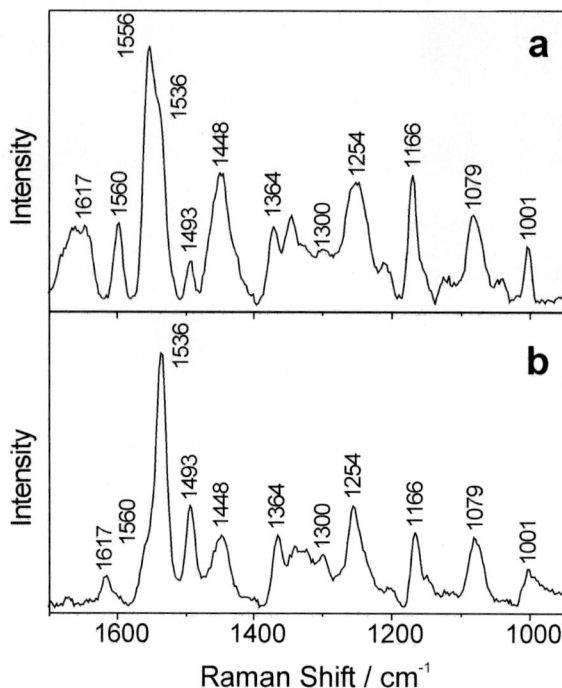

Fig. 3. The ensemble Raman spectrum of EGFP in (**a**) acetate buffer (pH = 5.0) and in (**b**) phosphate buffer (pH = 7.4)

the chromophore [30], the peak at $1556\,\mathrm{cm}^{-1}$ corresponding to the protonated form of the chromophore and the peak at $1536\,\mathrm{cm}^{-1}$ corresponding to the deprotonated form. Both forms of the chromophore can, therefore, be distinguished on the basis of these bands.

Figure 4a shows an averaged SERRS spectrum for EGFP obtained by adding up the spectra of 80 different single molecules. Most of the peaks in the spectrum are in agreement with the solution Raman spectrum of EGFP (Fig. 3) within an error of $\pm 10\,\mathrm{cm}^{-1}$, confirming that the Raman signal of individual EGFP molecules was measured. The relative amplitude of the Raman band of the protonated form in the averaged spectrum is higher than the amplitude measured for that band at the same pH in the solution [27, 29], indicating the acid-base equilibrium might be slightly modified due to the adsorption of EGFP on the Ag particles.

The time evolution of the SERRS spectra of a single EGFP molecule adsorbed on an Ag colloid is depicted in Fig. 4b. As can be seen in the sequence of spectra, the signals abruptly changed in both frequency and intensity, typical features of SM-SERRS spectra. The most important observation within the series of the spectra depicted in Fig. 4b is the sudden frequency jump from $1540\,\mathrm{cm}^{-1}$ to $1560\,\mathrm{cm}^{-1}$ between spectra 3 and 4. In view of the ob-

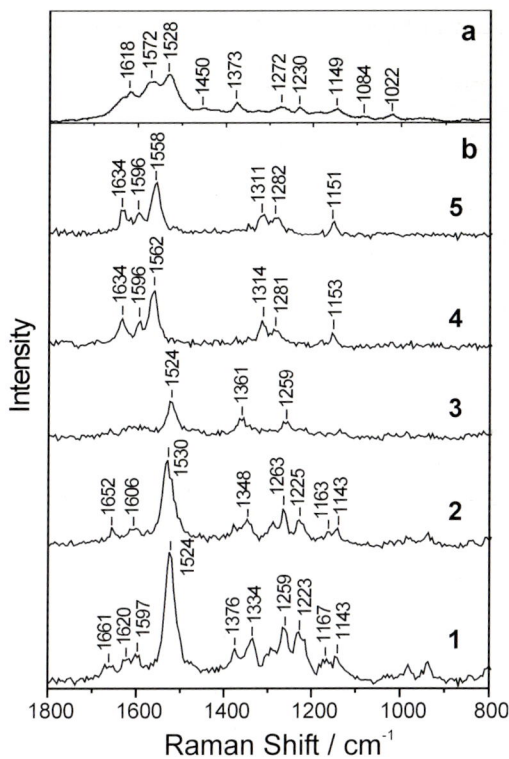

Fig. 4. SERRS spectra of a single EGFP molecule (5 s integration time) adsorbed on a Ag particle. (**a**) The averaged spectrum built up by adding the spectra of 80 different individual molecules. (**b**) The time series of the Raman spectra from an individual EGFP molecule

servation for the ensemble Raman spectra, we attribute this frequency jump to a conversion of the chromophore from the deprotonated to the protonated form in a timescale of approximately 10 s [16].

Other time series of SERRS spectra for individual EGFP molecules substantiate this hypothesis (Fig. 5). The spectral series clearly display the reversibility of the conversion between the two forms. Figure 5a shows the conversion of the deprotonated form of the chromphore of EGFP to its protonated form between spectra 1 and 2, and of the protonated form back to the deprotonated form between spectra 4 and 5. A similar conversion can be seen in Fig. 5b. In this case, the EGFP molecule shows a conversion from the protonated to the deprotonated form and then back to the protonated form. A third time series of the Raman spectrum (Fig. 5c) shows the conversion from the protonated to the deprotonated form. Within this series, both the protonated and the deprotonated bands were observed simultaneously in spectra 2 and 3. This finding points to the fact that reversible conversion takes

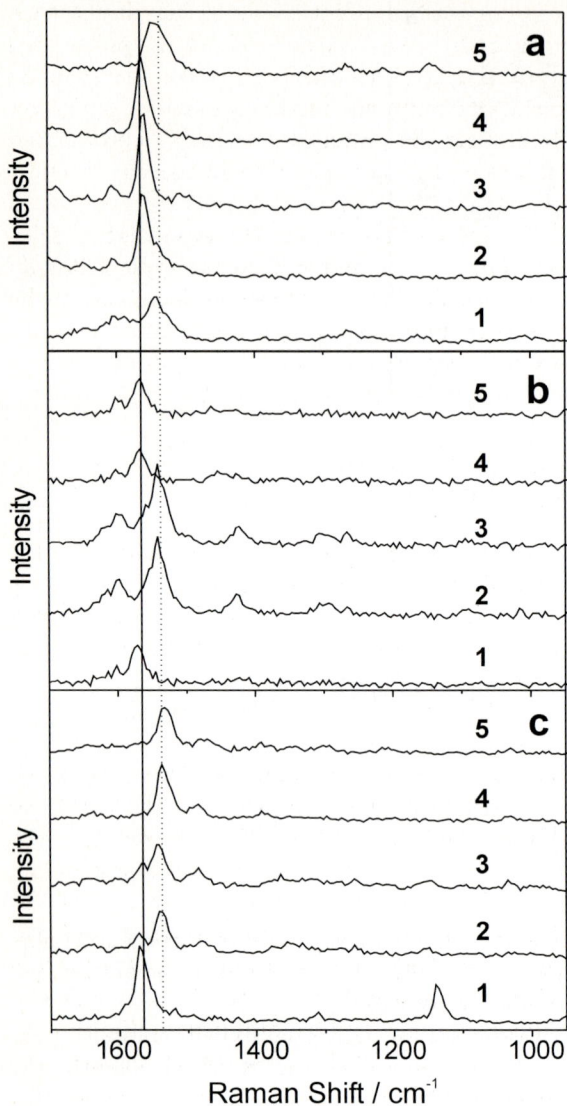

Fig. 5. Time series of SERRS spectra from three individual EGFP molecules (1 s integration time) adsorbed on an Ag particle. The *solid* and *dashed lines* show the vibrational peak positions of the protonated and deprotonated form of the chromophore, respectively

place in a timescale faster than or corresponding to the accumulation time used (1 s) for this particular molecule. The Raman bands of the protonated and the deprotonated form obtained from the SM-SERRS measurements do not perfectly agree with the values determined based on ensemble measurements ($1556 \, \mathrm{cm}^{-1}$ and $1536 \, \mathrm{cm}^{-1}$ for the protonated and the deprotonated form, respectively). However, it is well known that Raman spectra obtained from a single molecule show spectral fluctuations, especially in the case of protein molecules [12, 31, 32, 33]. Taking into account the wavelength resolution of our experimental setup ($10 \, \mathrm{cm}^{-1}$) together with the typical Raman spectral fluctuations for single molecules [34, 35], the small deviations of the Raman peaks obtained for the single-molecule measurements from the peaks in the ensemble spectrum can be interpreted as yet another typical phenomena in single-protein/molecule Raman spectroscopy.

3.3 Discussion on the Spectral Jumps Observed in the SM-SERRS Spectra of EGFP

In the previous section, the observed spectral jumps in the Raman spectra were attributed to a protonation/deprotonation reaction of the chromophore. Let us first consider another possible origin of the spectral jumps observed in the SM-SERRS spectra. An obvious possibility is the denaturation of the protein. Denaturation of heme proteins on Ag particles was reported previously [36]. However, the possibility of the denaturation is unlikely in our system for the following reasons. First, the Ag colloidal particles prepared by the Lee and Meisel citrate reduction method have almost no influence on the protein structure because the citrate ions were proposed to act as a coating that prevents denaturation of the protein [37]. Second, the reversibility of the conversion in the SM-SERRS spectra is expected only when the protein has its intact tertiary structures. If the protein denatures on the Ag surface, the chromophore would have the protonated form at the measurement condition (pH = 7.4) because the pKa of the chromophore not stabilized by the protein barrel is 8.1 [38, 39]. Consequently, the spectral jump from the protonated to the deprotonated form should correspond to the renaturation of the protein in that case. However, a denaturation/renaturation cycle within the measurement timescale (seconds) is quite unlikely. For these reasons, the possibility of the denaturation of the protein is unlikely.

Let us now return to the hypothesis of protonation/deprotonation as the origin for the spectral jumps in the Raman spectra of individual EGFP molecules. The protonation/deprotonation dynamics of GFP and its mutants have been extensively studied by fluorescence techniques. Upon excitation in the aborption band corresponding to the protonated form of the chromophore, fast (picosecond) excited-state proton transfer has been demonstrated [25, 40]. It has been shown for wild-type GFP that a photoconverted sample relaxes back toward initial equilibrium in the dark [40]. Photoinduced conversion between the deprotonated and the protonated form has also been demonstrated by hole-burning spectroscopy [41]. The latter technique also revealed that the ground-state energy level of the protonated form is close to that of the deprotonated form for EGFP, suggesting a potential ground-state conversion between the two forms. It has to be stressed that the relative positions of the energy levels are strongly dependent on the GFP mutant that is considered. Finally, the presence of the protonated and the deprotonated form of the chromophore has been suggested at the single-molecule level, albeit only indirect evidence was provided. Indeed, the long off-times in the fluorescence intensity trajectory of individual GFP molecules on 488 nm excitation have been attributed to the molecule switching to the protonated form and in that way shifting out of resonance with the excitation source [42,43]. Simultaneous two-color excitation in the absorption bands of the protonated and the deprotonated form leads to an enhancement of the fluorescence intensity, strongly suggesting that the chromophore can convert between the protonated and the deprotonated form [44]. It should be mentioned that the long off-times in the fluorescence trajectories were still observed even when exciting both the protonated and the deprotonated form simultaneously [44].

Taking into account these findings, conversion between the protonated and the deprotonated form is principally possible, either via a photoinduced process or an acid–base equilibrium in the ground state. In our case, the excited state is quenched rapidly by resonance energy transfer from the molecule to the Ag particles. Consequently, the conversion between the protonated and the deprotonated form observed in our SM-SERRS measurements will most probably be the result of an acid–base equilibrium in the ground state rather than the result of photoinduced protonation/deprotonation. The timescale of the conversion, ranging from seconds to tens of seconds and evidenced by the vibrational fingerprints of the protonated and deprotonated forms observed in the SM-SERRS experiments, corresponds to the duration of the long off-times observed in SM fluorescence trajectories of immobilized EGFP molecules (see next section). This timescale is also orders of magnitude longer than the timescale on which photoinduced excited-state proton transfer occurs [25]. However, a photoinduced process via an unknown dark state [45] can not be completely excluded at this point. SM-SERRS experiments on a GFP mutant in which the ground-state energy levels between the different forms of the chromophore are much larger could shed light on this issue.

3.4 The Relation between the Fluorescence and SERRS Intensity Trajectories of EGFP

Figure 6a shows a fluorescence intensity trajectory of a single EGFP molecule embedded in a polyvinylalcohol (PVA) thin film, excited at 488 nm. The trajectory displays several bright bursts separated by dark states. The bright bursts result from the fluorescence of the deprotonated form of the chromophore. The survival time of the bright state (on-time) ranges from subseconds to seconds. A similar behavior is observed for many of the GFP

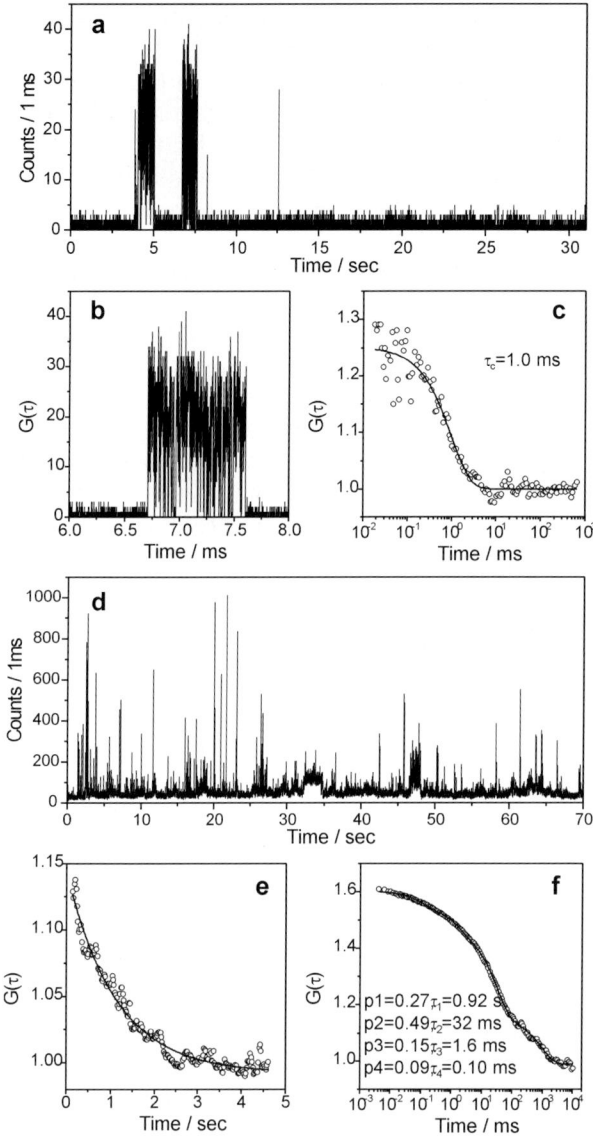

Fig. 6. The intensity time traces of the (**a,b**) fluorescence signal detected from single EGFP molecule embedded in polyvinyl alcohol and (**d**) integrated SERRS signal detected from a single EGFP molecule adsorbed on an Ag particle. (**c**) The autocorrelation curve obtained from the fluorescence burst displayed in Fig. 6b. (**e,f**) The autocorrelation curves (linear and logarithmic scales) obtained from the complete intensity trajectory of the SM-SERRS signal of the individual EGFP molecule displayed in Fig. 6d

mutants. Within a burst, an additional fast on-off blinking process is observed (Fig. 6b). An autocorrelation curve calculated from the burst in Fig. 6b can be fitted with a single-exponential function, giving an off-time of 1.0 ms (Fig. 6c), which is assigned to triplet blinking [46]. The single-exponential behavior of the correlation curve points to the fact that the intensity fluctuation can be attributed solely to the triplet blinking. Note that off-times attributed to conversion to the triplet state in solution as observed by fluorescence correlation spectroscopy (FCS) are much shorter (usually of the order of a few microsecond). The survival time of the dark state (off-time) ranges from seconds to several tens of seconds. As mentioned above, the origin of the dark state is still debated in the literature although this is a common phenomenon of GFP mutants. Possible processes mentioned in the literature to result in the observed dark states are the protonation of the chromophore, shifting the molecule out of resonance with the excitation source, a transition to an ionic state such as a zwitter ionic form, or an isomerization taking place in the chromophore. The findings reported here, e.g., the spectral fingerprints of both protonated and deprotonated forms of the chromophore as well as the observed timescale of interconversion, strongly suggest the protonation of the chromophore to be responsible for at least part of the long off-times observed in SM-fluorescence trajectories of EGFP.

In Fig. 6d an intensity trajectory of the integrated SERRS signal (including the broad background discussed in a number of publications) of a single EGFP molecule adsorbed on an Ag colloid is shown. The contribution of the fluorescence of EGFP to the transient is excluded because no fluorescence was detected in the spectra measured simultaneously (results not shown). This indicates efficient excitation transfer to the metal. The efficient quenching of the excited state results in a low probability of photobleaching, which leads to a long survival time of the SERRS signal as compared with the fluorescence signal. The count rate of the SERRS signal is one to two orders of magnitude higher than that of the fluorescence signal (see scales in Fig. 6a,d). The intensity of the signal, together with its long survival time, suggests that EGFP is not only an excellent fluorescence probe but also a potentially powerful SERRS probe. From the intensity ratio of the SERRS signal and the fluorescence signal of single EGFP molecules, one can calculate a SERRS cross section of roughly 3×10^{-15} cm^2, about an order of magnitude smaller than the SERRS cross sections reported for R6G [7]. The difference in cross section between EGFP and R6G could be explained by the fact that the chromophore of EGFP is separated by the β-barrel from the surface of the Ag particle. Since the electromagnetic enhancement factor decreases with increasing distance between the particle and the molecule, the larger distance between the chromophore and the particle could explain the lower cross section. Nevertheless, the value of the cross section of EGFP is actually surprising as the chromophore can not directly bind to the Ag colloids. At first sight, this excludes the contribution of the chemical enhancement to the SERRS signal since this contribution is believed to occur for molecules that adsorb directly on the metal. Such binding and subsequent contribution to the enhancement, if contributing at all, should in our case occur via the β-barrel amino acid residues.

The time trace of the SERRS signal displays intensity fluctuations, similar to those observed for the fluorescence time trace (Fig. 6d) of single EGFP molecules. However, the SERRS intensity time trace shows much more dynamics as compared with the fluorescence time trace. The fluctuation in the SERRS intensity time trace is the result of a complicated combination of factors including both the nature of the chromophore and the interaction between the protein and the Ag particles. Furthermore, all the nonfluorescent states of the chromophore are principally SERRS active, which further complicates the observed SERRS dynamics. The autocorrelation curve obtained from the SERRS intensity time trace in Fig. 6d is shown in Fig. 6e,f with a linear and logarithmic timescale, respectively. The correlation curve shows a multiexponential behavior (Fig. 6f). The slowest correlation time for this trajectory is 0.9 s (Fig. 6e), although other trajectories gave a slow correlation time of several seconds to tens of seconds (data not shown). A second to subsecond correlation time has been observed for single-molecule SERRS intensity trajectories of organic dye molecules too [11, 23, 47] as well as for protein molecules [32]. This slow fluctuation has been interpreted as lateral electron-induced desorption followed by lateral displacements [23, 48] or as the change in the orientation of the molecule in a gap region between two particles [11, 49]. Indeed, a recent report demonstrated that the blinking behavior of a R6G molecule adsorbed on Ag particles is absent at 77 K, where molecular motions, both rotation and translation, should be much slower [22]. In our case, the conversion between the deprotonated and the protonated form of the chromophore also occurs in this time range (vide supra). Since the conversion between the two forms leads to a change in the resonance condition, the slow fluctuation in the SERRS trajectory for EGFP is probably a convolution of protonation–deprotonation dynamics and motion of the protein in the gap between Ag particles.

In addition to the slow fluctuation, the autocorrelation trace shows shorter correlation times from hundreds of microseconds to tens of milliseconds (Fig. 6f). Although the origin of these fast fluctuations is not clear at this stage, some possibilities can be pointed out. Fluorescence correlation spectroscopy studies on GFP mutants showed that most of the GFP mutants display a fluctuation in the fluorescence intensity of the order of hundreds of microseconds to milliseconds [26, 50]. These fluctuations have been partially attributed to cis–trans isomerization of the chromophore [45]. While both isomers should be SERRS active, the isomerization will lead to a change in the orientation of the chromophore. Most probably, the EGFP molecules detected by SM-SERRS measurements are located in the gap region of the Ag particles (vide supra). In this configuration, the SERRS intensity is strongly dependent on the orientation of the molecule [49]. Consequently, the shorter correlation times from hundreds of microseconds to tens of milliseconds obtained in the SM-SERRS traces could reflect the isomerization dynamics of the chromophore. Although the mechanisms responsible for the fluctuations in the SERRS signal from single EGFP molecules are still not fully understood, we believe the SM-SERRS study of EGFP presented here demonstrates the potential of the technique to gain structural evidence for photophysical processes thought to occur in fluorescent proteins.

4 Summary

We have investigated SM-SERRS spectra and intensity trajectories of individual EGFP molecules adsorbed on silver colloids. Spectral jumps, in a timescale of seconds, were observed in the SM-SERRS spectral trajectories. These jumps were attributed to a slow, reversible conversion between the protonated and deprotonated form of the chromophore in EGFP based on comparison with an ensemble Raman spectra of EGFP at different pH. The timescale of the conversion seems to match the timescale of the (long) off-periods observed in SM-fluorescence intensity trajectories. SM-SERRS intensity trajectories show, beside the slow fluctuations of the order of seconds, several faster timescales of fluctuations. We present a possible rational for those fluctuations based on literature data. We demonstrate that SM-SERRS data can be used to elucidate some of the complex photophysical processes occurring in GFPs.

Acknowledgements

This work was supported by the Fonds voor Wetenschappelijk Onderzoek Vlaanderen, the Flemish Ministry of Education (GOA01-2), the Federal Science Policy of Belgium (IUAP-V-03), the Fonds National de la Recherche Scientifique and the European networks Sisitomas and Bionics.

References

[1] S. M. Nie, S. R. Emory: Science **275**, 1102 (1997)
[2] K. Kneipp, Y. Wang, H. Kneipp, L. T. Perelman, I. Itzkan, R. Dasari, M. S. Feld: Phys. Rev. Lett. **78**, 1667 (1997)
[3] M. Moskovits: Rev. Mod. Phys. **57**, 783 (1985)
[4] A. Otto: J. Phys. Condens. Matter **4**, 1143 (1992)
[5] A. Otto: Phys. Stat. Sol. A **188**, 1455 (2001)
[6] M. Moskovits: J. Raman Spectrosc. **36**, 485 (2005)
[7] A. M. Michaels, M. Nirmal, L. E. Brus: J. Am. Chem. Soc. **121**, 9932 (1999)
[8] A. M. Michaels, J. Jiang, L. E. Brus: J. Phys. Chem. B **104**, 11965 (2000)
[9] W. E. Doering, S. Nie: J. Phys. Chem. B **106**, 311 (2002)
[10] J. Jiang, K. Bosnick, M. Maillard, L. Brus: J. Phys. Chem. B **107**, 9964 (2003)

[11] Z. Wang, L. J. Rothberg: J. Phys. Chem. B **109**, 3387 (2005)
[12] H. X. Xu, E. J. Bjerneld, M. Käll, L. Börjesson: Phys. Rev. Lett. **83**, 4357 (1999)
[13] A. Otto: J. Raman Spectrosc. **33**, 593 (2002)
[14] R. Y. Tsien: Annu. Rev. Biochem. **67**, 509 (1998)
[15] P. C. Lee, D. J. Meisel: J. Phys. Chem. **86**, 3391 (1982)
[16] S. Habuchi, M. Cotlet, R. Gronheid, G. Dirix, J. Michiels, J. Vanderleyden, F. C. D. Schryver, J. Hofkens: J. Am. Chem. Soc. **125**, 8446 (2003)
[17] M. Cotlet, J. Hofkens, S. Habuchi, G. Dirix, M. V. Guyse, J. Michiels, J. Vanderleyden, F. C. D. Schryver: Proc. Natl. Acad. Sci. USA **98**, 14398 (2001)
[18] M. Futamata, Y. Maruyama, M. Ishikawa: J. Mol. Struct. **735–736**, 75 (2005)
[19] E. Hao, G. C. Schatz: J. Chem. Phys. **120**, 357 (2004)
[20] H. X. Xu: Appl. Phys. Lett. **85**, 5980 (2004)
[21] M. Käll, H. X. Xu, P. Johansson: J. Raman Spectrosc. **36**, 510 (2005)
[22] M. Futamata, Y. Maruyama, M. Ishikawa: J. Phys. Chem. B **107**, 7607 (2003)
[23] A. Weiss, G. Haran: J. Phys. Chem. B **105**, 12348 (2001)
[24] K. A. Bosnick, J. Jiang, L. E. Brus: J. Phys. Chem. B **106**, 8096 (2002)
[25] M. Cotlet, J. Hofkens, M. Maus, T. Gensch, M. V. der Auweraer, J. Michiels, G. Dirix, M. V. Guyse, J. Vanderleyden, A. J. W. G. Visser, F. C. D. Schryver: J. Phys. Chem. B **105**, 4999 (2001)
[26] U. Haupts, S. Maiti, P. Schwille, W. W. Web: Proc. Natl. Acad. Sci. USA **95**, 13573 (1998)
[27] A. Bell, X. He, R. M. Wachter, P. J. Tonge: Biochemistry **39**, 4423 (2000)
[28] V. Tozzini, A. R. Bizzarri, V. Pellegrini, R. Nifosì, P. Giannozzi, A. Iuliano, S. Cannistraro, F. Beltram: Chem. Phys. **287**, 33 (2003)
[29] S. G. Kruglik, V. Subramaniam, J. Greve, C. Otto: J. Am. Chem. Soc. **124**, 10992 (2002)
[30] X. He, A. F. Bell, P. J. Tonge: J. Phys. Chem. B **106**, 6056 (2002)
[31] A. R. Bizzarri, S. Cannistraro: Appl. Spectrosc. **56**, 1531 (2002)
[32] I. Delfino, A. R. Bizzarri, S. Cannistraro: Biophys. Chem. **113**, 41 (2005)
[33] A. R. Bizzarri, S. Cannistraro: Chem. Phys. Lett. **395**, 222 (2004)
[34] T. Vosgrone, A. J. Meixner: J. Lumin. **107**, 13 (2004)
[35] T. Vosgrone, A. J. Meixner: Chem. Phys. Chem. **6**, 154 (2005)
[36] G. Smulevich, T. G. Spiro: J. Phys. Chem. **89**, 5168 (1985)
[37] B. N. Rospendowski, K. Kelly, C. R. Wolf, W. E. Smith: J. Am. Chem. Soc. **113**, 1217 (1991)
[38] W. W. Ward, C. W. Cody, R. C. Hart, M. J. Cormier: Photochem. Photobiol. **31**, 611 (1980)
[39] O. Shimomura: FEBS Lett. **104**, 220 (1979)
[40] M. Chattoraj, B. A. King, G. U. Bublitz, S. G. Boxer: Proc. Natl. Acad. Sci. USA **93**, 8362 (1996)
[41] T. M. H. Creemers, A. J. Lock, V. Subramaniam, T. M. Jovin, S. Völker: Proc. Natl. Acad. Sci. USA **97**, 2974 (2000)
[42] R. M. Dickson, A. B. Cubitt, R. Y. Tsien, W. E. Moerner: Nature **388**, 355 (1997)
[43] M. F. Garcia-Parajo, G. M. J. Segers-Nolten, J. A. Veerman, J. Greve, N. F. van Hulst: Proc. Natl. Acad. Sci. USA **97**, 7237 (2000)
[44] G. Jung, J. Wiehler, B. Steipe, C. Bräuchle, A. Zumbusch: Chem. Phys. Chem. **6**, 392 (2001)

[45] P. Schwille, S. Kummer, A. A. Heikal, W. E. Moerner, W. W. Webb: Proc. Natl. Acad. Sci. USA **97**, 151 (2000)

[46] M. Cotlet, J. Hofkens, F. Köhn, J. Michiels, G. Dirix, M. V. Guyse, J. Vanderleyden, F. C. D. Schryver: Chem. Phys. Lett. **336**, 415 (2001)

[47] J. T. Krug, G. D. Wang, S. R. Emory, S. Nie: J. Am. Chem. Soc. **121**, 9208 (1999)

[48] G. Haran: Israel J. Chem. **44**, 385 (2004)

[49] H. X. Xu, M. Käll: Chem. Phys. Chem. **4**, 1001 (2003)

[50] J. Windengren, B. Terry, R. Rigler: Chem. Phys. **249**, 259 (1999)

Index

Surface-Enhanced Vibrational Spectroelectrochemistry: Electric-Field Effects on Redox and Redox-Coupled Processes of Heme Proteins

Daniel Murgida and Peter Hildebrandt

Technische Universität Berlin, Institut für Chemie, Max-Volmer-Laboratorium für Biophysikalische Chemie, Sekr. PC14, Strasse des 17. Juni 135, D-10623 Berlin, Germany
dh.murgida@tu-berlin.de, hildebrandt@chem.tu-berlin.de

1 Introduction

Soon after its discovery, the surface-enhanced Raman (SER) effect was exploited to study adsorbed biomolecules including proteins and nucleic acids [1, 2]. In their pioneering work on heme proteins adsorbed on Ag electrodes, *Van Duyne* and *Cotton* demonstrated that the SER effect and the molecular resonance Raman (RR) effect can be combined (surface-enhanced resonance Raman SERR) to selectively probe the protein cofactor [1]. Although at that time the origin of the surface enhancement was far from being fully understood, SER spectroscopy was proposed to become a valuable tool in studying biological systems. This optimistic view, however, was damped by later findings on adsorption-induced denaturation of proteins [3, 4], casting serious doubts on the general applicability of this technique to study biological systems.

It took eventually more than fifteen years to establish strategies that are appropriate to overcome this problem. Using biocompatible coatings of the metal surface now allows immobilization of proteins under preservations of their native structure [5, 6]. Additional methodological achievements including the development of time-resolved SERR spectroscopy [7,8,9] as well as the application of surface-enhanced infrared absorption spectroscopy (SEIRA) [10] have substantially broadened the scope and the potential of studies on immobilized proteins.

This account is dedicated to summarize recent results on the redox processes of heme proteins on coated electrodes using potential-dependent stationary and time-resolved SER, SERR and SEIRA spectroscopy. The first part of this work is devoted to outline various strategies for immobilizing proteins on metal electrodes, the adaptation of stationary and time-resolved surface-enhanced vibrational spectroscopic techniques to biological systems, and the control and determination of the interfacial electric-field. In the second and third parts, we summarize results obtained for proteins peripherally bound to and integrated in membrane models on electrodes. Special empha-

K. Kneipp, M. Moskovits, H. Kneipp (Eds.): Surface-Enhanced Raman Scattering – Physics and Applications, Topics Appl. Phys. **103**, 313–334 (2006)
© Springer-Verlag Berlin Heidelberg 2006

sis is laid on electric-field effects on redox and redox-coupled processes of immobilized cytochrome c and cytochrome c oxidase.

2 Strategy and Methodological Approach

The analysis of electric-field effects on interfacial processes of heme proteins requires, first, an appropriate strategy to immobilize proteins that provides a biomimetic environment and avoids irreversible denaturation (Sect. 2.1). Second, surface-enhanced vibrational spectroscopic techniques have to be adapted to the investigation of sensitive biological materials such that proteins are not degraded during the experiments. Moreover, the techniques must be capable of providing information about the structure and the dynamics of the adsorbed species (Sect. 2.2). Third, the devices must allow a controlled variation and the quantification of the electric-field strength (Sect. 2.3).

2.1 Protein Immobilization on SER/SEIRA Active Metal Electrodes

The most widespread approach for biocompatible coating of SER/SEIRA-active electrodes is based on the chemisorption of ω-functionalized alkanethiols or disulfides on Ag and Au surfaces [5,6]. Depending on their chain length and ω-substituents, these compounds can form densely packed self-assembled monolayers (SAMs) that have very negative potentials of reductive desorption (typically $< -0.6\,V$ vs. NHE) [11, 12, 13, 14]. The potential window for SER/SERR and SEIRA studies is, however, narrowed by reversible or irreversible potential-dependent phase transitions depending on the specific SAM [15].

The chemical variability of the SAM tail groups provides the basis for different modes of protein immobilization, particularly suitable for soluble proteins. Proteins having cationic or anionic binding domains can form stable electrostatic complexes with SAMs that contain acidic or basic tail groups, e.g., $-CO_2H$, $-PO_4H_3$, or $-NH_2$ [5, 6, 11, 12, 14, 16, 17, 18, 19, 20]. Electrostatic adsorption is particularly useful for proteins with a high molecular dipole moment that leads to a largely uniform orientation. NH_2- or COOH-terminated SAMs also serve as a starting point for covalent attachment by crosslinking to anionic or cationic amino acid side chains through carbodiimide reagents [21]. In this case, however, one usually obtains a broader distribution of orientations depending on the reactivity of the individual amino acid side chains. Alternatively, proteins can be crosslinked to the SAM via cysteines, which, if required, have to be introduced to the protein surface by site-directed mutagenesis. The thiol side chain can then be bound to NH_2-terminated SAMs previously treated with N-succinimidyl iodoacetate or N-β-maleimidopropionic acid and carbodiimide. Crosslinking of cysteine residues to OH-terminated SAMs is also possible using N-(p-maleimidophenyl)-isocyanate.

Using methyl-terminated SAMs, hydrophobic interactions also may allow a firm immobilization as has been demonstrated for cytochrome c (Cyt-c) [22] and azurin [11, 23]. Mixed monolayers containing protonable, hydrophobic and/or polar tail groups can also be constructed, although lateral phase separation may be difficult to avoid. In some cases, the redox center of the protein can be directly wired to the electrode using SAMs with tail groups that are capable of coordinating to the cofactor such as to the heme group in Cyt-c [24,25,26], or that may serve as a cofactor such as the flavin group in glucose oxidase [27]. In addition to the rather well-ordered SAMs, electrodes can be coated with polyelectrolytes or clays that allow for the embedment of proteins in relatively large quantities, albeit not in a uniform orientation [28, 29, 30].

Membrane proteins represent a special case since they require a hydrophobic environment that mimics the membrane core as a prerequisite for maintaining the native structure and avoiding protein aggregation. This requirement can be fulfilled either by solubilization in detergents or by reconstitution in phospholipid bilayers. Thus, the simplest approach for immobilizing membrane proteins consists in the direct adsorption of the detergent-solubilized protein onto the metal surface. This strategy has been recently employed for the immobilization of the heme-Cu quinol oxidase from *Acidianus ambivalens* on Ag electrodes coated with specifically adsorbed phosphate anions [31]. It was shown that, under the specific conditions of that study, adsorption occurs without displacement of the detergent, which then provides a biocompatible interface. Albeit straightforward, this mode of immobilization may lead to a distribution of different enzyme orientations. Alternatively, protein-containing lipid bilayers can be deposited on solid supports using Langmuir–Blodgett methods [32]. More recently, various types of solid-supported lipid bilayer membranes have been developed, including bilayers floating freely on quartz, indium tin oxide or gold surfaces, as well as polymer-, polyelectrolyte-supported and tethered bilayer lipid membranes, and some of these devices have been successfully employed in SERR and SEIRA spectroscopic studies [33, 34, 35, 36, 37, 38].

One of the most elegant approaches is based on binding the solubilized protein via a histidine tag (his-tag) to the electrode coated with Ni (or Zn) nitrilo triacetate (Ni–NTA). The high affinity of the Ni–NTA monolayer towards the his-tag guarantees a large surface coverage of uniformly oriented proteins even at the relatively high ionic strengths encountered under physiological conditions. The immobilized enzyme is then incubated in the presence of lipids and biobeads in order to remove the solubilizing detergent and allow the formation of a lipid bilayer. This method has been successfully applied for immobilization of cytochrome c oxidase (CcO) on Ag and Au electrodes [34, 36].

2.2 Surface-Enhanced Vibrational Spectroelectrochemistry of Heme Proteins

The electronic absorption spectrum of porphyrins is characterized by an intense band at ca. 410 nm (Soret band) and a weaker band at ca. 550 nm (Q band). Excitation in resonance with these transitions leads to a strong and selective enhancement of the Raman-active modes of the porphyrin such that the RR spectrum of a heme protein exclusively displays vibrational bands of the heme, regardless of the size of the optically transparent protein matrix. Due to the relatively high symmetry of the heme, which to a first approximation can be considered as D_{4h}, excitation in resonance with the allowed Soret and the forbidden Q band leads to specific enhancement patterns. Upon Soret-band excitation, the RR spectrum is dominated by the totally symmetric A_{1g} modes of the porphyrin, and displays those bands that are particularly diagnostic for specific structural properties of the porphyrin such as the oxidation, spin, and ligation state of the heme iron ("marker bands"; 1300 cm^{-1} to 1700 cm^{-1}) and the interactions of the heme with the surrounding protein ("fingerprint"; 200 cm^{-1} to 500 cm^{-1}) [39, 40]. For this reason and due to the distinctly larger resonance enhancement, Soret-band excitation is usually preferred compared to Q-band excitation.

When the protein is immobilized on SER-active metal surfaces RR and SER combine (SERR), provided that laser excitation simultaneously meets the resonance conditions of the electronic transition of the heme and the surface plasmons of the metal. This condition can only be fulfilled for Ag surfaces that exhibit maximum surface enhancement typically at ca. 500 nm. The highest SERR intensity is obtained with 413 nm or 407 nm excitation since the strong molecular RR enhancement clearly overrides the poorer surface enhancement in the violet region as compared to Q-band excitation at ca. 550 nm (strong SER, but low RR). Indeed, Soret-band excitation yields SERR signals of good quality even for heme proteins that are separated from the metal surface by spacers of up to 5 nm thickness [35, 37].

The SERR spectra that are obtained in this way allow different states of oxidation, spin and coordination of the heme iron [5, 6] to be distinguished. For instance, the SERR spectra of Cyt-c electrostatically immobilized on SAMs of mercaptohexadecanoic acid exclusively display the characteristic band patterns of the reduced and oxidized forms when measured 100 mV below and above the redox potential, respectively (Figs. 1a,b) [41].

Band positions and intensities of the ferrous and ferric form are sufficiently different to allow for quantitative spectral component analysis of redox mixtures. In this analysis, the complete spectra of the individual components are fitted to the experimental spectra of the mixture varying only the relative contributions of the component spectra [42]. The method is also applicable for spectra involving additional components such as those in which the heme iron exhibits different spin configurations and coordination patterns (Figs. 1c,d). On the basis of the component analysis, thermodynamic constants of redox

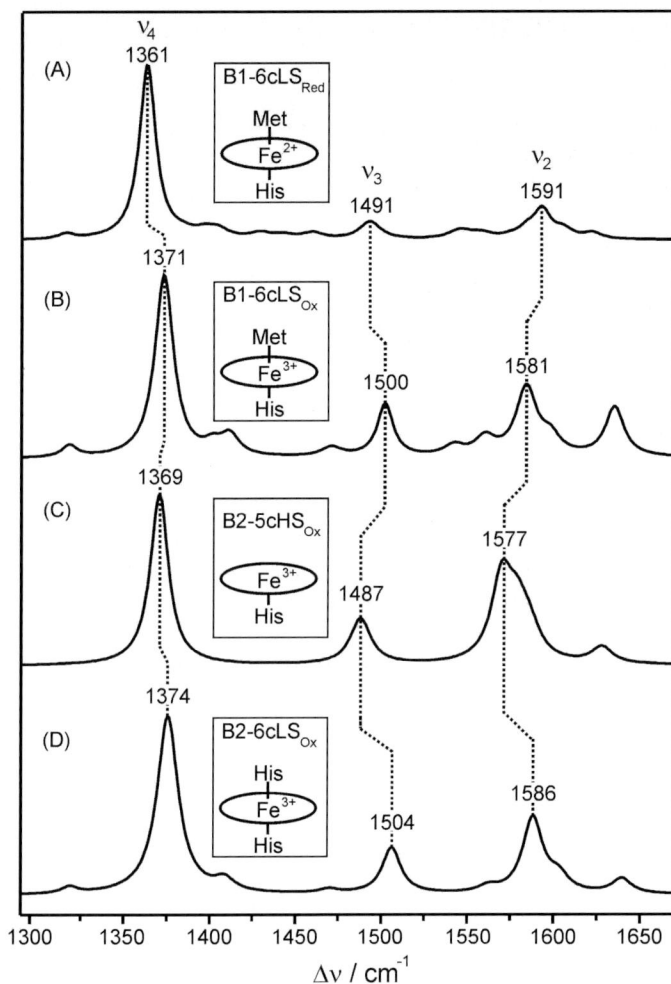

Fig. 1. SERR spectra of Cyt-c in different redox, coordination and spin states

and conformational equilibria of adsorbed heme proteins can be evaluated from potential-dependent SERR measurements as is illustrated in Fig. 2a.

SERR spectroelectrochemistry can be operated in the time-resolved (TR) mode by synchronizing fast potential jumps with short measuring events [6, 7, 9]. The most versatile TR SERR approach is based on electro-optically gated continuous wave (cw) excitation that generates laser pulses of low photon flux and variable duration down to the long nanoseconds timescale. These pulses define the measurement intervals that follow the potential jump after a variable delay time δ (Fig. 3). Subsequent to the measurement interval, the electrode potential is reset to the initial value to restore the original equilibrium. The sequence of potential jumps, delay times, and measurement

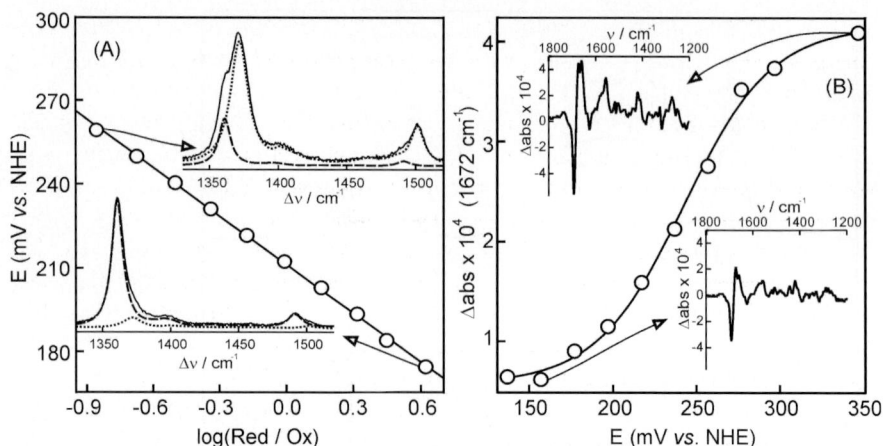

Fig. 2. Redox titrations of Cyt-c electrostatically adsorbed on a COOH-terminated SAM. (**a**) Nernstian plot based on potential-dependent SERR measurements. The *insets* show the experimental SERR spectra at two selected potentials with the reduced and oxidized components represented in *dashed* and *dotted lines*, respectively. (**b**) SEIRA titration curve based on the potential dependence of the amplitude of the $1672\,cm^{-1}$ band with respect to a reference potential of $100\,mV$. The *insets* show the difference SEIRA spectra at two selected potentials

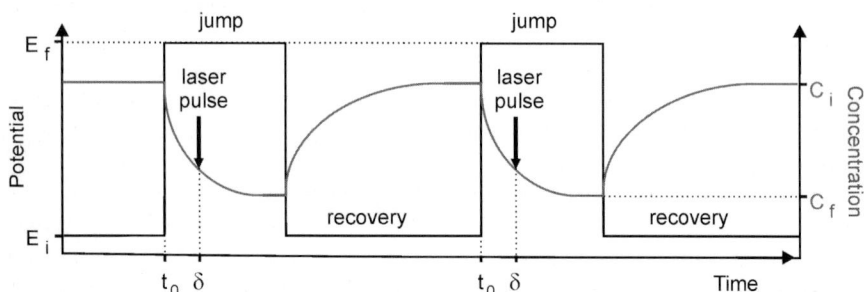

Fig. 3. Schematic representation of a TR SERR measurement. The sequence of potential jumps is represented with *black lines* and the corresponding variation of concentrations is indicated in *gray*

intervals, which are synchronized by a pulse delay generator, is repeated until the signal accumulated on a low-noise CCD detector of the spectrograph is of sufficient quality. A prerequisite for this approach is the full reversibility of the potential-dependent processes. The temporal resolution is mainly limited by the response time of the electrochemical cell, which is usually within a few tens of microseconds.

The main advantage of SERR spectroelectrochemistry, compared to conventional electrochemical techniques, is that it allows identifying and monitoring the molecular structures of the species that are involved in the interfacial redox reactions. This is particularly important for complex interfacial redox processes that involve various redox sites (e.g., multiheme proteins) or that are coupled to non-Faradaic processes. The main drawback of the method is that the SER effect is limited to microscopically roughened surfaces and only to a few metals, particularly Ag and Au. Laser-induced damage of the adsorbed proteins due to local heating or photochemical processes may also impose severe limitations. In most cases the damage can be avoided or minimized using low laser powers and continuous movement of the electrode surface with respect to the laser focus, for example using rotating electrodes [43, 44].

The information about the structure and dynamics of the adsorbed heme proteins that can be obtained by SERR spectroscopy is restricted to the redox site. To also probe structural changes of the protein matrix, SEIRA spectroscopy has to be employed. This technique traces spectral changes of the entire protein, specifically of the amide I modes that are sensitive indicators of the secondary structure. SEIRA measurements are most conveniently combined with the attenuated total reflection (ATR) technique [10]. For that purpose, a silicon ATR crystal is covered with a thin layer of SAM-coated Au or Ag that serves as the working electrode such that in-situ measurements in an electrochemical cell are possible without significant interference by water bands. Residual contributions from the solvent are largely cancelled since SEIRA measurements are carried out in the difference mode, which also significantly simplifies the SEIRA spectra. Since the enhancement of the IR bands strongly depends on the orientation of underlying dipole moments as well as on the distance with respect to the electrode surface, the absolute SEIRA spectra do not necessarily reflect all IR-active bands of the protein. Thus, unlike conventional IR spectroscopy in solution, determination of the protein secondary structure is not possible in SEIRA spectroscopy. However, potential-dependent processes of the adsorbed protein can be studied by subtracting a spectrum obtained at a reference potential from each measured spectrum. The difference spectra exclusively display those bands that reflect potential-dependent structural and orientational changes of the immobilized protein (Fig. 2b). Time-resolved SEIRA measurements can be performed by operating the FTIR spectrometer in the step-scan or rapid scan modes to monitor the dynamics of the protein structural changes subsequent to a potential jump.

2.3 Theoretical Description and Experimental Determination of the Electric-Field

The surface coatings described in Sect. 2.1 provide a biocompatible environment for the immobilization of several proteins. In addition, they offer the possibility of controlling the electric-field strength at the protein binding site.

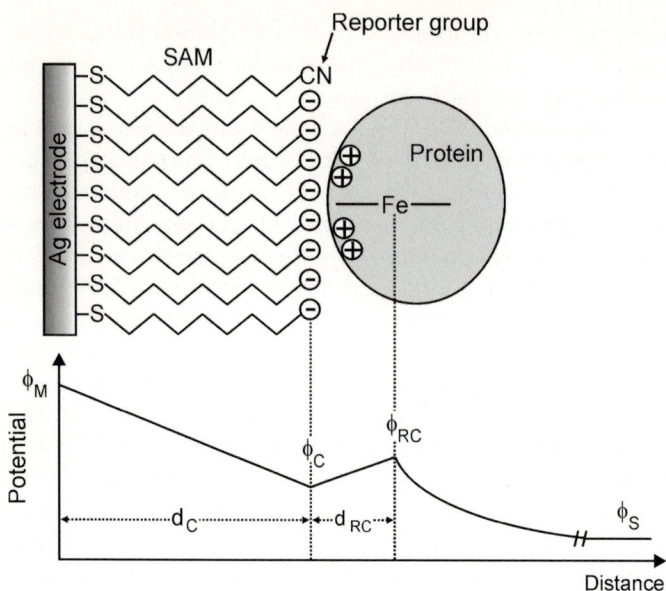

Fig. 4. Schematic representation of the interfacial potential distribution for Cyt-c electrostatically adsorbed on an Ag electrode coated with a COOH-terminated SAM

The interfacial potential distribution across the electrode/SAM/protein/solution interface can be described in terms of electrostatic theory as schematically shown in Fig. 4 [41]. The local electric-field at the protein binding site is controlled via the chain length and the tail functional group of the SAM, the electrode potential, and the pH and ionic strength of the solution. The predictions for SAMs of carboxylate-terminated thiols on Ag indicate local electric-field strengths at the SAM surface that are of the order 0.1 V/nm to 1 V/nm, and are, therefore, comparable to the values expected in the vicinity of charged phospholipid head groups of biological membranes [45].

Local electric-field strengths can also be determined experimentally by utilizing the vibrational Stark effect of appropriate reporter groups that are incorporated in the SAMs. For the nitrile function, the electric-field dependence of the vibrational frequency has been analyzed in detail by *Boxer* and coworkers [46, 47, 48]. On the basis of these results, determination of the C–N stretching frequency in SAMs including nitrile-terminated thiols by SER and SEIRA spectroscopy allows an estimation of the electric-field strength in the SAM/solution interface to be obtained (Fig. 4). Recent studies by *Harris* and coworkers and in our lab have shown that this approach provides results that are in good agreement with the values obtained by electrostatic theory [49, 50].

3 Electric-Field Effects on the Interfacial Processes of Heme Proteins

Electric-field strengths between $0.1\,\mathrm{V/nm}$ and $1\,\mathrm{V/nm}$ are sufficiently strong to alter protonation equilibria in proteins, to align and to induce molecular dipoles, to cause a redistribution of ions and solvent molecules, and even to perturb molecular energy levels [46]. These primary effects are likely to have consequences for the protein structure and the thermodynamics and kinetics of interfacial processes particularly when charge-transfer processes or changes of dipoles moments are involved.

3.1 Conformational Changes of Cytochrome c

Cyt-c is a small soluble monoheme protein that possesses a well-defined patch of positively charged lysine residues around the partially exposed heme edge that serves as the binding domain to its membrane-bound partner proteins [51]. Thanks to the heterogeneous distribution of surface charges and the resulting large dipole moment [52], it binds to negatively charged surfaces with high affinity and a largely uniform orientation. The binding domain of Cyt-c is likely to be the same for immobilization to electrodes coated with anionic SAMs as for complex formation with the natural reaction partners CcO or cytochrome c reductase. The SERR spectra of Cyt-c adsorbed to SAMs of carboxylate-terminated alkylthiols (C_n-SAMs with "n" denoting the number of methylen groups) on Ag with $n \geqslant 10$ are identical to the RR spectra in solution [41]. Also, the peak positions in the "reduced-minus-oxidized" IR difference spectra (SEIRA) obtained for Cyt-c adsorbed on Au coated with these C_n-SAMs and for the dissolved protein agree very well [10, 53, 54]. Furthermore, the redox potentials determined for the immobilized Cyt-c by SERR and SEIRA spectroscopy and corrected for the interfacial potential drop (see Sect. 3.2) were found to be the same as in solution. Thus, both the heme pocket and protein secondary structure remain unaffected upon adsorption under these conditions.

On the other hand, the SERR and SEIRA spectra of Cyt-c measured on shorter C_n-SAMs ($n < 10$) reveal additional contributions from non-native species denoted as B2 conformational state [41]. Similar B2 species have also been observed for electrostatic complexes of Cyt-c with micelles, phospholipid vesicles and polyanions in solution. The soluble complexes have been studied by a variety of spectroscopic methods, which revealed that the formation of the B2 state is associated with changes of the tertiary structure and the loss of the native axial ligand Met-80 [40,55,56,57]. This coordination site remains either vacant (five-coordinated high spin; 5cHS) or is occupied by a histidine residue (six-coordinated low spin; 6cLS), which has been identified as His-33 or His-26 (Fig. 5a). The transition from the native state B1 to the B2 state is reversible and involves the movement of the peptide segment 30(20)-49 to bring His-33 (His-26) into proximity with the heme iron.

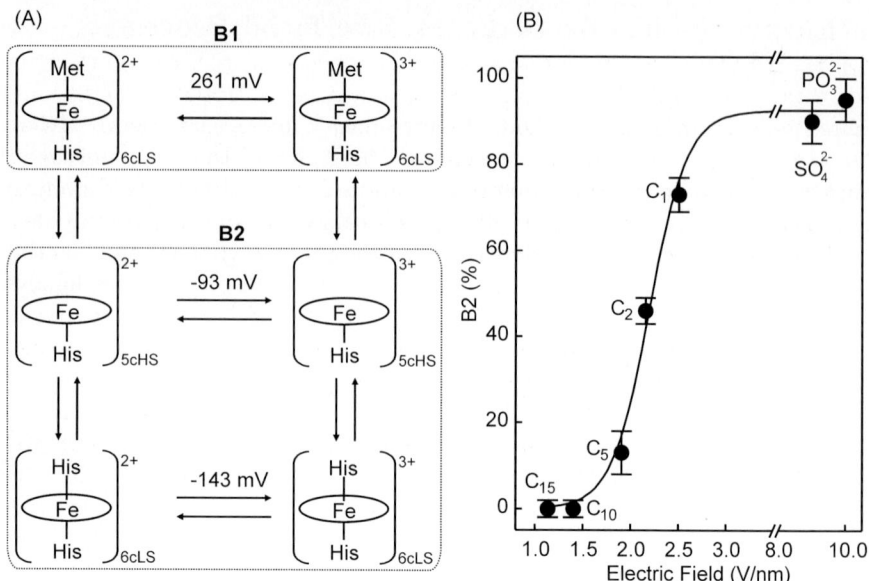

Fig. 5. (a) Schematic representation of the redox and conformational equilibria of Cyt-c in complexes with model systems. (b) Variation of the total amount of the B2 species of Cyt-c on electrodes with different coatings as a function of the electric field. See text for details

Formation of B2 is favored in the oxidized state due to the intrinsically lower stability of the Fe–Met bond in the ferric as compared to the ferrous form. On short C_n-SAMs, the B2 species can therefore only be observed for the ferric protein and reduction leads to the complete (re-)conversion to the B1 form [41]. The total amount of B2 species measured at the redox potential of the native state B1 shows a steady increase from 0 % at $n = 15$ and 10 up to 75 % at $n = 1$. This shift of the B2/B1 equilibrium is attributed to the increasing electric-field strength upon decreasing the SAM thickness. In agreement with this interpretation, a further increase of the electric field, which is achieved by coating the Ag with SAMs of phosphonate-terminated alkylthiols or with specifically adsorbed sulfate ions, shifts the conformational equilibrium almost completely towards B2 (Fig. 5b) [58]. At these latter two coatings, the electric-field is sufficiently strong to destabilize the Fe–Met bond also in the ferrous form. On sulfate-coated electrodes it was, therefore, possible to probe the redox equilibria of the B2 species leading to redox potentials of ca. -0.09 V and -0.14 V for the 5cHS and 6cLS couple, respectively [59]. This ca. 400 mV downshift of the redox potentials compared to the native B1 state can be readily attributed to the ligation and spin-state change as well as to the perturbation of the heme pocket structure.

A conformational transition to a B2-like conformational state is also observed upon hydrophobic binding both in solution (e.g., monomeric sodium dodecyl sulfate) [40, 55, 56] or on electrodes covered with CH_3-terminated SAMs [22]. Hydrophobic interactions primarily lead to the displacement of the peptide segment 81–85 that is directly linked to the Met-80 and thus may cause the dissociation of this ligand from the heme. This segment is located in the center of the ring-like arrangement of lysine residues that are involved in electrostatic binding to anionic surfaces, implying that both modes of binding not only cause the same changes in the ligation pattern of the heme but also lead to approximately the same gross orientation of the protein with respect to the surface.

In contrast to the striking similarities of the kind of conformational changes, electrostatic and hydrophobic binding are associated with dramatic differences in the dynamics of the conformational transitions as probed in TR SERR experiments. Upon electrostatic binding the B1 → B2 transition occurs on the timescale of seconds but the rate increases by ca. five orders of magnitude for hydrophobically adsorbed Cyt-c [22, 58, 59].

3.2 Modulation of Redox Potentials

Potential-dependent SERR and SEIRA measurements of Cyt-c on C_n-SAMs indicate small downshifts of the measured redox potential upon increasing the chain length from $n = 1$ to 15 [41]. On the basis of electrostatic theory, these shifts can be quantitatively ascribed to the potential drops at the electrode/SAM/protein interfaces and thus do not reflect an alteration of the redox potential of the heme. Cyt-c immobilized on electrodes coated with polyelectrolyte multilayers, however, displays a different picture since the observed negative shifts of the redox potentials are larger by up to a factor of five [28]. Such electrode coatings are obtained by sequential adsorption of polycations (poly[ethylene imine] – PEI; polyallylamine hydrochloride – PAH) and polyanions (poly[styrene sulfonate] – PSS). Efficient Cyt-c binding only occurs when the terminating layer is PSS, which forms tight complexes with the cationic Cyt-c. In these complexes, ca. 40 % of the Cyt-c molecules are converted to the B2 state. However, the high negative charge density that is provided by the surrounding sulfonate groups also affects the remaining native state B1 by destabilizing the ferrous form such that the redox potential is lowered by more than 200 mV. This shift is not associated with structural changes of the heme site implying a direct perturbation of the redox potential by the electric-field.

This electric-field dependent modulation of the redox potential has been studied in detail in the case of the tetraheme cytochrome c_3 (Cyt-c_3), a small soluble electron/proton shuttle in the respiratory chain of *Desulfovibrio gigas* [17]. NMR studies of Cyt-c_3 in solution have revealed reduction potentials that increase in the order heme I $(-287\,mV)$, heme II $(-279\,mV)$, heme III $(-262\,mV)$ and heme IV $(-183\,mV)$ [60]. When electrostatically adsorbed to Ag electrodes coated with C_{10}-SAM, the SERR spectra of both the fully

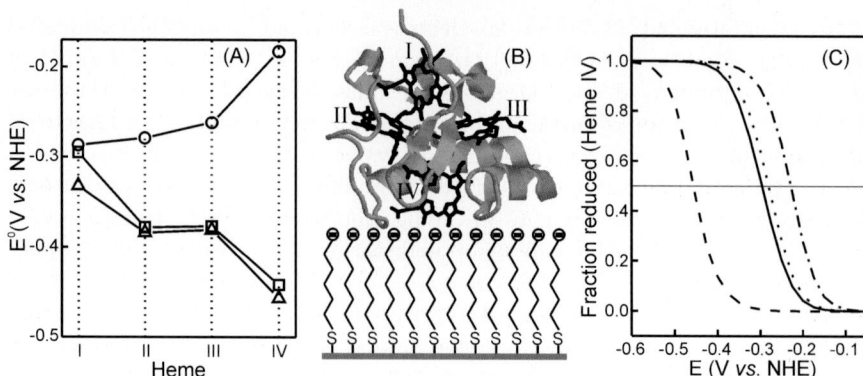

Fig. 6. (a) Reduction potentials for the individual hemes of Cyt-c_3 under different conditions. *Circles:* NMR titration in solution. *Squares:* SERR titration for Cyt-c_3 on an Ag electrode coated with C_{10}-SAM at pH 7. *Triangles:* theoretical prediction for Cyt-c_3 on a C_{10}-SAM at pH 7. (**b**) Orientation of Cyt-c_3 on a C_{10}-SAM as determined by molecular dynamics simulations. (**c**) Theoretical titration of heme IV of Cyt-c_3 computed under different conditions. *Dashed line:* charged C_{10}-SAM (pH 7). *Solid line:* solution. *Dotted line:* after removal of the C_{10}-SAM. *Dashed-dotted line:* neutral C_{10}-SAM

oxidized and fully reduced sample do not differ from the respective RR spectra in solution, suggesting that the protein retains its native structure upon immobilization. Electrochemical SERR titrations, however, indicate that the reduction potentials of the individual hemes are substantially downshifted with respect to the values in solution (Fig. 6a) such that the order of reduction is reversed with heme IV having the most negative redox potential ($-390\,\text{mV}$), followed by heme III ($-350\,\text{mV}$), heme II ($-350\,\text{mV}$) and heme I ($-270\,\text{mV}$).

Molecular dynamics simulations confirm that Cyt-c_3 docks to the C_{10}-SAM via the lysine-rich patch around heme IV under preservation of the protein structure (Fig. 6b). The redox potential shifts of the individual hemes determined in the SERR experiments are well reproduced by electrostatic calculations, with the largest downshift predicted for heme IV ($161\,\text{mV}$) that is in closest contact to the anionic SAM surface. Electrostatic calculations performed on the structure optimized for the immobilized Cyt-c_3 but in the absence of the charged SAM yield redox potentials very similar to those calculated for the protein in solution (Fig. 6c). The calculations further predict a positive shift of the redox potentials for Cyt-c_3 immobilized on a fully neutralized SAM [17]. These results suggest that the redox potentials of the bound Cyt-c_3 are controlled by the interplay of two effects that tend to shift the redox potentials in opposite directions. On the one hand, the distance dependent electrostatic interactions between the individual hemes and the carboxylate groups of the SAM stabilize the oxidized forms, specifically of heme IV that is in direct contact with the SAM, and, to a much lesser extent, of the more remote hemes. On the other hand, the low dielectric constant of the SAM with respect to the bulk aqueous medium tends to destabilize the formally monopositive ferric forms with respect to the formally neutral ferrous hemes.

A similar explanation holds for the behavior observed for cytochrome $P450_{cam}$ on Ag electrodes coated with C_n-SAMs of variable chain lengths [19]. SERR spectroscopy indicates that upon immobilization this enzyme does not retain the thiolate axial ligand and instead a P420 species is formed that is spectroscopically similar to the pressure-induced P420 in solution. The redox potential of P420 immobilized on a C_{15}-SAM is ca. 400 mV more positive than in solution but it shows a continuous downshift upon increasing the electric-field strength through shortening the chain length of the SAM or increasing the negative charge density on the electrode coating.

3.3 Electron-Transfer Dynamics of Cytochrome c and Other Soluble Electron Carriers

The heterogeneous electron-transfer (ET) of redox proteins immobilized on SAM-coated electrodes is expected to occur in the nonadiabatic regime due to the large separation between the redox center and the metal surface. Under these conditions, the formal ET rate constant $k_{ET}(0)$ is controlled by the weak electronic coupling that decays exponentially with the distance [61]:

$$k_{ET}(0) \propto H^0 \exp[-\beta(x - x_0)],\qquad(1)$$

where H^0 is the electronic coupling at the distance of closest approach x_0 and β is the tunneling decay parameter.

The distance dependence of $k_{ET}(0)$ has been studied for a number of proteins immobilized on Ag and Au electrodes coated with different types of SAMs, using a variety of electrochemical and spectroelectrochemical methods, including TR SERR [11, 16, 62, 63, 64, 65, 66]. A graphical summary for representative examples of these investigations is presented in Fig. 7a. At sufficiently long SAMs, the measured apparent rate constant $k_{app}(0)$, for all proteins exhibits the behavior predicted by (1) with approximately the same value for β corresponding to ca. 1.1 per CH_2 group, consistent with a redox reaction controlled by long-range tunneling. In contrast, $k_{app}(0)$ becomes distance independent at shorter SAMs suggesting either a change of mechanism or the coupling of ET with another process that becomes rate limiting.

For Cyt-c on a C_{15}-SAM the reduction rate shows a clear overpotential dependence, which can be treated according to Marcus-DOS theory for nonadiabatic ET reactions (Fig. 7b) yielding a reorganization energy $\lambda = 0.24$ eV [67, 68]. This value, which was also confirmed by temperature-dependent measurements of $k_{ET}(0)$, is distinctly lower than that determined for Cyt-c in solution (ca. 0.6 eV). The discrepancy can be understood taking

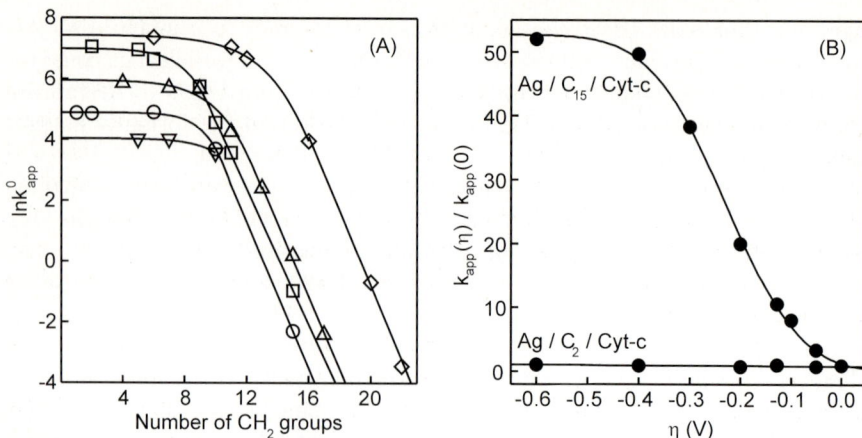

Fig. 7. (a) Distance dependence of the apparent ET rate for different systems. *Down-triangles:* cytochrome b_{562} on Ag coated with NH_2-terminated SAMs [16]. *Circles:* Cyt-c on Ag coated with COOH-terminated SAMs [62]. *Up-triangles:* azurin on Au coated with CH_3-terminated SAMs [11]. *Squares:* Cyt-c on Au coated with COOH-terminated SAMs [63]. *Diamonds:* Cyt-c on Au coordinatively bound to pyridine-terminated SAMs [64, 65, 66]. (b) Overpotential dependence of the apparent ET rate for Cyt-c on Ag electrodes coated with two different COOH-terminated SAMs

into account that the largest contribution to the reorganization energy results from the solvent and protein matrix rearrangement [69, 70]. In the adsorbed state, solvent reorganization is likely to be reduced as compared to the solution due to the exclusion of water molecules from the binding domain and due to the lower dielectric constant, which is presumably smaller than 10 in the electrochemical interface compared to 78 in the bulk solution.

On the other hand, the measured rate for Cyt-c on a C_1- or C_2-SAM is independent of the overpotential, indicating that at shorter distances ET does not constitute the rate-limiting step [58, 67]. In this region, $k_{app}(0)$ decreases upon increasing the viscosity of the solution as determined by TR SERR and electrochemical techniques [58, 63]. *Avila* et al. [63] proposed a reaction scheme that involves the adsorption of Cyt-c in a thermodynamically stable orientation that is redox-inactive and requires a transient reorientation before electron tunneling can take place. Since only the latter step is distance dependent, the overall reaction rate reaches a limiting value when ET approaches the rate of the distance independent reorientation. However, this model cannot account for all experimental observations, like the decrease of the apparent ET rate observed upon H_2O/D_2O exchange, which originally was attributed to proton-transfer steps becoming rate limiting [62]. The increasing ratio of the rate constants measured in H_2O and D_2O from 1.0 at C_{15}-SAMs to 4.0 at C_1-SAMs can be reinterpreted in terms of the ca. 20 % higher viscosity of D_2O with respect to H_2O; but it remains puzzling why the apparent rate constant in H_2O is constant at C_n-SAMs with $n < 10$ while it decreases in D_2O.

On the other hand, reorientation of Cyt-c must involve the rupture and formation of salt bridges between the protein surface and the electrode coating, and thus should be affected by the charge densities of the protein and the C_n-SAM as well as by the realignment of the protein dipole moment in the interfacial electric-field. Both the charge density of the C_n-SAM and the strength of the interfacial electric-field increase at short chain lengths such that protein reorientation should be slowed down when the field becomes sufficiently strong.

This prediction was in fact confirmed in TR SERR experiments with Cyt-c immobilized on Ag electrodes covered with specifically adsorbed sulfate ions and phosphonate-terminated SAMs [58, 59]. Although the strong electric-fields at these coatings lead to a substantial portion of the B2 state (vide supra) and even cause a slow irreversible denaturation, TR SERR still allows determination of the apparent ET rate for the fraction of immobilized protein that remains in the native state B1. The measured rates for Cyt-c adsorbed on SO_4–Ag and PO_3–SAM are significantly lower than for C_n-SAMs at comparable distances. Indeed, a plot of the apparent rates as a function of the electric-field strength (Fig. 8) yields a uniform picture for all the different coatings consistent with a heterogeneous redox reaction that is controlled by electron tunneling at long distances and by electric-field-dependent reorientation at short distances. Qualitatively similar results were obtained for cytochrome c_6 (Cyt-c_6), a soluble monoheme protein that acts as an electron carrier between the b_6f complex and photosystem I (PSI) in photosynthetic redox chains [20].

A different mechanism holds, however, for Cyt-c that is attached to the electrode via coordination of the heme iron by the tail group of pyridinyl-terminated SAM [24, 25, 26, 64, 65, 71]. Also in this case, the experimentally determined rate constants approach a limiting value at short distances and then become viscosity dependent. However, in contrast to electrostatic adsorption, the ET rate constant in the distance independent regime exhibits a strong increase with the overpotential [72]. This behavior is consistent with an adiabatic or friction-controlled ET mechanism, as originally suggested by *Waldeck* and coworkers [25, 26, 64, 65, 71].

3.4 Electric-Field Effects on the Electroprotonic Energy Transduction of Heme-Cu Oxidases

Heme-Cu oxidases constitute a superfamily of terminal enzymes in respiratory chains that catalyze the reduction of molecular oxygen to water and use the energy provided by exergonic electron-transfer events to pump protons trough the membrane against a gradient [73, 74]. The best-known members of this group of enzymes are the mitochondrial-type CcOs in which a binuclear

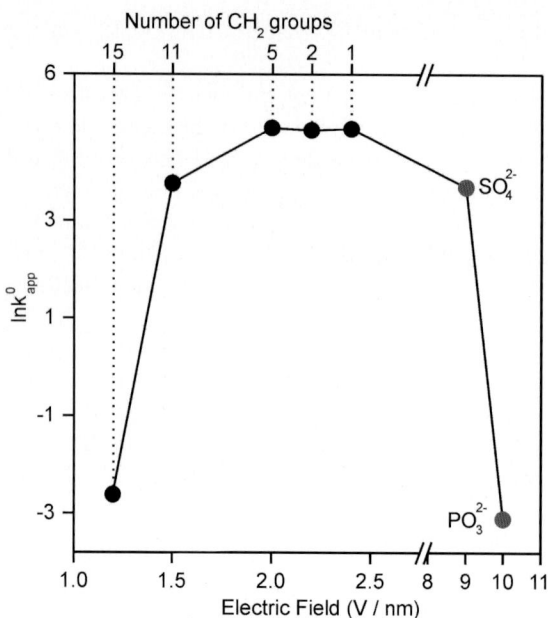

Fig. 8. Electric-field dependence of the apparent ET rate constant of Cyt-c on coated Ag electrodes. The *black points* represent COOH-terminated SAMs of different chain lengths. The *gray points* refer to Ag coated with either specifically adsorbed sulfate anions or with 11-thioundecyl-1-phosphonic acid

center, a high-spin heme a_3 and a copper B center (Cu_B), constitute the catalytic site. The entry site for the electrons delivered by Cyt-c is a binuclear copper A center (Cu_A) from which the electrons are further transferred to a low-spin heme a and subsequently to the catalytic site where bound oxygen is reduced.

We have recently studied CcO integrated into a lipid bilayer and anchored to a Ni(Zn)–NTA-coated electrode via a his-tag introduced at the C-terminus of either subunit I or II, such that the primary electron acceptor, the Cu_A site, is either facing the electrode or the solution, respectively [35, 37]. The immobilized enzymes retain the native structure at the level of the hemes a and a_3 as judged from the SERR spectra that do not differ from the corresponding RR spectra in solution. In both orientations the immobilized CcO can be reversibly reduced and oxidized under anaerobic conditions, as indicated by the potential-dependent changes of the oxidation marker bands (Fig. 9). However, a careful inspection of the SERR spectra in the $1500\,cm^{-1}$ to $1700\,cm^{-1}$ region demonstrates that only heme a is redox-active, while heme a_3 remains oxidized even at $-0.45\,V$, that is at very negative overpotentials.

The heterogeneous ET rate constant was found to be nearly identical for both orientations, ca. $0.002\,s^{-1}$, and independent of the electrode potential

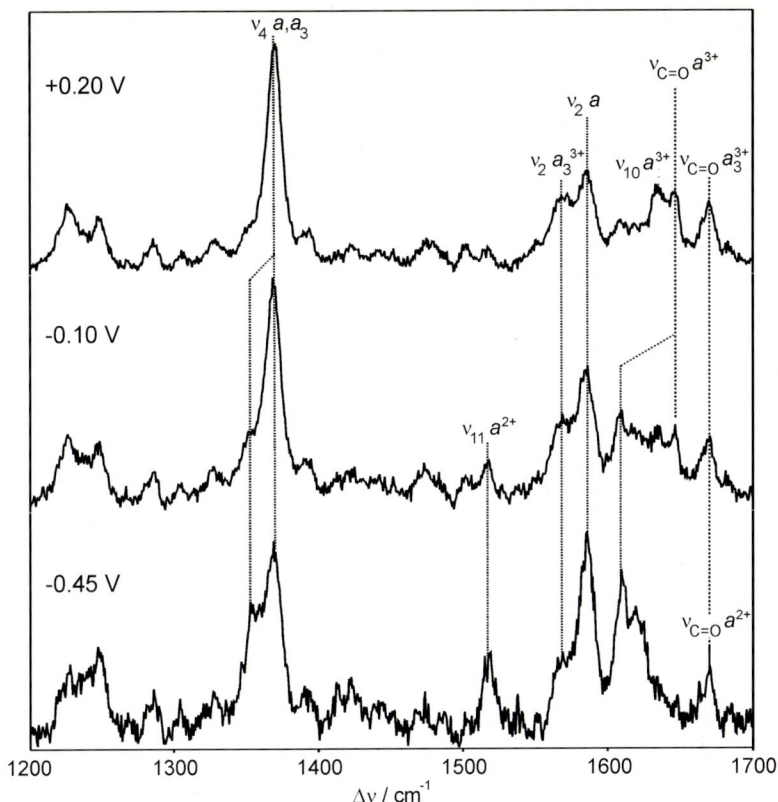

Fig. 9. Potential dependence of the SERR spectra of CcO immobilized on Ag electrodes using the Ni–NTA/his-tag approach. See text for details

between 200 mV and −160 mV. Thus, it was concluded that the natural electron entry site, i.e., the Cu_A center, does not participate in the electronic pathways involved in this model system. The through-space distance from the electrode surface to heme a is ca. 50 Å in both CcO orientations. For such long distances, electron tunneling is highly unlikely and cannot be reconciled with a rate constant of $0.002 s^{-1}$. Most likely the redox mechanism involves multistep electron tunneling or hopping via the amide functions of the NTA spacer and the polypeptide backbone as proposed by *Morita* and *Kimura* for helical peptides [75].

Irrespective of the details of the mechanism of electronic communication between the electrode and heme a in this artificial system, it is particularly remarkable that heme a_3 is not reduced. Intramolecular ET from heme a to heme a_3 is expected to occur within nanoseconds [76] despite the fact that, in the fully oxidized state, the midpoint potentials of hemes a and a_3 of mammalian CcO are ca. +390 mV and +200 mV, respectively, which in principle would correspond to an uphill electron-transfer process. However, the midpoint potentials of the four redox centers in CcO are strongly modulated by the respective oxidation states and the protonation state of redox-Bohr sites. The underlying molecular mechanism that ensures an efficient downhill electron-transfer cascade coupled to uphill proton translocation is still a matter of debate, but it essentially consists of a complex network of cooperativities, including Coulombic interactions and mechanochemical components [77,78,79]. The failure to reduce heme a_3 in the immobilized CcO, directly through the electrode or via intramolecular ET from heme a, suggests that the internal H^+/e^- cooperativity network is disturbed possibly due to electric-field-induced alterations of pK_as such that crucial proton-transfer steps coupled to the ET reactions are blocked or slowed down.

Indirect evidence for electric-field-induced perturbation of proton-transfer events has also been obtained for the quinol oxidaze (QO) from *A. ambivalens* [31]. Like CcO, this enzyme contains a Cu_B-heme a_3 binuclear active center and a heme a as its immediate electron donor, but instead of a Cu_A a loosely bound quinone acts as the electron-entry site [80]. The direct adsorption of detergent-solubilized QO to a "bare" Ag electrode occurs without displacement of the detergent that, hence, acts as a biocompatible interface. Indeed, the peak positions in the SERR spectra of the solubilized enzyme are identical to those in the RR spectra of QO in solution, indicating that the structures of the heme pockets are largely preserved [31]. Unlike CcO, potential-dependent SERR measurements of the immobilized QO clearly reveal a reversible electrochemistry for both hemes. Moreover, the spectral-component analysis yields nearly ideal Nernstian plots (Fig. 10) pointing to rather weak Coulombic interactions between the redox centers of QO that is also in sharp contrast to mitochondrial CcO. The redox potentials determined in the SERR experiments are $+320\,mV$ and $+390\,mV$ for hemes a and a_3, respectively, and thus are in the reversed order with respect to CcO. Both the reversed order of redox potentials as well as the lack of substantial Coulombic interactions represent a unique case among the superfamily of oxygen reductases. While CcO requires a complex network of cooperativity effects to guarantee a downhill ET sequence, the exergonicity in QO is already ensured by its inverted redox potentials.

Reduction of heme a_3 of QO in solution is followed by a weakening of the hydrogen-bond interactions of the formyl substituent as reflected by the frequency shift of the C=O stretching from $1661\,cm^{-1}$ to $1669\,cm^{-1}$ in the RR spectrum [80]. This process is believed to constitute a key step in the proton-pumping activity of the enzyme. On the other hand, the SERR spectrum of the reduced immobilized QO only displays the band at $1661\,cm^{-1}$ and no time-dependent increase of intensity at $1669\,cm^{-1}$ is noted within 30 min, implying that the redox-linked reorganization of the hydrogen-bond interactions in the catalytic center is blocked or at least drastically slowed down in the presence of high electric fields [31].

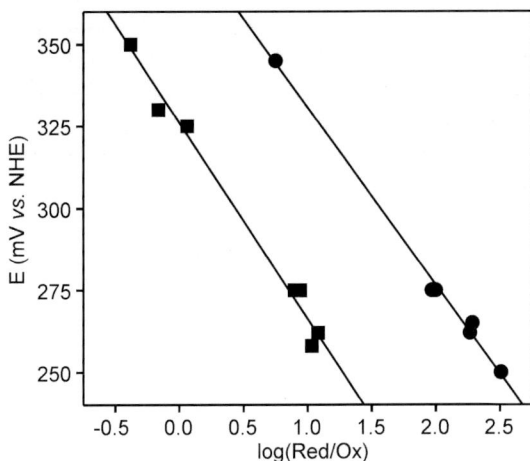

Fig. 10. Nernstian plots for the two hemes of QO immobilized on detergent-coated Ag electrodes. *Squares:* heme *a*. *Circles:* heme a_3

4 Concluding Remarks

Surface-enhanced vibrational spectroelectrochemistry constitutes a powerful approach for elucidating the reaction mechanisms and dynamics of immobilized redox proteins. In contrast to traditional electrochemical methods, the combination of SERR and SEIRA provides a direct and detailed molecular picture of all potential-dependent processes of the adsorbed species, including those that are non-Faradaic reactions. In this respect, these techniques may gain increasing importance in both fundamental and applied science. On the one hand, a profound knowledge of the molecular structure and dynamics of immobilized proteins and enzymes that can be obtained by these methods is a prerequisite for the rational design of bioelectronic devices of technological importance. On the other hand, surface-enhanced vibrational spectroscopies open new possibilities to study complex biomimetic systems, thereby providing novel insight into fundamental biological processes.

In this contribution we have presented a simple way for studying the influence of electric-fields on the various parameters that control electron-and proton-transfer reactions of soluble and membrane-bound proteins like Cyt-c and CcO. We have shown that electric fields of a magnitude comparable to those at the interfaces of biomembranes, i.e., the natural reaction environment, may have a substantial influence on redox potentials, protein structure, redox-linked conformational changes and orientation/reorientation of electrostatic complexes. These studies do not directly prove a functional role of the electric-field effects in vivo, but they clearly indicate that such effects might be present and deserve further investigation. The application of surface-enhanced vibrational spectroscopic methods in combination with more realistic model systems appears to be a promising strategy towards that goal.

References

[1] T. M. Cotton, S. G. Schultz, R. P. Van Duyne: J. Am. Chem. Soc. **102**, 7960 (1980)

[2] K. M. Ervin, E. Koglin, J. M. Sequaris, P. Valenta, H. W. Nurnberg: J. Electroanal. Chem. **114**, 179 (1980)

[3] R. E. Holt, T. M. Cotton: J. Am. Chem. Soc. **111**, 2815 (1989)

[4] G. Smulevich, T. G. Spiro: J. Phys. Chem. **89**, 5168 (1985)

[5] D. H. Murgida, P. Hildebrandt: Acc. Chem. Res. **37**, 854 (2004)

[6] D. H. Murgida, P. Hildebrandt: Phys. Chem. Chem. Phys. **7**, 3773 (2005)

[7] C. T. Shi, Z. Wei, R. L. Birke, J. R. Lombardi: J. Phys. Chem. **94**, 4766 (1990)

[8] D. H. Murgida, P. Hildebrandt: Angew. Chem. Int. Ed. **40**, 728 (2001)

[9] H. Wackerbarth, U. Klar, W. Gunther, P. Hildebrandt: Appl. Spectrosc. **53**, 283 (1999)

[10] K. Ataka, J. Heberle: J. Am. Chem. Soc. **125**, 4986 (2003)

[11] Q. J. Chi, J. D. Zhang, J. E. T. Andersen, J. Ulstrup: J. Phys. Chem. B **105**, 4669 (2001)

[12] N. Mohtat, M. Byloos, M. Soucy, S. Morin, M. Morin: J. Electroanal. Chem. **484**, 120 (2000)

[13] K. Shimazu, T. Kawaguchi, T. Isomura: J. Am. Chem. Soc. **124**, 652 (2002)

[14] C. A. Widrig, C. Chung, M. D. Porter: J. Electroanal. Chem. **310**, 335 (1991)

[15] M. Schweizer, H. Hagenstrom, D. M. Kolb: Surf. Sci. **490**, L627 (2003)

[16] T. Albrecht, P. D. Barker, D. H. Murgida, P. Hildebrandt: unpublished results

[17] L. Rivas, C. M. Soares, A. M. Baptista, J. Simaan, R. D. Paolo, D. H. Murgida, P. Hildebrandt: Biophys. J. **88**, 4188 (2005)

[18] A. J. Simaan, D. H. Murgida, P. Hildebrandt: Biopolymers **67**, 331 (2002)

[19] S. Todorovic, C. Jung, P. Hildebrandt, D. H. Murgida: J. Biol. Inorg. Chem. **11**, 119 (2006)

[20] A. Kranich, M. A. De la Rosa, P. Hildebrandt, D. H. Murgida: unpublished results

[21] D. H. Murgida, P. Hildebrandt: J. Mol. Struct. **565**, 97 (2001)

[22] L. Rivas, D. H. Murgida, P. Hildebrandt: J. Phys. Chem. B **106**, 4823 (2002)

[23] L. J. C. Jeuken, J. P. Mcevoy, F. A. Armstrong: J. Phys. Chem. B **106**, 2304 (2002)

[24] D. H. Murgida, P. Hildebrandt, J. Wei, Y. F. He, H. Y. Liu, D. H. Waldeck: J. Phys. Chem. B **108**, 2261 (2004)

[25] J. J. Wei, H. Y. Liu, A. R. Dick, H. Yamamoto, Y. F. He, D. H. Waldeck: J. Am. Chem. Soc. **124**, 9591 (2002)

[26] H. Yamamoto, H. Y. Liu, D. H. Waldeck: Chem. Commun. p. 1032 (2001)

[27] I. Willner, V. HelegShabtai, R. Blonder, E. Katz, G. L. Tao: J. Am. Chem. Soc. **118**, 10321 (1996)

[28] I. Weidinger, D. H. Murgida, W. F. Dong, H. Möhwald, P. Hildebrandt: J. Phys. Chem. B **110**, 522 (2006)

[29] M. K. Beissenhirtz, F. W. Scheller, F. Lisdat: Anal. Chem. **76**, 4665 (2004)

[30] M. K. Beissenhirtz, F. W. Scheller, W. F. M. Stocklein, D. G. Kurth, H. Mohwald, F. Lisdat: Angew. Chem. Int. Ed. **43**, 4357 (2004)

[31] S. Todorovic, M. M. Pereira, T. M. Bandeiras, M. Teixeira, P. Hildebrandt, D. H. Murgida: J. Am. Chem. Soc. **127**, 13561 (2005)

[32] J. A. He, L. Samuelson, L. Li, J. Kumar, S. K. Tripathy: Adv. Mater. **11**, 435 (1999)

[33] W. M. Mirsky (Ed.): *Ultrathin Electrochemical Chemo- and Biosensors, Technology and Performance* (Springer, Berlin, Heidelberg 2004)

[34] K. Ataka, F. Giess, W. Knoll, R. Naumann, S. Haber-Pohlmeier, B. Richter, J. Heberle: J. Am. Chem. Soc. **126**, 16199 (2004)

[35] M. G. Friedrich, F. Giess, R. Naumann, W. Knoll, K. Ataka, J. Heberle, J. Hrabakova, D. H. Murgida, P. Hildebrandt: Chem. Commun. p. 2376 (2004)

[36] F. Giess, M. G. Friedrich, J. Heberle, R. L. Naumann, W. Knoll: Biophys. J. **87**, 3213 (2004)

[37] J. Hrabakova, K. Ataka, J. Heberle, P. Hildebrandt, D. H. Murgida: Phys. Chem. Chem. Phys. **8**, 759 (2006)

[38] E. Sackmann: Science **271**, 43 (1996)

[39] S. Z. Hu, I. K. Morris, J. P. Singh, K. M. Smith, T. G. Spiro: J. Am. Chem. Soc. **115**, 12446 (1993)

[40] S. Oellerich, H. Wackerbarth, P. Hildebrandt: J. Phys. Chem. B **106**, 6566 (2002)

[41] D. H. Murgida, P. Hildebrandt: J. Phys. Chem. B **105**, 1578 (2001)

[42] S. Döpner, P. Hildebrandt, A. G. Mauk, H. Lenk, W. Stempfle: Spectrochim. Acta Part A: Mol. Biomol. Spectrosc. **52**, 573 (1996)

[43] P. Hildebrandt, M. Stockburger: Biochem. **28**, 6710 (1989)

[44] A. Bonifacio, D. Millo, C. Gooijer, R. Boegschoten, G. van der Zwan: Anal. Chem. **76**, 1529 (2004)

[45] R. J. Clarke: Adv. Coll. Int. Sci. **89**, 263 (2001)

[46] I. T. Suydam, S. G. Boxer: Biochem. **42**, 12050 (2003)

[47] S. S. Andrews, S. G. Boxer: J. Phys. Chem. A **106**, 469 (2002)

[48] S. S. Andrews, S. G. Boxer: J. Phys. Chem. A **104**, 11853 (2000)

[49] V. Oklejas, C. Sjostrom, J. M. Harris: J. Phys. Chem. B **107**, 7788 (2003)

[50] V. Oklejas, J. M. Harris: Langmuir **19**, 5794 (2003)

[51] R. A. Scott, A. G. Mauk (Eds.): *Cytochrome c - A Multidisciplinary Approach* (University Science Books, Sausalito 1995)

[52] W. H. Koppenol, J. D. Rush, J. D. Mills, E. Margoliash: Mol. Biol. Evol. **8**, 545 (1991)

[53] K. Ataka, J. Heberle: J. Am. Chem. Soc. **126**, 9445 (2004)

[54] N. Wisitruangsakul, I. Zebger, P. Hildebrandt, D. H. Murgida: unpublished results

[55] S. Oellerich, H. Wackerbarth, P. Hildebrandt: Eur. Biophys. J. Biophys. Lett. **32**, 599 (2003)

[56] S. Oellerich, S. Lecomte, M. Paternostre, T. Heimburg, P. Hildebrandt: J. Phys. Chem. B **108**, 3871 (2004)

[57] H. Wackerbarth, D. H. Murgida, S. Oellerich, S. Dopner, L. Rivas, P. Hildebrandt: J. Mol. Struct. **563**, 51 (2001)

[58] L. Rivas, M. Marti, P. Hildebrandt, D. H. Murgida: unpublished results

[59] H. Wackerbarth, P. Hildebrandt: Chem. Phys. Chem. **4**, 714 (2003)

[60] R. O. Louro, T. Catarino, D. L. Turner, M. A. Picarra-Pereira, I. Pacheco, J. LeGall, A. V. Xavier: Biochem. **37**, 15808 (1998)

[61] A. M. Kuznetsov, J. Ulstrup: *Electron Transfer in Chemistry and Biology. An Introduction to the Theory* (Wiley, Chichester 1999)

[62] D. H. Murgida, P. Hildebrandt: J. Am. Chem. Soc. **123**, 4062 (2001)

[63] A. Avila, B. W. Gregory, K. Niki, T. M. Cotton: J. Phys. Chem. B **104**, 2759 (2000)

[64] D. E. Khoshtariya, J. J. Wei, H. Y. Liu, H. J. Yue, D. H. Waldeck: J. Am. Chem. Soc. **125**, 7704 (2003)

[65] J. J. Wei, H. Y. Liu, D. E. Khoshtariya, H. Yamamoto, A. Dick, D. H. Waldeck: Angew. Chem. Int. Ed. **41**, 4700 (2002)

[66] J. J. Wei, H. Y. Liu, K. Niki, E. Margoliash, D. H. Waldeck: J. Phys. Chem. B **108**, 16912 (2004)

[67] P. Hildebrandt, D. H. Murgida: Bioelectrochem. **55**, 139 (2002)

[68] D. H. Murgida, P. Hildebrandt: J. Phys. Chem. B **106**, 12814 (2002)

[69] G. Basu, A. Kitao, A. Kuki, N. J. Go: J. Phys. Chem. B **102**, 2085 (1998)

[70] I. Muegge, P. X. Qi, A. J. Wand, Z. T. Chu, A. Warshel: J. Phys. Chem. B **101**, 825 (1997)

[71] T. D. Dolidze, D. E. Khoshtariya, D. H. Waldeck, J. Macyk, R. van Eldik: J. Phys. Chem. B **107**, 7172 (2003)

[72] J. Hrabakova, Y. Hongjun, D. H. Waldeck, P. Hildebrandt, D. H. Murgida: unpublished results

[73] M. M. Pereira, M. Teixeira: Biochim. Biophys. Acta Bioenerg. **1655**, 340 (2004)

[74] M. M. Pereira, M. Santana, M. Teixeira: Biochim. Biophys. Acta Bioenerg. **1505**, 185 (2001)

[75] T. Morita, S. Kimura: J. Am. Chem. Soc. **125**, 8732 (2003)

[76] A. Jasaitis, F. Rappaport, E. Pilet, U. Liebl, M. H. Vos: Proc. Natl. Acad. Sci. USA **102**, 10882 (2005)

[77] M. Brunori, A. Giuffre, P. Sarti: J. Inorg. Biochem. **99**, 324 (2005)

[78] M. Wikstrom: Biochim. Biophys. Acta Bioenerg. **1655**, 241 (2004)

[79] A. V. Xavier: Biochim. Biophys. Acta Bioenerg. **1658**, 23 (2004)

[80] T. K. Das, C. M. Gomes, T. M. Bandeiras, M. M. Pereira, M. Teixeira, D. L. Rousseau: Biochim. Biophys. Acta Bioenerg. **1655**, 306 (2004)

Index

Nanosensors Based on SERS for Applications in Living Cells

Janina Kneipp[1,2]

[1] Federal Institute for Materials Research and Testing, FG I.3,
Richard-Willstätter-Str. 11, D-12489, Berlin, Germany
[2] Wellman Center for Photomedicine, Harvard University, Medical School,
55 Fruit St., Boston, MA 02114, USA
jkneipp@janina-kneipp.de

1 Reasons for Intracellular SERS Approaches

1.1 Raman Spectra from Biological Samples

Over the past decades, the growing demand for noninvasive diagnostic and nondestructive structure analysis tools to identify and investigate tissues and cells has led to great progress in the development of Raman spectroscopic methods. Thanks to new solutions for experiments and data evaluation, Raman spectroscopy-based approaches have evolved into robust, reliable techniques to study the chemical composition of complex biosamples, and to shed light on their molecular make-up in health and disease. Since more than a decade ago, it has repeatedly been proven that the fingerprint-like information in Raman microspectra can add to our understanding of the biochemical background of various regular, induced or pathological changes in eukaryotic cells [1,2,3], which is, aside from being a basic research goal, a major prerequisite for progress in areas such as molecular medicine and nanobiotechnology.

In the numerous studies on cells, tissues, and micro-organisms, two major parameters were shown to influence the meaning and usefulness of the vibrational information: 1. the lateral resolution and hence the size, number, amount, and heterogeneity of biological object(s) the information is obtained from, and 2. the detection sensitivity, which is of consequence to the duration of the measurements and therefore to sample integrity, time resolution, and in-vivo applicability. As outlined in the chapters of this book, surface-enhanced Raman scattering (SERS) has its origin in the favorable optical properties of metal nanostructures, and benefits from the enhanced local optical fields in their immediate vicinity. The enhancement of the Raman signals in SERS leads to very sensitive probing of chemical structure and composition, and the nanometer-scaled spatial confinement of the effect enables highly localized measurements. Therefore, SERS can be utilized to get Raman spectral information from complex biosystems rapidly and at high lateral resolution, thereby addressing both issues named above. In this Chapter we will illustrate how SERS probing of the intrinsic biochemistry of

K. Kneipp, M. Moskovits, H. Kneipp (Eds.): Surface-Enhanced Raman Scattering – Physics and Applications, Topics Appl. Phys. **103**, 335–350 (2006)
© Springer-Verlag Berlin Heidelberg 2006

living cells can be accomplished and discuss the application of SERS probes in the context of the specific requirements posed by cellular systems.

1.2 Advantages of SERS for Studies in Cells

Although the field of vibrational biospectroscopy has made enormous progress and is evolving rapidly, building on recent developments in instrumentation and data analysis, the study of live systems remains one of its greatest challenges. The advantages of Raman scattering experiments over infrared (IR) absorption studies for in-vivo measurements are obvious not only in potential diagnostic procedures, since a Raman method as opposed to IR, e.g., enables use of fiber-optic probes, but in particular in microspectroscopic experiments with individual cells, where the diffraction-limited lateral resolution of visible or near IR-excited Raman experiments is higher than that of mid-IR absorption experiments, and hence permits investigations of subcellular structures [2].

Many studies have been performed with fixed or dried cells so far [3, 4], while investigations in living cells have proved challenging. The possibility to conduct Raman experiments on individual living cells in a microspectroscopic setup was shown by *Puppels* et al. [5] and later also other groups [6]. Raman investigations of living cells are not necessary complication free, as the applicable maximum intensity of the laser used to excite the Raman scattering in small probe volumes with low analyte concentrations is limited if damage to the cells is to be avoided [7]. Raman studies using laser powers low enough for living cells to withstand usually require signal accumulation times on the order of a few minutes or even longer, even under well-optimized experimental conditions [5,6]. In order to be able to measure Raman spectra on the timescales of biological processes taking place in a living cell, such as trafficking or transport events, where short sampling duration is essential, a dramatic decrease in measurement times is required.

To overcome the limitations posed by the relatively low efficiency of the Raman process under experimental conditions suitable for living cells, several studies have proposed to employ approaches that result in stronger signals, such as resonance Raman [8, 9, 10, 11], enhancement due to molecular aggregates [12], coherent anti-Stokes Raman scattering (CARS) [13], and also SERS [14].

It is clear that the gain in sensitivity resulting from an enhancement of Raman scattering always comes at a price. We pay for it with a higher selectivity concerning important experimental parameters: The process of resonance enhancement is operative only for specific molecules or molecular groups (those with an electronic transition matching the energy of the excitation light), and limits the experiment to a specific excitation wavelength. In the case of SERS, the enhancement is restricted to the immediate vicinity of a metal nanostructure. However, the independence from a specific laser wavelength for excitation of the Raman scattering and the ability to probe many different kinds of molecules make SERS techniques extremely attractive for efficient live-cell Raman approaches. When considering the development of such an approach, it is worthwhile to keep in mind the local confinement of the SERS effect, and how to make use of it to one's advantage. In-vivo SERS studies are conducted with individual probes positioned at discrete locations in the cell. The spectral information is obtained only from the nano environment of these SERS probes. This is different from the spectral information obtained in a "normal" Raman microspectroscopic experiment, where all positions in a whole cell are probed. The ability of very sensitive detection, along with the confinement of the spectral enhancement in SERS to the immediate vicinity of the nanoprobes makes SERS probes ideal tools for the investigation of small morphological structures in cells. The maximum lateral resolution in such a Raman experiment is no longer limited by the excitation wavelength; it is influenced by the metal nanostructure used to provide the enhancement.

Most studies on eukaryotic cells that exploit SERS, so far use the effect to detect the enhanced Raman signals of reporter molecules such as dyes [15, 16, 17, 18] or of extracellular compounds [19]. The idea to create SERS labels provides several advantages over other labels regarding sensitivity, label stability, and specificity. The latter is caused by the characteristic, fingerprint-like pattern of many narrow Raman bands of the label molecule rather than, e.g., a broad fluorescence band. This enables the construction of label arsenals comprising large numbers of structurally very similar molecules with yet very different SERS signatures [20,21]. The high specificity has been used in multiplexing experiments unprecedented so far by fluorescence studies [22]. In addition, for a number of common fluorescent dye labels, the sensitivity for SERS is higher than in fluorescence detection [20].

However, as shown numerous times by promising applications in the area of vibrational biospectroscopy, and in particular the Raman spectra of cells, no spectrum of a label in a cell can replace the information of a spectrum from the cellular molecules themselves. Therefore, an even greater advantage of SERS measurements in biological samples than the construction of new optical labels based on extrinsic reporter molecules is the ability to give us *intrinsic* chemical information from a cell or cellular compartment. It is evidenced by the successful SERS characterization of micro-organisms [23,24], that the application of SERS improves the sensitivity and lateral resolution of current biospectroscopic applications. In the next paragraphs, the application of SERS nanoprobes in living cells will be described.

2 SERS Nanosensors for Probing of Intrinsic Cellular Chemistry

2.1 Requirements of Cellular Systems

Unlike measurements in a homogeneous material or a solution, the retrieval of meaningful SERS spectroscopic information from systems that are relatively heterogeneous on the nanometer-scale such as cells requires experimental designs that allow control of the position of the SERS probes. As has been demonstrated, several different methods are in principle applicable to accomplish precise positioning of a SERS probe, such as the utilization of sharp probe tips and atomic force techniques or the combination with electron microscopy [24, 25], but most in-vivo and in-situ biospectroscopic approaches will require mobile metal nanostructures and if necessary active targeting of specific locations in a heterogeneous system. In addition to being controllable with respect to topology, uptake, and retention characteristics, the metal nanostructures need to be compatible with the biological system; the material should be inert and if possible not influence its surroundings. Due to its universal biocompatibility, gold is ideal for optical probing in biology. For decades, gold nanoparticles have been tools of the trade in cell biology because of their favorable physical and chemical properties. Delivery of nanoparticles into the cellular interior, as well as routing of the particles or targeting of cellular compartments can be achieved in a number of different ways, depending on the nature of the experiment or practical application, but also on the type of cell line and physicochemical particle parameters, such as size, shape, and surface functionalization [26, 27, 28, 29, 30]. We have been able to show that nanoparticulate structures from gold, while fulfilling the requirements posed by a cellular system, also provide the local optical fields necessary for ultrasensitive intracellular SERS probing [31].

2.2 SERS Nanosensors in Endosomal Structures

In our effort to develop Raman-based optical nanosensors to probe live cells, we have started to employ the fact that many cells take up nanostructures by themselves without further induction. This method can be used to selectively probe the cellular substructures involved in vesicular transport in the cells. Nonphagocytic cells can internalize structures of less than 1 μm in size, with clathrin-mediated endocytosis and transport into late endosomes and lysosomes occurring for sizes up to 200 nm and highest efficiency for particles of several tens of nanometers [27, 32]. Figure 1 shows a schematic of our delivery experiment and cells of a macrophage cell line, J774 (phagocytic) and a fibroblast cell line, NIH/3T3 (nonphagocytic), after incubation with gold nanoprobes with a mean particle size of 20 nm to 50 nm. Briefly, the subcultured cells are seeded on glass coverslips on which they were grown

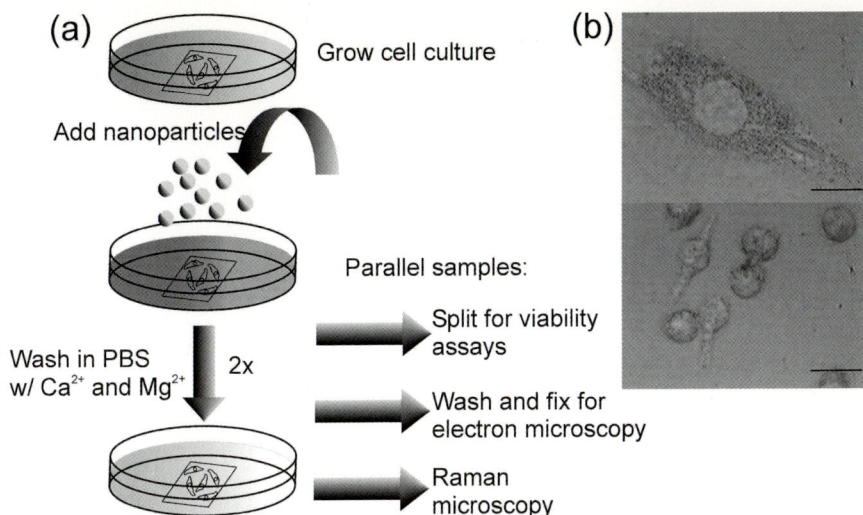

Fig. 1. (a) Schematic for the delivery of gold nanoprobes into eukaryotic cells by uptake from the culture medium. The cells are grown on flamed coverslips to the desired confluence. Gold nanoprobes are added to the cell-culture medium, and after defined periods of time removed, followed by thorough washing of the cells. Parallel samples are prepared for biological assays, electron microscopy, and Raman microspectroscopy. Raman measurements are carried out while the cells are alive in a PBS buffer solution. (b) Cells of cell lines NIH/3T3 (*upper panel*) J774 (*lower panel*) and after incubation with gold nanoparticles (*scale bar:* 20 μm). After several hours of incubation with the probes, numerous particle-containing endosomes have accumulated in these cell lines, therefore the particles become visible as black dots in a common light microscope. No particles are found in the area of the nuclei, since the size of the probes in this experiment (20 nm to 50 nm) exceeds the entry limit of the nuclear pore complex ($\leq$ 20 nm)

to confluence before the culture medium is replaced by culture medium containing the gold nanoparticles for delivery. After specified incubation times, varying from a few minutes to many hours, all excess particles are removed, and the cells are washed in buffer solution.

Raman spectra can be acquired from the living cells in the physiological environment by using a microspectroscopic setup. The applied laser intensities of $\sim$ 2 mW/ $\sim$ 1 μm spot size, accumulation time of 1 s and less, and excitation with 830 nm (out of resonance) preclude from the observation of non-SERS spectra from the cells. However, at positions in the cells where gold nanoparticles are present, surface-enhanced Raman spectra can be measured. At these positions, the Raman signatures from the cellular components in the immediate surroundings of the gold nanostructures experience surface enhancement, enabling sensitive chemical probing of the particles' vicinity in very short times. Figure 2 shows examples of such spectra. The

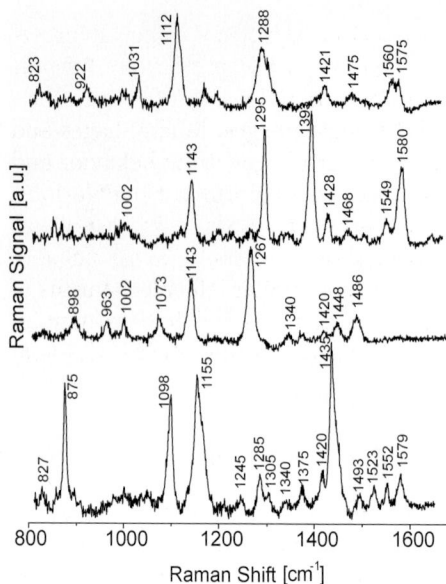

Fig. 2. Examples of SERS spectra acquired from NIH/3T3 cells after 3 h incubation with gold nanostructures. Excitation wavelength: 830 nm, collection time: 1 s

spectra display features of various typical cellular constituents, such as proteins ($1550 \, \mathrm{cm^{-1}}$ amide II, $1245 \, \mathrm{cm^{-1}}$, $1267 \, \mathrm{cm^{-1}}$ amide III, side chains Phe $1002 \, \mathrm{cm^{-1}}$, Tyr $825 \, \mathrm{cm^{-1}}$) and various nucleic acid constituents (e.g., 1580, 1575, $1098 \, \mathrm{cm^{-1}}$), to mention a few. Although the Raman shifts, relative peak intensities, and linewidths with SERS may differ slightly from those in normal Raman spectra due to the interaction of the molecule with the metal and due to large field gradients, many bands show similarities with the spectra collected from cell constituents in normal Raman measurements [4, 5, 6, 33]. In contrast to regular Raman experiments, however, each spectrum can be obtained in one second or less, which represents a great advantage with respect to probing speed. Such short acquisition times are on the order of the timescales of the processes occurring in a cell, enabling investigations that have been out of reach so far in other experiments in live cells.

Typically, the spectral sampling is done by raster scanning along a predefined set of points over each individual cell, until the whole area of a cell is covered. It is obvious that a cell is never completely filled with gold nanoparticles. Such a situation would represent a large perturbation from a cell's normal functional state and is not desirable. As a consequence, there are spectra with and without SERS signal in each of our raster scan measurements. The number of spectra that show a SERS signal in such a measurement is a function of the incubation time of the cells with the nanoprobes, of the uptake behavior of the specific cell line, and of the delivery regime. The observations in the spectroscopic experiments are supported by the ultrastructural information from electron microscopy obtained from parallel samples (compare the example in Fig. 3b). We have been studying various types of cell lines, among them macrophages, fibroblasts, and cells of epithelial origin. In agreement with the existing knowledge on the uptake behavior of these different cell types, the uptake in the epithelial cell line IRPT was much lower than in the fibroblast line NIH/3T3 and in J774 macrophages. Macrophages and fibroblasts were recently shown to exhibit similar phagocytotic behavior and form a phenotypic continuum, e.g., in wounds, where fibroblast populations, usually responsible for forming connective tissues, can also perform scavenger phagocytosis [28]. Based on very few biochemical markers, it was concluded that although the phenotype of both cell types is similar, the mechanisms of phagocytosis may be different. The SERS nanosensors will be a very good means to study the biochemistry of different phagosome and other endosomal structures in detail.

2.3 Gold Nanoparticles and the Cellular Environment

SERS nanoprobes for intracellular applications have to function with respect to both spectroscopic and cell biology aspects. Controlling the formation of gold nanoparticle aggregates is of great interest in constructing such probes for two obvious reasons. One is the size dependence of uptake or delivery efficiency and/or intracellular routing [26, 34]. The other reason is the influence of nanoaggregate formation on the SERS signals [20, 35]. In most cases, the nanoparticles are applied via the cell-culture medium or a buffer. Under many experimental circumstances, the required ionic conditions, pH and other parameters can cause flocculation of the particles. Therefore, different methods have been proposed to protect gold nanoparticles before adding them to cell-culture media, many using the protein bovine serum albumin [29,36,37]. The status of the particle colloidal solutions can be monitored by different methods, among them extinction spectra and transmission electron microscopy (see Fig. 3a and Fig. 4b). We found that under culture conditions that require addition of fetal calf serum (FCS), flocculation in the medium is prevented and the colloidal particles are stable for days. Results from TEM studies of nanoparticles after immersing them in FCS-containing culture medium suggest that apart from the individual particles, formation of nanoaggregates with an aggregate diameter of ~100 nm can occur [31]. From the first TEM experiments in cells after uptake, we have also learned that these particles and nanoaggregates are taken up by the cells individually. This shows that the apparent limitations posed by gold nanoaggregate formation in cell-culture systems could be used to an advantage, provided that particle properties are controlled before nanostructure delivery and also later, inside the cell. Controlled generation of nanoaggregates will be one of the major prerequisites to employ the large enhancement factors of such nanoaggregate structures in living systems in a systematic way. To achieve this, the combination of ultrastructural information from electron microscopy of parallel samples with the new in-vivo SERS sensors inside cells is of great help (Fig. 3b). Figure 3b shows an example of gold nanoparticles included in endosomes in the cell line IRPT.

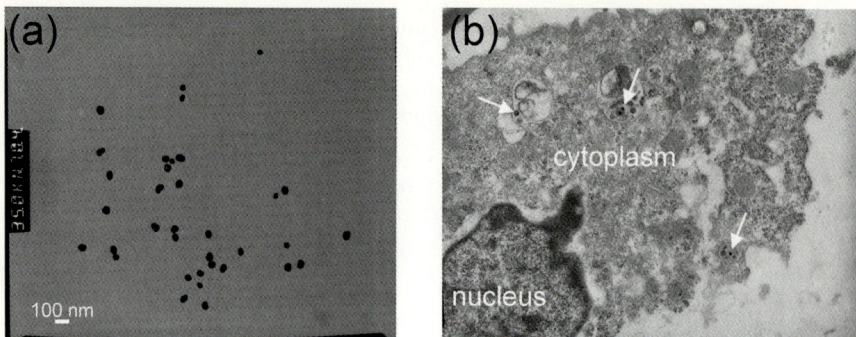

Fig. 3. (a) Electron micrograph of gold nanoparticles immersed in cell-culture medium. (b) Electron micrograph of an epithelial cell displaying individual particles in an endosome after 2 h incubation with gold nanoparticles (*arrows*) [31]

3 A Labeled Nanosensor Based on the SERS Signal of Indocyanine Green

3.1 Characterization of the SERS Label

As was mentioned in the first part of this Chapter, the utilization of a SERS approach to create optical labels with high sensitivity, specificity, and spatial localization is very promising also for intracellular probing. A label that could deliver chemical structure information from the molecules in its environment would be of even greater interest for studies in cells. We were able to show that it is possible to construct a SERS nanosensor that contains a label molecule and at the same time enhances the Raman signals from the molecules it is surrounded by when delivered into a cell [38]. The nanosensor is based on the SERS signature of the biocompatible dye indocyanine green (ICG). This dye is frequently used and approved as a fluorescence label for biomedical applications [39, 40]. In accordance with this, we observed a fluorescence signal rather than a Raman spectrum of ICG in measurements without metal nanoparticles. Even at saturation concentration of $\sim 10^{-3}$ M or from higher-concentrated dry samples, no Raman spectrum could be measured. However, in silver and gold colloidal solutions, we obtain a strong and characteristic ICG SERS signature at an ICG concentration of 10^{-7} M (compare traces A and B in Fig. 4a).

The extinction spectrum of ICG bound to the protein human serum albumin (HSA) in gold and silver colloidal solution is shown in Fig. 4b. Aside from the characteristic plasmon bands of the gold and silver nanoparticles around 540 nm and 420 nm, respectively, an absorption maximum around 800 nm of ICG is seen. This is an indication that the dye mainly exists as monomer. For comparison, the inset in Fig. 4b shows the same absorption maximum of 5×10^{-6} M ICG bound to HSA in aqueous solution in the absence of colloid.

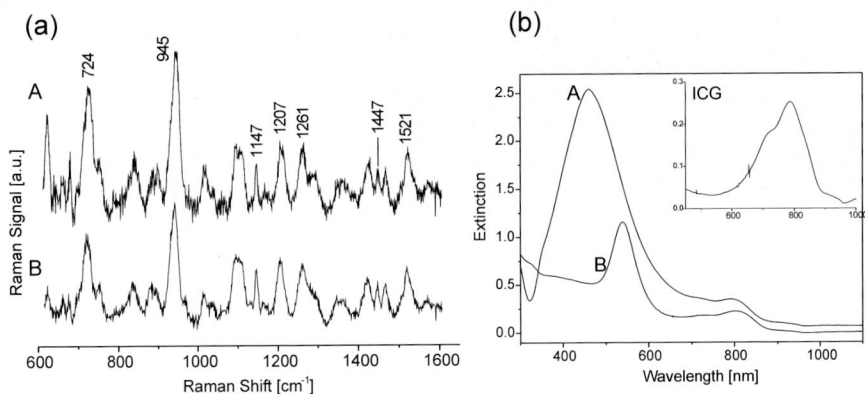

Fig. 4. (a) SERS spectra of 10^{-7} M indocyanine green in silver (*trace A*) and gold (*trace B*) colloidal solutions obtained using excitation wavelength 786 nm. (b) Extinction spectra of silver and gold colloidal sample solutions (*spectra A* and *B*, respectively) containing 5×10^{-6} M indocyanine green (ICG) bound to HSA. The inset shows the absorption of 5×10^{-6} M ICG bound to HSA in water. Reprinted with permission from [38], Copyright 2005 American Chemical Society

Serum albumins such as HSA prevent aggregation of the ICG dye [41], which is known to occur under certain conditions [42, 43, 44]. Although the absolute dye concentration in our solutions is too low ($< 10^{-6}$ M) for ICG to form aggregates [42], the local concentration on the surface of the gold and silver particles may be higher, and the presence of ions could increase its rate of aggregation [42]. As discussed above, the serum albumin, which is used to prevent ICG aggregation, also stabilizes the gold nanoparticles [29, 36, 37] and can therefore serve two purposes in our intracellular experiments.

To investigate the influence of HSA on the SERS spectra, we measured the spectra of pure ICG and ICG bound to HSA in gold colloidal solution (Fig. 5). Both spectra show very similar features. A control, consisting of 10^{-5} M HSA complexed with gold nanoparticles, did not produce a SERS signal. These findings are in good agreement with recent observations that serum albumin does not interfere with sensitive SERS detection of glucose [45]. The small differences between the spectra are very likely associated with formation of dye aggregates in the HSA-free samples. The high similarity between the spectra measured in the presence and absence of HSA (Fig. 5) indicates that the SERS signature is very robust and can be distinguished independent of the presence of ICG aggregation, making it applicable in different physiological environments (e.g., different cellular compartments).

3.2 Application of ICG-SERS Nanoprobes in Living Cells

A SERS nanoprobe consisting of 10^{-7} M ICG complexed with HSA on 60 nm gold nanoparticles was delivered into cultured cells of a metastatic Dun-

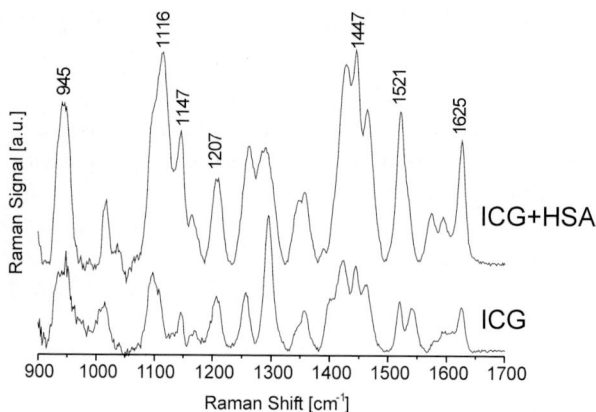

Fig. 5. SERS spectra measured from ICG (10^{-7} M) bound to HSA (*upper spectrum*) and from the pure dye (*lower spectrum*) in gold colloidal solution (excitation wavelength 680 nm). Reprinted with permission from [38], Copyright 2005 American Chemical Society

ning R3327 rat prostate carcinoma line (MLL) (donated by Dr. W. Heston, Memorial Sloan-Kettering Cancer Center, New York, NY). After overnight incubation, nanoparticles that were endocytosed must be included in lysosomes [46,47,48]. In accordance with this assumption we found gold accumulations in the range of 100 nm to 1000 nm in lysosomes by light and electron microscopy after incubation for 20 h and longer. The presence of gold particles in the cells could also be verified by the appearance of SERS signals collected from the cells. Microscopic inspection provided evidence that the cells were dividing after incubation with the ICG-SERS probes. While incubated with the nanosensors, the cells were visibly growing, and no evidence of cell death was found. A slightly lower density after 20 h incubation with the diluted medium was observed when compared with control cells growing in undiluted medium, probably resulting from the dilution of nutrients. Controls in diluted medium with ICG/NaCl and NaCl-diluted culture medium showed growth rates similar to those incubated with the gold particles and with the ICG-SERS nanoprobes.

Figure 6 shows examples of SERS spectra measured in single living cells that had been incubated with the ICG-gold hybrid probe at 830 nm excitation. Trace A in Fig. 6 displays an ICG SERS spectrum measured from a nanosensor in the physiological medium, trace B is an example of a spectrum measured inside a cell. The difference spectrum between both is shown in trace C and represents the Raman spectral contributions from the molecules in the cell. More examples are displayed in Fig. 6. In accordance with the localization of the ICG-SERS probes in lysosomes, where many different types of biomolecules are found, many different SERS signatures originate from the nanosensors' environment. This variety of combinations of spectral features can be assessed best by multivariate methods. The probes can also be localized inside the cells by image reconstruction based on their specific spectral information. Such images are reconstructed from Raman data collected by raster scanning over single living cells in a defined step width. As our data indicate, the SERS spectrum of ICG consists of more than ten characteristic bands distributed over a broad frequency range. For imaging of the label, this offers the advantage that spectral correlation methods can be used to enhance the contrast between the label and the cellular background. Figure 7 shows a spectral map of the probe in a cell based on the intensity of the $1147 \, \mathrm{cm}^{-1}$ ICG SERS line and on the product of two ICG SERS lines at $1147 \, \mathrm{cm}^{-1}$ and $945 \, \mathrm{cm}^{-1}$. As illustrated by the example in Fig. 7, using two bands already increases the contrast and enables better localization of the probe.

Comparing the ICG SERS spectra in Figs. 4, 5, and 6 (trace A), which were obtained with excitation wavelengths of 786 nm, 680 nm, and 830 nm, respectively, we observe that they exhibit the same Raman frequencies. The differences in relative Raman signals in the spectra are a consequence of altered resonance conditions for ICG due to the different excitation wavelengths, which is plausible if one also takes into account the differences in the absorption of ICG (inset in Fig. 4b).

4 Conclusions and Outlook

We have demonstrated the application of SERS in robust and sensitive optical probes for intracellular measurements that deliver intrinsic chemical information from subcellular structures and compartments. As was shown, SERS-based nanosensors for use in living cells and other small-scaled, complex biological materials provide a high content of molecular information and high sensitivity. Due to the large effective scattering cross section, they fulfill the requirements of dynamic in-vivo systems – the use of very low laser powers and very short data acquisition times. The utilization of gold nanoparticles for the development of SERS nanoprobes enable us to make use of well-established cell biology methods for probe delivery. In particular, functionalization of the gold nanoparticles will be very useful for further developments of targeted probe delivery and for control of the probes' positions.

The aspect of introducing reporter molecules as labels with the particles is a prerequisite for the development of further SERS-based cell biology tools, since it provides the basis for tracing multiple particle species in multilabel experiments. In such experiments, after controlled transfer of probes with different SERS reporters to different cellular substructures, the spectral signature of the reporter that appears along with the spectrum of the cell components indicates which part of the cell or biosample the spectrum is from. The structure sensitivity and the nature of the Raman process itself are major advantages of SERS over other methods such as fluorescence and enable the specific detection of a probe in addition to the molecules from

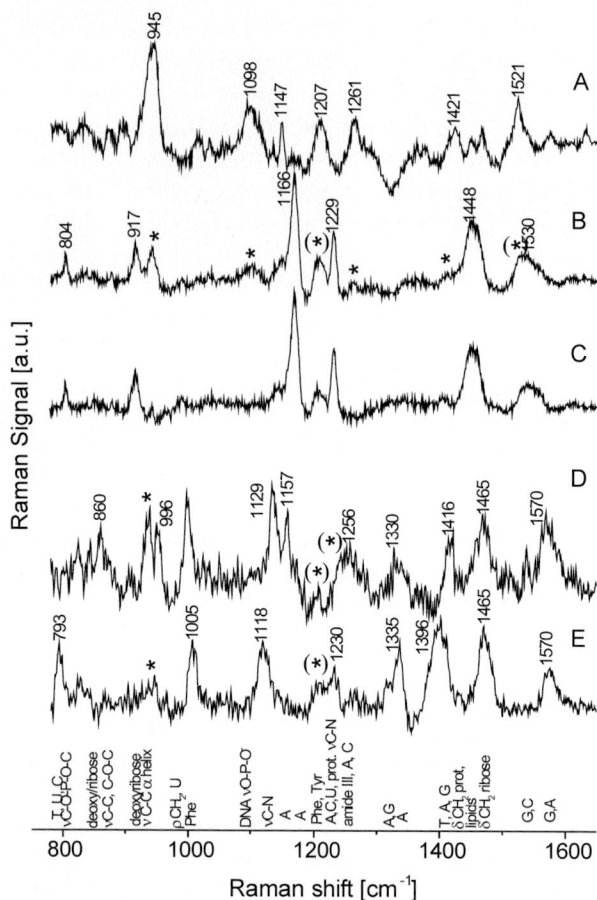

Fig. 6. Examples of SERS spectra measured in single living cells incubated with the ICG-SERS nanosensor at 830 nm excitation. *Trace A* represents the ICG signature. *Trace B* was measured in the cell. *Trace C* is the difference between spectrum B and A and displays only Raman lines of the cell. *Traces D* and *E*: More examples of spectra that contain contributions from ICG and cell components. Assignments of major bands are given below the spectra (based on [4, 5, 6, 33]). ICG bands are marked with an asterisk, while an asterisk in parentheses indicates the contribution of both ICG and cell. The relative contributions of ICG and cell components depend on the coadsorption of both kinds of molecules. Abbreviations: A, adenine, G, guanine, C, cytosine, T, thymine, U, Uracil, Phe, phenylalanine, Tyr, tyrosine, prot, proteins, ν, stretching, δ, deformation, ρ, rocking. Reprinted with permission from [38], Copyright 2005 American Chemical Society

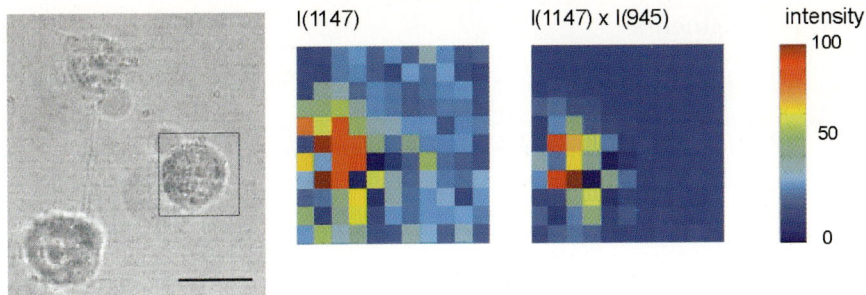

Fig. 7. Spectral map of ICG in a cell based on the $1147 \, \text{cm}^{-1}$ ICG SERS line and on the product of two ICG SERS lines at $1147 \, \text{cm}^{-1}$ and $945 \, \text{cm}^{-1}$. Intensities are scaled to the highest value in each area. A photomicrograph of the cell, indicating the studied area, is shown for comparison. *Scale bar*: $20 \, \mu\text{m}$. Reprinted with permission from [38], Copyright 2005 American Chemical Society

the cellular "background". As was illustrated here, biocompatible dyes such as ICG are promising candidates for the construction of SERS probes that not only highlight targeted biological structures, but simultaneously deliver molecular structure information from them as well.

With and without additional label molecules, SERS nanosensors will enable targeted molecular probing of subcellular structures and allow the study and imaging of molecular phenomena in-situ and in-vivo that can not be addressed otherwise. The ability of localized probing by SERS will be used to study parallel events in a cell, e.g., monitor transport processes that affect different cellular compartments. This will be achieved by directing gold nanoparticles to different locations in a cell in one experiment.

The size of the nanoprobes determines their delivery into cells and subcellular structures, but also SERS enhancement, confinement of the local fields, and spatial confinement of the collected information. Therefore, further research in this area is currently directed toward influencing the physicochemical and spectroscopic properties of noble metal nanoparticles inside cells and physiological media. The current proof-of-principle that cellular chemistry can be probed very selectively by means of intracellular SERS nanosensors is being followed by systematic SERS studies of well-characterized model systems, such as isolated cell compartments and compounds, which are being used to decipher the spectral information from the cells.

The concept of using modern vibrational spectroscopic methods along with exploiting nanoprobes and their local optical fields provides far-reaching perspectives and exciting potential capabilities for our understanding of cellular processes on the molecular level. In particular, the ability to detect local heterogeneity in large biomolecules and molecular structure in small subcellular units will be useful in future clinical diagnostic and therapeutic applications.

Acknowledgments

This work is supported in part by DOD grant # AFOSR FA9550-04-1-0079, NIH grant # PO1CA84203, and by the generous gift of Dr. and Mrs. J. S. Chen to the optical diagnostics program of the Massachusetts General Hospital, Wellman Center for Photomedicine.

References

[1] T. Bakker Schut, M. J. H. Witjes, H. J. C. M. Sterenborg, et al.: Anal. Chem. **72**, 6010 (2000)

[2] J. Kneipp, T. C. Bakker Schut, M. Kliffen, et al.: Vibrational Spectrosc. **32**, 67 (2003)

[3] H.-J. Van Manen, Y. M. Kraan, D. Roos, et al.: PNAS **102**, 10159 (2005)

[4] N. Uzunbajakava, A. Lenferink, Y. Kraan, et al.: Biophys. J. **84**, 3968 (2003)

[5] G. J. Puppels, F. D. Mul, C. Otto, et al.: Nature **347**, 301 (1990)

[6] W. L. Peticolas, T. W. Patapoff, G. A. Thomas, et al.: J. Raman Spectrosc. **27**, 571 (1996)

[7] K. Ramser, E. J. Bjerneld, C. Fant, et al.: J. Biomed. Opt. **8**, 173 (2003)

[8] N. M. Sijtsema, C. Otto, G. M. J. Segers-Nolten, et al.: Biophys. J. **74**, 3250 (1998)

[9] A. V. Feofanov, A. I. Grichine, L. A. Shitova, et al.: Biophys. J. **78**, 499 (2000)

[10] B. R. Wood, B. Tait, D. McNaughton: Biochim. Biophys. Acta **1539**, 58 (2001)

[11] H. J. Van Manen, N. Uzunbajakava, R. Van Bruggen, et al.: J. Am. Chem. Soc. **125**, 12112 (2003)

[12] B. R. Wood, S. J. Langford, B. M. Cooke, et al.: Febs. Lett. **554**, 247 (2003)

[13] J. X. Cheng, Y. K. Jia, G. F. Zheng, et al.: Biophys. J. **83**, 502 (2002)

[14] K. Kneipp, A. S. Haka, H. Kneipp, et al.: Appl. Spectrosc. **56**, 150 (2002)

[15] K. Nithipatikom, M. J. McCoy, S. R. Hawi, et al.: Anal. Biochem. **322**, 198 (2003)

[16] C. E. Talley, L. Jusinski, C. W. Hollars, et al.: Anal. Chem. **76**, 7064 (2004)

[17] M. B. Wabuyele, F. Yan, G. D. Griffin, et al.: Rev. Sci. Instrum. **76**, 063710 (2005)

[18] T. Vo-Dinh, F. Yan, M. B. Wabuyele: J. Raman Spectrosc. **36**, 640 (2005)

[19] G. Breuzard, J. F. Angiboust, P. Jeannesson, et al.: Biochem. Biophys. Res. Commun. **320**, 615 (2004)

[20] K. Kneipp, H. Kneipp, I. Itzkan, et al.: Chem. Rev. **99**, 2957 (1999)

[21] Y. W. C. Cao, R. C. Jin, C. A. Mirkin: Science **297**, 1536 (2002)

[22] J. M. Nam, C. S. Thaxton, C. A. Mirkin: Science **301**, 1884 (2003)

[23] S. Efrima, B. V. Bronk: J. Phys. Chem. B **102**, 5947 (1998)

[24] R. M. Jarvis, A. Brooker, R. Goodacre: Anal. Chem. **76**, 5198 (2004)

[25] B. Pettinger, B. Ren, G. Picardi, et al.: Phys. Rev. Lett. **92**, 96 (2004)

[26] L. K. Limbach, Y. Li, R. N. Grass, et al.: Environ. Sci. Technol. **39**, 9370 (2005)

[27] J. Rejman, V. Oberle, I. S. Zuhorn, et al.: Biochem. J. **377**, 159 (2004)

[28] W. J. Arlein, J. D. Shearer, M. D. Caldwell: Am. J. Physiol. Regulatory Integrative Comparative Physiol. **44**, R1041 (1998)

[29] A. G. Tkachenko, H. Xie, Y. L. Liu, et al.: Bioconjugate Chem. **15**, 482 (2004)

[30] C. Feldherr, E. Kallenbach, N. Schultz: J. Cell Biol. **99**, 2216 (1984)

[31] J. Kneipp, R. Peteranderl, M. McLaughlin, et al.: in preparation

[32] H. J. Gao, W. D. Shi, L. B. Freund: Proc. Natl. Acad. Sci. USA **102**, 9469 (2005)

[33] G. Thomas Jr., B. Prescott, D. Olins: Science **197**, 385 (1977)

[34] V. Levi, Q. Ruan, E. Gratton: Biophys. J. **88**, 2919 (2005)

[35] M. Moskovits: J. Raman Spectrosc. **36**, 485 (2005)

[36] C. Feldherr, D. Akin: J. Cell Biol. **111**, 1 (1990)

[37] H. Xie, A. G. Tkachenko, W. R. Glomm, et al.: Anal. Chem. **75**, 5797 (2003)

[38] J. Kneipp, H. Kneipp, W. L. Rice, et al.: Anal. Chem. **77**, 2381 (2005)

[39] A. Becker, C. Hessenius, K. Licha, et al.: Nature Biotechnol. **19**, 327 (2001)

[40] B. Ebert, U. Sukowski, D. Grosenick, et al.: J. Biomed. Opt. **6**, 134 (2001)

[41] J. Malicka, I. Gryczynski, C. D. Geddes, et al.: J. Biomed. Opt. **8**, 472 (2003)

[42] R. Weigand, F. Rotermund, A. Penzkofer: J. Phys. Chem. A **101**, 7729 (1997)

[43] M. Landsman, G. Kwant, G. Mook, et al.: J. Appl. Physiol. **4**, 575 (1976)

[44] K. Urbanska, B. Romanowska-Dixon, Z. Matuszak, et al.: Acta Biochim. Pol. **49**, 387 (2002)

[45] C. R. Yonzon, C. L. Haynes, X. Y. Zhang, et al.: Anal. Chem. **76**, 78 (2004)

[46] J. Yi, X. M. Tang: Cell Res. **9**, 243 (1999)

[47] C. T. Okamoto: Adv. Drug Delivery Rev. **29**, 215 (1998)

[48] N. A. Bright, B. J. Reaves, B. M. Mullock, et al.: J. Cell Sci. **110**, 2027 (1997)

Index

Biomolecule Sensing with Adaptive Plasmonic Nanostructures

Vladimir P. Drachev and Vladimir M. Shalaev

Purdue University, West Lafayette, IN 47907
{vdrachev,shalaev}@purdue.edu

1 Introduction

One of the challenges of biomolecule sensing with surface-enhanced Raman scattering (SERS) is to preserve all of the advantages of Raman spectroscopy applications for structural biology. There are many examples where Raman spectroscopy provides important information on large, macromolecular structures as a whole and in defining small regions of large complexes through ligand–macromolecule recognition (reviewed in [1, 2, 3, 4, 5, 6]). The Raman scattering process involves interplay between atomic positions, electron distribution, and intermolecular forces. Hence, Raman spectroscopy potentially can be one of the techniques used to reveal correlations between structure and function.

It is a common belief that the protein–metal surface interaction may lead to structural changes of proteins and the loss of protein functionality to some extent. To what extent this occurs is a question that needs to be addressed for any particular type of SERS-active substrates. The applicability of SERS to molecular biology has been under extensive study since the 1980s [7]. Despite some unknowns in the SERS process, the molecular mechanisms of biomolecule–metal surface interactions and the distance dependence of the Raman enhancement, SERS has been widely used in biomolecular spectroscopy [8, 9, 10, 11, 12, 13, 14, 15, 16].

In this Chapter we demonstrate several examples of protein sensing with our SERS substrate employing a new, adaptive property of well-known vacuum-evaporated silver films. The deposition of protein solutions on such a film results in the rearrangement of the initial metal nanostructures. Such protein-mediated restructuring leads to the formation of aggregates of metal particles naturally covered and matched with molecules of particular sizes and shapes. This procedure optimizes the SERS signal and, in parallel, stabilizes the metal surface with proteins. Such a substrate, which is referred to as an *adaptive silver film (ASF)*, provides a large SERS enhancement and hence allows protein sensing at monolayer protein surface densities, while enabling the adsorption of proteins without significant changes in their conformational states [17, 18, 19, 20]. For example, we showed that spectral differences in SERS spectra of human insulin and its analog insulin lispro can be detected and assigned to their difference in conformational states. An interesting op-

K. Kneipp, M. Moskovits, H. Kneipp (Eds.): Surface-Enhanced Raman Scattering – Physics and Applications, Topics Appl. Phys. **103**, 351–366 (2006)
© Springer-Verlag Berlin Heidelberg 2006

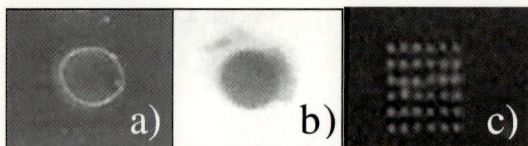

Fig. 1. Typical protein spot on ASF substrate before (**a**) and after (**b**) washing with a Tris-buffered saline solution (the 2 μl aliquot of 0.5 μM bacterial alkaline phosphatase/C-terminal FLAG-peptide fusion (fBAP) was deposited manually), spot size is about 2 mm; (**c**) – antibody array deposited with a quill-type spotter, typical size of about 4 × 4 mm, individual spot size is about 100 μm

portunity is enabled with SERS detection and a pseudotyping procedure for Ebola virus, a potential biowarfare agent. Antibody–antigen binding results in distinct spectral changes in SERS spectra of the first layer protein, either antibody or antigen, depending on the chosen binding scheme. While this Chapter addresses SERS applications, we should mention the excellent performance of adaptive surfaces as a solid support for antibody–antigen binding reactions tested with a microarray protocol and fluorescence detection and compared with commercial substrates [19, 21].

2 Adaptive Plasmonic Nanostructures

Both SERS enhancement mechanisms – electromagnetic and molecular (chemical) – work effectively for molecules in immediate proximity to a rough metal surface. Hence molecule adsorption is an important component of SERS and is always accompanied by surface-chemistry processes. One of the key ideas behind adaptive plasmonic nanostructures is to provide the needed flexibility under protein deposition for surface-chemistry to form particle aggregates naturally covered and stabilized with the proteins of interest.

Figure 1a shows a typical protein spot after manual deposition of proteins in a Tris-buffered saline (TBS) solution on an ASF with subsequent drying. In the example in the figure, the concentration of the deposited antibody was 1 μM and the volume was about 2 μl. After washing with TBS/Tween-20 for 15 min to 30 min, the nonadherent metal is removed from the substrate except in the areas where protein (antibody or antigen) has been deposited (Fig. 1b). An example of an array deposition with a quill-type spotter is shown in Fig. 1c. The capture proteins (a set contains antihuman interleukin 6 monoclonal antibody and other) were dissolved in phosphate-buffered saline (PBS) (300 μg/ml) for spotting on the ASF substrate. The approximate spotting volume was 0.7 nl, yielding spots of about 100 μm in diameter. The substrates were custom printed by Tele-Chem International.

The nanostructures of the ASF before and after protein deposition and washing are shown in Fig. 2a and b. The initial film was fabricated with an e-beam evaporator at high vacuum (10^{-7} Torr); clean glass slides were covered

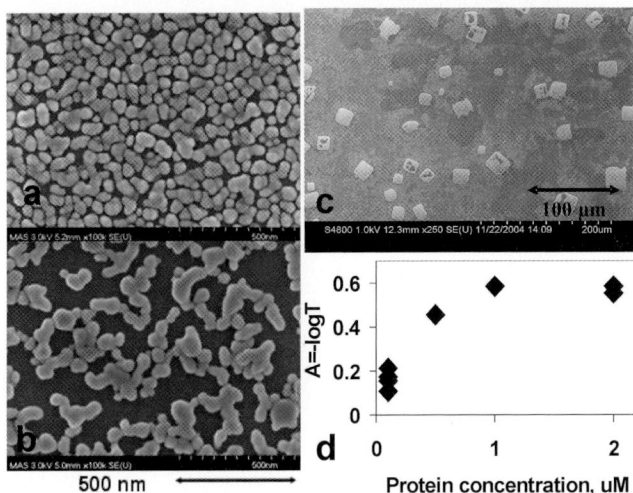

Fig. 2. FE SEM images of ASF substrates: (**a**) initial structure of 11 nm silver film (**b**) same substrate as in (**a**) but inside antibody spot after deposition of 2 μl, 0.5 μM in TBS solution and washing; (**c**) substrate after deposition of TBS without proteins; (**d**) absorbance inside protein spot (antihuman interleukin 10) at 568 nm versus protein concentration. (**a**) and (**b**) are adapted with permission from [19] © 2005 American Chemical Society

first by a sublayer of 10 nm of SiO_2 followed by an 8 nm to 11 nm Ag layer deposited at a rate of 0.05 nm/s. As illustrated by the field emission scanning electron microscope (FE SEM) images, protein-mediated restructuring results in the formation of aggregates of silver particles covered with proteins (Fig. 2b), as opposed to the relatively disintegrated but closely spaced particles of the initial film before protein deposition (Fig. 2a). Depending on the mass thickness of the initial film, small or large fractal-like aggregates can be formed during the nanoscale restructuring process.

A lower concentration of protein results in lower metal coverage (the ratio of white area to total area in the FE SEM images) and lower extinction at a particular wavelength. A decrease of metal coverage correlates with decreasing optical absorption. The absorbance of the metal film inside a protein spot increases linearly with protein concentration and then saturates above a certain concentration, which can be considered to be optimal (Fig. 2d). The concentration dependence shows almost no change after 30 min of washing in TBS/Tween 20 solution, which confirms the stabilization of the film by the proteins.

The transparent areas outside of a typical protein spot certainly contain no silver particles, as indicated by absorption measurements in those areas. The metal-particle coverage inside the protein spot is also reduced relative to the initial film. To determine the chemical form of the silver remaining in the transparent areas, TBS (Tris: 0.05 M, NaCl: 0.138 M, KCl: 0.0027 M) was deposited on the ASF and dried. Then the deposition region was studied with FE SEM and X-ray diffraction analysis. The transparent area contains $40 \times 40\,\mu m$ particles as seen from FE SEM images (Fig. 2c). X-ray diffraction results indicate a reduced Ag 101 peak and two peaks from NaCl and AgCl. This implies that silver particles are transformed to silver salt in the transparent area.

An estimate of the thermodynamics of the (redox) reaction of Ag with oxygen shows that Ag oxidation is a downhill reaction (with negative free energy) under the experimental conditions. When the silver film is exposed to a TBS buffer (pH 7.4), metal silver tends to be oxidized by oxygen and form AgCl due to the low solubility of AgCl in water ($K_{sp} = 1.8 \times 10^{-10}$) [22]. The total reaction and its standard potential under standard conditions [22] will be:

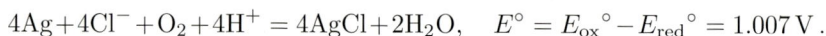

$$4Ag + 4Cl^- + O_2 + 4H^+ = 4AgCl + 2H_2O, \quad E^\circ = E_{ox}^\circ - E_{red}^\circ = 1.007\,V\,.$$

For calculating the actual reaction potential under the experimental condition at 298 K, we use the Nernst equation:

$$E = E^\circ - (0.059/4) \log\{1/([Cl^-]^4[H^+]^4 P_{O_2}\}\,,$$

where $[Cl^-]$ is about 0.14 M, $[H^+]$ is about $10^{-7.4}$ M and the partial pressure of oxygen in air is 0.21 atm. So the experimental value of E is estimated to be 0.51 V, which means that the reaction is thermodynamically favorable (downhill in free energy). The positive potential is the driving force for the oxidation of Ag to Ag^+.

Thus a deposition of proteins in a buffer solution results in the competition of two processes: etching through the oxidation of silver surface in buffer, and stabilization with protein interaction. Etching of the particle surface makes particles movable and leads to the protein-mediated rearrangement of the initial particle nanostructure.

3 SERS Features of Conformational States: Insulin

Insulin represents an interesting example for Raman spectroscopy. It consists of 51 aminoacids distributed in two chains (the A and B chains), which are linked by two disulfide bonds. Insulin typically exists in the hexameric form, although its monomeric form is the active form of the hormone. The size of an insulin hexamer is about 3 nm to 5 nm. The problem of insulin oligomerization has stimulated development of a number of recombinant insulin analogs. The first of these molecules, insulin *lispro,* is engineered as a rapidly acting, blood-glucose-lowering agent. ASFs were used to examine the differences in Raman spectra of two insulin isomers: human insulin and insulin lispro. Both of these molecules have the same set of amino acid side chains, and differ only in the two amino acid residue locations, causing a slight change in the molecules' conformational states. Specifically, the lysine and proline residues on the C-terminal of the B chain are interchanged in their positions in lispro molecules as compared to human insulin. This small difference, however, causes an important clinical effect for diabetes treatment, making insulin lispro a fast-acting agent in the bloodstream while human insulin is slower to act. The differences in SERS spectra between these two insulin isomers can be detected with ASFs and assigned to the α-helix Raman markers and Phe ring-breathing mode.

The Raman system used in this study comprises an Ar/Kr ion laser (Melles Griot), a laser bandpath holographic filter, two Super-Notch Plus filters (Kaiser Optical Systems), focusing and collection lenses, an Acton Research 300i monochromator with a grating of 1200 grooves per millimeter, and a liquid-nitrogen-cooled CCD system (1340 × 400 pixels, Roper Scientific). SERS spectra were typically collected using an excitation laser wavelength of 568.2 nm with normal incidence and 45° scattering. An objective lens (f/1.6) provided a collection area of about 180 μm^2. The spectral resolution was about 3 cm^{-1}. Normal Raman spectra were collected in a backscattering geometry using a micro-Raman system, which consists of a He–Ne laser (632.8 nm) with 12 mW power focused to a spot of about 2 μm diameter on the sample using a 80× microscope objective.

3.1 SERS Versus Normal Raman

A SERS spectrum collected from the central part of an insulin spot is shown in Fig. 3a after linear polynomial background subtraction and normalization. The spot was deposited from a 3 μl drop of 1 μM insulin in a 0.1 mM HCl solution. The extinction coefficient ε_{280} of 5.7 mM^{-1} · cm^{-1} [23] and Raman band assignments are known from the published literature [24, 25].

A comparison of the insulin SERS spectra with our normal Raman spectra of insulin on quartz (Fig. 3b) and with insulin in solution [24] suggests that all the Raman fingerprints of insulin are enhanced by approximately the same factor in our study. Phenylalanine (Phe) and tyrosine (Tyr) often contribute to the protein Raman spectra along with the amide I and amide III bands of the peptide backbone vibrations. Insulin contains phenylalanine located at the B1, B24, B25 residues of the B chain and tyrosine at A14, A19, B16, and B26 [25]. The spectra contain "indicators" of the phenylalanine-to-tyrosine ratio: two peaks at 624 (Phe) and 643 (Tyr) cm^{-1}; the pair of tyrosine Raman makers at 832 and 850 cm^{-1}, the backbone-related C_α–C stretching mode at 945 cm^{-1}; and phenylalanine ring modes at 1003 cm^{-1} and 1030 cm^{-1}. In addition there are a group of peaks between 1150 cm^{-1} and 1500 cm^{-1} that can be constituted from Tyr and Phe peaks at 1176 cm^{-1} and 1206 cm^{-1}, amide III bands at 1242 cm^{-1} and 1267 cm^{-1} (they give the peak at 1248 cm^{-1}, Fig. 3b), –CH deformation modes at 1308 cm^{-1} and 1342 cm^{-1} (peak at 1323 cm^{-1} in our case), a –COO$^-$ symmetrical stretching

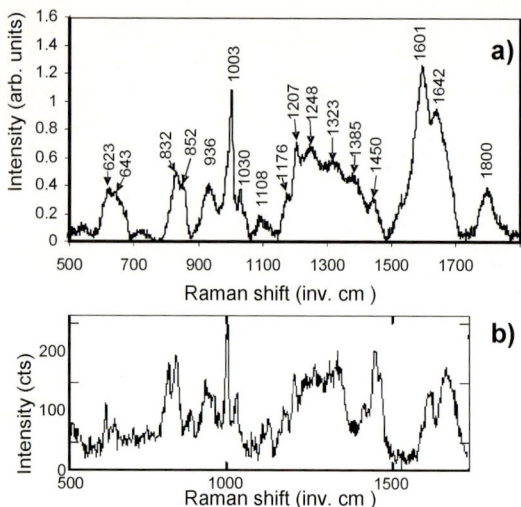

Fig. 3. Recombinant human insulin: (**a**) SERS spectrum after background subtraction and normalization with constant = 7 counts per milliwatt second, (**b**) Raman spectrum of insulin on quartz substrate

mode at $1422 \, \text{cm}^{-1}$, and a $-CH_2$ deformation mode at $1450 \, \text{cm}^{-1}$. The spectra also contain overlapped peaks around $1594 \, \text{cm}^{-1}$ that are usually assigned to the Phe and Tyr modes of an aromatic ring (at $1590 \, \text{cm}^{-1}$, $1605 \, \text{cm}^{-1}$, and $1615 \, \text{cm}^{-1}$), and peaks around $1640 \, \text{cm}^{-1}$ and higher that are amide I modes of the α-helical $(1662 \, \text{cm}^{-1})$ and random coil $(1680 \, \text{cm}^{-1})$ structures [25]. The relative intensities of the Raman peaks are slightly different for SERS of insulin on ASF and for normal Raman of insulin on quartz. The intensity ratios of the Raman peaks $(828 \, \text{cm}^{-1} : 850 \, \text{cm}^{-1} : 1003 \, \text{cm}^{-1} : 1450 \, \text{cm}^{-1} : 1600 \, \text{cm}^{-1})$ are equal to $(0.58 : 0.57 : 1 : 0.32 : 0.35\text{--}1.2)$ for SERS, and $(0.54 : 0.64 : 1 : 0.38 : 0.52)$ for normal Raman on quartz. Note that the $1600 \, \text{cm}^{-1}$ peak intensity is very sensitive to the experimental conditions; in particular, it grows at high incident light intensity.

The Phe(B1) and Tyr(A14, A19) residues of insulin are exposed to the hexamer surface (Protein Data Bank, www.pdb.org), which could be the reason for the similarity between the SERS and normal Raman spectra. It was suggested also in [24] that the Phe peak at $1003 \, \text{cm}^{-1}$ is mostly composed from one of the three phenylalanine residues, Phe(B1).

The average macroscopic SERS enhancement for insulin on an ASF as measured in our experiments (using the same Raman system for SERS and NR) is about 3×10^6, which is among the largest reported for island or percolation Ag films (10^5 for nitrobenzoate [26] and 5.3×10^5 for *trans*-1,2-bis(4-pyridyl)-ethylene [27]). The enhancement was calculated as an average over the substrate surface probed by the laser beam without normalization on the metal surface coverage, which is about 0.5–0.6 in our case. What is important

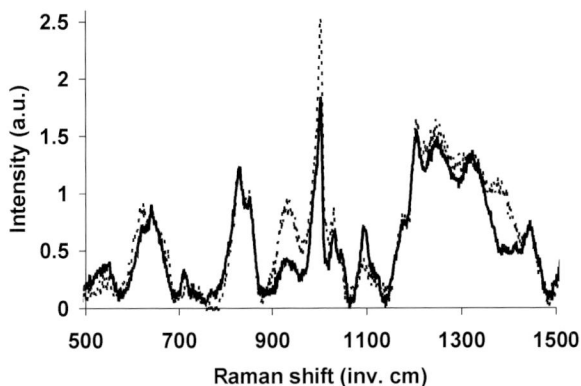

Fig. 4. SERS spectra for human insulin (*dashed gray*) and its analog insulin lispro (*solid black*). Adapted with permission from [17] © 2005 American Chemical Society

for bioapplications is that this enhancement is large enough to detect a sub-monolayer of proteins, namely $80\,\mathrm{fmol/mm^2}$ surface density for insulin. The protein-mediated aggregation provides a beneficial effect to the sensitivity. It is known that the aggregation of even a few particles makes a large difference for SERS enhancement and can allow single-molecule detection [28, 29, 30].

3.2 Human Insulin Versus Insulin Lispro

Representative SERS spectra for human insulin and insulin lispro are shown in Fig. 4 (dashed and solid lines are human insulin and lispro, respectively, and both spectra were subjected to linear polynomial background subtraction). The spectra were normalized to the Tyr peaks at $832\,\mathrm{cm^{-1}}$ and $853\,\mathrm{cm^{-1}}$, since they show approximately equal average intensity over three spots for both human insulin and lispro. Human insulin spectra, as compared to insulin lispro, indicate more intense peaks at $1387\,\mathrm{cm^{-1}}$, $945\,\mathrm{cm^{-1}}$, and for the phenylalanine (Phe) ring-breathing mode at $1003\,\mathrm{cm^{-1}}$. These differences hold at low laser intensity and power ($1\,\mathrm{mW}$, laser beam size is about $80\,\mu\mathrm{m}$ to $100\,\mu\mathrm{m}$) and become less pronounced under exposure to higher laser power.

The zinc–insulin hexamer is an allosteric protein that exhibits two different protein conformations referred to as extended (T) and α-helical (R) conformation states [31]. The T–R state allosteric transitions result in the conversion of B1–B8 residues from a random coil structure to an α-helix conformation along with about $30\,\text{Å}$ displacement for the Phe(B1) residue, which can be detected by Raman spectroscopy [24]. Depending on the conformational state of the displaced Phe(B1), the latter can be at different distances and thus have different orientations with respect to the metal surface, enabling the observed increase in Phe peak intensities by a factor of 1.4 for human insulin as compared to insulin lispro.

The C–C_α–H bending band at $1385\,\mathrm{cm}^{-1}$ and the N–C–C_α skeletal band at $940\,\mathrm{cm}^{-1}$ are stronger in human insulin than in insulin lispro. The $890\,\mathrm{cm}^{-1}$ to $945\,\mathrm{cm}^{-1}$ band is a characteristic spectral line for an α-helix structure and is known to be sensitive to structural changes [32, 33, 34, 35]. This spectral line is typically centered at $940\,\mathrm{cm}^{-1}$ and disappears or displays weak intensity upon conversion to β-sheet or random coil structures. It is also known that the C–C_α–H bending band at $1371\,\mathrm{cm}^{-1}$ appears for the R_6 conformation of hexameric human insulin, which has the longest sequence (B1–B19) of α-helix, and disappears in the spectra of $T_3R_3^f$ and T_6 conformations, with (B4–B19) of α-helix for a "frayed" R^f state and (B9–B19) for a T state [32]. Thus the spectral differences in the SERS spectra of the two insulins can be attributed to the larger α-helix content for human insulin and PheB1 displacement. This conclusion is in agreement with X-ray crystallographic studies [36], where it was observed that insulin lispro crystallizes as a $T_3R_3^f$ hexamer. Specific orientations of molecular bonds on the silver surface emphasize the SERS spectral difference between the two insulins, making the differences much stronger than for conventional Raman. Since comparative SERS spectra reveal characteristic features of the insulin and its analog that are correlated with normal Raman and X-ray studies, one can conclude that there is no significant alteration in the conformational states of insulin on the ASFs.

4 Ebola Virus after Pseudotyping

Biological agents are difficult to detect or protect against; they are invisible, odorless, and tasteless, and their dispersal can be accomplished silently. Early detection of a biological agent in the environment allows for early, specific treatment and advanced warning time during which prophylaxis would be effective. Ebola virus is a Class IV biological agent, has a mortality rate up to 90 % and cannot be used safely outside of highly specialized laboratories. The typical size of the Ebola virus is $\sim 80\,\mathrm{nm}$ to $100\,\mathrm{nm}$ in diameter with a shell of about $20\,\mathrm{nm}$; the length can vary from $0.5\,\mathrm{\mu m}$ to $14\,\mathrm{\mu m}$. The problem of whole-organism detection is the issue of safety, which makes it difficult to test a sensor for such a harmful sample. SERS provides an attractive possibility to detect harmful viruses using the pseudotyping procedure [37]. Through pseudotyping, the shell of the Ebola virus can be attached to the core of a harmless virus [38]. The initial virus loses its capacity to multiply and its activity is completely lost through exposure to UV light. This brings the virus down to a Class I biological agent, which is much easier to handle in typical laboratory environments. At the same time, however, the shell is specific to the given virus and is a glycoprotein that can be detected with SERS since SERS probes the protein layers near the metal surface. Hence

SERS can be considered a label-free whole-organism fingerprinting technique that can be trained with a harmless sample and then be applied to the real, harmful sample.

Our initial studies show spectra of monolayers of the Ebola virus after pseudotyping.[1] Spots of different volumes and different concentrations of the sample were manually deposited on ASF substrates and allowed to dry. SERS spectra were collected from each of the spots on the ASFs and normal Raman spectra were collected from the spots on Teflon-coated ($\sim 15\,nm$) stainless steel substrates. The SERS substrates were then washed for 5 min with deionized water and dried, after which the spectra were recollected. The background was subtracted from the spectra using a Fourier method. All spectra are normalized. A typical normalization constant is about 10–15 counts per milliwatt second. The SERS spectra of Ebola virus after deposition were obscured by noise-type peaks. The same sample shows good-quality spectra after washing in DI water for 5 min. Washing the substrate after deposition helps by keeping only a monolayer of the virus on the spot, while the extra layers of virus and possible contaminants are washed away. The estimated average surface density for a normal Raman spot is 160 cells per microsquaremeter and that for the SERS spot is 25 cells per microsquaremeter. The normal Raman spectrum was collected from the outer spot ring where the virus density is higher than that of the center, while the SERS spectrum was collected from the center of the spot after washing.

The SERS spectrum of Ebola (Fig. 5a) looks similar to a typical protein spectrum, implying that the outer shell (glycoprotein) signal is enhanced. The normal Raman spectrum (Fig. 5b) is very different since the virus as a whole is probed in this case, including the shell (Ebola) and the core (not Ebola).

5 Tag-Free Antibody–Antigen Binding Detection

The detection of protein–ligand binding events most commonly involves labeling strategies with a variety of schemes, including the use of surface-plasmon resonance (SPR) [39, 40, 41, 42] or surface-enhanced Raman scattering [43, 44, 45]. One of the SERS-based approaches takes advantage of the binding properties of antibody and antigen molecules (or DNA strands) and uses metal nanoparticles coated with Raman-active chromophores as tags [14, 15, 46]. To provide the required level of enhancement, silver clusters are added to form a complex sandwich structure that includes a nanoparticle labeled with a Raman-active dye, complementary biomolecules and the metal clusters. The detection of SERS events without inclusion of exogenous labels has been previously employed for an immune reaction using colloidal

---

[1] The samples were obtained by courtesy of Prof. D. Sanders and Department of Biological Sciences, Purdue University.

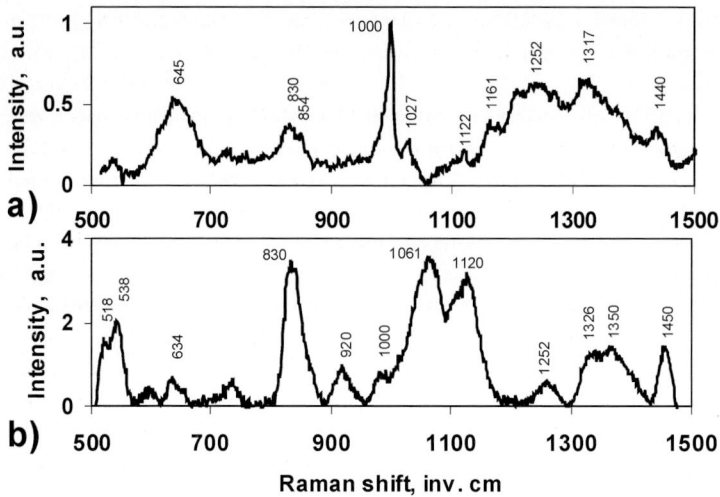

Fig. 5. Spectra of Ebola virus after pseudotyping: (**a**) SERS at $\lambda_{exc} = 568$ nm and (**b**) normal Raman at $\lambda_{exc} = 633$ nm

gold particles [47]. Other applications of SERS detection for binding to proteins include aflatoxins to RNA polymerase and specific organisms to their respective antibodies [48].

We demonstrated that SERS substrates based on nanostructured adaptive silver films make it possible to detect the formation of specific antigen–antibody complexes at a monolayer level by SERS, fluorescence and chemiluminescence techniques. The antibodies rearrange and stabilize the silver film to make the complex resistant to washing and incubation procedures, while preserving the protein activity as recognition agents. The ASF substrates can potentially be used for routine laboratory analyses in ways similar to membrane-based immunoblotting protocols. One advantage of the ASF substrate is that it does not require chromophore-tagged secondary antibodies since direct detection of SERS spectral signatures associated with the antigen–antibody binding event are observed. These early-stage demonstrations illustrate opportunities for the use of nanoparticle surface systems for analysis of biomolecular interactions relevant for diagnostic applications.

To probe the antigen–antibody binding event using SERS, a general protocol was devised that involves first the deposition and immobilization of a monoclonal antibody or a corresponding antigen on an ASF substrate. Typically, $2\,\mu$l of $0.5\,\mu$M Ab solution forms a spot of about $2\,$mm after drying overnight, similar to that shown in Fig. 1. The nonadherent metal particles were then removed by washing with a buffered solution and deionized water to reveal immobilized protein-adapted aggregates representing antibody (or antigen) in a small array. The specific proteins used in our studies include the anti-FLAG M2 monoclonal antibody (fAb) and the bacterial alkaline phosphatase/C-terminal FLAG-peptide fusion (fBAP). Proteins for control experiments included the bacterial alkaline phosphatase (BAP) without the FLAG peptide that was generated by enterokinase cleavage. Subsequent incubation of the protein-adapted aggregates with antigen (or antibody) was conducted, and the nonspecifically bound material was removed by washing with a standard buffer solution for Western blotting (TBS/Tween-20) followed by rinsing five times with deionized water. SERS spectra of the immobilized fAb (or fBAP) were compared before and after reaction with the cognate antigen (or antibody) partner.

An estimated average protein density was about $D_{av} = 120$ fmol/mm^2, or 0.072 molecules per nanosquaremeter. If a single protein molecule occupies an average area of approximately 20 nm^2, then 1.4 layers of protein would be expected in each cluster. This value would represent an upper limit since any loss of protein during the washing steps would reduce the surface coverage; the post-washing surface coverage can be estimated to be monolayer or sub-monolayer.

The immobilized fAb/metal clusters yielded reproducible SERS results, and representative spectra are shown in Fig. 6a (black). Upon incubation with fBAP (0.75 nM in TBS), spectral changes were observed with the most dominant features appearing in the 1200 cm^{-1} to 1400 cm^{-1} region (Fig. 6a, gray). To establish the specificity of the spectral changes, a second array was probed with BAP (0.67 nM) as a control experiment. The spectra of the fAb remained unchanged after incubation with BAP.

In order to further validate our observations regarding the specific interactions on the protein-adapted aggregates, antigen (fBAP at 0.5 μM in TBS) was first deposited on an ASF substrate followed by antibody (fAb at 4 nM) incubation. The results of the SERS measurements are shown in Fig. 6b, and the spectra of fBAP/metal clusters (black trace) undergo changes after incubation with the fAb solution (gray trace). These changes are most evident in the group of peaks centered at 900 cm^{-1}, where a triple peak appears instead of the double peak present before incubation with antibody. The intensity of the peak at 1000 cm^{-1} is also significantly decreased, while that around 1280 cm^{-1} is increased.

For both experiments, the signal from the first layer of the protein dominates the SERS spectra observed from the immune complex. Therefore, the spectra shown in Figs. 6a and 6b are significantly different, with the former being primarily the SERS spectrum of the antibody fAb and the latter representing the antigen fBAP spectrum. In both cases, binding with proteins of the second layer results in detectable and reproducible changes in the SERS spectrum of the first layer, revealing strong protein–protein interaction. The fact that the first layer dominates the observed SERS spectra indicates the sharp drop of the enhancement on the scale of several nanometers.

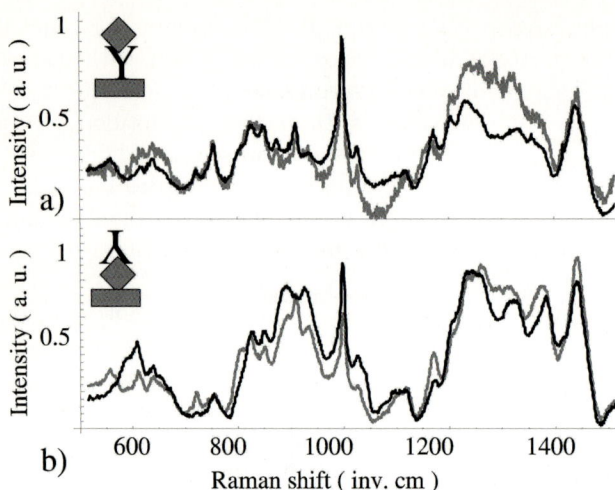

Fig. 6. SERS features of antibody(fAb)–antigen(fBAP) binding reaction on ASF: (a) spectra of fAb (*black*), the same fAb-modified surface incubated with fBAP (*gray*), (b) inverse order: spectra of fBAP on the ASF substrate (*black*) and the same fBAP-modified surface after incubation with fAb solution (*gray*), $\lambda_{\mathrm{exc}} = 568\,\mathrm{nm}$. Adapted with permission from [19] © 2005 American Chemical Society

6 Protein-Binding Detection with Dye Displacement

SERS spectra of dye molecules and dye conjugate proteins always contain a fluorescence contribution that can be quenched due to the presence of the metal surface, depending on the dye–surface distance. The SERS signal of some molecule placed on the same substrate along with an antibody should be sensitive to the changes in the environment caused by antigen binding, since we detect this signal from roughly a single layer of the molecules. This is the idea of the experiment with rhodamine 6G (R6G) illustrated in Fig. 7.

If we chose a molecule with a high SERS cross section, we have a chance to work with a strong SERS signal while the procedure would be as simple as that of label-free detection described above.

The procedure involves several steps: 1. Manual deposition of $2\,\mu\mathrm{l}$, $1\,\mu\mathrm{M}$ solution of antihuman interleukin 10 (IL10) in PBS. 2. Incubation of the substrate with $1\,\mathrm{nM}$ R6G in TBS Tween 20 for $40\,\mathrm{min}$ followed by five DI water rinses. 3. Substrate incubation with $1\,\mathrm{nM}$ antigen cocktail containing nine antigens including IL10 in TBS/Tween 20. The antibody concentration was varied from $0.1\,\mu\mathrm{M}$ to $2\,\mu\mathrm{M}$ under deposition from spot to spot. SERS spectra were collected after each step of the procedure shown in Fig. 7 without background subtraction using an excitation wavelength of $568\,\mathrm{nm}$. The spectrum is the antibody, antihuman IL10 spectrum in the first case (Fig. 7a), while the R6G spectrum is observed in the second case (Fig. 7b). After incubation with antigen IL10, the spectrum shows some combination of the decreased

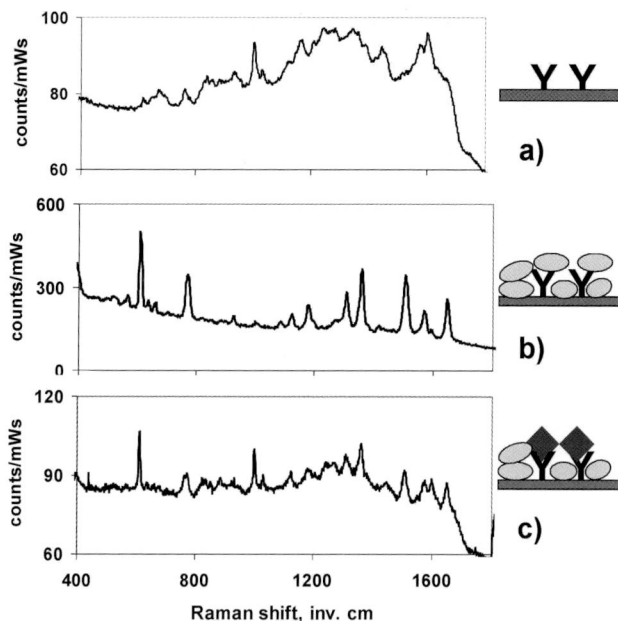

Fig. 7. Sketch of the procedure and corresponding SERS spectra for the experiment on R6G displacement caused by antibody–antigen binding: (**a**) antihuman interleukin 10 manually deposited on ASF ($2\,\mu$l, $1\,\mu$M solution in PBS), (**b**) after incubation with $1\,$nM R6G in TBS/tween 20 solution (rinsed in DI water 5 times), (**c**) after incubation with antigen cocktail containing $1\,$nM human interleuikin 10. Spectra are shown without background subtraction

R6G peaks and a slightly modified antibody spectrum. The dependence of the dye signal on the antibody concentration is very different before and after incubation with the antigens providing the binding reaction.

The intensity of the dye SERS signal is highest at the highest concentration ($2\,\mu$M) initially and becomes highest at the lowest concentration after antigen incubation. The relative change of the R6G signal caused by antigen incubation is approximately inversely proportional to the antibody surface density, as seen in Fig. 8. This suggests a correlation between the displacement of R6G molecules and antigen–antibody binding. The effect of antigen incubation on the displacement of R6G molecules can potentially be applied to binding detection and thus merits further study.

7 Summary

The examples provided demonstrate the appropriateness of nanostructured adaptive silver films (ASFs) to enhance the strength of Raman spectroscopy

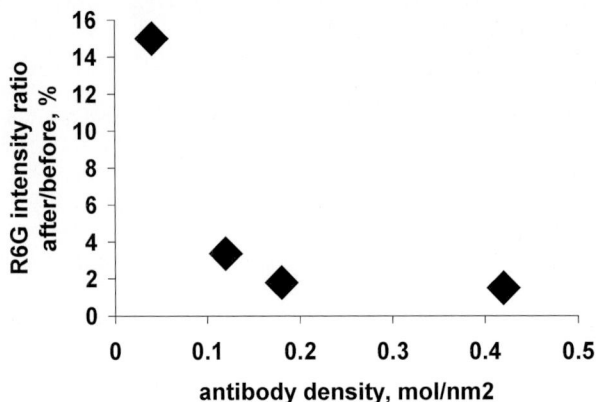

Fig. 8. Intensity of R6G SERS signal (peak at $613\,\mathrm{cm}^{-1}$, background is subtracted) after incubation with antigen solution relative to the intensity before incubation as a function of the average antibody surface density

in biological applications. The restructuring under biomolecule deposition allows one to match the metal-particle aggregate geometry with large molecules such that the conformation and functionality are preserved. This produces excellent conditions for SERS enhancement. Note here that there is a promising way to improve further the sensitivity of SERS-based biosensors by including a bulk metal layer as a sublayer between the adaptive silver film and a glass substrate [20]. The interaction of the silver film with the biomolecule solution acts to stabilize the system, making it applicable for all possible protocols for bioarray treatment and detection. The most straightforward application of SERS in immunoassays appears to be the active signaling reporter for binding reactions. However, we believe the most attractive benefits would be in the direct detection of Raman features of protein–ligand interaction, including a follow-up comprehensive analysis of the vibrational fingerprints.

Acknowledgements

We thank our coauthors M. D. Thoreson, E. N. Khaliullin, V. C. Nashine, H. K. Yuan, Y. Xie, T. Goyani, D. Ben-Amotz, M. L. Narasimhan, and V. J. Davisson who contributed in this work. This work was supported in part by a grant from Inproteo, LLC.

References

[1] T. G. Spiro, B. P. Garber: Ann. Rev. Biochem. **46**, 553–572 (1977)
[2] R. Callender, H. Deng: Ann. Rev. Biophys. Biomol. Struct. **23**, 215 (1994)
[3] W. L. Peticolas: Methods Enzymol. **246**, 389 (1995)

[4] P. R. Carey: J. Raman Spectrosc. **29**, 7 (1998)

[5] R. Callender, H. Deng, R. Gilmanshin: J. Raman Spectrosc. **29**, 15 (1998)

[6] R. Tuma: J. Raman Spectrosc. **36**, 307 (2005)

[7] G. D. Chumanov, R. G. Efremov, I. R. Nabiev: J. Raman Spectrosc. **21**, 43 (1990)

[8] T. Vo-Dinh: Trends Anal. Chem. **17**, 557–582 (1998)

[9] G. Bauer, N. Stich, T. G. M. Schalkhammer: *Methods and Tools in Biosciences and Medicine: Analytical Biotechnology* (Birkhäuser, Basel 2002) pp. 253–278

[10] M. S. Sibbald, G. Chumanov, T. M. Cotton: J. Electroanal. Chem. **438**, 179–185 (1997)

[11] T. Vo-Dinh, D. L. Stokes, G. D. Griffin, M. Volkan, U. J. Kim, M. I. Simon: J. Raman Spectrosc. **30**, 785–793 (1999)

[12] K. R. Brown, A. P. Fox, M. J. Natan: J. Am. Chem. Soc. **118**, 1154–1157 (1996)

[13] K. E. Shafer-Peltier, C. L. Haynes, M. R. Glucksberg, R. P. V. Duyne: J. Am. Chem. Soc. **125**, 588–593 (2003)

[14] Y. W. C. Cao, R. Jin, C. A. Mirkin: Science **297**, 1536–1540 (2002)

[15] Y. C. Cao, R. Jin, J. M. Nam, C. S. Thaxton, C. A. Mirkin: J. Am. Chem. Soc. **125**, 14676–14677 (2003)

[16] D. S. Grubisha, R. J. Lipert, H. Y. Park, J. Driskell, M. D. Porter: Anal. Chem. **75**, 5936–5943 (2003)

[17] V. P. Drachev, M. D. Thoreson, E. N. Khaliullin, V. J. Davisson, V. M. Shalaev: J. Phys. Chem. B **108**, 18046 (2004)

[18] V. P. Drachev, M. D. Thoreson, E. N. Khaliullin, A. K. Sarychev, D. Zhang, D. Ben-Amotz, V. M. Shalaev: Proc. SPIE **5221**, 76 (2003)

[19] V. P. Drachev, V. C. Nashine, M. D. Thoreson, D. Ben-Amotz, V. J. Davisson, V. M. Shalaev: Langmuir **21**, 8368 (2005)

[20] V. P. Drachev, M. D. Thoreson, V. C. Nashine, E. N. Khaliullin, D. Ben-Amotz, V. J. Davisson, V. M. Shalaev: J. Raman Spectrosc. **36**, 648 (2005)

[21] V. P. Drachev, M. L. Narasimhan, H.-K. Yuan, M. D. Thoreson, Y. Xie, V. J. Davisson, V. M. Shalaev: Proc. SPIE **5703**, 13 (2005)

[22] A. J. Bard, R. Parsons, J. Jordan: *Standard Potentials in Aqueous Solution* (Marcel Dekker, New York 1985)

[23] R. R. Porter: Biochem. J. **53**, 320–328 (1953)

[24] D. Ferrari, J. R. Diers, D. F. Bocian, N. C. Kaarsholm, M. F. Dunn: Biopolymers (Biospectroscopy) **62**, 249–260 (2001)

[25] N.-T. Yu, C. S. Lu: J. Am. Chem. Soc. **94**, 3250 (1972)

[26] D. A. Weitz, S. Garoff, T. J. Gramila: Opt. Lett. **7**, 168 (1982)

[27] R. P. V. Duyne, J. C. Hulteen, D. A. Treichel: J. Chem. Phys. **99**, 2101 (1993)

[28] K. Kneipp, Y. Wang, H. Kneipp, L. T. Perelman, I. Itzkan, R. R. Dasari, M. Feld: Phys. Rev. Lett. **78**, 1667–1670 (1997)

[29] S. Nie, S. R. Emory: Science **275**, 1102–1106 (1997)

[30] H. Xu, E. J. Bjerneld, M. Kall, L. Borjesson: Phys. Rev. Lett. **83**, 4357 (1999)

[31] N. C. Kaarsholm, H. C. Ko, M. F. Dunn: Biochemistry **28**, 4427 (1989)

[32] T.-J. Yu, J. L. Lippert, W. L. Peticolas: Biopolymers **12**, 2161–2176 (1973)

[33] M. C. Chen, R. C. Lord, R. Mendelson: Biochem. Biophys. Acta **328**, 252–260 (1973)

[34] B. G. Frushour, J. L. Koenig: Biopolymers **13**, 1809–1819 (1974)

[35] M. C. Chen, R. C. Lord, R. Mendelson: J. Am. Chem. Soc. **96**, 3038–3042 (1976)
[36] E. Ciszak, J. M. Beals, B. H. Frank, J. C. Baker, N. D. Carter, G. D. Smith: Structure **3**, 615–622 (1995)
[37] T. Goyani: MS thesis, School of Electrical and Computer Engineering, Purdue University (2004)
[38] S. A. Jeffers, D. A. Sanders, A. Sanchez: J. Virology **76**, 12463–12472 (2002)
[39] L. A. Lyon, M. D. Musik, M. J. Natan: J. Anal. Chem. **70**, 5177 (1998)
[40] W. Knoll, M. Zizlsperger, T. Liebermann, S. Arnold, A. Badia, M. Liley, D. Piscevic, F. J. Schmitt, J. Spinke: J. Colloid. Surf. A **161**, 115 (2000)
[41] B. P. Nelson, T. E. Grimsrud, M. R. Liles, R. M. Goodman, R. M. Corn: Anal. Chem. **73**, 1 (2001)
[42] J. C. Riboh, A. J. Haes, A. D. MacFarland, C. R. Yonzon, R. P. V. Duyne: J. Phys. Chem. B **107**, 1772 (2003)
[43] T. E. Rorh, T. Cotton, N. Fan, P. J. Tarcha: J. Anal. Biochem. **182**, 388 (1989)
[44] X. Dou, T. Takama, Y. Yamaguchi, H. Yamamoto, Y. Ozaki: Anal. Chem. **69**, 1492 (1997)
[45] J. Ni, R. J. Lipert, G. B. Dawson, M. D. Porter: Anal. Chem. **71**, 4903 (1999)
[46] T. Vo-Dinh, F. Yan, M. B. Wabuyele: J. Raman Spectrosc. **36**, 640 (2005)
[47] X. Dou, Y. Yamguchi, H. Yamamoto, S. Doi, Y. Ozaki: J. Raman Spectrosc. **29**, 739 (1998)
[48] A. E. Grow, L. L. Wood, J. L. Claycomb, P. A. Thomson: J. Microbiol. Methods **53**, 221 (2003)

Index

Glucose Sensing with Surface-Enhanced Raman Spectroscopy

Chanda Ranjit Yonzon, Olga Lyandres, Nilam C. Shah, Jon A. Dieringer, and Richard P. Van Duyne

Department of Chemistry and Department of Biomedical Engineering, Northwestern University, Evanston, Illinois 60208-3113
vanduyne@chem.northwestern.edu

1 Introduction

Diabetes mellitus is a chronic disorder that requires careful regulation of blood-glucose levels in order to maintain the health of diabetic patients. Failure to regulate these levels within tight limits leads to severe secondary health complications to the diabetics' retina, kidneys, nerves, and circulatory system [1]. Most commonly, diabetics measure blood-glucose levels four to six times per day with an electrochemical-based finger-stick method. The finger-stick method is not capable of continuous monitoring and suffers from low patient compliance due to the pain and discomfort associated with blood sampling from the capillaries. Such intermittent testing can fail to detect significant fluctuations in blood-glucose levels and places the patient in dangerously hypo- or hyperglycemic conditions [2]. The development of a continuous monitoring device for glucose with as low a degree of invasiveness as possible will clearly have an enormous impact on the long-term health on the 171 million diabetics worldwide [3].

Because of the importance of this healthcare challenge, several groups are developing methods for minimally invasive, biologically compatible, quantitative glucose detection. The most advanced and commercially available glucose sensors measure glucose indirectly by electrochemical detection [4, 5]. One of the disadvantages of this indirect detection method is that glucose oxidase, the enzyme that catalyzes the oxidation, needs to be replenished. Therefore, the lifetime of the sensor is limited. Another serious drawback inherent to enzymatic glucose sensors is the lack of stability due to the intrinsic nature of enzymes [6].

Optical methods for direct glucose detection have been explored of which vibrational spectroscopies show great promise. Vibrational spectroscopic methods applied to date for glucose sensing comprise of infrared absorption [7], normal Raman spectroscopy (NRS) [8, 9], and surface-enhanced Raman spectroscopies (SERE) [10,11,12,13,14,15]. The implementation of near- and midinfrared spectroscopic methods has fundamental limitations due to the competing absorption by water and spectral congestion. However, the application of multivariate calibration models presents a possible remedy. The innate property of Raman spectroscopy, in all its forms, allows it to distin-

K. Kneipp, M. Moskovits, H. Kneipp (Eds.): Surface-Enhanced Raman Scattering – Physics and Applications, Topics Appl. Phys. **103**, 367–379 (2006)
© Springer-Verlag Berlin Heidelberg 2006

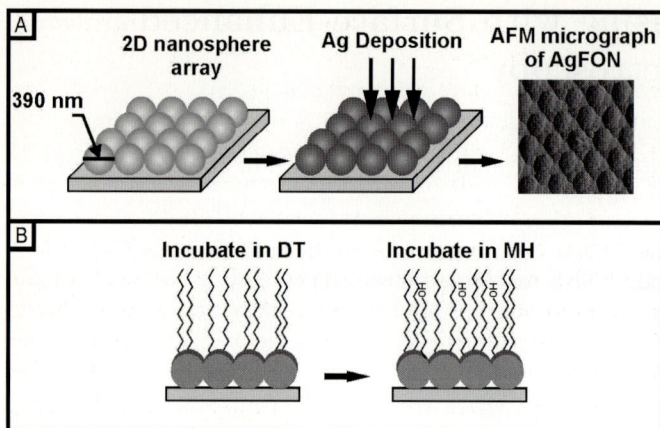

Fig. 1. Schematic of the glucose-sensor fabrication. (**a**) Polystyrene nanospheres were drop-coated on a supporting substrate followed by 200 nm of Ag deposition on the polystyrene mask. The AFM micrograph shows the topography of the AgFON. (**b**) The AgFON surface is first immersed in a 1 mM DT for 45 min. Then, the surface is placed in a 1 mM MH for an overnight incubation

guish between molecules with great structural similarity (e.g., glucose and fructose) [16]. Moreover, normal Raman spectroscopy has been shown to be able to detect physiological concentrations of glucose *in vitro* from a simulated aqueous humor solution [8]. However, high laser powers and long acquisition times are required due to the inherently small NRS cross section of glucose, $5.6 \times 10^{-30}\,\mathrm{cm}^2 \cdot \mathrm{molecule}^{-1} \cdot \mathrm{sr}^{-1}$ [17]. Higher-intensity Raman signals and lower detection limits can be achieved by using SERS. In comparison with infrared and NRS, SERS enjoys both the advantage of application in aqueous media because of the small Raman scattering cross section of water [17] and the sensitivity for trace-level detection [18].

SERS glucose sensing was performed on a silver film over nanospheres (AgFON) surfaces (Fig. 1a). AgFON surfaces were fabricated by first assembling polystyrene nanosphere masks on clean copper substrates. About 200 nm of silver was deposited on the mask in an e-beam deposition chamber [19]. Bare AgFON surfaces display extremely stable SERS activity when challenged with high potentials excursions [20] and high temperatures in ultrahigh vacuum [21]. In addition, AgFON surfaces left in ambient conditions in the absence of light for over a month have been shown to demonstrate SERS activity [22].

The only published SER spectrum of glucose uses a two-step surface-preparation technique using electrochemically roughened electrodes and colloidal nanoparticles [10]. In the experiment performed by *Weaver* and co-workers, the glucose molecules must be trapped in the junction between the roughened electrode and the colloidal nanoparticle. Although this substrate has potential for future work in glucose detection, substrate stability remains to be demonstrated. Electrochemically roughened electrodes are known to have metastable nanostructures; their enhancement factors are strongly potential dependent and, at sufficiently negative potentials, experience irreversible loss of SERS activity. Also, colloidal nanoparticles aggregate when exposed to media with high ionic strength such as would be encountered in glucose sensing. The normal Raman cross section should provide sufficient signal for glucose detection; however, all efforts to observe glucose on a bare AgFON surface using SERS were unsuccessful. The inability to observe glucose signal on a bare AgFON is attributed to the weak or nonexistent affinity to the silver surface. To bring glucose within the range of electromagnetic enhancement of the AgFON surface, a self-assembled monolayer (SAM) can be formed on its surface to partition the analyte of interest (Fig. 1b), in a manner analogous to that used to create the stationary phase in high-performance liquid chromatography [23, 24]. Implementing a partition layer has three advantages: 1. the SAM protects the Ag surface from oxidation; 2. the SAM is exceedingly stable; and 3. the surface structure can be tailored by choosing appropriate SAMs in accordance with the analyte of interest.

Several SAMs were tested for their ability to partition glucose efficiently to the AgFON surface. Of these, straight-chain alkanethiols and ethylene-glycol-terminated alkanethiols partitioned glucose most effectively [11, 12]. However, decanethiol (DT) produced a hydrophobic surface, and partitioning glucose in an aqueous phase was not feasible. On the other hand, although tri(ethylene glycol)-terminated alkanethiol (EG3) partitioned glucose in a phosphate buffer environment, the intricate synthesis of this compound limited its availability. A new-mixed SAM, based on two commercially available components, DT and mercaptohexanol (MH), produced an efficient partition layer for the SERS-based glucose-monitoring device (Fig. 1b) [14].

The exact mechanism of mixed SAM formation has not been well characterized. However, according to a space-filling computer model, combining the longer DT component with the shorter MH component creates a pocket. It is hypothesized that this pocket improves glucose partitioning, bringing glucose even closer to the SERS-active surface than was possible with EG3. In addition, the DT/MH SAM has dual hydrophobic/hydrophilic functionality analogous to EG3, which allows partitioning and departitioning of glucose.

To ensure that the lengths of SAMs do not exceed the electromagnetic enhancement of the SERS surface, experiments were performed to map the distance decay function of SERS in order to determine the length scale of electromagnetic enhancement. Alumina was deposited on the AgFON surface by a custom-fabricated atomic layer deposition (ALD) system in thicknesses of 0.0 nm, 1.6 nm, 3.2 nm and 4.8 nm. The SER spectrum of 50 mM pyridine in 100 mM NaCl/water was measured over the AgFON and alumina AgFON samples. The excitation was performed in an epi configuration through a 20× objective (Nikon, NA = 0.5) on an inverted microscope. The SERS signal was collected through the same objective in a 180° geometry.

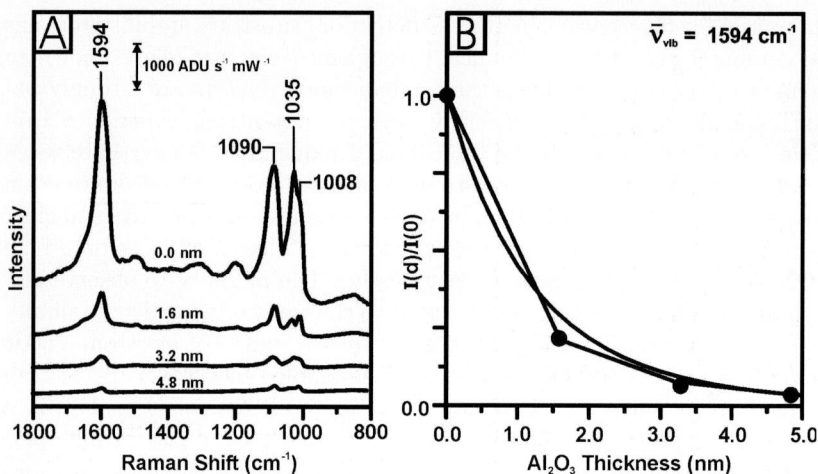

Fig. 2. (a) SER spectra of pyridine adsorbed on AgFON samples treated with various thicknesses of alumina (0.0 nm, 1.6 nm, 3.2 nm, 4.8 nm), $\lambda_{ex} = 532$ nm, $P_{laser} = 1.0$ mW, $t = 300$ s. (b) Plot of SERS intensity as a function of alumina thickness for the 1594 cm^{-1} band (*filled circles* and *straight line* segments). The *solid curved line* is a fit of this data to (1)

The distance dependence of SERS has been theoretically approximated as:

$$I = \left(1 + \frac{r}{a}\right)^{-10}, \qquad (1)$$

where I is the intensity of the Raman mode, a is the average size of the field-enhancing features on the surface and r is the distance of the analyte from the surface [25]. Figure 2a shows the SER spectra of pyridine adsorbed on AgFON surfaces with various thicknesses of alumina. Figure 2b depicts a plot of the relative intensity of the 1594 cm^{-1} band as a function of alumina thickness. Fitting the experimental data to (1) leads to the average size of the enhancing particle $a = 12.0$ nm. The term d_{10} defines the distance away from the nanoparticle required to decrease the SERS intensity by a factor of ten. Figure 2 depicts a d_{10} value of 2.8 nm for an AgFON, which is larger than the DT/MH SAM thickness (a monolayer of DT on silver is 1.9 nm thick) [26]. Therefore, all glucose molecules partitioned in the SAM are electromagnetically enhanced.

2 SERS of Glucose

The DT/MH-functionalized AgFONs were made by incubating the AgFONs in 1 mM DT in ethanol for 45 min and then transferring the substrates to

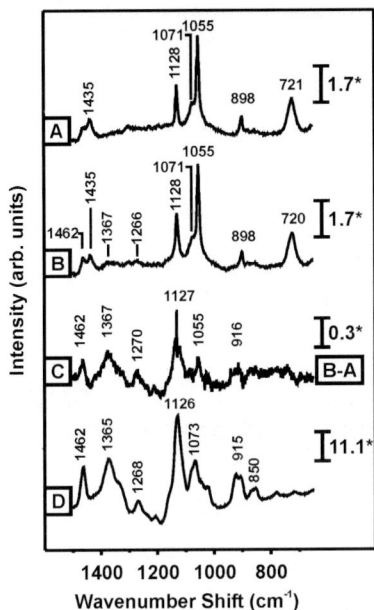

Fig. 3. (a) DT/MH monolayer on AgFON surface, $\lambda_{ex} = 532$ nm, $P_{laser} = 10$ mW, $t = 20$ min. (b) Mixture of DT/MH monolayer and glucose partitioned from a 100 mM solution. (c) Residual glucose spectrum produced by subtracting (a) from (b). (d) Normal Raman spectrum of 4 M aqueous glucose for comparison, $\lambda_{ex} = 532$ nm, $P_{laser} = 30$ mW, $t = 2$ min. *a.d.u. s$^{-1} \cdot$ mW^{-1}

1 mM MH in ethanol for at least 12 h (Fig. 1b). The DT/MH-functionalized AgFONs were then placed in an environment-controlled flow cell.

Figure 3 shows example spectra from the different stages of glucose detection on a DT/MH surface. Figure 3a shows the SER spectrum of 1-DT/MH on a AgFON surface. After 2 min incubation in 100 mM aqueous glucose solution, the SER spectrum in Fig. 3b was observed. This spectrum is the superposition of the SER spectra for the partition layer and glucose. Figure 3b clearly shows vibrational features from both the analyte glucose (1462, 1367, 1266 and 1128 cm^{-1}) and 1-DT (1435, 1128, 1071, 1055, 898, and 721 cm^{-1}) constituents. The SERS difference spectrum resulting from subtraction of spectrum 3 A from spectrum 3 B is shown in Fig. 3c. The difference spectrum can be compared directly to the normal Raman spectrum of 4 M glucose solution shown in Fig. 3d.

3 Reversibility and Real-Time Glucose Sensing

An implantable glucose sensor must also be reversible in order to successfully monitor fluctuations in glucose concentration throughout the day. To demonstrate the reversibility of the sensor, the DT/MH-modified AgFON surface was exposed to cycles of 0 and 100 mM aqueous glucose solutions (pH $\sim$ 7) without flushing the sensor between measurements to simulate real-time sensing (Fig. 4 inset). Nitrate was used as an internal standard in all the experiments (1053 cm^{-1} peak) to minimize the effective laser power

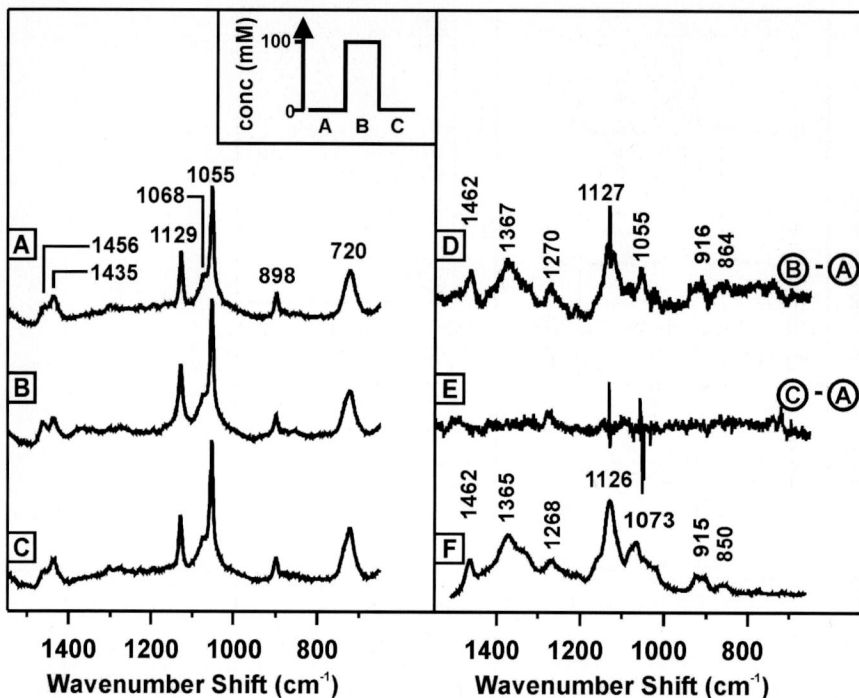

Fig. 4. Glucose pulsing sequence on the DT/MH modified AgFON surface (*inset*). SER spectra of the sample cycled between 0 and 100 mM aqueous glucose solutions (**a**), (**b**), (**c**), $\lambda_{ex} = 532$ nm, $P_{laser} = 10$ mW, $t = 20$ min, pH ~ 7. Normal Raman spectrum of 4 M aqueous glucose solution (**f**). Difference spectra showing partitioning/departitioning of glucose (**d**), (**e**). *a.d.u. $s^{-1} \cdot mW^{-1}$

fluctuations. The $1053\,\mathrm{cm}^{-1}$ band corresponds to a symmetric stretching vibration of NO_3^- and was used to normalize the spectra [27]. SER spectra were collected for each step (Fig. 4a–c).

Figure 4f shows the normal Raman spectrum of a 4 M aqueous glucose solution for comparison. In the normal Raman spectrum of a concentrated aqueous glucose solution, peaks at 1462, 1365, 1268, 1126, 915, and $850\,\mathrm{cm}^{-1}$ correspond to crystalline glucose peaks [14]. The difference spectrum (Fig. 4d) represents partitioning of glucose in DT/MH SAM, which shows the glucose features at 1461, 1371, 1269, 1131, 916, and $864\,\mathrm{cm}^{-1}$. This corresponds to the peaks in the normal Raman spectrum of glucose in aqueous solution (Fig. 4f). The literature has shown that SERS bands can shift up to $25\,\mathrm{cm}^{-1}$ when compared to normal Raman bands of the same analyte [28]. The absence of glucose spectral features in the difference spectrum (Fig. 4e) represents complete departitioning of glucose. The DT/MH mixed SAM presents a completely reversible sensing surface for optimal partitioning and departitioning of glucose.

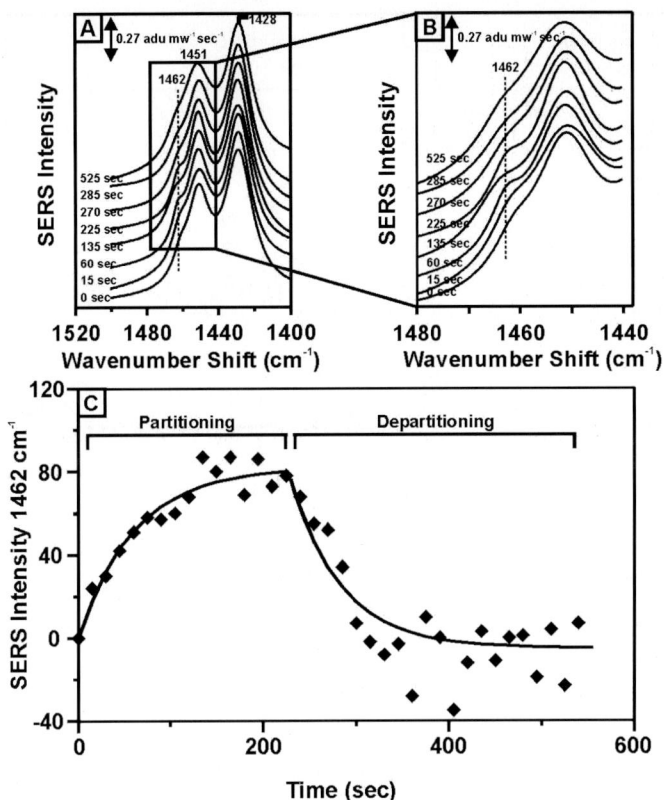

Fig. 5. Real-time SERS response to a step change in glucose concentration in bovine plasma. (**a**) SER spectra of the SAM and glucose at various times. Peaks at 1451 and 1428 cm^{-1} are features of SAM and 1462 cm^{-1} indicates glucose. Glucose was injected at $t = 0$ s, and the cell was flushed with bovine plasma at $t = 225$ s. (**b**) Expanded scale version of (**a**). (**c**) Partitioning and departitioning of glucose. $\lambda_{ex} = 785$ nm, $P_{laser} = 100$ mW, and $t = 15$ s. The 1/e time constants were calculated to be 28 s for partitioning and 25 s for departitioning. Reproduced with permission from Ref. [14]. Copyright 2005 Anal. Chem.

In addition to reversibility, which is an important characteristic for a viable sensor, the sensor should be able to partition and departition glucose on a biologically relevant timescale. The real-time response was examined in two systems: aqueous medium and bovine plasma simulating the *in vivo* environment. To evaluate the real-time response of the sensor, the 1/e time constant for both partitioning and departitioning in aqueous medium was determined to be 8 s. Due to interfering analytes in bovine plasma, the 1/e time constant increased to 28 s for partitioning and 25 s for departitioning (Fig. 5).

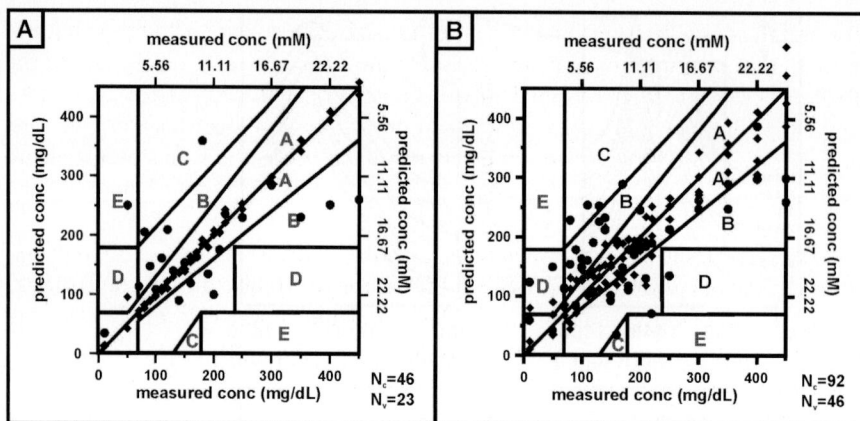

Fig. 6. Calibration (♦) and validation (•) for (a) multiple analyte system and (b) bovine plasma system. (a) PLS calibration plot was constructed and validated over a range of glucose concentrations (10 mg/dl to 450 mg/dl) in 1 mM lactic acid and 2.5 mM urea at pH ∼ 7 with RMSEC = 9.89 mg/dl (0.55 mM) and RMSEP = 92.17 mg/dl (5.12 mM) using 7 loading vectors (λ_{ex} = 785 nm, P_{laser} = 8.4 mW, t = 2 min). (b) PLS calibration was constructed and validated over a range of glucose concentrations (10 mg/dl to 450 mg/dl) in bovine plasma with RMSEC = 34.3 mg/dl (1.9 mM) and RMSEP = 83.16 mg/dl (4.62 mM) using 7 loading vectors (λ_{ex} = 785 nm, P_{laser} = 10 mW to 30 mW, t = 2 min)

4 Quantitative Aspects of Glucose Sensing with SERS

A viable glucose biosensor must be capable of detecting glucose in the physiologically relevant concentration, 10–450 mg/dl (0.56 mM to 25 mM), pH and in the presence of interfering analytes. The DT/MH-modified AgFON surface was exposed to various concentrations of glucose in water (pH ∼ 7) containing lactate (1 mM) and urea (2.5 mM) in physiological concentrations, which are potential interferents for glucose detection (Fig. 6a). Glucose solutions ranging from 10 mg/dl to 450 mg/dl with lactate and urea were randomly introduced in the cell and incubated for 2 min to ensure complete partitioning. SER spectra were collected using two substrates and multiple spots. A calibration model was constructed using partial least-squares leave-one-out (PLS-LOO) analysis with 46 randomly chosen independent spectral measurements of known glucose concentrations (Fig. 6a). Based on the calibration, 95 % of the data is represented by seven latent variables.

The number of latent variables can be interpreted as the inherent dimensionality of the system, in other words, the number of variables present including the concentration of the analyte of interest. These variables can include, and are not limited to, the temperature and humidity conditions in the laboratory on the day of the experiment, the focusing of the optical elements, the enhancement of the sensing surface at different locations, and the laser power and mode fluctuations, as well as noise in the data. Although using too many latent variables can cause overmodeling of the data, including all of the above-mentioned variation in the experimental design is necessary to build a robust calibration model [29]. For example, the training set needs to be able to accurately predict glucose concentrations at more than one temperature to account for thermal fluctuations *in vivo* and to still function if subject movement alters the position of the optical focus.

The use of seven latent variables resulted in a model with a root-mean-square error of calibration (RMSEC) of 9.89 mg/dl (0.549 mM). The RMSEC describes the accuracy of the model itself. Because real-world applications of the sensor will include a number of independent and uncontrolled variables, a versatile mathematical model is needed, and it is reasonable that the dimensionality of the system is large.

A low RMSEC is necessary for, but does not in itself ensure, accurate prediction of concentrations based on measurements from samples outside the training set. Therefore, a separate set of spectra consisting of 23 independent data points was used to validate the model. Validation tests the ability of the model to predict the concentration of samples not used in the calibration, and more precisely reflects the accuracy of the sensor. The root-mean-square error of prediction (RMSEP) was calculated to be 92.17 mg/dl (5.12 mM).

After data analysis using PLS-LOO, the results are presented in a Clarke error grid (Fig. 6). *Clarke* and coworkers established the Clarke error grid as the metric for evaluating glucose-sensor efficacy in the clinical concentration range [30]. The Clarke error grid is divided into five major zones: zone A predictions lead to clinically correct treatment decisions; zone B predictions lead to benign errors or no treatment; zone C predictions lead to overcorrecting acceptable blood glucose concentrations; zone D predictions lead to dangerous failure to detect and treat; and zone E predictions lead to further aggravating abnormal glucose levels. Figure 6a depicts that 98 % of the calibration points and 87 % of the validation points fall in the A and B range of the Clarke error grid.

To transition from the *in vitro* sensor to an *in vivo* sensor, the sensor should also demonstrate quantitative detection in a more complex medium. Bovine plasma was used to simulate the *in vivo* environment of an implantable glucose sensor, which will eventually be implanted under the skin in the interstitial fluid. Prior to use, bovine plasma was passed through a 0.45 µm diameter pore size filter. The filtered plasma was then spiked with glucose concentrations ranging from 10 mg/dl to 450 mg/dl. DT/MH-functionalized AgFON substrates were placed in the flow cell and exposed to glucose-spiked bovine plasma. SER spectra were collected at each concentration using multiple samples and multiple spots in random order to construct a robust calibration model. Calibration was constructed using PLS-LOO analysis described above using seven latent variables and presented in a Clarke error grid (Fig. 6b). To construct the calibration, 92 randomly chosen data points were used, resulting in an RMSEC of 34.3 mg/dl (1.90 mM).

For the validation, 46 data points were used with an RMSEP of 83.16 mg/dl (4.62 mM). In the Clarke error grid, 98 % for calibration and 85 % for validation fall in the A and B range. The errors in both experiments can be reduced by using more data points for the calibration. In addition, error can also be attributed to variation in SERS enhancement at different spots and different substrates [31]. The results show that the DT/MH-modified AgFON glucose sensor is capable of making accurate glucose measurements in the presence of many interferring analytes.

5 Temporal Stability of the SERS Glucose Sensor

An implantable glucose sensor must be stable for at least 3 days [32]. Herein, the stability of the DT/MH-functionalized AgFON surface is studied for 10 days in bovine plasma (Fig. 7). SER spectra were captured every 24 h from three different samples and three spots on each sample. Figure 7a represents the DT/MH spectrum acquired on day 2. Figure 7b shows the average intensity of the 1119 cm^{-1} peak for DT/MH on the AgFON for each day as a function of time. The 1119 cm^{-1} band corresponds to a symmetric stretching vibration of a C–C bond [33]. Only a 2 % change in intensity of the 1119 cm^{-1} peak was observed from the first day to the last day with a standard deviation of 1216 counts, indicating that it did not vary significantly during the 10-day period. This small change in intensity can be attributed to the rearrangement of the SAM during the incubation in bovine plasma [34]. The temporal stability of the 1119 cm^{-1} peak intensity indicates that the DT/MH SAM was intact and well ordered, making this SAM-functionalized surface a potential candidate for an implantable sensor.

6 Conclusions

Since the discovery of SERS nearly thirty years ago, it has progressed from model-system studies of pyridine to state-of-the-art surface-science studies coupled with real-world applications. We have demonstrated a SERS-based glucose sensor as an example of the latter. A SERS-active surface functionalized with a mixed SAM was shown to partition and departition glucose efficiently. The two components of the SAM, DT and MH, provide the appropriate balance of hydrophobic and hydrophilic groups. The DT/MH-functionalized SERS surface partitioned and departitioned glucose in less than 1 min, which indicates that the sensor can be used in real-time, continuous sensing. Furthermore, quantitative glucose measurements, in the physiological concentration range, in a mixture of interfering analytes and in bovine plasma were also demonstrated. Finally, the DT/MH-functionalized SERS surface showed temporal stability for at least 10 days in bovine plasma, making it a potential candidate for implantable sensing.

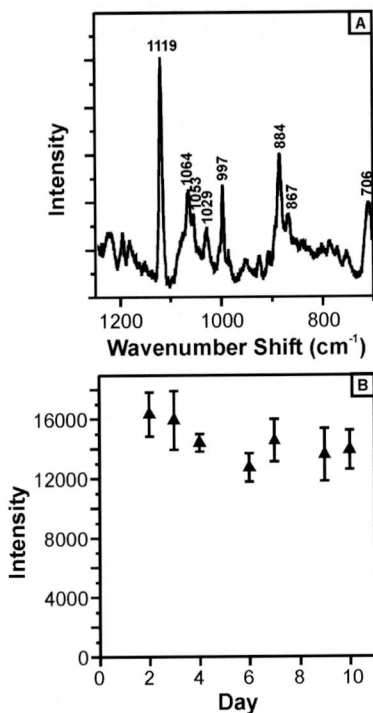

Fig. 7. Stability of the DT/MH-functionalized AgFON. (**a**) SER spectrum of DT/MH-functionalized FON. (**b**) Time course of intensity of the $1119\,\mathrm{cm}^{-1}$ peak. $\lambda_{\mathrm{ex}} = 785\,\mathrm{nm}$, $P_{\mathrm{laser}} = 55\,\mathrm{mW}$, $t = 2\,\mathrm{min}$

Acknowledgements

The authors acknowledge Alyson V. Whitney for making ALD samples. This work was supported by the National Institutes of Health (DK066990-01A1), the US Army Medical Research and Materiel Command (W81XWH-04-1-0630), the National Science Foundation (CHE0414554), and the Air Force Office of Scientific Research MURI program (F49620-02-1-0381).

References

[1] S. A. Ross, E. A. Gulve, M. Wang: Chem. Rev. **104**, 1255 (2004)
[2] E. Boland, T. Monsod, M. Delucia, C. A. Brandt, S. Fernando, W. V. Tamborlane: Diabetes Care **24**, 1858 (2001)
[3] Diabetes programme
URL http://www.who.int/diabetes/en/
[4] A. Heller: Ann. Rev. Biomed. Eng. **1**, 153 (1999)
[5] G. S. Wilson, Y. Hu: Chem. Rev. **100**, 2693 (2002)
[6] S. Park, T. C. Chung, H. C. Kim: Anal. Chem. **75**, 3046 (2003)
[7] J. T. Olesberg, L. Liu, V. V. Zee, M. A. Arnold: *Anal. Chem.* (2005) web released: November 30

[8] J. L. Lambert, J. M. Morookian, S. J. Sirk, M. S. Borchert: J. Raman Spectrosc. **33**, 524 (2002)

[9] A. M. Enejder, T. G. Scecina, J. Oh, M. Hunter, W.-C. Shih, S. Sasic, G. L. Horowitz, M. S. Feld: J. Biomed. Opt. **10**, 031114 (2005)

[10] M. F. Mrozek, M. J. Weaver: Anal. Chem. **74**, 4069 (2002)

[11] K. E. Shafer-Peltier, C. L. Haynes, M. R. Glucksberg, R. P. Van Duyne: J. Am. Chem. Soc. **125**, 588 (2003)

[12] C. R. Yonzon, C. L. Haynes, X. Y. Zhang, J. T. Walsh, R. P. Van Duyne: Anal. Chem. **76**, 78 (2004)

[13] D. A. Stuart, C. R. Yonzon, X. Zhang, O. Lyandres, N. C. Shah, M. R. Glucksberg, J. T. Walsh, R. P. Van Duyne: Anal. Chem. **77**, 4013 (2005)

[14] O. Lyandres, N. C. Shah, C. R. Yonzon, J. T. Walsh, M. R. Glucksberg, R. P. Van Duyne: Anal. Chem. **77**, 6134 (2005)

[15] C. L. Haynes, C. R. Yonzon, X. Zhang, R. P. Van Duyne: J. Raman Spectrosc. **36**, 471 (2005)

[16] S. Soderholm, Y. H. Roos, N. Meinander, M. Hotokka: J. Raman Spectrosc. **30**, 1009 (1999)

[17] R. L. McCreery: *Raman Spectroscopy for Chemical Analysis* (John Wiley & Sons, New York 2000)

[18] J. M. Sylvia, J. A. Janni, J. D. Klein, K. M. Spencer: Anal. Chem. **72**, 5834 (2000)

[19] R. P. Van Duyne, J. C. Hulteen, D. A. Treichel: J. Chem. Phys. **99**, 2101 (1993)

[20] L. A. Dick, A. D. McFarland, C. L. Haynes, R. P. Van Duyne: J. Phys. Chem. B **106**, 853 (2002)

[21] M. Litorja, C. L. Haynes, A. J. Haes, T. R. Jensen, R. P. Van Duyne: J. Phys. Chem. B **105**, 6907 (2001)

[22] X. Zhang, M. A. Young, O. Lyandres, R. P. Van Duyne: J. Am. Chem. Soc. **127**, 4484 (2005)

[23] P. Freunscht, R. P. Van Duyne, S. Schneider: Chem. Phys. Lett. **281**, 372 (1997)

[24] D. Blanco Gomis, J. Muro Tamayo, M. Alonso.: Anal. Chim. Acta **436**, 173 (2001)

[25] B. J. Kennedy, S. Spaeth, M. Dickey, K. T. Carron: J. Phys. Chem. B **103**, 3640 (1999)

[26] M. M. Walczak, C. K. Chung, S. M. Stole, C. A. Widrig, M. D. Porter: J. Am. Chem. Soc. **113**, 2370 (1991)

[27] P. A. Mosier-Boss, S. H. Lieberman: Appl. Spectrosc. **54**, 1126 (2000)

[28] A. M. Stacy, R. P. Van Duyne: Chem. Phys. Lett. **102**, 365 (1983)

[29] K. R. Beebe, R. J. Pell, M. B. Seasholtz: *Chemometrics: A Practical Guide* (Wiley Interscience, New York 1998)

[30] W. L. Clarke, D. Cox, L. A. Gonder-Frederick, W. Carter, S. L. Pohl: Diabetes Care **10**, 622 (1987)

[31] C. L. Haynes, R. P. Van Duyne: J. Phys. Chem. B **107**, 7426 (2003)

[32] F. R. Kaufman, L. C. Gibson, M. Halvorson, S. Carpenter, L. K. Fisher, P. Pitukcheewanont: Diabetes Care **24**, 2030 (2001)

[33] M. A. Bryant, J. E. Pemberton: J. Am. Chem. Soc. **113**, 8284 (1991)

[34] H. A. Biebuyck, C. D. Bain, G. M. Whitesides: Langmuir **10**, 1825 (1994)

Index

Quantitative Surface-Enhanced Resonance Raman Spectroscopy for Analysis

W. E. Smith, K. Faulds, and D. Graham

Centre for Molecular Nanometrology, WestCHEM, Department of Pure and
Applied Chemistry, University of Strathclyde, 295 Cathedral Street,
Glasgow, G1 1XL, Scotland
w.e.smith@strath.ac.uk

1 Introduction

SERS/SERRS has huge potential for use as a detection technique for the development of sensitive and quantitative analytical methods. These methods have been shown to have single-molecule detection capability. The sharp signals obtained are molecularly specific giving easy discrimination from any broad background fluorescence from nonadsorbed material and providing in-situ identification of specific analytes in a mixture.

The technique is not widely used, in part because of uncertainty regarding the nature of the effect and in part because only a few methods have been developed specifically using SERS/SERRS detection. A further problem, until recently, was the cost, complexity, reliability and availability of the equipment. However, the recent huge advances in Raman spectroscopy have made reliable and more affordable spectrometers more widely available.

The continuing debate about the nature of the effect can be a problem for new users of the technique for quantitative analysis. The theory should not matter as long as the effect is reproducible but an understanding would greatly increase confidence. Here we give a practical account of the SERRS effect to aid the analyst to design reliable experiments rather than to understand the fundamentals. Other chapters in this book will deal with the theory.

SERS requires that the analyte is adsorbed on a roughened metal surface usually of silver or gold. Signals can be obtained from molecules on the surface or, with reduced sensitivity, at a short distance away from it (10 Å to 20 Å). The roughness required on the surface is in the nanoscale region and many effective substrates are known, including aggregated colloid, cold-deposited or chemically deposited silver, and suitably roughened electrodes. There are also proprietary substrates that give a more even distribution of the roughness features. The effect of the roughening is to create local features in the surface plasmon and provide a significant component perpendicular to the surface that enables scattering. The nature of this roughness affects the resonant frequencies at which the surface plasmon will absorb or scatter light and for SERS the most effective scattering will be with excitation frequencies near to the resonant frequency [1]. For SERRS, both the range of resonance

K. Kneipp, M. Moskovits, H. Kneipp (Eds.): Surface-Enhanced Raman Scattering – Physics and
Applications, Topics Appl. Phys. **103**, 381–396 (2006)
© Springer-Verlag Berlin Heidelberg 2006

frequencies of the plasmon and the molecular absorption of the dye used as an analyte has to be considered. The usual metals used to obtain SERS or SERRS are silver and gold. These metals have plasmons in the visible region and have a reasonable combination of robustness and chemical activity. In addition, when a photon interacts with a metal surface it can either interact with the surface to cause scattering or it can be absorbed. The ratio of scattering to absorption is dependent on the metal and the frequencies used. In the visible region, this ratio, particularly with blue or green excitation, is more favorable for silver than for gold. Other metals also give SERS/SERRS. However, they are either very reactive or give relatively small enhancements with visible excitation. Therefore, to build good quantitative methods, silver and gold have the right combination of robustness and high activity to make them metals of choice for quantitative method development.

The large enhancements in Raman scattering created by the use of SERS can be substantially improved by using analytes that are dyes to give both a surface enhancement and a molecular resonance enhancement (SERRS). With visible excitation, enhancements for SERS are often quoted as about 10^6 or slightly more whereas enhancements for SERRS are quoted as 10^{14} or 10^{15}. Similar enhancement factors have been reported for SERS with infrared excitation [2]. Thus, either SERS or SERRS can be used to give sensitive quantitative analysis but it is SERRS that should provide the ultimate sensitivity.

SERRS has other advantages for quantitative analysis. SERS signals are sensitive to the orientation of the molecule on the enhancing surface. This is an advantage if a study of the surface chemistry is intended since it can provide additional information on the nature of the surface. However, for quantitation, it is a problem since it can make the signals variable due to chemical effects such as pH and to reorientation due to molecular packing on surfaces with more than one-tenth of a monolayer coverage. Additionally, in some molecules, particularly those with a center of symmetry, new bands appear in the SERS spectra on absorption making it difficult to compare the surface and bulk spectra. Fortunately, both these problems are much reduced with SERRS. Usually, the SERRS signal from the adsorbate has a strong resemblance to the resonance Raman spectra in solution or the solid state, making the species immediately and positively identifiable. A further advantage of SERRS is that the added sensitivity reduces the possibility of interfering signals from adsorbed impurities since they will require both to adsorb and to possess a chromophore reducing the number of effective impurities considerably. This does not exclude interference from a colored impurity but the ready recognition of the spectrum of the preferred analyte normally alerts the operator to the problem immediately. Overall, SERRS is easier to use than SERS in quantitative analysis but, as we shall show in the examples within this Chapter, both can be used effectively.

For the purpose of this Chapter, it is sufficient to regard an effective analyte as adsorbed close to a metal surface and bathed in a cloud of surface electrons that when excited on a roughened surface, cause the electrons in the molecule to undergo a huge polarization change causing SERS/SERRS. We know that the enhancement can be much greater in "hot spots" [3], areas with particularly intense electric fields such as the point of contact or near contact of two colloidal particles. There is therefore a problem with SERRS in detection at the single-molecule level in that one molecule could be much more enhanced than another.

2 Experimental Approach

2.1 Choice of Wavelength

For SERS, the wavelength should be chosen to excite the surface plasmon. In some cases this is not important since the plasmons are very broad. Examples of these include surfaces made by cold or chemical deposition of metal from the vapor phase, or from aggregated colloid with a wide range of particle and cluster sizes. Other surfaces made by deposition in a controlled way onto structured substrates do require more consideration of the wavelength. Figure 1 shows the absorption spectrum of silver colloid that is close to monodisperse. It shows the profile wavelength dependence of the absorption of light by the surface plasmon for unaggregated colloid. This is chosen as the example of a substrate because the plasmon spans a relatively narrow range of wavelengths compared to many solid-state substrates or aggregated colloid and so the effects of wavelength dependence are more critical. The scattering depends on both the absorption and emission frequencies but these are relatively close together in terms of the wavelength ranges discussed here and so they are not differentiated. For unaggregated colloid, excitation close to the peak will give the most effective results.

It is possible to aggregate the colloid in a controlled manner to bring particles together. This shifts the plasmon absorption to the red and broadens it. It can be inferred from studies of immobilized colloidal clusters that this process may create hot spots due to the interaction between particles and this will increase the scattering very significantly [4]. Often, external aggregating agents such as sodium chloride are used to do this. Zeta-potential measurements indicate that this reduces the charge on the surface making the colloidal particles less stable and causing aggregation. The result of this is that a series of different-sized clusters are formed, each with a different absorption maximum and bandwidth. The absorption spectrum is now a sum of all these giving a much broader redshifted spectrum compared to the unaggregated colloid. It is likely that some of these clusters will be in resonance with any excitation wavelength chosen in the visible region. In this situation, the wavelength chosen is less critical but the actual enhancement depends on

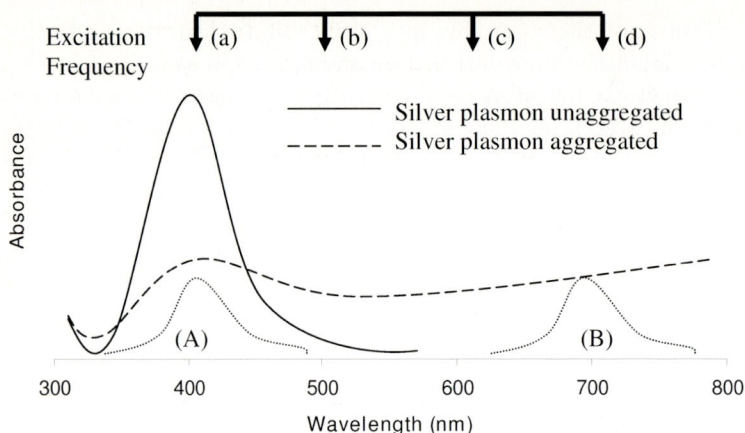

Fig. 1. The plasmon-resonance absorption profile of unaggregated silver colloid together with four possible choices of excitation frequency. Laser excitation (**a**) is ideal for SERS with unaggregated colloid. On aggregation, the plasmon resonance moves to the red and broadens so (**b**), (**c**) and (**d**) will give effective SERS. The absorption spectrum of two dyes, one with an absorption maximum at 400 nm (A) and the other with an absorption maximum at 700 nm (B) are shown. For SERRS, (**a**) is ideal for an unaggregated colloid with the 400 nm dye and (**d**) should be effective for the 700 nm dye if an aggregated colloid is used. Diagram reproduced with modification from [5]

the number of clusters present at the wavelength chosen. In SERRS, the situation is somewhat more complex since both the plasmon resonance and the molecular resonance contribute. Figure 1 shows examples of possible choices of excitation frequencies for a dye that absorbs at 400 nm and one that absorbs at 700 nm.

At low concentrations of analyte with no added aggregating agent, where the amount of analyte is insufficient to cause appreciable aggregation, the molecular-resonance contribution from the dye is crucial. Figure 2 shows the intensity dependence on wavelength for the largest peak in the SERRS spectrum of two dyes and the molar absorptivity at these dyes at the excitation wavelengths [5]. The two plots are similar. Only enough dye was added to reach about one hundredth of a monolayer in coverage. There was no change in the zeta-potential during this process, indicating that the colloid is stable and that there is little to no aggregation.

By contrast, when more dye is added to cause aggregation by altering the surface charge of the colloidal particles, the relationship to the absorption spectrum is lost and a much broader profile is obtained. Note also that the enhancement 50–100 times greater indicating that particle–particle interactions cause most of the enhancement.

Thus, with SERRS and without the application of very intense local fields caused by aggregating the colloid, the frequency chosen should match that of

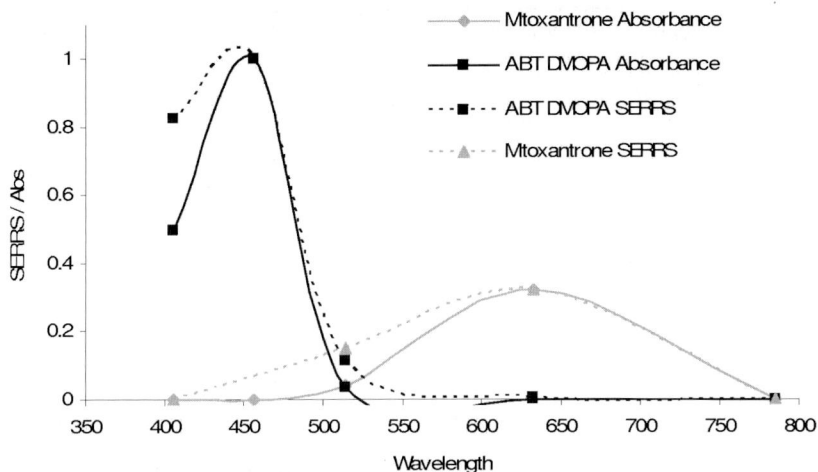

Fig. 2. Wavelength dependence of SERRS intensities and the molar absorptivity for the azo dye ABT DMOPA (dye B in Fig. 3) and the drug mitoxantrone at the excitation frequencies chosen. There is a close dependence that is completely lost upon aggregation, when the SERRS enhancement covers the whole frequency range with a maximum at about 500 nm. Reproduced from [5]

the absorption peak. However, when a colloid is aggregated, and this is the condition under which the greatest enhancement will be obtained, there is no clear dependence on the absorption maximum. Excitation frequencies in the middle of the visible region will work well with this colloid and these analytes. The effective frequencies may well be different for other dyes, other sizes of silver particles, or particles from another metal such as gold. In general, the use of aggregated colloid for analytical procedures can be advantageous since there is a greater enhancement and the dependence of the scattering intensity on the wavelength of excitation is less pronounced.

2.2 Types of Assay

SERRS/SERS enhancements arise from nanoscale features on a roughened surface or on a particle or cluster of particles. For many systems, particularly if "hot spots" created by nanofeatures such as particle–particle interactions are involved, the scattering intensity from single nanoscale events is difficult to reproduce quantitatively and methods that average over many nanoscale events are likely to be more effective. Fortunately, with standard Raman spectrometers the optical limit of resolution is far larger than the nanoscale features so that averaging is inherent in many measurements.

In colloidal suspensions, either aggregated or unaggregated, many thousands of nanoparticles will exist within the smallest interrogation volume possible using a Raman microscope and they are constantly moving due to

Brownian motion. In addition, the Raman process is very fast with the result that with a 1 s accumulation time, signals from many different nanoparticles may be accumulated and many opportunities are available for the Raman scattering event to occur on any one particle. Thus, a large number of signals from individual scattering events can be averaged, enabling quantitative detection at very low levels in a simple and reliable manner.

Another feature of SERRS enhancement, which is crucial for its use as a quantitative detection technique, is that once the dye is adsorbed on the silver surface, fluorescence is effectively quenched. The nature of this process is not often discussed but it probably involves the reduction in lifetime of the fluorophore due to energy transfer from the dye-excited state to the plasmon or metal surface. This process can be very effective. In addition, SERRS is effective with nonfluorescing as well as fluorescing dyes allowing for a wider and more effective labeling chemistry. However, it must be remembered that only fluorophores adsorbed on or close to the surface will be quenched and fluorescence from nonadsorbed material such as plasma or other biological media can still be a problem.

2.3 Choice of Analytes

For the development of a sensitive and quantitative procedure, it is preferable that the analyte should adhere strongly to the surface. This is not essential, and an example of a procedure using a weakly adsorbing analyte is given later, but such procedures are more prone to error and interference and the sensitivity is reduced. Some analytes adsorb very effectively without the need for any modification. For example, the anticancer agent mitoxantrone adheres strongly to silver and is used later in this Chapter as an example of an effective analytical procedure [6].

One great advantage of SERRS assays is that there is very little interference from other molecules. For example, diluted blood plasma can be used directly with the adsorbing proteins and the nonadsorbing material contributing very little to the spectra against the highly effective Raman scattering that can be obtained from an effective SERRS chromophore such as mitoxantrone. However, this very high selectivity illustrates one of the reasons why overexpectations from SERRS analysis can occur. In the case of mitoxantrone, no other molecule in the plasma with any appreciable concentration gave strong enough SERRS/SERS to compete with mitoxantrone despite the fact that they were present in much higher concentrations. Thus, SERS/SERRS will be very effective where a strongly adhering and strongly scattering molecule is the analyte of interest. It will be very difficult to use where a weakly absorbing and weakly scattering molecule is the analyte. In the latter case, the recommendation in this Chapter is that derivitization of the analyte or modification of the surface is carried out to provide good surface attachment.

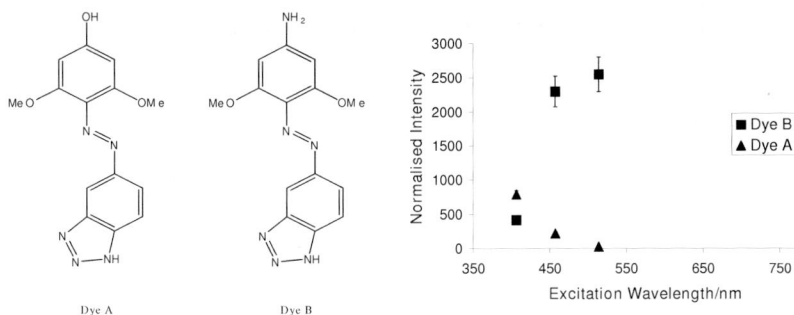

Fig. 3. Wavelength dependence of SERRS for two dyes adsorbed onto the surface of silver colloid for which the plasmon absorption spectrum is given in Fig. 1. Dye A does not cause aggregation and dye B causes aggregation above 1/10 monolayer coverage giving greater enhancement at longer wavelengths. Reproduced from [12]

SERRS has great potential for the development of methods that label the analyte to obtain sensitive and selective detection, a common approach in fluorescence assays. The same advantages of sensitivity and selectivity are obtained as is the case for fluorescence but with added advantages. The labels are more molecularly specific, making for easier in-situ identification of the label and they are less sensitive to interlabel quenching than fluorophores. A series of dyes with functional groups has been designed specifically for SERRS using azo chemistry. All contain, either a benzotriazole group [7, 8, 9], an 8-hydroxy quinoline [10] or some similar grouping to complex to the surface. These dyes are designed for silver and are based on the likelihood that surface complexing will occur, probably to silver (I) and silver (II) ions present on the surface. Similar strategies for gold use complexing groups such as thiols for preference.

Figure 3 shows two of these dyes. Both are likely to replace the organic layer on the surface [11] and create a new surface complex by complexing to the silver ions on the surface through more than one of the nitrogen groups in the triazole ring. However, one dye (dye A) has a phenolic group that will ionize at neutral pHs to provide a negative charge on the surface. Since the surface of many of the colloids used is negative, when the dye adsorbs, the surface remains negatively charged and therefore no aggregation occurs. The other dye (dye B) on complexing will reduce the charge on the silver and cause aggregation. The effect of this can be seen in Fig. 3. Thus, the largest signals from dye A will be achieved with excitation at the frequency of the plasmon resonance of the unaggregated colloid (the absorption maxima of the dye is close to that of colloid). Dye B will give large enhancements with longer-wavelength excitation and the greatest enhancement will be where the largest number of effective clusters is present [12].

An extension of this approach is to make bifunctional dyes. Dye C shown in Fig. 4 contains both a benzotriazole group to complex to the surface and

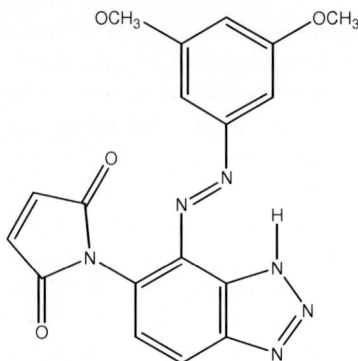

Fig. 4. Structure of a bifunctional dye with a benzotriazole group to bind to a silver surface and a maleimide group to attach to other molecules such as proteins

a maleimide group to attach to molecules such as thiols, or biomolecules such as proteins, antibodies and DNA [13]. In this way it can attach the protein to the surface and also provide a label for that attachment. A very similar method has been used to label rhodamine to provide more effective attachment for that dye to the surface [14].

For many types of analysis, chemical derivatization of the analyte is not an effective procedure. Positively charged molecules often adhere strongly to negatively charged surfaces in any case, but for some other molecules, special surfaces may be required. One effective surface used a particular combination of SAMs (surface-assembled monolayers) to adsorb glucose efficiently and create an effective SERRS assay. Two SAMs of different lengths were adsorbed on the surface and functionalities added to provide a specific surface on which the glucose could adsorb effectively [15]. This process is more general and could be extended to other molecules that cannot be easily derivatized.

2.4 Choice of Substrates

There are many ways of making effective substrates for quantitative analysis including lithography, colloid, and roughened electrodes. A full evaluation is outside the scope of this Chapter. Rowlen and coworkers evaluated several surfaces that are effective for SERRS/SERS [16]. The main advantages of these substrates are that they are simple, cheap and effective. The effectiveness of each substrate depends on properties such as the lifetime of the film, the robustness of the film, and the evenness of the enhancement across it. One recently reported film contained spin-coated TiO_2 in a polymer with silver nitrate added [17]. It is activated by UV light when the silver grows as silver metal to produce a roughened surface on top of the TiO_2. The film is quite robust and can stand up to a standard "sellotape" test. It also has the advantage that it can be activated at the time of use. In many cases these films have "hot spots" and surfaces that give very high SERS/SERRS intensities can often prove to give irreproducible signals when a microscope is used for interrogation since only very small areas are observed on the surface. Another problem with many of these systems is that simple delivery of samples to the point of analysis such as spotting the sample onto the surface, and drying it, does not provide control of the coverage of the surface by the analyte. The sample tends to form a ring as it dries out and the signal intensity across the surface is uneven. There are simple methods to overcome this. For example, a larger beam area can be used to cover a statistically meaningful part of the surface, the sample can be spun to average the roughness or it may be possible to measure the spectrum without drying the sample out.

Specifically designed surfaces can give good enhancement and very good sensitivity. Electron lithography is one effective technique. The use of microscale beads to form a pattern on the surface and then coating the beads with metal produces effective surfaces. The removal of the beads leaves reproducible pattered surfaces that make an effective and reliable substrate. These surfaces are compatible with microfluidics or robotics and this could improve reliable sample delivery.

A widely used alternative to solid-state substrates are colloidal suspensions of silver or gold. Correctly prepared, they can be close to monodisperse and will last for many years (Faraday's original gold colloid is still viable and is held at the Royal Institute in London). The most widely used colloids are made by reaction of silver nitrate with borohydride [18, 19], citrate [20] or EDTA to reduce the silver. One problem with evaluating much of the literature on colloids is that many papers do not provide information on particle size or charge. Often, the UV visible spectrum is given instead but the relationship to colloid particle size, size distribution and shape is complex. For example one rule of thumb for Lee–Meisel silver colloid is that about 30 nm diameter particles give an absorption maximum at about 400 nm to 406 nm and the bandwidth of the colloid should be about 60 nm if the suspension is close to monodisperse [11]. However, in-depth studies of this type of colloid indicate that even if these criteria are met, the colloid may contain an appreciable number of needles and a range of particle sizes. Thus, the UV-visible spectrum, although a useful way of characterizing a colloid quickly, is not sufficient for proper characterization. Colloidal suspensions with a wide range of particle sizes will give good SERRS but usually they have a maximum absorbance further into the visible region. The main advantage of preparing a colloid that has a narrow bandwidth and a band as far towards the ultraviolet as possible, is that this specification can be made more reproducibly and this helps with regard to developing quantitative methods.

3 Examples of Analytical Methods

3.1 DNA

DNA analysis is widely used in molecular biology and fluorescent labeling is the most commonly used method to analyze a specific molecule at high sensitivity in the presence of other molecules. Recently, a simple procedure to use SERRS to obtain similar advantages has been developed. To make this procedure effective, and to obtain strong adsorption onto the metal surface, the DNA has to be either labeled with a positively charged dye, such as rhodamine, which will adhere to SERRS-active substrates, or if a negatively charged dye label is used, bases modified with propargyl amine groups are incorporated close to the label to provide a surface-attachment group. In the latter case, these groups that are positively charged at neutral pH and adhere strongly to the surface pulling the label into proximity with it whatever the charge on the label [21].

However, these methods in themselves were not found to be sufficiently reliable for good quantitative application. A second development was required to make the method effective. As already discussed, the commonly used silver colloids are negatively charged and so is DNA due to the phosphate groups. To reduce the charge and cause controlled aggregation, one of the charge-neutralization agents known to be very effective with DNA, spermine, a poly amine, was added as an aggregating agent. When this was done, the DNA assay was extremely successful. Twelve commercially available labels have now been investigated. All gave quantitative behavior over a few orders of magnitude with very sensitive detection limits [22, 23].

The potential for quantitative assay development using SERRS in DNA analysis is shown in Fig. 5. Theoretically, SERRS and fluorescence should have approximately the same enhancement per molecule. However, the techniques differ greatly. In SERRS, the effective labels are attached to the metal surface, whereas for fluorescence, the labels are distributed in solution throughout the sample. Further, higher power densities can be used with SERRS than with fluorescence before photodegradation of the sample is caused. To compare the relative sensitivities of the two techniques in practice, two standard, state-of-the-art, DNA analyzers that use fluorescence detection and a Renishaw Raman microprobe were used in separate concentration-dependent studies with the same set of eight DNA oligonucleotides. They were labeled with fluorophores and deisgned to be effective for SERRS by adding propargyl amine groups where necessary. The results show a huge advantage for SERRS with lower detection limits by a factor of at least 10^3 in all cases except one [24]. For one label, SERRS was still more effective than fluorescence but the advantage was much smaller. This was where an infrared fluorophore was used with visible excitation. Effectively, this label is a SERS label and consequently the sensitivity is reduced. Thus, correctly designed labels with SERRS are more sensitive than the equivalent fluorescence labels in practise.

A major advantage of SERRS over fluorescence is that the signals from SERRS are more molecularly specific and much sharper. This provides much better potential for simultaneous detection of multiple analytes since more dyes can be individually identified within a mixture. The combination of multiple detection and sensitivity makes SERRS a method of choice for the de-

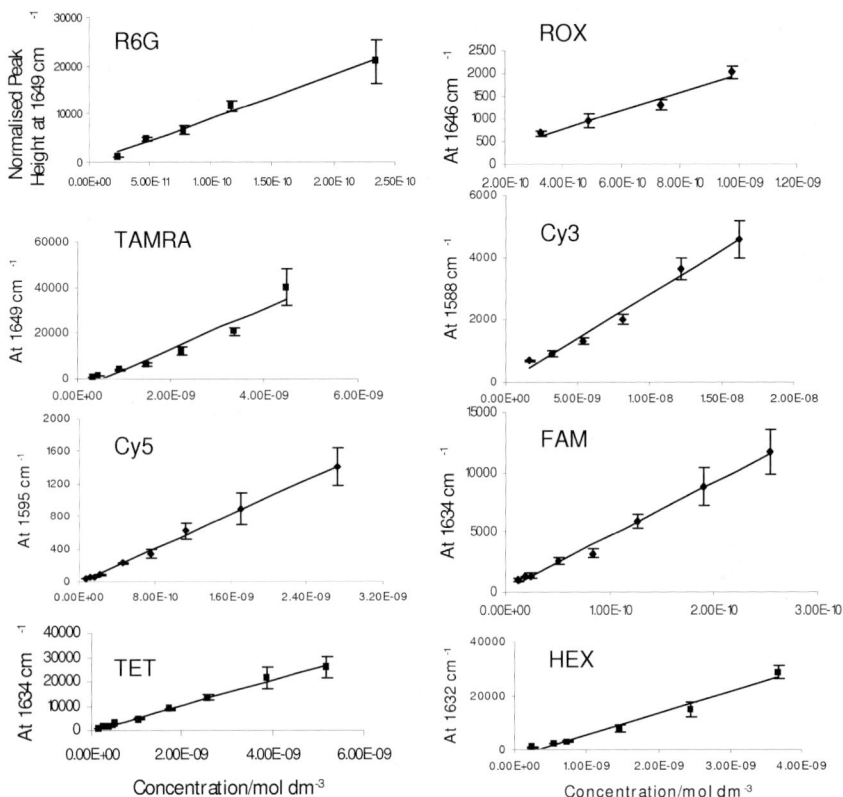

Fig. 5. The concentration dependence of the intensity of the major peak in the SERRS spectrum for an oligonucleotide labeled with 8 different labels. With the large enhancements obtained from SERRS, only peaks from the labels are observed

velopment of more advanced molecular biology detection techniques. SERRS has previously been used for the multiplex genotyping of the mutational status of the cystic fibrosis gene using two different labels [25]. Another format that has been used is that of lab-on-a-chip. In this example microfluidics chips were generated from PDMS and a DNA sequence labeled with a fluorophore, spermine and silver nanoparticles were introduced into the chip, the SERRS signals were measured at a point further down the channel [26]. The microfluidic chip allowed detection of three different DNA sequences corresponding to different variants of the *E-coli* bacterium. This indicates that DNA detection by SERRS can be carried out in a number of different ways that are compatible with modern molecular biological assay formats.

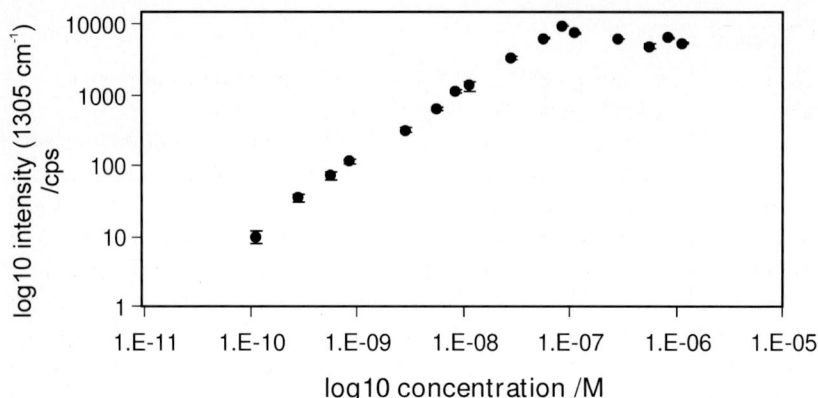

Fig. 6. Concentration dependence of the most intense peak in the SERRS spectrum of mitoxantrone. The flat region at high concentration is caused by saturation of the available silver surface with dye. Reproduced from [6]

3.2 Mitoxantrone

Mitoxantrone is a clinical drug used in the treatment of breast cancer. It is toxic and requires to be given at the correct dose. Instant feedback to obtain the level of mitoxantrone in an individual patient's blood would be of considerable value. However, the standard analytical method takes approximately 4 h. It requires the separation of the mitoxantrone from the plasma followed by HPLC analysis. As already discussed, mitoxantrone is a particularly effective drug for use with SERRS detection since it sticks readily to silver surfaces and gives strong signals. Using a flowcell to aid quantitation, a drop of plasma from a patient treated with mitoxantrone was determined by SERRS [6]. The signal from mitoxantrone could easily be detected in the presence of the plasma. The reason this works is that the flowcell dilutes out the plasma, removing some of the fluorescent background from proteins that do not absorb onto the silver surface and mitoxantrone is such a strong complexing agent for silver surfaces it complexes to the surface in the presence of adhering proteins. Thus, the mitoxantrone signal is enhanced and other interfering signals are very weak. A concentration-dependent plot of mitoxantrone covering the clinical relevant concentration region and going up to the point at which monolayer coverage of mitoxantrone is obtained, and the signal stops increasing is shown in Fig. 6. For this particular drug, this is a very effective and fast analysis taking under two minutes with a 10 s accumulation time. The main problem is that it will only work for those drugs that are effective SERRS labels, making it easy to oversell the method as a general technique for the analysis of drugs.

3.3 Drugs of Abuse

Examples so far are extremely sensitive but require labeling chemistry to create the selective enhancement. However, methods that use lower sensitivities may also be effective in some cases. Drugs of abuse could be modified to provide SERRS, however, labeling chemistry would need to be carried out on the actual sample causing time delays and complexity in analysis. There is a need in this particular area for simplicity, making it possible to work in the field with handheld equipment therefore analysis must rely on the drug adhering to the metal surface. The detection of amphetamine sulfate using SERS from various substrates including silver and gold colloid and vapor-deposited metal films has been reported [27]. This study highlighted the importance of ensuring the correct metal substrate and conditions are used to obtain SERS signals from analytes that have poor absorption to metal surfaces. Better SERS signals were obtained from gold colloid compared to silver colloid as long as the correct aggregating agent was used, suggesting that amphetamine adsorbs more easily onto gold rather than silver surfaces. Lower detection limits could be obtained from gold vapor-deposited films compared to silver vapor-deposited films. These are interesting assays because the drugs do not adhere well to the surfaces used. The drug was incubated with gold colloid and the gold removed by centrifugation. The supernatant was resuspended in fresh colloid and the signal was only slightly reduced. This suggests that only a small percentage of the drug is adhering onto the surface at any one time. However, the results are quantitative over limited ranges but in this range relatively effective detection is obtained. Signals were obtained from amphetamine that allowed the drug to be detected down to concentrations between $10^{-5}\,\mathrm{mol\,dm^{-3}}$ to $10^{-6}\,\mathrm{mol\,dm^{-3}}$.

3.4 Glucose

Glucose is a good example of a molecule for which there is a need for effective and fast methods of quantitative analysis. However, it does not adhere effectively to most standard substrates and as a result few SERS spectra had been reported till recently. However, a method has been developed to enable SERS to be obtained by forming a partition layer of SAMs on a carefully constructed silver substrate [15]. This layer captures the glucose and holds it on the surface in the electromagnetic field giving glucose spectra with significant enhancement. Quantitative results were obtained using a form of partial least squares to analyze the data and the root-mean-squared error of prediction is close to that required clinically. Developments of this type will greatly extend the range of analytes for which SERS/SERRS can be used quantitatively.

3.5 Derivatization Assays

The use of SERRS labels for DNA detection is one example of SERRS labeling of biomolecules. However, it is sometimes possible to derivatize a small

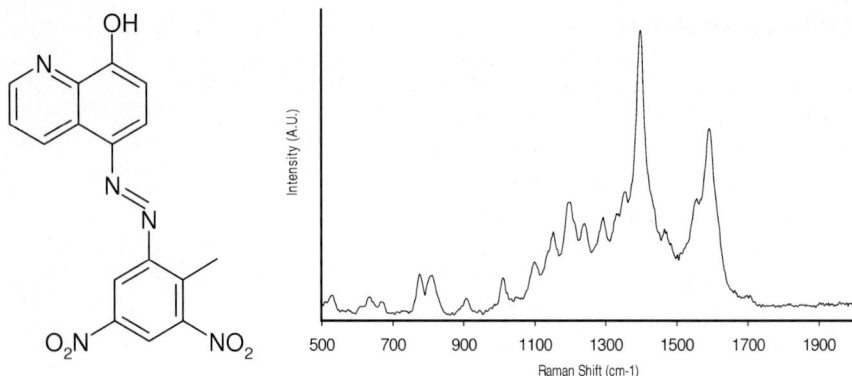

Fig. 7. A dye made by reducing and derivatizing TNT in a flow system. The use of the flow system enabled quantitative results to be obtained. Features of the parent molecule are still present and they affect the spectrum, giving confirmation that it was reaction with TNT that produced the SERRS spectrum

analyte to achieve an effective SERRS chromophore with good surface adhering characteristics and an effective chromophore. SERRS has an advantage for this type of work in that the specific nature of SERRS often enables spectral features of the original analyte to be identified in the derivatized product, reducing the chance of reaction between the wrong substances passing unnoticed. The requirements for detection of explosives are such that a very high degree of sensitivity is essential and methods such as those used with drugs of abuse will not be sufficiently effective for many targets. Therefore, a chemistry to derivatize explosives such as TNT, RDX and PETN, the components of Semtex, has been developed [28]. Figure 7 shows the spectrum of a derivatized TNT molecule that, in an assay using a flowcell, gave a detection limit in the picogram region.

A device has been made for the detection of TNT using a different chemistry where sodium hydroxide was added and the molecule dropped onto a gold surface. The drawback of SERRS detection is that a different chemistry may be required if the target molecule changes. Compare this to a mass spectrum that can measure all of a range of samples. However, under certain circumstances, for example if microfluidics are used and the chemistry is optimized, the detection limits for SERRS may be lower than those for mass spec, allowing low vapor pressure explosives such as RDX to be detected [29].

4 Analysis Using Individual Nanoparticles

So far, the methods described have used a suspension of colloid or a substrate to obtain a measurement that covers a wide number of individual SERRS events. However, in principle one of the best uses of SERRS would be at the single-nanoparticle level. Considerable progress has been made in this direction and individual nanoparticle labeling is now well established. One way to label these single nanoparticles effectively is to use a silica coating to trap the dye onto the surface of the particle. This appears to provide effective labeling in biological media. The problem with single-particle analysis in this particular field is the difficulty to obtain reliable and reproducible signals per nanoparticle [30]. For labeling of cellular events and for tracing particular antibodies, etc. into an animal, these particles can be attached to biomolecules. They can then be recognized in situ through the metal nanoparticle itself and the individual SERRS signal can then be detected. Another use of antibodies labeled in this way is in detection systems for diagnostics and it is anticipated that more quantitative methods for specific targets will be developed in this field.

5 Summary

In this Chapter, we adhere to a basic philosophy for obtaining quantitative SERRS. Effective quantitative SERRS is best obtained where the analyte is preselected either by choosing a suitable molecule such as mitoxantrone that has an intrinsic SERRS activity or by the use of labeling/derivatization chemistry. For the best results, the molecule should adhere strongly to the SERRS surface and have a chromophore close to the excitation frequency. Quantitation is easily obtained by averaging a number of events such as those that will occur if a suspension of colloid is used. The advantages in sensitivity and selectivity of SERRS detection make it very attractive for the next generation of bioanalytical equipment and for the solution to problems where extreme sensitivity is required.

References

[1] S. Nie, S. E. Emory: Science **275**, 1102 (1997)
[2] K. Kneipp, Y. Wang, H. Kneipp, L. T. Perelman, I. Itzkan, R. R. Dasari, M. S. Feld: Phys. Rev. Lett. **78**, 1667 (1997)
[3] V. A. Markel, V. M. Shalaev, P. Zhang, W. Huuynh, L. Tay, T. L. Haslett, M. Moskovits: Phys. Rev. B **59**, 10903 (1999)
[4] H. Xu, E. J. Bjerneld, M. Kall, L. Borjesson: Phys. Rev. Lett. **83**, 4357 (1999)
[5] D. Cunningham, R. E. Littleford, W. E. Smith, P. J. Lundahl, I. Khan, D. W. McComb, D. Graham, N. Laforest: Faraday Discuss. **132**, 111 (2005)
[6] C. McLaughlin, D. MacMillan, C. McCardle, W. E. Smith: Anal. Chem. **74**, 3160 (2002)
[7] G. McAnally, C. McLaughlin, R. Brown, D. C. Robson, K. Faulds, D. R. Tackley, W. E. Smith, D. Graham: Analyst **127**, 838 (2002)
[8] L. Fruk, A. Grondin, W. E. Smith, D. Graham: Chem. Commun. **18**, 2100 (2002)

[9] D. Graham, L. Fruk, W. E. Smith: Analyst **128(6)**, 692 (2003)
[10] C. McHugh: PhD Thesis, University of Strathclyde (2002)
[11] C. H. Munro, W. E. Smith, M. Garner, J. Clarkson, P. C. White: Langmuir **11**, 3712 (1995)
[12] K. Faulds, R. E. Littleford, D. Graham, G. Dent, W. E. Smith: Anal. Chem. **76**, 592 (2004)
[13] L. Fruk, D. Graham: Heterocycles **60**, 2305 (2003)
[14] A. F. McCabe, D. Graham, D. McKeown, W. E. Smith: J. Raman Spectrosc. **36**, 45 (2005)
[15] O. Lyandres, N. C. Shah, C. R. Yonzon, J. T. Walsh, M. R. Glucksberg, R. P. Van Duyne: Anal. Chem. **77(19)**, 6134 (2005)
[16] K. L. Norrod, L. M. Sudnik, D. Rousell, K. L. Rowlen: Appl. Spectrosc. **51**, 994 (1997)
[17] A. Mills, G. Hill, M. Stewart, D. Graham, W. E. Smith, S. Hodgen, P. J. Halfpenny, K. Faulds, P. Robertson: Appl. Spectrosc. **58**, 922 (2004)
[18] K. Cermakova, O. Sestak, P. Matejka, V. Baumruk, B. Vlckova: Collect. Czech. Chem. Commun. **58**, 2682 (1993)
[19] R. Keir, D. Sadler, W. E. Smith: Appl. Spectrosc. **56**, 551 (2002)
[20] P. C. Lee, D. Meisel: J. Phys. Chem. **86**, 3391 (1982)
[21] D. Graham, W. E. Smith, A. D. T. Linacre, C. H. Munro, N. D. Watson, P. C. White: Anal. Chem. **9**, 4703 (1997)
[22] K. Faulds, D. Graham, W. E. Smith: Anal. Chem. **76**, 412 (2004)
[23] K. Faulds, L. Stewart, W. E. Smith, D. Graham: Talanta **67**, 667 (2005)
[24] K. Faulds, R. P. Barbagallo, J. T. Keer, W. E. Smith, D. Graham: Analyst **129**, 567 (2004)
[25] D. Graham, B. J. Mallinder, D. Whitcombe, W. E. Smith: Chem. Phys. Chem. **12**, 746 (2001)
[26] F. T. Docherty, P. B. Monaghan, R. Keir, D. Graham, W. E. Smith, J. M. Cooper: Chem. Commun. **118** (2004)
[27] K. Faulds, W. E. Smith, D. Graham, R. J. Lacey: Analyst **127**, 282 (2002)
[28] C. J. McHugh, W. E. Smith, R. J. Lacey, D. Graham: Chem. Commun. **21**, 2514 (2002)
[29] R. Keir, E. Igata, M. Arundell, W. E. Smith, D. Graham, C. McHugh, J. M. Cooper: Anal. Chem. **74**, 1503 (2002)
[30] S. P. Mulvaney, M. D. Musick, C. D. Keating, M. J. Natan: Langmuir **19**, 4784 (2003)

Index

Rapid Analysis of Microbiological Systems Using SERS

Roger Jarvis[1], Sarah Clarke[2], and Royston Goodacre[1]

[1] The University of Manchester, School of Chemistry, P.O. Box 88, Sackville Street, Manchester, M60 1QD
[2] Institute of Biological Sciences, Edward Llwyd Building, The University of Wales Aberystwyth, Ceredigion, SY23 3DA
Roy.Goodacre@manchester.ac.uk

1 Introduction

Research in the biological sciences field is now moving forward at an unprecedented rate. Modern technological advances, particularly computational processing power, but also modern analytical methods, have been embraced by the life sciences. In combination these have developed to such an extent that it is now possible to begin to investigate complex systems using powerful techniques that in the past would most likely have been the preserve of the analytical chemist. There is still a wealth of highly challenging problems in biology that need to be solved, and this will necessitate the continual evolution of novel analytical strategies.

Genotyping technology revolutionized the study of biological systems; by using the polymerase chain reaction to amplify genetic material extracted from a cell, gel electrophoresis and more recently multiplex genotyping could be performed to map genetic sequences, and observe genetic differences between organisms. However, even this highly qualitative approach has yielded many more questions than it actually answers, since the function of genes of interest cannot normally be directly inferred from the "static" sequence blueprint. Thus, within the areas of functional genomics and systems biology many studies are actually aimed at investigating the organism's phenotype directly. The phenotype is defined as the result of the expressed genotype of the organism and is influenced by the environment which it inhabits. Whilst genetic analysis can be enlightening, it is often not as informative as phenotypic information, and because of this, in microbial research biochemical analyses have been adopted to measure quantitatively cellular components such as proteins and metabolites. However, more recently great steps have been made to introduce rapid spectroscopic approaches to study the phenotype.

So, exactly what are biologists interested in understanding about microbial systems? Broad areas of interest include: microbial classification and identification; mRNA, protein and metabolic profiling; functional genomics; systems biology; optimization, regulation and understanding of microbial bioprocesses; and more specifically with increased incidence of antibiotic resis-

K. Kneipp, M. Moskovits, H. Kneipp (Eds.): Surface-Enhanced Raman Scattering – Physics and Applications, Topics Appl. Phys. **103**, 397–408 (2006)

tance, the mode of action of hopefully antimicrobial drugs. To extract the maximum information from these experiments the best approach is to employ global analysis tools; which often indicates spectroscopic measurements, from which relevant knowledge can be extracted. This data-mining process uses mathematical tools to ask questions of the data that have much lower dimensionality than the spectral data and typically involve quantitative or categorical modeling. For example, a question might be; "based on the Raman fingerprint I have generated from this group of bacteria, which bands are discriminatory and allow objective identification?" In addition, many spectroscopic methods provide a real insight into the biochemistry of an organism by conventional interpretation of spectral bands, and these data are far more easily and rapidly obtained through this route as opposed to that of the traditional biochemical approach.

There are a wide range of spectroscopic methods encompassing the vibrational, nuclear magnetic resonance and hyphenated mass spectrometries that have been used to great effect on a range of biological problems. The purpose of this Chapter is to introduce some (micro-)biological applications for which surface-enhanced Raman scattering (SERS) has been recently employed. SERS is emerging as a very powerful tool in the biological sciences due to its excellent sensitivity and ability to quench fluorescence, which can plague Raman measurements of biological material excited in the visible to the near-infrared. Specifically we shall discuss, SERS for the characterization and identification of micro-organisms, the monitoring of industrial bioprocesses and finally, gene-function analysis.

2 Spectroscopic Characterization of Micro-Organisms

Traditionally, the task of classifying microbes has been performed by a comparison of macro- and micromorphological characteristics or by biochemical tests. In more recent times genomic analysis has been used as a means of identification or classification and 16S ribosomal RNA sequencing is now the "gold" standard used for this task [1]. However, there are drawbacks associated with all of these approaches. Naturally, comparison of morphology is not a wholly reliable means of classification; this particular phenotype of an organism can be repeatedly expressed amongst biochemically diverse species and therefore is not on its own a suitable means of differentiating between bacteria [2]. The API system (http://www.biomerieux.com/) is a popular biochemical method used in routine laboratory analysis, approximately 2000 research publications (since the early 1990s) refer to the use of API. With API, a series of biochemical tests is applied to an organism cultured in the laboratory, the response of the organism to these tests is then matched against a database of possible results to provide identification. Whilst biochemical detection methods are reliable, they do not always provide conclusive decisions at the species level, whilst strain-level characterization is generally impossible and analysis of environmental isolates uncertain. The process itself is also very time consuming; cell culturing, running the test and analysis of results often require several days. The analysis of rDNA also has the same issues since many steps are involved including cell culturing, DNA extraction, sequencing and results analysis. In addition, as the 16S rRNA is highly conserved it is only useful at the species level.

In contrast to these biochemical and molecular techniques another route has been taken through the use of vibrational spectroscopy to generate "fingerprints" of intact bacterial samples. As far back as 1911, Coblentz suggested that biological samples could be analyzed by infrared (IR) absorption spectroscopy [3]. IR spectroscopy is a vibrational technique that measures the absorbance of radiation by a sample. The two vibrational techniques of IR and Raman spectroscopy are useful complementary techniques, since they can be used to probe a broad range of molecular symmetries [4]. It was IR spectroscopy that was first applied to the identification and characterization of *Eubacteriales* and *Lactobacillus* isolates, with data published as early as the 1950s [3, 5]. Despite these early successes the application of IR spectroscopy to the microbial taxonomy field did not gain popularity. Unfortunately at this time, the engineering behind spectrometers was still not advanced enough to provide rapid, sensitive, reproducible and low-cost instrumentation. It was not until the last quarter of the 20th century that the development of the interferometer, the microprocessor and powerful wave-transformation algorithms led to a resurgence of literature reporting whole-organism fingerprinting studies using IR technology [6, 7].

Raman spectroscopy is a recent addition to the physicochemical spectroscopic technologies that have been applied to the problem of rapid characterization and identification of micro-organisms. Although, in 1974 *Spiro* first suggested that resonance Raman spectroscopy could have potential for biological analysis [8], the earliest literature suggesting that Raman had the capability to be used in categorical analysis of microbes did not come until the early 1980s, and even then the reports did not go so far as to demonstrate that this was possible in practice [9, 10, 11]. In the 1990s the first reports of NIR FT-Raman were published, which examined the chemical nature of both bacterial and fungal cells, but again did not go so far as to use those data for discrimination [12, 13]. It was not until the start of this century that significant work showing the ability of Raman spectroscopy to be used as a microbial characterization and discrimination tool was reported [14,15,16,17]. These studies showed that Raman spectroscopy at near-infrared frequencies could be used to characterize bacterial cells at the earliest stages in colony development. Using Raman spectroscopy, single bacterial cells have also been analyzed, both conventionally using Raman microscopy [18], and with more complex laser tweezers systems that can aid the reduction of fluorescence at NIR wavelengths [19, 20].

The major shortcoming of Raman spectroscopy is that given the weak Raman scattering achieved by many biological samples, the spectral acquisition

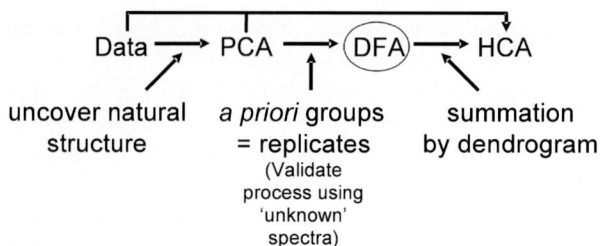

Fig. 1. The typical multivariate analysis methodology applied to the problem of biological characterization using SERS spectral fingerprints. The underlying theme is simplification or dimensionality reduction

time can be many minutes. However, enhancement methods such as SERS will provide a solution to this problem. Several investigations into SERS of bacteria have been undertaken [21, 22, 23, 24, 25], and more recently it has been shown that SERS can reliably differentiate between different bacteria [26, 27]. There is also much interest in detecting aerosols of spore-forming pathogens, with the obvious target being the identification of *Bacillus anthracis*. Preliminary work has already been carried out using SERS for this purpose and it has been shown that the dipicolinic acid biomarker, common to all spore-forming pathogens, can be detected [28, 29, 30, 31].

3 Introduction to Multivariate Cluster Analysis

By their very nature SERS spectra are multivariate. That is to say, each Raman wave number shift measured can be plotted against each other such that if we measured a meager 100 shifts then a sample can be said to reside somewhere in 100-dimensional space. Obviously this abstract space is very hard to visualize (!) and thus the underlying theme of multivariate cluster analysis is dimensionality reduction of the SERS spectra to a few new components that can be easily plotted and the relationship between the spectra visualized.

In this Chapter we are discussing the application of SERS to the problem of classification in highly complex biological systems. The major step that needs to be taken with large sets of multivariate data to achieve this objective, is a transformation of the original spectral domain into a reduced form that is both robust and interpretable in terms of the problem being studied.

Whilst there are many chemometric methods that can be applied to multivariate data (the reader is referred to [32, 33, 34]), the strategy that we have adopted for cluster analysis as reported in [35] is depicted in Fig. 1.

Initially, principal components analysis (PCA [36]) is used, which is an unsupervised method of data reduction where the original data matrix is projected onto a smaller variable subspace, and the resultant principal-component scores represent a majority of the variance in the data. This can be represented as:

$$T = XL,\tag{1}$$

where T is an $n \times d$ matrix of principal-component scores with a magnitude of d (dependent on the number of PCs to be extracted); X is an $n \times p$ matrix of independent variables (e.g., mean-centered spectra); and L is a $p \times d$ loadings matrix. In the situation where groups of SERS spectra from different bacteria are separated in the first PC, the loadings matrix from PC1 can be inspected to ascertain which SERS bands are most important for this separation.

However, often an unsupervised approach is insufficient to separate closely related bacterial classes based on their complex spectral profiles, and therefore one must use supervised analysis that can be used as predictive models. This involves proposing an a-priori class structure from which a determinative model is derived. One popular supervised method is discriminant function analysis (DFA [37]), which maximizes the within-group to between-group ratio (Fisher ratio) to differentiate between classes (groups). This approach has been used with great success for classification problems involving spectral fingerprinting [38,39,40]. DFA calculates a number of linear discriminant functions for separating groups by finding the eigenvalues and eigenvectors of the expression:

$$W^{-1}B,\tag{2}$$

where W is the within-sample matrix of sums of squares and crossproducts, and B is the between-sample matrix of sums of squares and crossproducts. As DFA is a supervised algorithm that optimizes the Fisher ratio, so as to separate different classes, it is necessary to avoid overfitting. In overfitting, the model has learnt the (training) data perfectly but is no longer able to predict the identity of new (test) data. That is to say it can not generalize. To avoid this we project test SERS data obtained from fresh cultures of bacteria of which we know the identity into a previously generated DFA cluster space. If the test data cocluster with the corresponding training data we are convinced that our model is valid.

For the identification of large numbers of different bacteria one often needs to inspect more than just the first 2 or 3 discriminant functions. In our strategy depicted above (Fig. 1) we reduce this problem by constructing a dendrogram based on DFA output using hierarchical cluster analysis (HCA). In this process, we can further summarize the multivariate outputs from DFA by taking the Euclidean distance between the a-priori group centers in PC-DFA space to construct a similarity matrix. These distance measures can then be processed by an agglomerative clustering algorithm to produce a dendrogram [37]. This can provide more lucid results than those obtained from plotting discriminant function scores as ordination plots.

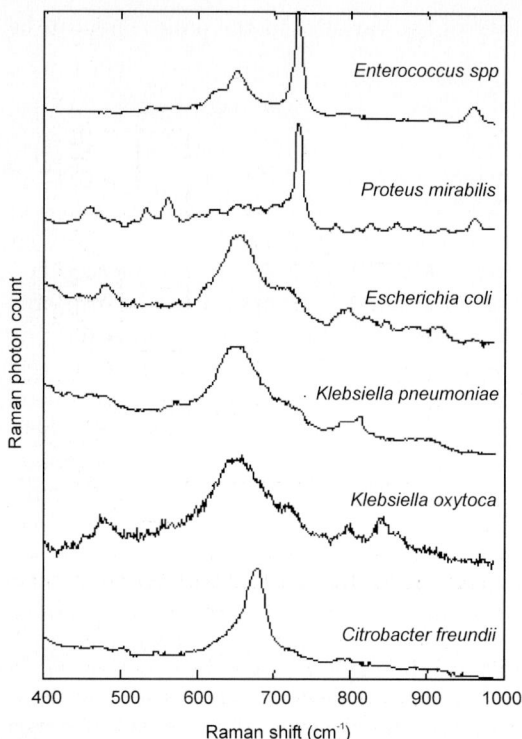

Fig. 2. Typical unprocessed SERS spectra showing an example from each species of UTI isolates studied. Each spectrum took 10 s to collect and the counts were in the thousands

4 Identification of Micro-Organisms Using SERS

It is possible to acquire extremely good SERS data from bacterial cells using minimal preparation. In the examples shown in Fig. 2, which are SERS spectra of urinary tract infection (UTI) isolates, using aggregated citrate-reduced colloid [41] strong spectral signals were obtained in only 10 s with 785 nm excitation and ~ 2 mW laser power at the sample. Importantly, these spectral fingerprints also have clear characteristic differences, which show great promise for application to the problem of bacterial characterization. In fact, using SERS fingerprints from a large number of UTI clinical isolates (courtesy of Bronglais Hospital, Aberystwyth), it has been possible to define a categorical model to discriminate between the major causative organisms of UTI [26].

One of the major benefits of taking a spectroscopic approach to bacterial characterization is that there is the potential to classify the microbial isolate to the subspecies or strain level. This is not possible using conventional biochemical methods; however, with the sensitivity of SERS it is possible

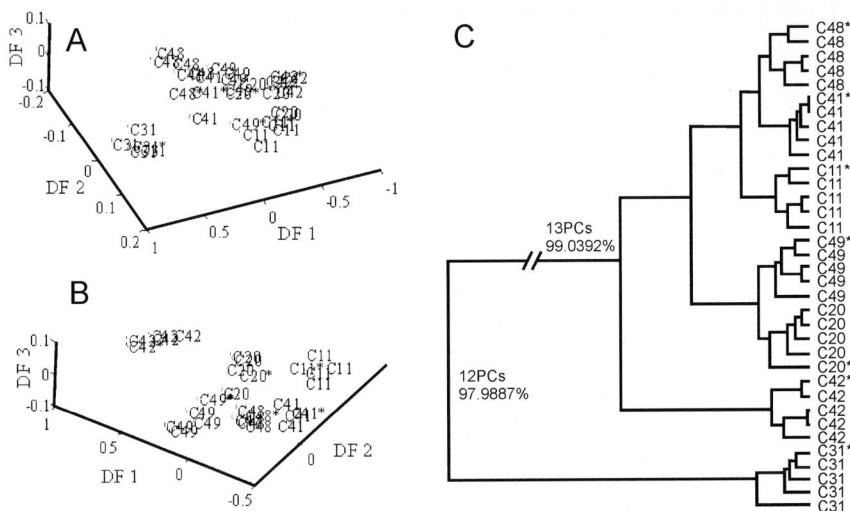

Fig. 3. (a) An ordination plot characterizing 7 clinical isolates of *Escherichia coli* from urinary tract infections. The items highlighted in *bold* with an asterisk are the validation samples. Isolate C31 is massively different from the other isolates, therefore C31 was extracted and the remaining data analyzed again. This result (b) clearly demonstrates that SERS can be applied to discrimination of microorganisms at the strain level. (c) A composite dendrogram generated by HCA using the combined PC-DFA space from the training and validation replicates used to generate the ordination plots in (a) and (b). This representation of the results shows how the validation replicates fall tightly within the clusters formed from the training data

to distinguish such closely related strains based on the organism phenotype. For isolates of *Escherichia coli*, obtained from patients with UTI, subspecies classification is shown in Fig. 3. This takes a stepwise approach to classification modeling, that is detailed fully in [26], and essentially involves removing groups from the analysis that are so clearly different that they prevent clear separation of other classes.

5 Monitoring Industrial Bioprocesses

The ability to control a bioprocess is paramount for product yield optimization, and it is imperative that the *concentration* of the fermentation product is assessed accurately [42]. Raman spectroscopy has historically been used to monitor the reaction of chemical processes and it was only a matter of time before this approach was used to analyze bioprocesses. Whilst chemical reactions involve few chemical species and vibrational modes more readily attributable to substrates and products, biological processes are more complex and rarely are vibrations directly related to bioproduct formation. In addition, for low-concentration products in a complex milieu the signal from the analyte of interest is often masked because of the weakness of the Raman effect, and this has perhaps discouraged analyzts from taking this approach. However, SERS provides an opportunity for off-line and at-line monitoring of bioprocesses that is both sensitive for the detection and quantification of low yield primary and secondary metabolites. Clearly SERS would not be performed on-line due to the poisonous nature of the silver colloid on the microbial process.

Our initial work has concentrated on the production of penicillin. In this industrially important bioprocess, *Penicillium chrysogenum* fermentations produce penicillin G as the major secondary metabolite of commercial interest. In conventional Raman spectroscopy using 785 nm excitation, the Raman spectra of penicillin G at high concentrations are dominated by the resonance enhancement of the aromatic ring vibration at $1005\,cm^{-1}$. By contrast, SERS penicillin G spectra, also collected at 785 nm, contained a greater number of peaks, with a much-improved signal-to-noise ratio and with significantly reduced fluorescence. With respect to the quantification of penicillin G it was shown that Raman spectroscopy could be used to quantify the amount of penicillin present in broths when relatively high levels of penicillin were analyzed ($> 50\,mM$). By contrast, using simple integration under SERS-enhanced peaks excellent quantification of penicillin G from considerably lower concentrations of the antibiotic were achieved.

6 Gene-Function Analysis

Whole genome sequencing has shown that there are many genes for which the function is unknown. There is thus a requirement to assign functions to these orphan genes and one approach to this is through "guilt by association" [43, 44]. By analyzing knockouts of a known function together with those of an *un*known function, cluster analysis on spectroscopic measurements can be used to infer metabolism classes based on the distances between groups [45]. The approach to this problem in terms of data analysis is the same as for bacterial characterization; the main differences are in the type of sample under analysis and the way in which the cluster analyzes are interpreted.

The "metabolic footprint" is a measure of the metabolites in the extracellular material, such as spent culture media, urine or blood [43, 44]. Spectroscopic fingerprints of such samples are seen as the best method for determining gene function by an inductive approach (i.e., mining data for knowledge, rather than testing a hypothesis) [46, 47, 48, 49]. However, detecting small quantitative or qualitative changes in growth media requires sensitive instrumentation, and therefore mass-spectral techniques have primarily been the method of choice. Consequently, sample preparation and spectral collection times are many minutes, which can be limiting when large libraries of knockout mutants need to be profiled. SERS is potentially a more

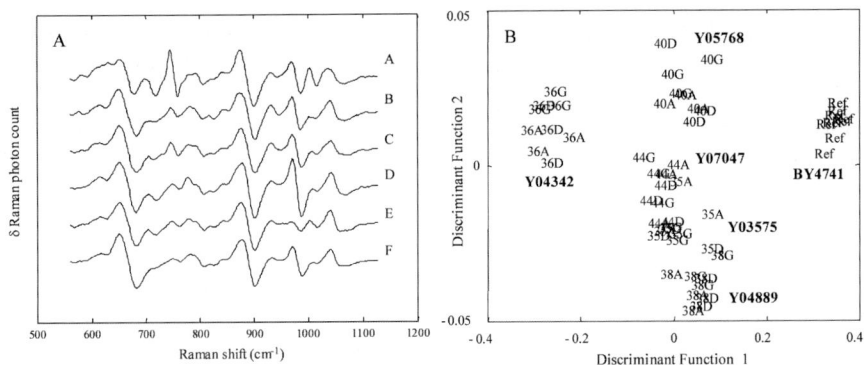

Fig. 4. (**A**) First-derivative SERS spectra (785 nm, $\sim$ 8 mW laser power at the sample) of depleted haploid yeast-cell culture media from (**A**) BY4741 Wild type; (**B**) Y07047; (**C**) Y03575; (**D**) YO5768; (**E**) Y04889; (**F**) Y04342. (**B**) PC-DFA model using SERS spectra, showing the haploid yeast reference strain and 5 haploid mutant strains resolved into separate clusters

rapid "holistic" fingerprinting method than MS. The low limit of detection for SERS measurements also makes the method an ideal candidate for detecting small differences in extracellular metabolites between gene knockout mutants.

Table 1. Haploid yeast knockout mutant strains used in a SERS metabolic footprinting study

Experiment Ref.	ID	Metabolism descriptor
Reference	BY4741	Wild type
35 A, 35D, 35G	Y03575	Zinc-finger transcription factor, controls expression of ADH2, peroxisomal, ethanol, glycerol and fatty acid genes
36 A, 36D, 36G	Y04342	Broad-specificity amino-acid permease, high-affinity glutamine permease
38 A, 38D, 38G	Y04889	Iron homeostasis
40 A, 40D, 40G	Y05768	Involved in manganese homeostasis
44 A, 44D, 44G	Y07047	A transcriptional repressor for allantoin and GABA catabolic genes, a negative regulator of multiple nitrogen catabolic genes

As a preliminary example, a small subset of the 7000 eukaryotic *Saccharomyces cerevesiae* single-gene knockout mutants (courtesy of Prof. Stephen G. Oliver, The University of Manchester) are listed in Table 1 and these were analyzed by SERS. The putative metabolism classes for these mutants suggest that the deleted genes relate to a broad range of biochemical pathways within the cell. In Fig. 4a examples of first-derivative SERS spectra acquired from the metabolic footprint of these samples are shown, with a total spectral integration time of only 1 min for each sample.

Whilst some obvious quantitative differences can be seen between these spectral fingerprints, for the discovery of the relationships between the wild type and the various gene knockouts analyzed, it is necessary to take a multivariate approach using cluster analysis. PC-DFA clearly separates these mutants from wild type and separate groups were recovered for each of the knockouts (Fig. 4b). This preliminary result demonstrates the potential for SERS to be used as a rapid screening technique in gene-function analysis and will be explored more fully in the future.

7 Concluding Remarks

Raman and SERS spectroscopy clearly presents itself as a highly versatile tool that provides complex chemical fingerprints from a wide range of biological materials. For microbial investigations, these generally require multivariate cluster analysis, or more advanced machine-learning techniques [50], for clear microbial characterization, in terms of elucidating the relationship between bacteria, and for robust unequivocal identification of infectious agents. We have recently demonstrated that SERS has the exquisite sensitivity required for the classification of micro-organisms and is reproducible enough for the identification of a wide variety of bacteria [26, 27, 31]. It is noteworthy that in our studies we have not just used the "stare and compare" analysis adopted by others, but have employed multivariate methods that show reproducibility across the full spectral range collected. Finally, we have also demonstrated single bacterial cell analysis using SERS [27] and that this approach is reproducible enough to be used for the quantification of microbial fermentations [51].

References

[1] A. Griffiths, W. Gelbart, R. Lewontin, S. Wessler, D. Suzuki, J. Miller: *An Introduction to Genetic Analysis* (WH Freeman, New York 2004)
[2] E. Leifson: Bacteriol. Rev. **30**, 257 (1966)
[3] J. Riddle, P. Kabler, B. Kenner, R. Bordner, S. Rockwood, H. Stevenson: J. Bacteriol. **72**, 593 (1956)
[4] N. B. Colthup, L. H. Daly, S. E. Wiberley: *Introduction to Infrared and Raman Spectroscopy* (Academic, London 1990)
[5] J. Goulden, M. Sharpe: J. General Microbiol. **19**, 76 (1958)
[6] D. Ellis, G. Harrigan, R. Goodacre: *Metabolic Profiling: Its Role in Biomarker Discovery and Gene Function Analysis* (Kluwer Academic, Boston 2003)
[7] D. Naumann: Appl. Spectrosc. Rev. **36**, 239 (2001)
[8] T. G. Spiro: Acc. Chem. Res. **7**, 339 (1974)

[9] R. A. Dalterio, M. Baek, W. H. Nelson, D. Britt, J. F. Sperry, F. J. Purcell:
     Appl. Spectrosc. **41**, 221 (1987)
[10] R. A. Dalterio, W. H. Nelson, D. Britt, J. F. Sperry: Appl. Spectrosc. **41**, 417
     (1987)
[11] K. A. Britton, R. A. Dalterio, W. H. Nelson, D. Britt, J. F. Sperry: Appl.
     Spectrosc. **42**, 782 (1988)
[12] A. C. Williams, H. G. M. Edwards: J. Raman Spectrosc. **25**, 673 (1994)
[13] H. G. M. Edwards, N. C. Russell, R. Weinstein, D. D. Wynnwilliams: J. Raman
     Spectrosc. **26**, 911 (1995)
[14] K. Maquelin, C. Kirschner, L. P. Choo-Smith, N. van den Braak, H. P. Endtz,
     D. Naumann, G. J. Puppels: J. Microbiol. Met. **51**, 255 (2002)
[15] K. Maquelin, L. P. Choo-Smith, T. van Vreeswijk, H. P. Endtz, B. Smith,
     R. Bennett, H. A. Bruining, G. J. Puppels: Anal. Chem. **72**, 12 (2000)
[16] K. Maquelin, L. P. Choo-Smith, H. P. Endtz, H. A. Bruining, G. J. Puppels:
     J. Clinical Microbiol. **40**, 594 (2002)
[17] L. P. Choo-Smith, K. Maquelin, T. van Vreeswijk, H. A. Bruining, G. J. Pup-
     pels, N. A. G. Thi, C. Kirschner, D. Naumann, D. Ami, A. M. Villa, F. Orsini,
     S. M. Doglia, H. Lamfarraj, G. D. Sockalingum, M. Manfait, P. Allouch,
     H. P. Endtz: Appl. Environm. Microbiol. **67**, 1461 (2001)
[18] K. C. Schuster, I. Reese, E. Urlaub, J. R. Gapes, B. Lendl: Anal. Chem. **72**,
     5529 (2000)
[19] C. G. Xie, Y. Q. Li: J. Appl. Phys. **93**, 2982 (2003)
[20] C. G. Xie, M. A. Dinno, Y. Q. Li: Opt. Lett. **27**, 249 (2002)
[21] A. A. Guzelian, J. M. Sylvia, J. A. Janni, S. L. Clauson, K. M. Spencer: *Vi-
     brational Spectroscopy-Based Sensor Systems* (SPIE-INT Society Optical En-
     gineering, Bellingham 2002) p. 182
[22] L. Zeiri, B. V. Bronk, Y. Shabtai, J. Eichler, S. Efrima: Appl. Spectrosc. **58**,
     33 (2004)
[23] L. Zeiri, B. V. Bronk, Y. Shabtai, J. Czege, S. Efrima: Colloids and Surfaces
     A – Physicochemical and Engineering Aspects **208**, 357 (2002)
[24] N. F. Fell, A. G. B. Smith, M. Vellone, A. W. Fountain: *Vibrational
     Spectroscopy-Based Sensor Systems* (SPIE, SanJose, CA 2002) p. 174
[25] S. Efrima, B. V. Bronk: J. Phys. Chem. B **102**, 5947 (1998)
[26] R. M. Jarvis, R. Goodacre: Anal. Chem. **76**, 40 (2004)
[27] R. M. Jarvis, R. Goodacre: Anal. Chem. **76**, 5198 (2004)
[28] W. Premasiri, D. Moir, M. Klempner, N. Krieger, G. Jones, L. Ziegler: J. Phys.
     Chem. B **109**, 312 (2005)
[29] X. Zhang, M. Young, O. Lyandres, R. V. Duyne: J. Am. Chem. Soc. **127**, 4484
     (2005)
[30] A. E. Grow, L. L. Wood, J. L. Claycomb, P. A. Thompson: J. Microbiol. Met.
     53, 221 (2003)
[31] R. M. Jarvis, A. Brooker, R. Goodacre: Faraday Discuss. **132**, 281 (2006)
[32] M. Otto: *Chemometrics: Statistics and Computer Application in Analytical
     Chemistry* (Wiley, New York 1999)
[33] D. L. Massart, B. G. M. Vandeginste, S. N. Deming, Y. Michotte, L. Kauf-
     man: *Chemometrics: A Textbook* (Elsevier Science Publishers B V, Amsterdam
     1988)
[34] R. Brereton: *Chemometrics: Data Analysis for the Laboratory and Chemical
     Plant* (John Wiley & Sons, Chichester 2003)

[35] R. Goodacre, E. M. Timmins, R. Burton, N. Kaderbhai, A. Woodward, D. B. Kell, P. J. Rooney: Microbiol. **144**, 1157 (1998)
[36] I. T. Jolliffe: *Principal Component Analysis* (Springer, New York, Heidelberg 1986)
[37] B. F. J. Manly: *Multivariate Statistical Met.: A Primer* (Chapman & Hall/CRC, New York 1994)
[38] E. Kinoshita, Y. Ozawa, T. Aishima: *Flavonoids in the Living System* (Plenum, New York 1998) p. 117
[39] B. S. Radovic, R. Goodacre, E. Anklam: J. Anal. Appl. Pyrolysis **60**, 79 (2001)
[40] E. M. Timmins, D. E. Quain, R. Goodacre: Yeast **14**, 885 (1998)
[41] P. C. Lee, D. Meisel: J. Phys. Chem. **86**, 3391 (1982)
[42] M. Pons: *Bioprocess Monitoring and Control* (Hanser, Munich 1991)
[43] N. Kaderbhai, D. Broadhurst, D. Ellis, R. Goodacre, D. Kell: Comp. Func. Genomics **4**, 376 (2003)
[44] J. Allen, H. Davey, D. Broadhurst, J. Heald, J. Rowland, S. Oliver, D. Kell: Nature Biotechnol. **21**, 692 (2003)
[45] D. Kell: Curr. Opin. Microbiol. **7**, 296 (2004)
[46] D. B. Kell, S. G. Oliver: Bioessays **26**, 99 (2004)
[47] J. Nicholson, J. Connelly, J. Lindon, E. Holmes: Nature Rev. Drug Discovery **1**, 153 (2002)
[48] J. Griffin: Philos. Trans. **359**, 857–871 (2004)
[49] D. Huhman, L. Sumner: Phytochem. **59**, 347 (2002)
[50] R. Jarvis, R. Goodacre: Bioinformatics **21**, 860 (2005)
[51] S. Clarke, R. Littleford, W. Smith, R. Goodacre: Analyst **130**, 1019 (2005)

Index

Surface-Enhanced Raman Scattering for Biomedical Diagnostics and Molecular Imaging

Tuan Vo-Dinh, Fei Yan, and Musundi B. Wabuyele

Center for Advanced Biomedical Photonics, Oak Ridge National Laboratory, Oak Ridge, TN 37830-6101, USA
vodint@ornl.gov (current address: tuan.vodinh@duke.edu)

1 Introduction

Raman spectroscopy, based on molecular vibrational transitions, has long been regarded as a valuable tool for the identification of chemical and biological samples as well as the elucidation of molecular structure, surface processes, and interface reactions. Despite such advantages, Raman scattering suffers the disadvantage of extremely poor efficiency or inherently small Raman cross section (e.g., $10^{-30}\,\mathrm{cm}^2$ per molecule) even when high laser power is used, thus precluding the possibility of analyte detection at low concentration levels without special enhancement processes. Nevertheless, there has been a renewed interest in Raman techniques in the past two decades, largely due to the discovery of the surface-enhanced Raman scattering (SERS) effect, which results from the adsorption of molecules on nanotextured metallic surfaces [1, 2, 3]. In 1974, it was first reported that a strong enhanced Raman scattering signal occurred with pyridine molecules adsorbed on silver electrode surfaces that had been roughened electrochemically by oxidation–reduction cycles [3]. This observation of an enhanced Raman signal, which was originally attributed to a high surface density produced by the roughening of the surface of electrodes [3], was reported in 1977 to be a result of a surface-enhancement process, hence the term surface-enhanced Raman scattering (SERS) effect [1, 2]. During a period between the mid-1970s and the early 1980s, this early enthusiasm for SERS decreased and did not lead to practical applications because the Raman enhancement effect had been observed for only a limited number of molecules (mostly pyridine) under very specific experimental conditions. In 1984, the general applicability of SERS as an analytical technique for trace detection of a wide variety of chemicals using solid substrates covered with silver-coated nanospheres was first demonstrated in our laboratory [4]. This SERS technique based on metallic nanostructures has been further developed and applied to the detection of environmental pollutants and health-effect biomarkers [5, 6, 7, 8, 9, 10, 11, 12, 13]; the technique has also been used in the investigation of adsorption and reaction process at electrochemical interfaces [14]. Several studies have even

K. Kneipp, M. Moskovits, H. Kneipp (Eds.): Surface-Enhanced Raman Scattering – Physics and Applications, Topics Appl. Phys. **103**, 409–426 (2006)
© Springer-Verlag Berlin Heidelberg 2006

shown that the sensitivity of SERS can rival that of fluorescence in detecting single molecules [15, 16, 17, 18, 19].

Over the last few years, there has been a great deal of effort in our laboratory in the development of SERS techniques for use in biomedical diagnostics, pathogen detection, gene identification, gene mapping, and DNA sequencing. The hybridization of a nucleic acid probe to its DNA target provides a very high degree of accuracy for identifying complementary DNA sequences. We have used this process to develop a new generation of DNA-based SERS probes for medical diagnosis [20, 21]. The SERS gene-probe system could offer a unique combination of performance capabilities and analytical features of merit.

SERS techniques also provide useful molecular probes for cell-based assays. Direct observation of molecular events inside single living cells could significantly improve our understanding of basic cellular processes as well as improving our knowledge of the intracellular transport and fate of therapeutic agents. Therefore, it is very important to develop techniques to identify and detect the reactions of individual molecules in living cells with improved spatial and temporal resolution. For example, labeling of specific proteins, membrane and genetic material by chemical or recombinant techniques has enabled tracking of the location of individual molecules within cells and tissues using fluorescence microscopy with great sensitivity [22, 23, 24]. By contrast, Raman imaging is emerging as a rapid and nondestructive analytical tool that yields highly compound-specific information for chemical analysis and has great potential for high-throughput analysis and direct imaging. Recent reports have shown an increasing trend in applying Raman spectroscopy in cellular and tissue imaging studies [25, 26, 27, 28]. The development of SERS techniques for biological analysis and medical diagnostics has been investigated in our laboratory [29, 30, 31, 32, 33, 34]. In addition, SERS studies on living cells using gold nanoparticles [35] and coherent anti-Stokes Raman scattering (CARS) for cellular imaging have been reported [36]. In this Chapter, we present an overview of the various SERS methods and instrumentation developed in our laboratory for use in biomedical applications: SERS gene probes for DNA detection, a hyperspectral surface-enhanced Raman imaging (HSERI) system that combines hyperspectral imaging capabilities with SERS for potentially identifying cellular components with high spatial and temporal resolution, near-field scanning optical microscopy (NSOM) for sub-wavelength Raman spectral acquisition, and a novel SERS-based "molecular sentinel" probe for medical diagnostics.

2 Methods and Instrumentation

2.1 Silver Nanoparticle Island Films for Biomedical Diagnostics

While the enhancement mechanism for SERS is most likely due to the intense localized fields arising from surface-plasmon resonance in nanostructures of various metals (e.g., Au, Ag, Cu, Li, Al, Na, Fe) with sizes on the order of tens of nanometers, the most intense SERS enhancement is found to be obtained from molecules adsorbed onto silver surfaces only. For this reason, our research has mainly focused on SERS studies using silver substrates. The main challenge with any SERS-active substrate is the attainment of reproducible spectra, without which quantitative analysis is difficult. One of the most successful substrates first developed in our group is a glass slide coated with silver-island films. Typically, a 9 nm mass thickness of silver is deposited on a cleaned glass surface prepared by a physical vapor deposition (PVD) method. Prior to deposition, glass slides are soaked in saturated KOH solution in 2-propanol (i-PrOH) overnight. Then the slides are rinsed with distilled water several times and air dried. An electron beam evaporation system (CVE 301 EB, Coke Vacuum Products, Norwalk, CT) is used for the PVD process. A 1 nm chromium layer is deposited on the glass surface before silver deposition to stabilize the silver layer on the glass surface. The deposition rate and PVD chamber pressure were $0.01 \, nm \cdot s^{-1}$ and $1.33 \times 10^{-4} \, Pa$, respectively. Freshly coated silver-island surfaces are kept under vacuum prior to further treatment.

For medical diagnostics, especially gene detection, a multistep procedure was involved in this pretreatment and hybridization. Typically, Ag surfaces are first immersed in 100 ml of an ethanolic solution containing a mixture of $0.4 \, \mu M$ 1-mercaptoundecanoic acid and $4.0 \, \mu M$ 1-mercapto-undecanol for 30 min. This is followed by extensive washing with the degassed EtOH and air drying. A solution containing O-(N-succinimidyl)-N, N, N', N'-tetramethyluronium tetrafluoroborate ($50.8 \, \mu mol$), $50 \, \mu l$ of diisopropylethylamine ($287 \, \mu mol$) and 1 ml of acetonitrile (MeCN) is distributed over the clean surfaces and kept in an air tight container overnight. The chemically treated silver surfaces are subsequently washed with MeCN and air dried. These substrates are then transferred to a Petri dish that is vapor saturated with 8 % H_2O in methanol (MeOH). A stock solution of the 5'-amino-labeled capture-probe sequence is prepared by dissolving 3.27 mg in $90 \, \mu l$ of H_2O and $950 \, \mu l$ of MeOH. Aliquots ($15 \, \mu l$) of this stock solution are applied to the substrates (except for the blank), and immediately exposed to broadband UV radiation for 60 s. The silver substrates are rehydrated with 8 % H_2O–MeOH, covered with a microscope cover slide and kept in an air tight container for 6 h. The silver-coated slides are washed with H_2O, covered with hybridization solution and placed in an incubator at 37.5 °C for 60 min. The slides are then washed with H_2O, air dried and then covered with $185 \, \mu l$ of hybridization buffer and $15 \, \mu l$ of DNA–SERS probe solution. Microscope coverslips are used to prevent evaporation of the oligonucleotide solution. The silver surfaces are placed in an H_2O–MeOH-saturated airtight container, which is placed in a 37.5 °C hybridization oven for 15 h. They are later repeatedly washed with H_2O and air dried. A control for nonspecific binding, consisting of modified silver surfaces with self-assembled monolayers (SAMs) of alkanethiols, without bound capture oligonucleotides, is also used in the experiments. The control is exposed to the DNA–SERS probe solution during hybridization in the same manner as the silver surfaces with immobilized capture oligonucleotides.

2.2 Silver Nanoparticles for Cellular Imaging

The use of nanoparticles for in-situ detection of specific reactions as well as intracellular molecular imaging has significant potential for living cells and tissues. SERS appears to have superior properties compared to fluorescence in terms of dye labeling of nanoparticles made of gold or silver metals. Recently, we have had considerable success in utilizing a novel type of SERS-active silver colloid, which can be easily prepared according to a method reported by *Leopold* and *Lendl* [37]. Briefly, 10 ml of AgNO$_3$ aqueous solution (10 mM) are added to 90 ml of a hydroxylamine hydrochloride solution (1.67 mM) containing 3.33 mM sodium hydroxide. The procedure yields a rather monodispersed size of silver particles with an average diameter in the range of 23 nm (as examined by the TEM measurements). Other advantages of this approach include its ease of preparation at room temperature and its immediate applicability for SERS measurements without any subsequent treatment.

Before Raman dye labeling, an aliquot of the above-prepared silver colloids is incubated with mercaptoacetic acid (MAA, $\sim$ 1.0 mM) for 3 h. Because –SH groups tend to bind to Ag (and Au) surfaces strongly, thus leaving the carboxylic groups hanging around the silver colloids and allowing the covalent binding of a variety of amine-containing Raman-active dye compounds, as well as many protein-like biomolecules. The MAA-labeled silver colloids can then be separated from the solution by centrifugation at 10 000 rpm for $\sim$ 10 min. The clear supernatant is discarded and the loosely packed silver sediments are resuspended in 100 μl of 0.1 M 4-morpholinoethane sulfonic acid (MES) buffer, pH 5.0. The colloids are washed four times with MES buffer to remove the excess MAA. Cresyl violet acetate (CV) is chosen as the Raman reporter due to its large SERS enhancement previously described [38]. Typically, the labeling is carried out as follows, 2 μl of 1 mM CV and 100 μl of 1-ethyl-3-(3 dimethylaminopropyl)-carbodiimide (EDC) solution (10 mg dissolved in 1 ml of ultrapure H$_2$O) are added to 100 μl of MAA-labeled silver colloids in MES buffer and allowed to react for 24 h. The reporter-labeled silver colloids are separated from the solution by centrifugation. After four rinses, the silver sediments are resuspended in 100 μl of sterile ultrapure water and stored at 4 °C.

Chinese hamster ovary (CHO) cells are used as model systems in our SERS studies. Cells are obtained from the American Type Culture Collection (ATCC, Manassas, VA) are grown in T-25 flasks (Corning, NY) using Ham's F-12 medium (Invitrogen, Carlsbad, CA) containing 1.5 g/l sodium bicarbonate and 2 mM L-glutamine, and supplemented with 10 %, fetal bovine serum (Gibco, Grand Island, NY). The stock cultures are kept in a 5 % CO$_2$ cell culture incubator at 37 °C with 95 % relative humidity. When cells reach 70 %

to 80 % confluence, they are subcultured at 1 : 20 ratio. For experiments, cells are seeded onto glass chamber slides from Nalge Nunc International (Naperville, IL). Cells are incubated with 10 μl of CV-labeled silver colloids for 24 h in Ham's F-12 medium in the CO_2 incubator. Prior to fixing, the cells were washed three times with phosphate-buffered saline (PBS) buffer. The Cl^- ions present in the PBS buffer induce the formation of colloidal aggregates, which result in clustering. The larger clusters formed by aggregated colloids are believed to be the most efficient site for Raman enhancement. The cells were fixed with 4 % paraformaldehyde for 5 min followed by multiple rinses with 100 % cold methanol.

2.3 SERS Instrumentation

Several Raman instrumentation systems are used for various SERS studies. The detection system for medical diagnostics is designed to allow spectral recording of individual spots at the hybridization sites. This detection system, composed of commercially available or off-the-shelf components, is illustrated in Fig. 1. In most of our studies, a 632.8 nm helium-neon laser with ∼ 5 mW excitation power is used. Signal collection is performed at 180° with respect to the incident laser beam, which is focused through the backside of the translucent substrate. A Raman holographic filter is used to reject the Rayleigh-scattered radiation from the collected Raman signal. The Raman signal is focused onto the entrance slit of a spectrograph, equipped with a red-enhanced intensified charge coupled device (ICCD) detector, having a total accumulation time of 60 s per spectrum.

The schematic diagram of a hyperspectral Raman imaging system is illustrated in Fig. 2. The system consists of an inverted microscope coupled with a 632.8 nm helium-neon laser. The light from the laser was passed through a set of diverging and collimating lenses (L1), an iris, and then diverted into a microscope objective (60×, 0.85NA) using a dichroic filter and focused on a sample mounted onto a translation stage. SERS signals were collected by the same objective, transmitted through the dichroic mirror, and then through a holographic notch filter (HNF) into an acousto-optic tunable filter (AOTF) device. The AOTF projected the diffracted (first-order) light at an angle different from the undiffracted (zero-order) light. The AOTF has a spectral operating range from 600 nm to 900 nm that corresponds to the relative wave number range (from $0\,cm^{-1}$ to $4691.7\,cm^{-1}$ with respect to a 632.8 nm excitation and a spectral resolution of $7.5\,cm^{-1}$ at 633 nm). The first-order beam exiting the AOTF was passed through a beamsplitter (BS) (70/30 ratio), then through a second iris and imaged onto a thermoelectrically cooled intensified charged coupled device (ICCD) containing a front-illuminated chip with a 512 × 512 two-dimensional array of pixels ($19 \times 19\,\mu m^2$). The ICCD was computer controlled with WinView software. The refracted beam was focused down onto the active area of an avalanche photodiode (APD). An

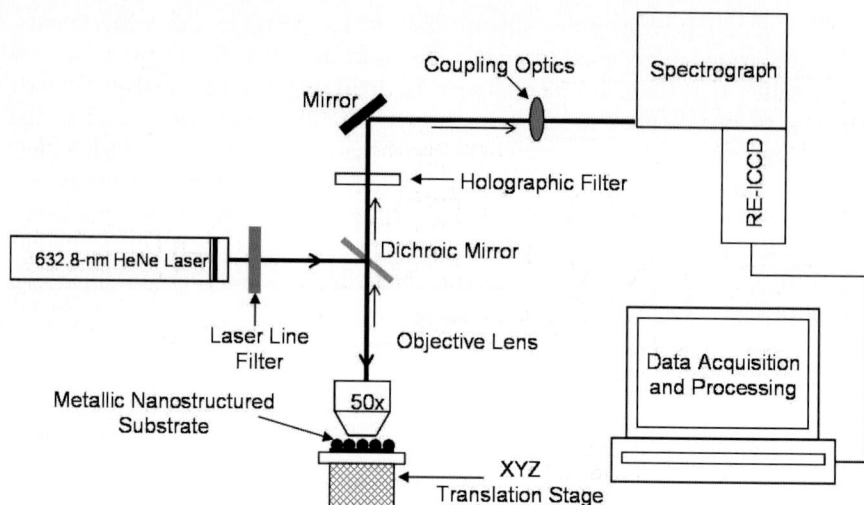

Fig. 1. Schematic diagram of conventional SERS system for SERS spectral acquisition. Laser excitation: 632.8 nm

APD is an ideal detector for several reasons: small size, high quantum efficiency (QE) and large amplification capabilities. The APD used has a QE of $\sim 70\,\%$ and very low dark count ($< 100\,\mathrm{c/s}$) thus reducing the possible noise arising from the use of amplifiers. A TTL pulse of 2.5 V from the APD is sent to a universal counter where the pulses are counted for a specified acquisition time. The APD detector is controlled by an integrated LabVIEW program developed in-house. The AOTF-based HSERI system is also integrated with an incandescent tungsten light for bright-field imaging and a mercury lamp for fluorescence imaging. SERS spectra and images were acquired after focusing the laser beam to an area of interest on the sample of cells adhered to the glass chamber slides. The images were acquired upon excitation with 15 mW laser power and an accumulation time of 6 s (0.6 s per frame). The SERS spectra were recorded with accumulation times of 25 s to 50 s. For specific instrumentation details, please refer to the original papers cited in this Chapter.

3 SERS Applications

3.1 SERS Gene Probe for Medical Diagnostics

The primary advantage of Raman scattering is the very narrow bandwidth of a typical Raman peak ($\leq 1\,\mathrm{nm}$). As shown in Figs. 3a and 3b, two Raman-active dyes, i.e., brilliant crystal blue (BCB) and crystal fast violet (CFV), have their most intense characteristic peaks at ~ 579 and $591\,\mathrm{cm}^{-1}$, respectively. The intensity ratios between these two peaks are proportional to the

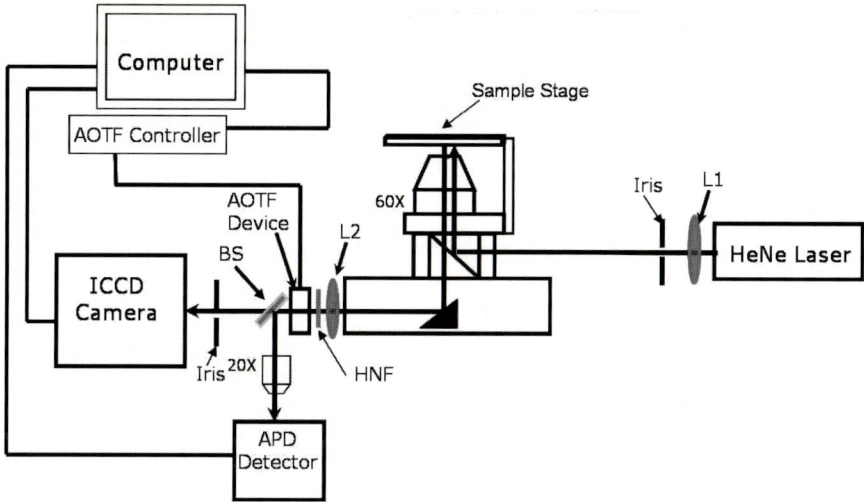

Fig. 2. Schematic diagram of a hyperspectral Raman imaging system. SERS images and spectrums are filtered with the acousto-optic tunable filter (AOTF) positioned between the microscope and the detectors. Optical elements, L1 and L2 are used to expand and collimate the laser beam. The SERS signal is filtered through a 633 nm holographic notch filter (HNF) into the AOTF and sent through a beamsplitter (BS) onto the detectors

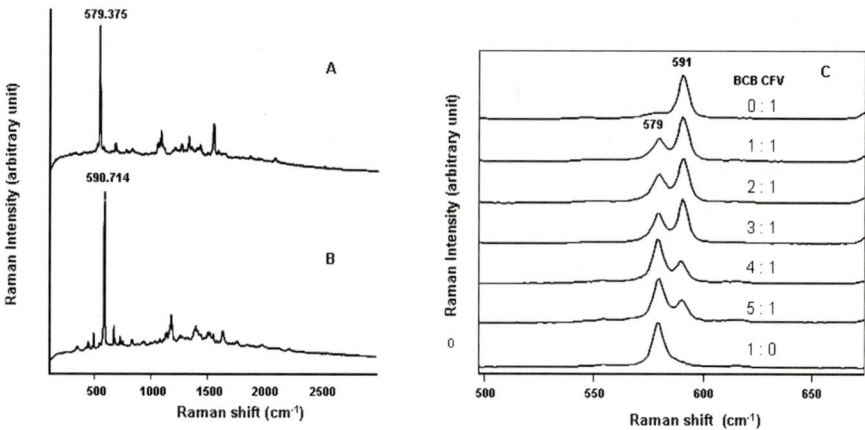

Fig. 3. Surface-enhanced Raman spectra of (**a**) Brilliant cresyl blue (BCB), (**b**) Cresyl fast violet (CFV), and (**c**) mixtures of BCB and CFV at different ratios of BCB to CFV: 0 : 1; 1 : 1; 2 : 1; 3 : 1; 4 : 1; 5 : 1; 1 : 0 (from *top* to *bottom*)

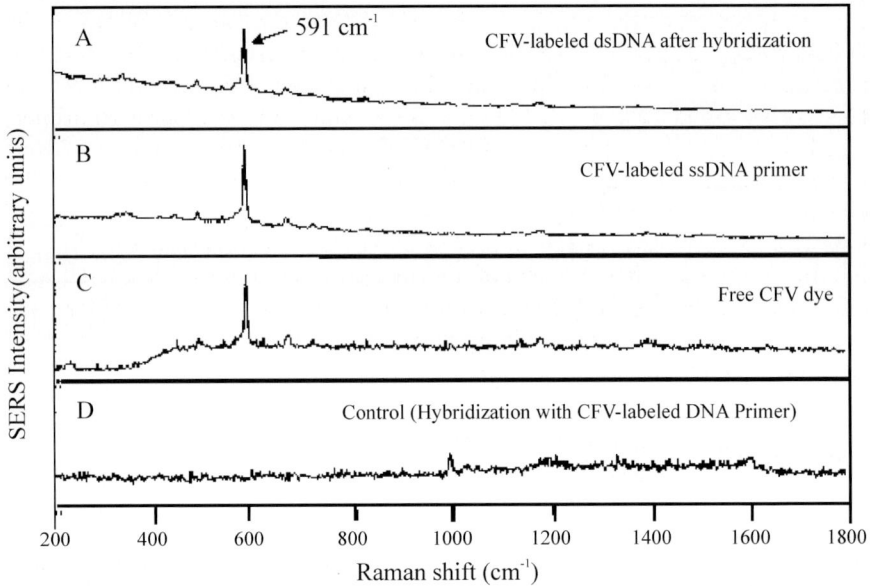

Fig. 4. Illustration of the SERS gene-probe technique for the detection of the HIV1 gag gene. (**a**) CFV-labeled DNA after hybridization; (**b**) CFV-labeled primer DNA; (**c**) Free CFV; and (**d**) hybridization control

concentration ratios of these two dyes from $1:1$ to $1:4$ (Fig. 3c), providing the opportunity for multiplexing in diagnostics applications [20, 21, 38, 39, 40].

To demonstrate the SERS gene-detection scheme, we used a silver-island-coated glass substrate, on which capture probes were bound and directly used for hybridization. A SERS gene probe has been developed for the selective detection of HIV DNA [21]. Infection with the human immunodeficiency virus Type 1 (HIV1) often results in a uniformly fatal disease if not detected and treated adequately. Unfortunately, standard HIV serologic tests, including the enzyme-linked immunosorbent assay (ELISA) and the Western blot assay, are not definitive in the diagnosis of HIV infection during early infancy because of the confounding presence of transplacentally derived maternal antibodies in the infant's blood. Direct nucleic acid-based tests that detect the presence of HIV viral sequences are required. Figure 4 illustrates the results of the HIV nucleotide hybridization experiments.

Figure 4a shows the SERS detection of the CFV-labeled target DNA strand that has been hybridized with the surface-bound capture probe. Detection of a single-strand DNA labeled by CFV is shown in Fig. 4b. In this case, the sample was simply spotted on the immobilization support that was coated with silver. Figure 4c corresponds to free-CFV detection. There does not appear to be any major alteration in the CFV spectrum as a result of being bound to single- or double-stranded DNA. As a further demonstration of selectivity (Fig. 4d), no CFV signals were observed following the hybridization and rinsing steps when no complementarity existed between the capture probe sequence and the single-stranded, CFV-labeled DNA. In this example, the CFV-labeled DNA was the single-stranded forward primer, which lacked any complementarity with the capture probe and was hence removed during the rinsing step. In summary, these results demonstrate the effectiveness of our SERS gene-probe technique for biomedical diagnostics. Our results demonstrate the potential of SERS as a practical tool for the identification and differentiation of multiple genes related to medical diseases and infectious pathogens, provided that the Raman shifts of a set of Raman-active dyes could be controllably adjusted and easily conjugated to any gene sequence of interest in the foreseeable future.

3.2 Hyperspectral Imaging of Raman Dye-Labeled Silver Nanoparticles in Single Cells

Applications of our Raman system to hyperspectral imaging are illustrated by imaging cresyl violet (CV)-labeled silver colloids spotted on a glass slide. It is worth noting that our labeling approach is different from other approaches reported in the literature [41, 42, 43], in which spontaneous adsorption of the thiol molecules on metal surfaces is used for dye labeling. Our approach involved developing labeled nanoparticles that could be covalently coupled to a larger number of molecules having the amine-moieties (these include Raman-active dyes such as CV, rhodamine 123, etc., as well as targeting biomolecules such as antibodies and amine-functionalized DNA). Figure 5 shows selected SERS images collected over the spectral range between $296\,\mathrm{cm}^{-1}$ to $3304\,\mathrm{cm}^{-1}$. We observe intense images due to the SERS signal from the CV-labeled silver nanoparticles at CV's characteristic peaks $(596\,\mathrm{cm}^{-1}, 1196\,\mathrm{cm}^{-1}$ and $1644\,\mathrm{cm}^{-1})$ indicated by arrows on the SERS spectra. The narrow spectral bandpass ($< 1\,\mathrm{nm}$) of the CV label is advantageous to high-throughput analysis where multiple Raman probes are required. Images shown in Fig. 5 were acquired with a $7\,\mathrm{cm}^{-1}$ spectral resolution AOTF scanned in steps of $12\,\mathrm{cm}^{-1}$ and $24\,\mathrm{cm}^{-1}$ from the most intense Raman peak of CV at $596\,\mathrm{cm}^{-1}$. The resulting images of CV-labeled silver colloidal particles were used to further demonstrate the hyperspectral-SERI concept where images were recorded over the spectral region of interest with the threshold value equivalent to the SERS signal intensity at $992\,\mathrm{cm}^{-1}$.

Figure 6 shows a bright-field image (left), a SERS image (center), and a composite image (right) of a fixed Chinese hamster ovary (CHO) cell after the passive uptake of CV-labeled silver nanoparticles. Two types of control cells were used in this study: Control-1 cells were not incubated with the CV-labeled particles and Control-2 cells were incubated with the CV-labeled particles for less that 1 h and rinsed four times with PBS buffer. Both types of control cells did not show any SERS signal under the measuring conditions used (data not shown). In a typical experiment, CHO cells were incubated

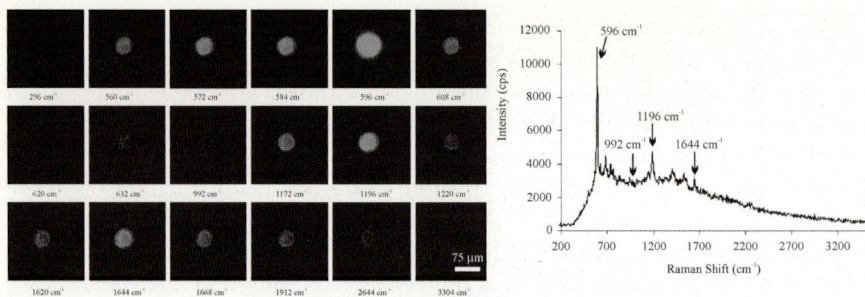

Fig. 5. *Left:* Raman images of CV-labeled silver colloidal particles collected at various Raman shifts. The images were acquired by scanning the AOTF over the entire spectrum using a 632.8 nm HeNe laser for excitation. *Right:* SERS spectrum of CV-labeled silver colloidal particles; *scale bar:* 75 µm

with 10 µl of CV-labeled silver colloids for 24 h in Ham's F-12 medium in the CO_2 incubator. Prior to fixing and imaging, the cells were rinsed three times with PBS buffer to ensure the removal of silver nanoparticles adsorbed outside the cells. The images indicate that the SERS signal is inhomogeneously distributed over the cell (Fig. 6b). This feature is likely due to the fact that some regions within the cell do not contain the CV-labeled silver clusters. Variation in the SERS signal intensities that are observed within the cell may be attributed to two things: 1. differing enhancement due to the SERS-active site resulting from the nanostructure (silver clusters); and 2. the heterogeneous distribution of the laser within the field of view. The image in Fig. 6 was obtained with an acquisition time of 25 s using the AOTF system set at $596\,cm^{-1}$ that correlated with the CV's characteristic peak shown in the SERS spectrum.

3.3 SERS Near-Field Scanning Optical Microscopy (SERS-NSOM)

Near-field scanning optical microscopy (NSOM) is becoming a useful tool for high-resolution optical imaging well below the diffraction limit. So far, the greatest advantage of this novel scanning probe technique, namely the addition of a spectral to the spatial resolution, has been mostly restricted to fluorescence and luminescence spectroscopy [19]. We have reported the combination of NSOM and SERS to obtain spectral, spatial, and chemical information of molecular adsorbates with subwavelength lateral resolution [18,19]. Briefly, a commercially available near-field scanning optical microscope (Aurora, Topometrix, Santa Clara, CA) is the basic unit of our setup, an optical fiber was etched by the protection layer method, in which the fiber was etched by a mixture of 40 % HF and an oil (e.g., squalane or p-xylene). The very end of the etched tip has dimensions smaller than 200 nm. The fiber was further coated with aluminum having a thickness of 120 nm in an electron

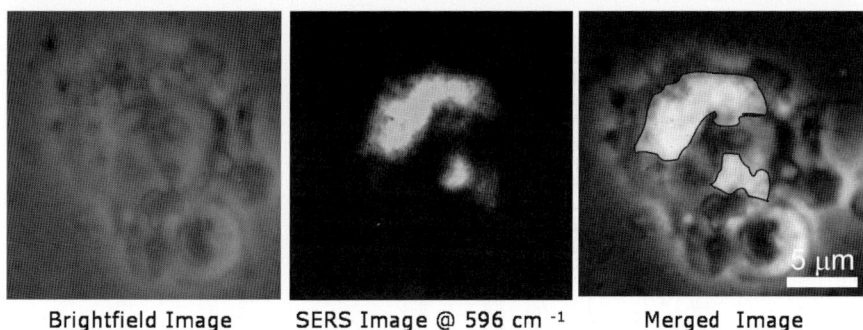

Fig. 6. SERS images and spectra of CHO cells incubated with cresyl violet-labeled silver colloidal particles. The bright-field image (*left*), the total SERS image (*middle*) and the composite image (right) of a fixed CHO cell are shown. The SERS image was acquired at a SERS intensity signal of 596 cm^{-1}. Laser excitation: 632.8 nm He–Ne; *scale bar:* 5 µm

gun evaporation system at a pressure of 10^{-5} mbar. During a typical run, the tip was kept in shear-force feedback, and set at the desired lateral position until the CCD camera finished data collection, before it was moved to the next position. The light was collected in reflection mode using a microscope objective (NA = 0.65) and focused onto a 100 µm glass fiber that acted as a confocal element. The lateral resolution with this setup was 80 nm to 100 nm. The laser power at the sample was 1 µW. All Raman spectra were measured with the 488 nm line of the argon laser. Using a 50 µm slit in the spectrometer gives a spectral resolution of < 5 cm^{-1}. The SERS-active substrate was prepared by using a 50 µl volume of a suspension of Teflon submicrometer-size spheres (diameter 200 nm) spin coated on the surface of a glass substrate at 800–2000 rpm for about 20 s. A silver-island film was deposited on the nanosphere-coated substrate in an electron beam evaporation system (CVE 301 EB, Cooke Vacuum Products, Norwalk, CT) at a deposition rate of 2 nm/s. The thickness of the silver layer deposited was 100 nm. This sub-

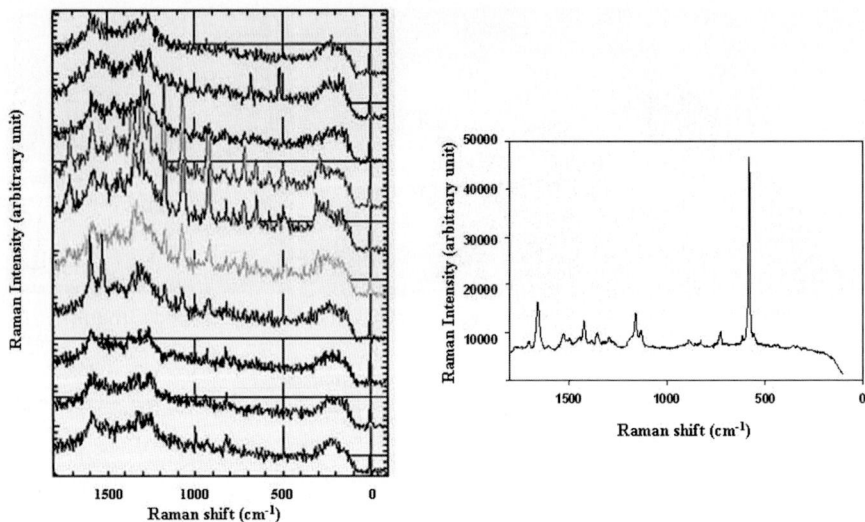

Fig. 7. Time evolution of the Raman spectrum of BCB-labeled DNA on microspheres: 1486th–1495th scans out of 1800 repeats (from *bottom* to *top*). Temperature: 23 °C; excitation wavelength: 488 nm with a slit of 25 μm; laser power: ~ 1.3 mW at sample; and acquisition interval: acq 1 s / pause 1 s

strate-preparation technique provides excellent reproducibility (5 % to 10 % standard deviation from batch to batch).

Near-field SERS spectra of BCB on silver substrates have been obtained. Spectra from as few as 300 molecules can be recorded. Figure 7a depicts ten progressions of Raman spectra obtained from the same spot, with each individual spectrum acquired with an integration time of 1 s. Figure 7b shows a far-field reference spectrum of the BCB label obtained by a Renishaw Raman microscope. The near-field and far-field Raman spectra of the BCB dye are almost identical, except that the peak intensity at 580 cm^{-1} appears to be more dominant in far-field, while in near-field, all the peak intensities are comparable [18]. The line at ~ 1590 cm^{-1} has generally been assigned to the typical Raman-active band of highly ordered macroscopic graphite crystals that is associated with inplane atomic displacement and has E$_{2g}$ symmetry. The line at ~ 1340 cm^{-1} has been attributed to the hydrocarbon CH bending modes and modes observed in disordered graphite [44], and it is the symmetry forbidden D-band that becomes active with the relaxation of the wavevector conservation in nanocrystalline and glassy carbon [45].

Surface studies of Ag-island substrates using AFM shows a regularly covered, grainy metal-island distribution. A maximum height of aggregated silver particles of 85 nm in the z-direction can be deduced from the data. The average particle size outside the agglomerates varies between 40 nm and 100 nm full-width-at-half-maximum (FWHM) laterally and 20 nm to 30 nm in height.

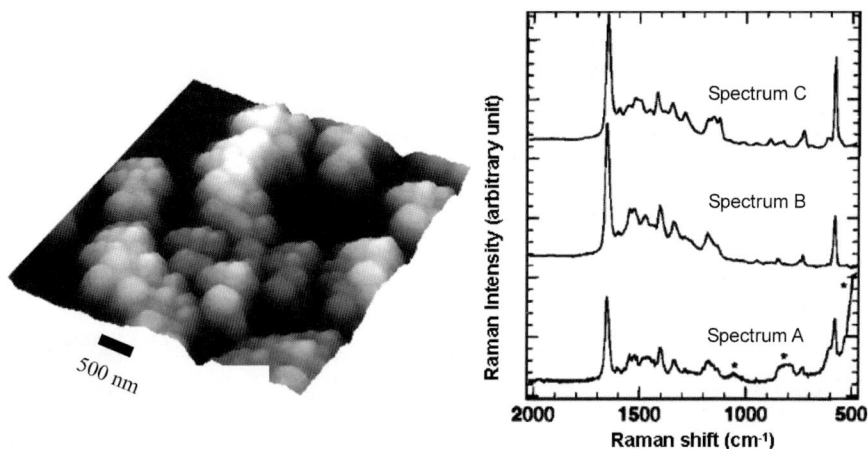

Fig. 8. *Left:* AFM image of a silver substrate coated with BCB-labeled DNA. The silver-coated Teflon nanospheres show a strong agglomeration on the glass cover slide. *Right:* Near-field and far-field surface-enhanced Raman spectra of BCB-labeled DNA adsorbed on silver SERS substrates: (**A**) near-field Raman spectrum of the BCB label (exposure time: 300 s), (**B**) micro-Raman spectrum of the BCB label (exposure time: 3 s), and (**C**) far-field reference spectrum of the BCB label. The bands marked with an *asterisk* in spectrum B are Raman bands from the glass in the fiber tip

Surface-enhanced Raman spectra of adsorbed BCB molecules on silver-island substrates, which have been recorded in the near-field mode with a ~ 200 nm aperture are shown in Fig. 8 [18]. In this figure, the spectra A, B, and C have been obtained at different tip positions on the SERS substrates. The exposure time of the CCD camera was 100 s. The broad peak at 800 cm^{-1} can be assigned to Raman scattering of the glass fiber due to SiO$_4$ vibrations. Local chemical identification of the BCB molecules can be performed by analyzing the group of peaks between 500 cm^{-1} and 1500 cm^{-1}, which can be assigned to different vibrational modes of the aromatic ring system. A local Raman enhancement factor of 10^{13} or greater can be derived from a comparison with fluorescence measurements [18, 19].

3.4 SERS Molecular Sentinels

Recently, we reported a new diagnostic approach that involves a plasmonics-based nanoprobe, referred to as a "molecular sentinel" (MS), consisting of a DNA hairpin loop having a Raman label molecule at one end and, a metal nanoparticle at the other end [46, 47]. The MS nanoprobe combines two basic features: 1. the modulation of the plasmonics effect to change the SERS intensity of the label, and 2. the specificity of a DNA hairpin loop sequence to recognize and discriminate a variety of molecular target sequences. The struc-

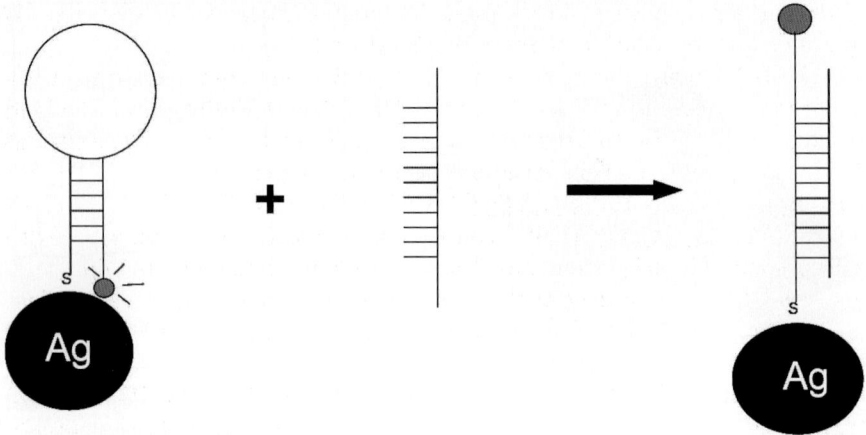

Fig. 9. Signaling concept of SERS molecular sentinels (SERS MS): SERS signal is detected when the MS probe is in the hairpin (closed-state) conformation, whereas the SERS signal is decreased in the open (hybridization) state

ture and operating principle of the SERS MS detection scheme is depicted in Fig. 9.

The MS basic structure consists of a DNA probe sequence that has a Raman label at one end and a metallic (e.g., Ag or Au) nanoparticle attached to the other end via a thiol group. The metal nanoparticle serves as a signal-enhancing system (nanoenhancer) for the SERS signal associated with the Raman label. The gene-diagnostic system consists of a DNA probe sequence having a middle section that contains a sequence complementary to the target sequence to be detected and two arms that have complementary sequences in order to form a hairpin-loop configuration under normal conditions. We design the hairpin-loop configuration such that the Raman label is in contact with or in close proximity ($< 1\,\mathrm{nm}$) to the nanoparticle that induces a strong SERS effect on the adjacent Raman label molecule. Under normal conditions, the hairpin-loop configuration is consistent with a strong SERS signal when the Raman label is excited with laser light. However, when hybridization occurs with a complementary target DNA, the hairpin loop opens and physically separates the Raman label from the nanoparticle. Since the Raman enhancement, E_R, depends strongly on the distance d, between the Raman label and the metal nanoparticle ($E_R = d^{-12}$), the hybridization process leads to a strong decrease in the SERS effect and produces a decrease in the SERS signal upon laser excitation. The plasmonics nanoprobes play the role of molecular sentinels patrolling the sample solution with their warning light "switched on" when no significant event occurs. Whenever a target species is identified and detected, the molecular sentinels extinguish their light, thus providing measurable optical signal changes. The usefulness of the MS approach in medical diagnostics is demonstrated in the detection of the *gag* gene sequence of the human immunodeficiency virus type1 (HIV1) in a homogenous solution at room temperature [46].

The HIV1 molecular sentinel nanoprobes that incorporated a partial sequence for a human immunodeficiency – HIV1 isolate Fbr020 from Thailand reverse transcriptase (pol) gene – were designed with a stem sequence that produced a stable hairpin structure at room temperature. The HIV1 MS nanoprobe (5′-HS-(CH$_2$)$_6$)-<u>GATCGC</u>CCTTTTCCATTTCCATACATAT TTCTGTTA<u>GCGATC</u>-R6G) consisted of a 42 bp DNA hairpin probe with a 6 bp stem. The 6 bp stem was designed to have rhodamine 6G on the 3′ end and a thiol substituent at the 5′ end that could then be used for covalent coupling to the surface of silver nanoparticles. The complementary arms of the MS are shown as the underlined sequence. Figure 10 shows the SERS spectra from the HIV1 SERS-MS nanoprobe targeted to the HIV1 gene in the absence and in the presence of the target DNA sequence. In the absence (blank) of the target DNA (Fig. 10, upper curve), the hairpin conformation of the HIV1 SERS-MS nanoprobes remains stable, resulting in a close proximity of the rhodamine 6G label with the surface of the silver nanoparticles (nanoenhancers). As a result, a strong SERS signal from rhodamine 6G is detected. However, in the presence of the complementary HIV1 target sequence (Fig. 10, lower curve) the HIV1 SERS-MS nanoprobes bind to the target DNA (positive recognition) thus resulting in the physical separation of the rhodamine 6G label from the surface of the silver nanoparticle and as a result the SERS signal from the MS nanoprobes becomes significantly quenched. It is noteworthy that the presence of a noncomplementary target sequence (Fig. 10, middle curve) did not significantly affect the SERS signal, indicating that the hairpin-loop structure of the HIV1 SERS-MS nanoprobes was not disrupted, thus indicating a negative recognition [46].

4 Conclusion

The development of diverse SERS probes and portable instrumentation has spurred renewed interest in Raman as a practical analytical tool in other areas [48, 49, 50, 51, 52]. For medical diagnostics, the SERS gene-probe technology has great potential for a wide variety of applications in areas where nucleic acid identification is involved. With the SERS gene-label multiplex technique, complex samples, which may contain different genes related to medical diseases and infectious pathogens, can be separated and directly analyzed using multiple SERS gene labels simultaneously. In biomedical, genomics and proteomics applications and in high-throughput analysis, the SERS gene-probe technology could lead to the development of detection methods that minimize the time, expense, and variability of preparing samples.

Furthermore, the use of AOTF-based hyperspectral surface-enhanced Raman imaging (HSERI) will allow the recording of multiple spectral images with both spatial and spectroscopic information of chemical components.

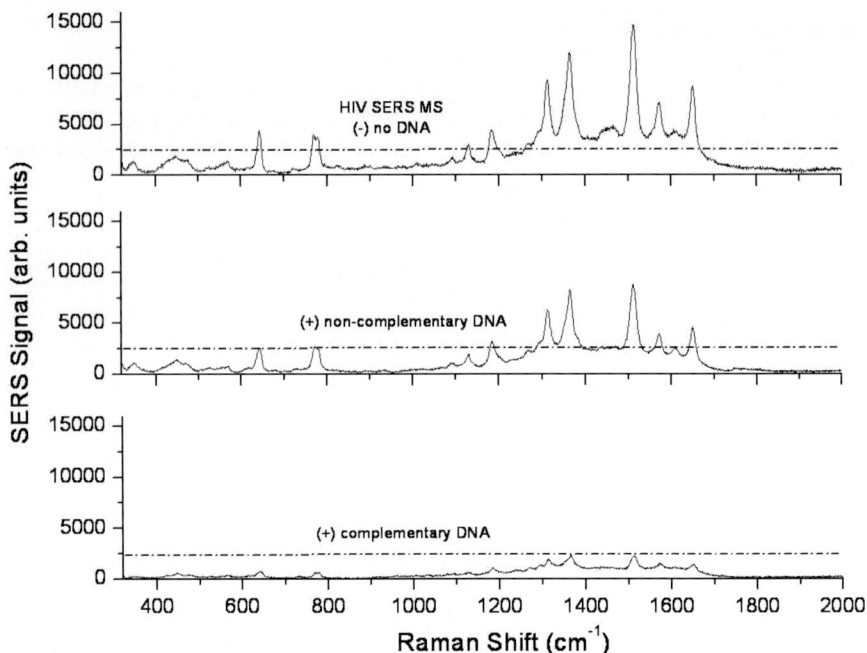

Fig. 10. Detection of HIV1 sequence using SERS "molecular sentinels" probes: SERS spectra of HIV1 SERS-MS nanoprobe with no target DNA sequence (*upper curve:* blank) and in the presence of a noncomplementary DNA target sequence (*middle curve:* negative diagnostic) and a complementary HIV1 DNA target (*lower curve:* Positive diagnostic). The threshold level of the SERS signal from the SER-MS nanoprobe is indicated by the *dotted line*

The HSERI technology has the capability of rapidly detecting the spatial distribution of molecules with high resolution without applying additional data-processing algorithms. By changing the surface functionalities of metal nanoparticles and controlling the aggregation, the variability in the SERS signal could be greatly reduced. The system allows rapid imaging and has the potential for high-throughput analysis.

Acknowledgements

This work was sponsored by the Office of Biological and Environmental Research, US Department of Energy, under Contract DE-AC05-00OR22725 with UT-Battelle, LLC, and by the Laboratory Directed Research and Development Program (Advanced Plasmonics Sensors Project) at Oak Ridge National Laboratory. Fei Yan and Musundi B. Wabuyele are also supported by an appointment to the Oak Ridge National Laboratory Postdoctoral Research Associates Program, administered jointly by the Oak Ridge National Laboratory and Oak Ridge Institute for Science and Education.

References

[1] D. L. Jeanmaire, R. P. V. Duyne: J. Electroanal. Chem. **84**, 1 (1977)

[2] M. G. Albrecht, J. A. Creighton: J. Am. Chem. Soc. **99**, 5215 (1977)

[3] M. Fleischman, P. J. Hendra, A. J. McQuillan: Chem. Phys. Lett. **26**, 163 (1974)

[4] T. Vo-Dinh, M. Y. K. Hiromoto, G. M. Begun, R. L. Moody: Anal. Chem. **56**, 1667 (1984)

[5] T. Vo-Dinh, G. H. Miller, J. Bello, R. Johnson, R. L. Moody, A. Alak, W. H. Fletcher: Talanta **36**, 227 (1989)

[6] M. Meier, A. Wokaun, T. V. Dinh: J. Phys. Chem. **89**, 1843 (1985)

[7] T. Vo-Dinh, M. Meier, A. Wokaun: Anal. Chim. Acta **181**, 139 (1986)

[8] P. D. Enlow, M. C. Buncick, R. J. Warmack, T. Vo-Dinh: Anal. Chem. **58**, 1119 (1986)

[9] A. L. Alak, T. Vo-Dinh: Anal. Chem. **59**, 2149 (1987)

[10] T. Vo-Dinh, M. Uziel, A. Morrison: Appl. Spectrosc. **41**, 605 (1987)

[11] J. M. Bello, D. L. Stokes, T. Vo-Dinh: Anal. Chem. **61**, 1779 (1989)

[12] J. M. Bello, D. L. Stokes, T. Vo-Dinh: Anal. Chem. **62**, 1349 (1990)

[13] A. Helmenstine, M. Uziel, T. Vo-Dinh: J. Tox. Environ. Health **40**, 195 (1993)

[14] Z. Q. Tian, B. Ren: Ann. Rev. Phys. Chem. **55**, 197 (2004)

[15] S. M. Nie, S. R. Emory: Science **275**, 1102 (1997)

[16] K. Kneipp, Y. Wang, H. Kneipp, L. T. Perelman, I. Itzkan, R. Dasari, M. S. Feld: Phys. Rev. Lett. **78**, 1667 (1997)

[17] K. Kneipp, H. Kneipp, R. Manoharan, E. B. Hanlon, I. Itzkan, R. R. Dasari, M. S. Feld: Appl. Spectrosc. **52**, 1493 (1998)

[18] V. Deckert, D. Zeisel, R. Zenobi, T. Vo-Dinh: Anal. Chem. **70**, 2646 (1998)

[19] D. Zeisel, V. Deckert, R. Zenobi, T. Vo-Dinh: Chem. Phys. Lett. **283**, 381 (1998)

[20] T. Vo-Dinh, K. Houck, D. L. Stokes: Anal. Chem. **66**, 3379 (1994)

[21] N. Isola, D. L. Stokes, T. Vo-Dinh: Anal. Chem. 1998, **70**, 1352 (1998)

[22] T. Enderle, T. Ha, D. F. Ogletree, D. S. Chemla, C. Magowan, S. Weiss: Proc. Natl. Acad. Sci. USA **94**, 520 (1997)

[23] D. J. Stephens, V. J. Allan: Science **300**, 82 (2003)

[24] J. P. Knemeyer, D. P. Herten, M. Sauer: Anal. Chem. **75**, 2147 (2003)

[25] K. E. Shafer-Peltier, A. S. Haka, M. Fitzmaurice, J. Crowe, J. Myles, R. R. Dasari, M. S. Feld: J. Raman Spectrosc. **33**, 552 (2002)

[26] S. W. E. Van de Poll, T. C. Bakker Schut, A. Van der Laarse, G. J. Pupples: J. Raman Spectrosc. **33**, 544 (2002)

[27] J. A. Timlin, A. Garden, M. D. Morris, R. M. Rajachar, D. H. Kohn: Anal. Chem. **72**, 2229 (2000)

[28] M. D. Schaeberle, H. R. Morris, J. F. Turner, P. J. Treado: Anal. Chem. **71**, 175A (1999)

[29] T. Vo-Dinh, A. Alak, R. L. Moody: Spectrochim. Acta B **415**, 605 (1988)

[30] T. Vo-Dinh: Trends Anal. Chem. **17**, 557 (1998)

[31] D. L. Stokes, T. Vo-Dinh: Sens. Actuat. B Chem. **69**, 28 (2000)

[32] L. R. Allain, T. Vo-Dinh: Anal. Chim. Acta **469**, 149 (2002)

[33] M. Culha, D. L. Stokes, L. R. Allain, T. Vo-Dinh: Anal. Chem. **75**, 6196 (2003)

[34] T. Vo-Dinh, L. R. Allain, D. L. Stokes: J. Raman Spectrosc. **33**, 511 (2002)

426   Tuan Vo-Dinh et al.

[35] K. Kneipp, A. S. Haka, H. Kneipp, K. Badizadegan, N. Yoshizawa, C. Boone, K. E. Shafer-Peltier, J. T. Motz, R. R. Dasari, M. S. Feld: Appl. Spectrosc. **56**, 150 (2002)
[36] A. Zumbusch, G. R. Holtom, X. S. Xie: Phys. Rev. Lett. **82**, 4142 (1999)
[37] N. Leopold, B. J. Lendl: J. Phys. Chem. B **107**, 5723 (2003)
[38] Y. C. Cao, R. Jin, J. Nam, C. S. Thaxton, C. A. Mirkin: J. Am. Chem. Soc. **125**, 14676 (2003)
[39] Y. C. Cao, R. Jin, C. A. Mirkin: Science **297**, 1536 (2002)
[40] L. Wang, C. Yang, W. Tan: Nano Lett. **5**, 37 (2005)
[41] K. Nithipatikom, M. J. McCoy, S. R. Hawi, K. Nakamoto, F. Adar, W. B. Campbell: Anal. Biochem. **322**, 198 (2003)
[42] W. E. Doering, S. Nie: Anal. Chem. **75**, 6171 (2003)
[43] J. Ni, R. J. Lipert, G. B. Dawson, M. D. Porter: Anal. Chem. **71**, 4903 (1999)
[44] C. E. Taylor, S. D. Garvey, J. E. Pemberton: Anal. Chem. **68**, 2401 (1996)
[45] P. J. Moyer, J. Schmidt, L. M. Eng, A. J. Meixner: J. Am. Chem. Soc. **122**, 5409 (2002)
[46] M. B. Wabuyele, T. Vo-Dinh: Anal. Chem, **77**, 7810 (2005)
[47] T. Vo-Dinh, M. B. Wabuyele: Nano. Biotechnol. in press
[48] D. L. Stokes, Z. Chi, T. Vo-Dinh: Appl. Spectrosc. **58**, 292 (2004)
[49] B. M. Cullum, J. Mobley, Z. Chi, D. L. Stokes, G. H. Miller, T. Vo-Dinh: Rev. Sci. Instrum. **71**, 1602 (2000)
[50] M. Volkan, D. L. Stokes, T. Vo-Dinh: Sens. Actuators B-Chem. **106**, 660 (2005)
[51] M. B. Wabuyele, F. Yan, G. D. Griffin, T. Vo-Dinh: Rev. Sci. Instrum. **76** (2005)
[52] F. Yan, M. B. Wabuyele, G. D. Griffin, A. A. Vass, T. Vo-Dinh: IEEE Sensors J. **5**, 665 (2005)

Index

Ultrasensitive Immunoassays Based on Surface-Enhanced Raman Scattering by Immunogold Labels

Hye-Young Park[1], Jeremy D. Driskell[1], Karen M. Kwarta[1], Robert J. Lipert[1], Marc D. Porter[1], Christian Schoen[2], John D. Neill[3], and Julia F. Ridpath[3]

[1] Institute for Combinatorial Discovery, Ames Laboratory-US DOE, and Departments of Chemistry and of Chemical and Biological Engineering, Iowa State University, Ames, Iowa, 50011
[2] Concurrent Analytical, Inc., Laramie, Wyoming, 82070
[3] Virus and Prion Diseases of Livestock Unit, National Animal Disease Center, United States Department of Agriculture, Ames, Iowa 50010
mporter@porter1.ameslab.gov

1 Introduction

Many immunoassays (e.g., cancer-marker screening) [1] require the concomitant determination of several analytes. There are two chip-based approaches often employed to address this need. One approach immobilizes different antibodies on a solid, chip-scale support at spatially separated addresses. Multiple antigens can then be detected using the same label, with identification based on address location [2, 3]. Alternatively, different labels can be used to detect different analytes simultaneously in the same spatial area [4, 5, 6, 7, 8, 9, 10, 11, 12, 13]. Depending upon the strategy, readout can be achieved by a variety of different techniques [1], including scintillation counting [14], fluorescence [6, 7], absorption [8], electrochemistry [9], chemiluminescence [11], Rayleigh scattering [12], and Raman scattering [13]. Of these techniques, fluorescence spectroscopy holds a dominant position, primarily because of its high sensitivity.

In the past few years, breakthroughs related to signal enhancement and instrumentation have rekindled interest in the potential of Raman spectroscopy as an analytical tool for chip readout. Scattering cross sections for conventional Raman spectroscopy are intrinsically weak, and are comparatively much lower than analogs for infrared spectroscopy, however, surface-enhanced Raman scattering (SERS) has demonstrated the ability to detect picomolar and even femtomolar amounts of biolytes. As discussed in several other contributions to this collection, this development results mainly from the large enhancements of electrical fields that arise upon excitation at the surface of several types of roughened metal surfaces [15, 16]. Among the metallic substrates employed by SERS are metal nanoparticles (e.g., gold and silver colloids). Impressively, recent reports have shown that some types of nanoparticle substrates can yield SERS intensities for adsorbed species comparable

K. Kneipp, M. Moskovits, H. Kneipp (Eds.): Surface-Enhanced Raman Scattering – Physics and Applications, Topics Appl. Phys. **103**, 427–446 (2006)
© Springer-Verlag Berlin Heidelberg 2006

to or even exceeding those from fluorescence [17, 18]. Nanoparticles also appear to provide a route to the reproducible fabrication of enhancing surfaces, potentially serving as a means for achieving a more reliable calibration of a SERS response. Furthermore, the analytical merits of metallic nanoparticles as labels in various bioassays are well established, with the best-known successes using colorimetry [19], photothermal deflection [20, 21], surface-plasmon resonance [22], and scanning electron microscopy [23] as readout tools.

Technological advances have transformed the instrumentation requisite for high-sensitivity Raman spectroscopy measurements. A little more than a decade ago, Raman spectroscopy typically employed large and expensive laboratory hardware that consisted of high-power lasers, double and triple monochromators, specialized detectors, and vibrational-isolation systems. Today, breakthroughs in optical filters, array detectors, and fiber optics have moved Raman spectroscopy to the industrial plant floor and other demanding environments. These developments have also markedly reduced equipment and labor costs, while maintaining a remarkably high level of performance in a rugged field-deployable instrument. As a consequence of these innovations, the next decade will, in all likelihood, witness the widespread deployment of Raman spectroscopy into areas previously not imagined.

Based on these precedents, we recently began to explore the use of SERS in the clinical arena by assessments of the utility of SERS in immunoassays [24]. Many other laboratories have initiated similar pursuits [24, 25, 26, 27, 28, 29, 30, 31, 32]. Along these lines, several additional attributes of SERS with respect to fluorescence should be noted [15, 33, 34].

1. Raman bands are generally 10–100 times narrower than most fluorescence bands. This difference minimizes the potential overlap of different labels in a given spectral region.
2. The optimum excitation wavelength for SERS is dependent on the size and composition of the nanoparticle. As a consequence, only one excitation wavelength is required for multiple analytes.
3. Raman scattering is not sensitive to humidity or affected by oxygen and other quenchers. This characteristic facilitates applications in a variety of environments.
4. The SERS signal is less prone to photobleaching, potentially enabling one to signal average for extended time periods to lower the limit of detection (LOD).

The purpose of this Chapter is to review our recent efforts in the development of SERS in sandwich immunoassays in which surface-functionalized gold colloids are employed as extrinsic Raman labels (ERLs), a strategy that requires neither resonance enhancement nor enzymatic amplification. The method utilizes the strong SERS signal from aromatic compounds (i.e., Raman reporter molecules (RRMs)) that are immobilized on gold nanoparticles and subsequently coupled to a molecular recognition element such as an antibody. The identity of each antigen selectively extracted by a capture antibody substrate is therefore determined from the characteristic SERS spectrum of the nanoparticle-bound RRM linked to the tracer antibody, whereas the amount of antigen is quantified by the spectral intensity of the reporter species.

To this end, we first examine key details related to the design, synthesis, and stabilization of ERLs, highlighting key components of this type of assay in terms of methodologies that minimize nonspecific adsorption and fundamental factors that maximize the magnitude of the enhancement of the SERS signal. This discussion is followed by a recent example of the application of ERLs to early disease detection in an assay for prostate cancer [24]. This Chapter then takes the first step to the potential extension of this concept to serve as a chip-scale readout methodology for fast, sensitive detection of foodborne pathogenic bacteria (i.e., *Escherichia coli* (*E. coli*) O157:H7) and simulants for biowarfare (BW) agents (i.e., the bacteria *Yersinia pestis* (plague)). Furthermore, we demonstrate the ultimate in sensitivity by the single organism detection of captured *E. coli* O157:H7. We conclude by presenting results that demonstrate the low-level detection of porcine parvovirus (PPV) in a clean sample solution to emphasize the potential sensitivity of this method. Importantly, all of these targets differ greatly in size (e.g., 1 nm for PSA to 3 μm for *E. coli* O157:H7). This study, therefore, demonstrates the flexibility in terms of analytes and assesses the merits of our assay platform to address needs (e.g., accuracy, sensitivity, and specificity) in the detection of BW agents and pathogens central to food and water security.

2 Assay Design

Figure 1 depicts the three-step, sandwich assay [24, 35]. It involves:

1. use of a capture antibody substrate to selectively extract and concentrate antigens (i.e., biolytes) from solution
2. selective tagging of captured antigens with ERLs
3. readout by Raman spectroscopy.

This section details the design of key components of our assay platform.

2.1 Preparation of Capture Antibody Substrates

The preparation of capture antibody surfaces begins with creation of the underlying gold substrate, followed by the chemisorption of a disulfide-based coupling agent and the subsequent covalent immobilization of a layer of the capture antibody [24, 35]. We currently employ template-stripped gold (TSG) as the underlying "biochip" surface, which is prepared via established protocols [36, 37]. TSG has an extremely flat gold surface, which is well suited for

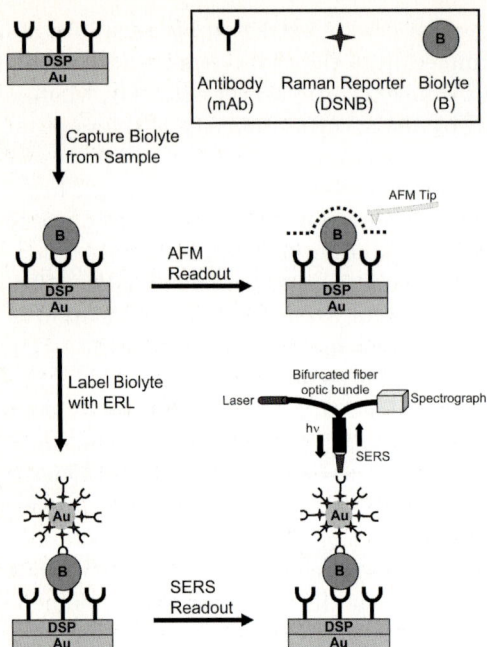

Fig. 1. Three-step sandwich assay and readout

imaging the presence of nanometer-sized objects (e.g., proteins and viruses) by atomic force microscopy (AFM) [36] and provides a means to validate performance by cross-correlation with the SERS response.

There are three more steps involved in the capture substrate preparation. First, a TSG chip is addressed by soft lithography [38]. In this step, a polydimethylsiloxane stamp with a centered, 3 mm hole, is immersed in a 2 mM solution of octadecanethiol (ODT) for 1 min and dried with nitrogen gas. The dried stamp is then gently pressed into conformal contact with a TSG chip for 30 s. This process leaves an uncoated gold area (3 mm diameter) that is surrounded by a hydrophobic ODT-derived monolayer that acts to confine small droplets of aqueous fluids.

Next, the substrate is submerged for ~ 12 h in a 0.1 mM ethanolic solution of either dithiobis(succinimidyl undecanoate) (DSU) or dithiobis(succinimidyl propionate) (DSP). This step chemisorbs a monolayer of the corresponding thiolate, which then serves as a coupling agent for the covalent immobilization of proteins (e.g., capture antibodies) and other amine-containing molecules via amide linkages. The construction of the capture antibody substrate is completed by pipetting a dilute buffered solution of the capture antibody onto the substrate and a 12 h incubation in a humidity chamber. The effectiveness of this step has been verified by several characterization techniques (e.g., atomic force microscopy (AFM) and infrared reflection spectroscopy (IRS)).

Lastly, the surface is treated for $\sim 1\,\mathrm{h}$ with a blocking agent (e.g., bovine serum albumin (BSA; 1% w/w in borate buffer at pH 8.3). In this step, unreacted succinimidyl groups of the adlayer of coupling agent are "capped" with BSA. Other blockers, such as human serum albumin, evaporated milk, and various other formulations, can also be used, with the selection often the result of a panel of trial-and-error tests.

There are, however, two issues that require consideration. The first deals with the overall efficiency of the coupling reaction. IRS has shown that a DSU-derived monolayer is exhaustively converted to an amide by reaction with a small molecule like ethylamine. These experiments have also demonstrated that the spectrum was devoid of evidence for a carboxylate functionality, a group that would be formed by the hydrolysis of the succinimidyl endgroup.

While the reaction of the endgroup with a small molecule ethylamine appears exhaustive, the same cannot be said for the reaction with a larger, more complex molecule like an IgG protein, which has 60–80 lysine residues having primary amines that can react with the succinimidyl terminus of the adlayer. IRS after such a coupling reaction indicates that bands corresponding to the succinimidyl endgroup are still present at 20% to 25% of their original intensities after treatment with IgG. The less-than exhaustive conversion of the endgroup is attributed, at least in part, to steric effects, recognizing the much larger cross section of IgG ($\sim 120\,\mathrm{nm}^2$) with respect to the succinimidyl endgroup ($\sim 2\,\mathrm{nm}^2$). We add that the distribution of lysine residues throughout the protein structure indicates that the immobilized antibody can adopt a distribution of orientations.

One more issue needs to be addressed. How viable are the immobilized antibodies? Recent findings by *Shannon* and coworkers provide a basis for assessment of this issue [39]. The results indicated that only $\sim 30\%$ of the immobilized IgG antibodies were effective in binding its antigen, a situation attributed to steric effects with respect to the spatial orientation of the capture antibody and/or to interactions with the adlayer causing the protein to denature. This finding suggests that synthetic routes that may afford a more favorable level of spatial orientation for the immobilized antibody may lead to a lower level of detection. However, the improvement would at best be only a factor of 2–3, and has therefore not been pursued.

2.2 ERL Preparation

The preparation of the Raman reporter-labeled immunogold particles, using commercially available gold nanoparticles, is shown in Fig. 2. In building these labels, various derivatives of dithiobis(benzoic acid), which could readily be converted to the corresponding succinimide ester with N-hydroxysuccinimide (NHS), were synthesized. Of those tested, 5,5′-dithiobis(succinimidyl-2-nitro-benzoate) (DSNB) proved particularly attractive because of the intrinsically strong Raman scattering cross-section of its symmetric NO_2 stretch. As such, treatment of colloidal gold with this derivative yields a coating of the thiolate

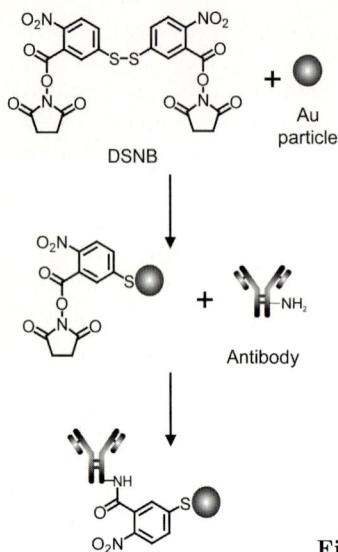

Fig. 2. Preparation of extrinsic Raman labels (ERLs)

of DSNB, which, paralleling the chemistry used to construct the capture antibody substrates, can readily couple to the primary amines of a tracer antibody by formation of an amide linkage. Moreover, this design strategy minimizes the distance between the gold surface and scattering center of the label. Minimization of this distance is of significance because a simplified electromagnetic model indicates that enhancement varies inversely with the 12th power of the distance between scattering center and metal particle [40, 41, 42]. Our labels are therefore constructed to place the scattering center as close to the surface as possible in order to fully exploit this prediction.

There are three more important advantages to our ERLs, all of which arise from the use of gold colloids in label development.

1. Gold is a strongly enhancing surface at long-wavelength excitation. Readout is therefore less susceptible to interference by native fluorescence, which is excited at shorter wavelengths.
2. Each nanoparticle is coated with a large number of RRMs (10^3–10^5). As a consequence, the response of an individual binding event is markedly amplified.
3. Gold surfaces are readily modified by thiols and disulfides [43]. This chemistry provides a versatile and easy-to-modify platform for label construction.

Prior to modifying with DSNB, the pH of the nanoparticle solution is adjusted with borate buffer (pI 8.5). This pI: 1. is above the pH of the antibody, which inhibits aggregation of the labeled nanoparticles; and 2. deprotonates the amines of the antibody, which favors the formation of an amide linkage

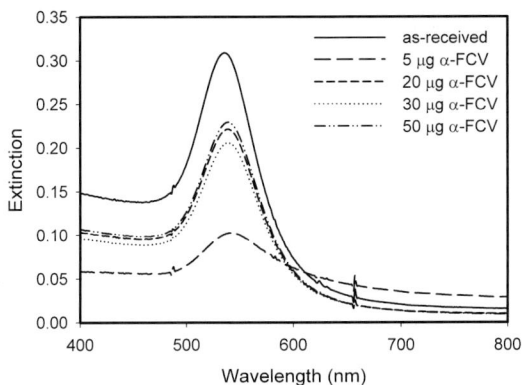

Fig. 3. UV-visible extinction spectra of colloidal gold (60 nm) before and after mixing with anti-FCV mAb (5 μg to 50 μg) for 1 h, followed by addition of NaCl to final concentration of 150 mM NaCl [38]

by reaction with the succinimidyl ester of DSNB. The next steps then modify the particles with DSNB and the tracer antibody [24].

The effectiveness of the antibody coupling step is often assessed using flocculation tests [44, 45]. A recent example of results is summarized by the extinction spectra in Fig. 3 from work in the development of an ERL for feline calicivirus (FCV), which is a simulant for the human calicivirus known as the Norwalk virus. This study systematically varied the amount of anti-FCV monoclonal antibody (mAb) added to the colloidal solution and monitored aggregation upon the addition of NaCl to physiological concentration (150 mM) [38].

As shown in Fig. 3, the as-received particle solution has a strong extinction maximum at 535 nm that is consistent with the location of the plasmon resonance of isolated gold particles with an average diameter of 60 nm [46]. The precipitation of particles from the solution modified with 5 μg of anti-FCV mAb is revealed by the large decrease in the strength of the plasmon band. The broadening and shift to longer wavelengths is also indicative of aggregate formation. The changes upon addition of larger quantities of anti-FCV are in reasonable agreement with expectations from the dilution that results from each of the solution manipulation steps, and are an indication of stable colloidal solutions. Moreover, there was no observable precipitate for samples stored for several weeks. All subsequent procedures therefore modified the particles by addition of 20 μg of mAb.

2.3 Immunoassay Protocols

Immunoassays are typically carried out by exposure of capture substrates to antigens for a predetermined time (3 h to 12 h) in a humidity chamber, followed by rinsing with buffer. The antigen binding step can be validated through optical microscopy and AFM investigations. These microscopy techniques are used to enumerate pathogens and cross-correlate pathogen surface concentration with the SERS intensity. Finally, a small amount of the ERLs is pipetted onto the substrate, incubated again (3 h to 12 h) in a humidity chamber, and rinsed with buffer. In addition to quantifying the analyte, the label imparts a high level of confidence in the identification of the analyte through the use of a second highly specific antibody. Assay results were used to optimize capture substrate preparation and ERL preparation with respect to minimization of nonspecific adsorption through the use of blocking agents (e.g., BSA) and surfactant additives (e.g., Tween 20 and Tween 80).

3 Spectroscopic Instrumentation

Raman spectroscopic measurements were performed using a NanoRaman I spectrophotometer (Concurrent Analytical). The NanoRaman I instrument is a portable, field-deployable instrument equipped with a fiber-optic probe, a Czerny–Turner imaging spectrometer ($f/2.0$, $6\,\mathrm{cm}^{-1}$ to $8\,\mathrm{cm}^{-1}$ resolution), thermoelectrically cooled CCD (Kodak 0401 E), and HeNe laser (632.8 nm). The incident laser light is focused to a $25\,\mu\mathrm{m}$ spot size on the substrate at normal incidence using an objective with a numerical aperture of 0.68. The power at the sample is $\sim 3\,\mathrm{mW}$. The same objective is used to collect the scattered radiation. Spectral integration times for the data herein were 1 s to 60 s. The system is environmentally sealed, allowing for extensive use in field applications. The footprint of the system is $16 \times 8 \times 8\,\mathrm{in}$, has a weight of $\sim 14\,\mathrm{lbs}$, and requires less than 100 W of power. The operational lifetime of the system is limited by the laser, which is $\sim 10\,000\,\mathrm{h}$. There are no moving parts. This setup enables "point-and-shoot" spectral acquisition, while performing at levels comparable to most laboratory-scale instruments but at much lower cost.

4 Maximization of Signal Strength

In addition to strategies that mitigate nonspecific adsorption, these assays can be greatly improved by considerations of the effect of nanoparticle size on the SERS enhancement factor. Since enhancement originates to a large extent from an amplification of the electromagnetic field due to excitation of the plasmon resonance, the dependence of the plasmon resonance on particle size and shape is central to understanding the fundamental factors that can be manipulated to maximize sensitivity. Current theory [47], as extensively detailed in other contributions to this text, predicts correlation of enhancement with particle size and excitation frequency. However, experimental studies at the individual particle level show a strong discrepancy with respect to theoretical expectations [48, 49, 50]. In each case, the discrepancy between theory and experiments was attributed to the presence of aggregation or facets on the particle surface, which may act as locations of sharp surface asperity with unusually large electric-field enhancements and/or as sites active for charge-transfer-based enhancements.

Collectively, these and other experiments demonstrate the need for more monodisperse nanoparticles that are void of these anomalies in order to construct a viable assay based on SERS. It is, however, important to note the changes in particle size can have contrasting effects on the assay. Size alters the plasmon-resonance frequency, which has an impact on the optimal excitation wavelength. Size also changes the number of RRMs coated on an individual particle. Moreover, size has a direct impact on the time required for the ERLs to bind to the captured antigen and on the stability of the colloidal solution with respect to particle precipitation. As a consequence, assay optimization reflects a compromise between maximization of SERS enhancement, the number of labels bound to a particle, particle stability, and minimization of incubation time.

Figure 4 presents some of the results from a study of enhancement as a function of the size of individual gold nanoparticles [51]. These experiments coupled the DSNB-labeled nanoparticles to the underlying gold substrate by an amine-terminated thiolate, enabling a determination of particle size by AFM. To facilitate the evaluation of individual particles, patterned substrates were used to define addresses that could be readily mapped by AFM. The response of individual particles was then measured using an optical/Raman microscope.

Figure 4a shows an example of the AFM-measured topographic image of the patterned substrate modified with DSNB-labeled gold nanoparticles having a nominal diameter of 80 nm. The bright spots in the image represent individual gold particles, with the circles designating areas where SERS measurements were made. Area 1, 2, and 3 contain single gold nanoparticles, but area 4 is devoid of particles. Cross-sectional analysis from the AFM images indicate that the particles in areas 1–3 have respective sizes of 80.2 nm, 67.6 nm, and 56.9 nm but are irregularly (i.e., nonspherical) shaped. A histogram of particle sizes reveal that the average particle size is 80 nm. However, the occasional discovery of smaller sizes allowed for comparison of signals from different-sized nanoparticles. The spectroscopic results from each of these regions, which represent a small portion of the collected data, are shown in Figs. 4b–e. Spectral features characteristic of DSNB-coated gold particles are evident in areas 1–3. Note that the SERS intensity is observed to decrease with decreasing particle sizes over this size range. No SERS signal was detected in area 4.

As detailed elsewhere [51], plots of the SERS intensities show that the surface area-normalized response, which accounts for differences in the number of RRMs, undergoes a gradual increase as the size of the particles decreases from 130 nm, reaching a maximum at ~ 70 nm. This finding led to the investigation of detecting labeled particles upon dispersion in aqueous solution.

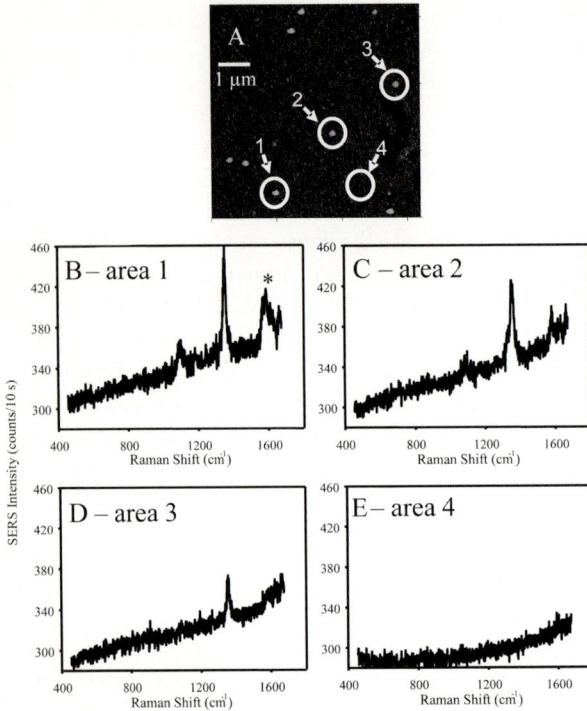

Fig. 4. (**a**) Topographic AFM image of DSNB-coated Au nanoparticles immobilized on aminoethane thiol modified gold. (**b–d**) Single-particle Raman spectra of particle 1, 2, and 3 respectively. (**e**) Blank spectrum (from spot 4). Peak labeled with * due to room lights [51]

We have not yet, however, detected a response in any experiments with the particle dispersions, even though a much larger number of particles were present in the focused laser beam than in the experiments with the immobilized particles. Importantly, the ability to readily measure SERS from particles in close proximity to a gold surface suggests that the coupling between the particle and surface has shifted the plasmon resonance to a wavelength more favorable for excitation with 632.8 nm light [52, 53, 54, 55, 56, 57].

This possibility was tested by obtaining extinction spectra for a dispersion of 60 nm DSNB-labeled nanoparticles and for labeled nanoparticles coupled to a gold and a silver substrate. This design follows the theoretical prediction that the extent of plasmon coupling is dependent on the separation between particle and surface, the size of the particle, and the composition of the two materials [58, 59, 60, 61]. Of particular relevance is the prediction that the dielectric function of the underlying substrate significantly influences the plasmon-resonance frequency of the metal particle. Established theory pre-

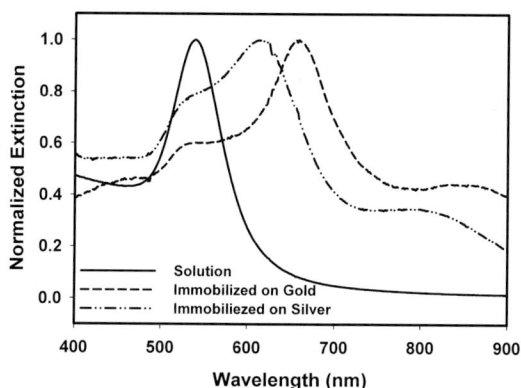

Fig. 5. The extinction spectra for DSNB-labeled 60 nm gold nanoparticles in solution, immobilized on gold with aminoethanethiol, and immobilized on silver with aminoethanethiol. Each spectrum is normalized with respect to its extinction maximum for presentation

dicts that coupling will redshift the surface plasmon for the gold substrate more than for the silver substrate [62, 63, 64].

The data in Fig. 5 show that the aqueous dispersion of 60 nm gold particles has an extinction maximum of 539 nm. To probe the influence of the underlying substrate, DSNB-labeled particles were coupled to gold and silver substrates previously modified with a 2-aminoethanethiol-derived monolayer. The surface-plasmon resonances for these substrates were measured using UV-vis reflection spectroscopy with an incident angle of 58°. For the particles tethered to the gold and silver substrates, the surface plasmon is located at longer wavelengths, 657 nm and 617 nm, respectively.

SERS spectra for these substrates were then collected and compared. The symmetric nitro stretch at $1336\,cm^{-1}$ for DSNB-labeled 60 nm nanoparticles on the gold substrate was three times stronger than the labeled nanoparticles bound to silver. This result follows the accepted theory that maximized surface enhancement occurs when the surface plasmon is located between the excitation wavelength and the location of the scattered band [65, 66, 67, 68]. These studies stress the importance of choosing the appropriate substrate composition and nanoparticle size for optimal performance.

5 Assays

5.1 Early Disease Detection

In the area of human health care, the ability to detect a diagnostic, subclinical change in biomarker concentration is central to mounting an effective treatment of a disease [69]. Prostate-specific antigen (PSA), a 33 kDa

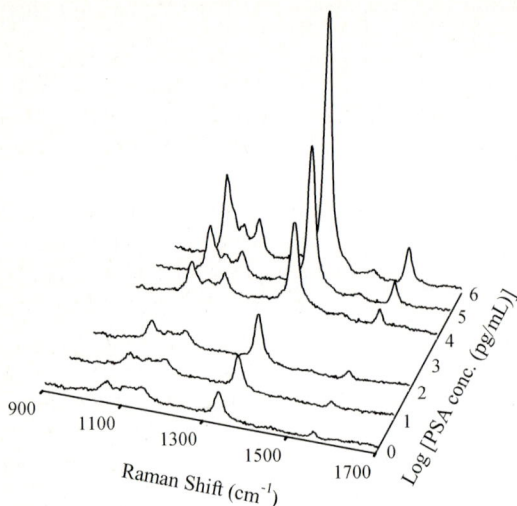

Fig. 6. Representative SERS spectra for the assay of different concentrations of PSA

glycoprotein has been used clinically as a marker for prostate cancer since 1988 [70]. The dominant form of PSA is a complex, however, the distinction between complexed PSA and free-PSA (f-PSA) is clinically relevant because the probability of prostate cancer occurrence increases as the percentage of f-PSA decreases. Figure 6 presents a series of spectra obtained in a recent evaluation of the performance of ERLs in the low-level detection of f-PSA directly from spiked human serum. The data in Fig. 6, obtained at 60 s integrations, used a matched pair of mAbs and 30 nm ERLs (assay performed prior to size optimization). As is evident, the SERS spectra reveal the presence of the ERLs, having features (e.g., the symmetric nitro stretch, ν_s (NO$_2$), at 1336 cm^{-1} and an aromatic ring mode at 1558 cm^{-1}) consistent with the DSNB-derived adlayer. Moreover, the intensity of these features tracks with increases in f-PSA levels, and span a working range of six orders of magnitude. This level of performance not only encompasses concentrations critical to prostate-cancer diagnosis, but also demonstrates the ability of this strategy to detect an exceedingly low level of f-PSA. Analysis estimates that the LOD is $\sim$ 1 pg/ml ($\sim$ 30 fM), and is defined by the nonspecific adsorption of the ERLs in the blank spectrum. This limit compares favorably with commercially available kits based on radiometric, chemiluminescent, and ELISA methods, which have LODs of 5 pg/ml to 100 pg/ml f-PSA.

The ability to detect small amounts of analyte using our assay format is underscored by a rough estimate of the number of PSA molecules responsible for the observed response at the LOD. At an LOD of 1 pg/ml, the 40 µl sample of f-PSA contains 40 fg of f-PSA or $\sim 0.8 \times 10^6$ proteins. If we assume that 1. the capture surface exhaustively binds all of the proteins in the 40 µl

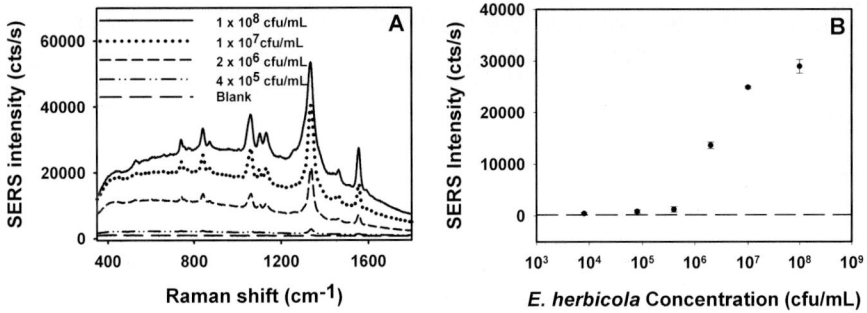

Fig. 7. (a) Representative SERS spectra for the assay of different concentrations of *E. herbicola*. (b) Dose–response curve for *E. herbicola* constructed using the intensity of the symmetric nitro stretch ($1336\,\mathrm{cm}^{-1}$) from the ERLs

sample, and 2. the capture antigens are uniformly distributed across the $5\,\mathrm{mm}$ diameter surface of the capture substrate, then there are only ~ 15 PSA molecules in the $25\,\mu\mathrm{m}$ diameter area irradiated by the laser on the substrate.

5.2 Assay of *Erwinia Herbicola* (BW Stimulant)

The tragic events of September 11, 2001 have placed the development of techniques for the rapid, low-level detection of BW agents at the highest possible levels of emphasis [71]. Traditional detection methods for this purpose include culture- [72, 73, 74, 75], ELISA- [76], and fluorescence- [77, 78] based assays. While generally effective, these methods often have a low level of throughput, can be difficult to multiplex, or suffer from photobleaching or shelf-life issues. More recently, research to address these needs has been realized by breakthroughs like the polymerase chain reaction [79], surface-plasmon resonance [80], quartz-crystal microbalance [81, 82], immunofiltration [83], immunomagnetic technology [84], chemiluminescence [85], and SERS [31]. Nevertheless, the demands of homeland security continue to place a premium on portability, ease of use, and cost, all of which drive the research agenda in a large number of analytical research laboratories.

Figure 7a shows representative spectra from an assay for *Erwinia herbicola* (*E. herbicola*), the simulant for the bacteria *Yersinia pestis* (plague) [86, 87], using the NanoRaman I system. This assay employed 80 nm ERLs, which offer maximized signal based on particle-size studies, and used polyclonal anti-*E. herbicola*. Features diagnostic of a DSNB-derived monolayer formed on the gold nanoparticles are again evident. The presence of these features, coupled with the increase in their strength with increasing antigen concentration, demonstrates the general effectiveness of the assay.

A dose–response curve was constructed based on the intensity of the symmetric nitro stretch, averaged over five different locations on each capture surface. This curve is shown in Fig. 7b. The LOD for *E. herbicola* is

$\sim 4000\,\mathrm{cfu/ml}$. Sample-to-sample variation of the signals was about $10\,\%$, adding that this LOD is about five times better than conventional ELISA methods [87]. The spot-to-spot variation was $3\text{--}10\,\%$, except for the very low concentrations ($0\,\mathrm{cfu/ml}$ to $10^5\,\mathrm{cfu/ml}$), in which the variation in signal strength was much greater ($15\text{--}100\,\%$). For example, the signal on one spot was several hundred counts, whereas that at another spot was only a few tens of counts. This situation is probably because only a small number of captured cells are present on the substrates at low concentrations, leading to a relatively inhomogeneous spatial distribution of the analytes with respect to the size of the focused laser spot. These results begin to demonstrate the exciting potential for the extension of this assay format to the needs of homeland security.

6 Showcasing Performance

6.1 Single *E. coli* O157:H7 SERS

The SERS signal from a single *E. coli* O157:H7 cell was measured using a SERS microscope [51]. After completing the sandwich immunoassay utilizing DSNB-based ERLs, the laser beam, focused to a spot $2.5\,\mu\mathrm{m}$ to $3\,\mu\mathrm{m}$ in diameter, was placed onto a single *E. coli* O157:H7 cell tagged with ERLs. Since the size of the laser spot is comparable to that of *E. coli* O157:H7, the observed signal originates primarily from the irradiated cell and not other portions of the capture substrate. A strikingly large signal from a single bacterium is evident (Fig. 8). On the other hand, no signal was observed on a sample location devoid of *E.coli* O157:H7 (Fig. 8), further demonstrating the selectivity of our ERLs and the absence of background typically due to nonspecific binding. The lack of a background signal facilitates ultralow levels of detection.

In an earlier single-particle SERS study [51], $80\,\mathrm{nm}$, DSNB-coated particles gave a SERS signal of ~ 6 counts per second per particle using the same instrumental setup. The signal of ~ 600 counts per second from a single cell, therefore, suggests that the cell is covered with many particles. Moreover, given the large size of *E.coli* O157:H7 cells, it is highly likely that ERLs captured on the top surface of the cells will not contribute strongly to the SERS signal, based on the importance of particle–substrate electromagnetic coupling in producing the enhanced Raman scattering in these experiments and the rapid decay of this coupling as the particle–substrate separation distance increases [51,64]. In our earlier study, we estimated that the ERLs were located $10\,\mathrm{nm}$ to $20\,\mathrm{nm}$ above the metal substrate. Here, ERLs on the upper surface of captured cells could be removed by as much as $1000\,\mathrm{nm}$ from the substrate. We suspect that some ERLs are localized around the periphery of the cells and near the substrate. It is also possible that the SERS signal in bacterial assays is further enhanced by particle–particle coupling of ERLs

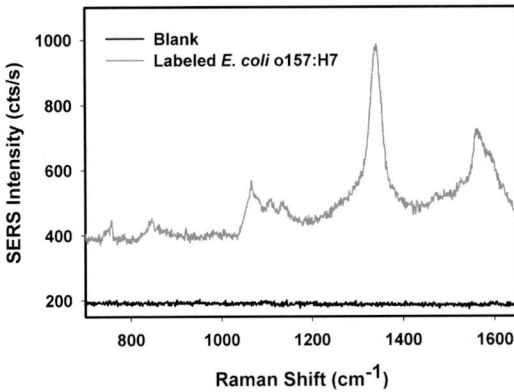

Fig. 8. SERS spectra of a single *E. coli* O157:H7 cell labeled with ERLs and a blank area

on the cell surface. Although not fully investigated, the drying process may have caused the dehydration of the cells, bringing the particles closer to the surface and closer to each other, both contributing to the enhancement of the signal. Nevertheless, this result highlights the applicability of ERLs in single-bacterium detection.

6.2 Assay of PPV

As a last example of our recent findings, Fig. 9 and Fig. 10 present results from a SERS-based investigation of PPV. PPV was chosen because of its importance to veterinary diagnostics and to demonstrate the universality of the SERS-based detection scheme by the successful detection of a viral pathogen. PPV is an important cause of swine reproductive failure that is commonly referred to as SMEDI (stillbirth, mummification, embryonic death, and infertility) [88]. The virion has a 25 nm hydrated diameter and the spherically shaped capsid is composed of 60 repeating protein subunits that encapsulate its 5.2 kb single-stranded DNA. Additionally, PPV is stable over a wide range of pH and temperature. This stability and morphology make PPV an ideal target for assay validation with AFM.

Figure 9a shows the spectra (1 s integration) for the SERS-based detection of PPV in phosphate-buffered saline (PBS) using the NanoRaman I instrument. This assay is performed with an optimum mAb selected through AFM analysis from a panel of mAbs supplied by the National Animal Disease Center. The same mAb was used for the extraction of PPV onto the capture substrate and for the labeling of the PPV with 60 nm ERLs because the epitope is repeated 60 times per virus. Much like the assay for *E. herbicola*, the spectra display bands diagnostic of the RRM, DSNB. Again, the dose–response curve was constructed from the intensity of the symmetric nitro stretch of the RRM, and is shown as part of Fig. 10. The LOD for PPV

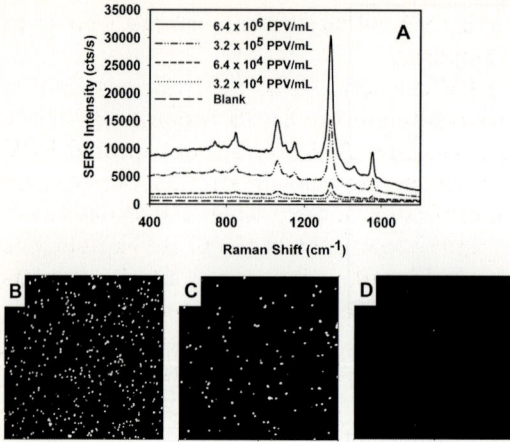

Fig. 9. (a) Representative SERS spectra for the assay of different concentrations of PPV. (b) AFM micrographs ($5 \times 5\,\mu m$) of PPV bound to capture substrates at three different concentrations. The images represent the number of PPV bound after exposure to a virus solution of (b) 2×10^7, (c) 3×10^6, (d) 0 HA units per milliliter

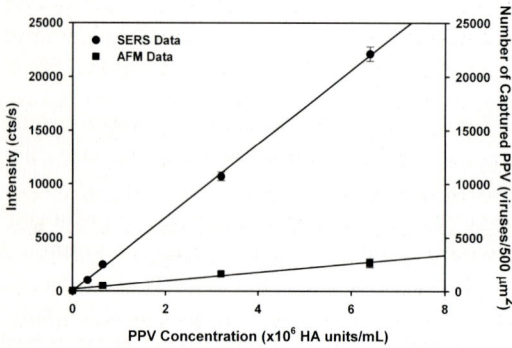

Fig. 10. Dose–response curves for SERS-based (*circles*) and AFM-based (*squares*) detection of PPV

is $\sim 10\,000$ hemagglutination (HA) units per milliliter as calculated from three times the standard deviation of the noise in the Raman measurement because the blank gave no signal. The SERS response is linear from 0 to 6.4×10^6 HA units per milliliter; higher concentrations result in a plateau of the SERS signal. The lack of detectable nonspecific binding from the blank is most likely due to optimization of the blocking agent, the cleanliness of the suspending virus solution (PBS), and the use of mAbs. Although the sample solution for this assay is not a typical biological matrix, it was chosen to emphasize the potential sensitivity of this technique with respect to minimization of the background response.

A parallel investigation into PPV extraction onto the capture substrate was performed with AFM to cross-correlate the SERS responses with the number of captured viruses. Representative AFM images of captured PPV at different concentrations are shown in Figs. 9b–d. The number of PPV in each image ($5 \times 5\,\mu m$) was enumerated and extrapolated to determine the number of PPV interrogated by the laser spot in the SERS experiments ($500\,\mu m^2$). These results are shown in Fig. 10, which enables a facile comparison of the SERS and AFM dose–response curves. The correlation between the curves shows that for each captured virus, the SERS response is $\sim 7\,\text{cts/s}$. The noise in the blank signal is $10.5\,\text{cts/s}$; therefore, the LOD is defined as the concentration of PPV that yields $32\,\text{cts/s}$. These data suggest that at the LOD, only five viruses are in the area irradiated by the laser.

7 Conclusions

This Chapter has shown that the combination of surface enhancement with respect to the close proximity of the scattering site to the particle surface, the amplification due to the large number of RRMs coating each particle and the binding affinity of monoclonal antibodies leads to an extremely low level of detection for sandwich-based immunoassays [89]. By exploiting the facile ability to manipulate the surface architecture of ERLs, high-sensitivity assays can be designed for pathogenic bacteria, e.g., *E. coli* O157:H7, and BW agents like the plague with its simulant *Erwinia herbicola*. Results were also reported that showcased the extreme levels of sensitivity that can be realized in the absence of nonspecific adsorption and other possible contributors to a background spectral response. To fully realize this potential, efforts to markedly reduce incubation times and to concentrate samples will be needed. It will be the successful integration of such processes that drive the widespread movement of SERS-based assays from research to clinical laboratories.

Acknowledgements

Heat-killed *E. coli* O157:H7 was the generous gift from Dr. Nancy Cornick (Department of Veterinary Medicine, Iowa State University). This work was supported by a grant from the USDA-NADC, Concurrent Analytical, Inc. through a grant from the DARPA CEROS program and by the Institute for Combinatorial Discovery of Iowa Sate University. The Ames Laboratory is operated for the US Department of Energy by Iowa State University under contract W-7405-eng-82.

References

[1] E. P. Diamandis, T. K. Christopoulos: *Immunoassay* (Academic, New York 1996)
[2] J. Donohue, M. Bailey, R. Gray, et al.: Clin. Chem. **35**, 1874 (1989)
[3] B. Baslund, J. Wieslander: J. Immunol. Meth. **169**, 183 (1994)
[4] C. Blake, M. N. Al-Bassam, B. J. Gould, et al.: Clin. Chem. **28**, 1469 (1982)
[5] S. Gutcho, L. Mansbach: Clin. Chem. **23**, 1609 (1977)
[6] J. Vuori, S. Rasi, T. Takala, et al.: Clin. Chem. **37**, 2087 (1991)
[7] Y. Y. Xu, K. Pettersson, K. Blomberg, et al.: Clin. Chem. **38**, 2038 (1992)
[8] A. Varenne, A. Vessieres, M. Salmain, et al.: Anal. Biochem. **242**, 172 (1996)
[9] F. J. Hayes, H. B. Halsall, W. R. Heineman: Anal. Chem. **66**, 1860 (1994)
[10] A. J. Beavis, K. J. Pennline: Cytometry **15**, 371 (1994)
[11] C. R. Brown, K. W. Higgins, K. Frazer, et al.: Clin. Chem. **31**, 1500 (1985)
[12] F. K. Fotiou: Anal. Chem. **64**, 1698 (1992)
[13] T. E. Rohr, T. Cotton, N. Fan, et al.: Anal. Biochem. **182**, 388 (1989)
[14] S. Gutcho, L. Mansbach: Clin. Chem. **23**, 1609 (1977)
[15] R. L. Garrell: Anal. Chem. **61**, 401 A (1989)
[16] E. J. Zeman, G. C. Schatz: J. Phys. Chem. **91**, 634 (1987)
[17] K. Kneipp, H. Kneipp, R. Manoharan, et al.: Appl. Spectrosc. **52**, 1493 (1998)
[18] S. Nie, S. R. Emory: Science **275**, 1102 (1997)
[19] R. Elghanian, J. J. Storhoff, R. C. Mucic, et al.: Science **277**, 1078 (1997)
[20] H. Kimura, S. Matsuzawa, C.-Y. Tu, et al.: Anal. Chem. **68**, 3063 (1996)
[21] C. Y. Tu, T. Kitamori, T. Sawada, et al.: Anal. Chem. **65**, 3631 (1993)
[22] L. A. Lyon, M. D. Musick, M. J. Natan: Anal. Chem. **70**, 5177 (1998)
[23] K. Park, S. R. Simmons, R. M. Albrecht: Scan. Microsc. **1**, 339 (1987)
[24] D. S. Grubisha, R. J. Lipert, H.-Y. Park, et al.: Anal. Chem. **75**, 5936 (2003)
[25] J.-M. Nam, C. S. Thaxton, C. A. Mirkin: Science **301**, 1884 (2003)
[26] T. E. Rohr, T. Cotton, N. Fan, et al.: Anal. Biochem. **182**, 388 (1989)
[27] X. Dou, T. Takama, T. Yamaguchi, et al.: Anal. Chem. **69**, 1492 (1997)
[28] J. Ni, R. Lipert, B. Dawson, et al.: Anal. Chem. **71**, 4903 (1999)
[29] S. Xu, X. Ji, W. Xu, et al.: The Analyst **129**, 63 (2004)
[30] S. P. Mulvaney, M. D. Musick, C. D. Keating, et al.: Langmuir **19**, 4784 (2003)
[31] X. Zhang, M. A. Young, O. Lyandres, et al.: J. Am. Chem. Soc. (2005)
[32] D. O. Ansari, D. A. Stuart, S. Nie: Proc. SPIE **5699**, 82 (2005)
[33] L. A. Lyon, C. D. Keating, A. P. Fox, et al.: Anal. Chem. **70**, 341R (1998)
[34] T. Vo-Dinh: Trends Anal. Chem. **17**, 557 (1998)
[35] J. Ni, R. J. Lipert, G. B. Dawson, et al.: Anal. Chem. **71**, 4903 (1999)
[36] P. Wagner, P. Kernen, M. Hegner, et al.: FEBS Lett. **356**, 267 (1994)
[37] P. Wagner, M. Hegner, H.-J. Guentherodt, et al.: Langmuir **11**, 3867 (1995)
[38] J. D. Driskell, K. M. Kwarta, R. J. Lipert, et al.: Anal. Chem. **77**, 6147 (2005)
[39] Y. Dong, C. Shannon: Anal. Chem. **72**, 2371 (2000)
[40] S. L. McCall, P. M. Platzman: Phys. Rev. B **22**, 1660 (1980)
[41] S. L. McCall, P. M. Platzman, P. A. Wolff: Phys. Lett. A **77A**, 381 (1980)
[42] J. Gersten, A. Nitzan: J. Chem. Phys. **73**, 3023 (1980)
[43] R. G. Nuzzo, D. L. Allara: J. Am. Chem. Soc. **105**, 4481 (1983)
[44] W. D. Geoghegan: J. Histochem. Cytochem. **36**, 401 (1988)
[45] W. D. Geoghegan, G. A. Ackerman: J. Histochem. Cytochem. **25**, 1187 (1977)

[46] R. K. Chang, T. E. Furtak: *Surface Enhanced Raman Scattering* (Plenum, New York 1982) p. 315
[47] E. J. Zeman, G. C. Schatz: J. Phys. Chem. **91**, 634 (1987)
[48] S. R. Emory, S. Nie: Anal. Chem. **69**, 2631 (1997)
[49] J. T. Krug, II, G. D. Wang, S. R. Emory, et al.: J. Am. Chem. Soc. **121**, 9208 (1999)
[50] R. G. Freeman, R. M. Bright, M. B. Hommer, et al.: J. Raman Spectrosc. **30**, 733 (1999)
[51] H.-Y. Park, R. J. Lipert, M. D. Porter: Proc. SPIE-Nanosensing: Mater. Dev. **5593**, 464 (2004)
[52] K. Kim, J. K. Yoon: J. Phys. Chem. B **109**, 20731 (2005)
[53] W. R. Holland, D. G. Hall: Phys. Rev. B **21**, 7765 (1983)
[54] T. Kume, N. Nakagawa, S. Hayashi, et al.: Solid State Commun. **93**, 171 (1995)
[55] A. V. Shchegrov, I. V. Novikov, A. A. Maradudin: Phys. Rev. Lett. **78**, 4269 (1997)
[56] J. Zheng, Y. Zhou, X. Li, et al.: Langmuir **19**, 632 (2003)
[57] C. J. Orendorff, A. Gole, T. K. Sau, et al.: Anal. Chem. **77**, 3261 (2005)
[58] S. Link, M. A. El-Sayed: J. Phys. Chem. B **103**, 8410 (1999)
[59] T. R. Jensen, G. C. Schatz, R. P. Van Duyne: J. Phys. Chem. B **103**, 2394 (1999)
[60] A. D. McFarland, R. P. Van Duyne: Nano Lett. **3**, 1057 (2003)
[61] M. D. Malinsky, K. L. Kelly, G. C. Schatz, et al.: J. Am. Chem. Soc. **123**, 1471 (2001)
[62] P. K. Aravind, H. Metiu: Surf. Sci. **124**, 506 (1983)
[63] M. M. Wind, J. Vlieger, D. Bedeaux: Physica **143A**, 164 (1987)
[64] T. Okamoto, I. Yamaguchi: J. Phys. Chem. B **107**, 10321 (2003)
[65] K. Kneipp, K. Kneipp, I. Itzkan, et al.: J. Phys.: Condens. Matter **14**, R597 (2002)
[66] D. A. Weitz, S. Garoff, T. Gramila: Opt. Lett. **7**, 168 (1982)
[67] N. Felidj, J. Aubard, G. Levi, et al.: Appl. Phys. Lett. **82**, 3095 (2003)
[68] S. J. Oldenburg, J. B. Jackson, S. L. Westcott, et al.: Appl. Phys. Lett. **75**, 2897 (1999)
[69] B. B. Chomel: J. Vet. Med. Educ. **30**, 145 (2003)
[70] T. J. Polascik, J. E. Oesterling, A. W. Partin: J. Urol. **162**, 293 (1999)
[71] D. A. Henderson: Science **283**, 1279 (1999)
[72] J. Bordet, E. Renaux: Ann. inst. Pasteur **45**, 1 (1930)
[73] H. Depla: Arch. Int. Pharm. Ther. **28**, 223 (1923)
[74] R. J. Hewlett, G. N. Hall: J. Hyg. **2**, 473 (1912)
[75] B. Moore, S. Williams: Biochem. J. **5**, 181 (1911)
[76] T. Vo-Dinh, G. D. Griffin, K. R. Ambrose: Appl. Spectrosc. **40**, 696 (1986)
[77] H. P. Anne, H. J. Linwood, J. F. Peruski Leonard: J. Immunol. Meth. **263**, 35 (2002)
[78] M. Seaver, J. D. Eversole, J. J. Hardgrove, et al.: Aerosol Sci. Technol. **30**, 174 (1999)
[79] P. Belgrader, W. Benett, D. Hadley, et al.: Clin. Chem. **44**, 2191 (1998)
[80] R. Slavik, J. Homola, E. Brynda: Biosens. Bioelectron. **17**, 591 (2002)
[81] B. Koenig, M. Graetzel: Anal. Lett. **26**, 1567 (1993)
[82] E. Prusak-Sochaczewski, J. H. Luong, G. G. Guilbault: Enzyme Microb. Technol. **12**, 173 (1990)

[83] J. M. Libby, H. G. Wada: J. Clin. Bicrobiol. **27**, 1456 (1989)

[84] H. Yu, J. W. Raymonda, T. M. McMahon, et al.: Biosens. Bioelectron. **14**, 829 (2000)

[85] R. Vidziunaite, N. Dikiniene, V. Miliukiene, et al.: J. Biolumin. Chemilumin. **10**, 193 (1995)

[86] C. A. Rowe, L. M. Tender, M. J. Feldstein, et al.: Anal. Chem. **71**, 3846 (1999)

[87] M. T. McBride, S. Gammon, M. Pitesky, et al.: Anal. Chem. **75**, 1924 (2003)

[88] F. A. Murphy, E. P. J. Gibbs, M. C. Horzinek, et al.: *Veterinary Virology*, 3rd ed. (Academic, San Diego 1999)

[89] E. Delamarche, B. Michel, C. Gerber, et al.: Langmuir **10**, 2869 (1994)

Index

Detecting Chemical Agents and Their Hydrolysis Products in Water

Stuart Farquharson[1], Frank E. Inscore[1], and Steve Christesen[2]

[1] Real-Time Analyzers, Middletown, CT 06457
[2] US Army Edgewood, APG-EA, MD 21010
    Stu@rta.biz

1 Introduction

Chemical agents were introduced as weapons during World War I [1], and have, unfortunately, become part of the terrorist's arsenal in the past decade [2]. There are many possible deployment scenarios that terrorist can use, and the deliberate poisoning of drinking water is one scenario that is of great concern. Countering such an attack could be greatly aided by a portable analyzer that could detect poisons in water at $\mu g \cdot l^{-1}$ concentrations, within minutes, and with few or no false-positive responses. A number of analytical methods have been investigated for this purpose, and include gas chromatography coupled mass spectrometry (GC/MS) [3, 4, 5, 6, 7, 8, 9], ion mobility spectrometry (IMS) [10, 11], infrared spectroscopy (IR) and Raman spectroscopy [12, 13, 14, 15, 16, 17].

However, all of these analytical methods suffer from various limitations. Creasy et al. measured $0.02 \, \mathrm{mg} \cdot l^{-1}$ nerve agents in water using GC/MS, but the analysis required as much as an hour to perform. Steiner at al. required the addition of electrospray ionization and a time-of-flight MS as a second detector for their IMS analyzer, and only achieved $10 \, \mathrm{mg} \cdot l^{-1}$ sensitivity for chemical-agent simulants in water. Although both portable IR and Raman spectrometers have been developed, Braue and Pannella only measured $0.1 \, \mathrm{g} \cdot l^{-1}$ nerve agents in water using IR, while *Christesen* et al. only measured $5 \, \mathrm{g} \cdot l^{-1}$ sulfur-mustard in chloroform using Raman spectroscopy [13, 17].

The primary limitation of Raman spectroscopy, sensitivity, can be overcome through surface-enhancement of the Raman-scattering mechanism [18]. The interaction of surface-plasmon modes of metal particles with target analytes can increase scattering efficiencies by as much as 14 orders of magnitude, although 6 orders of magnitude are more common. The details of surface-enhanced Raman spectroscopy (SERS) can be found in the beginning of this book. The utility of SERS to measure chemical agents was first demonstrated by *Alak* and *Vo-Dinh* by measuring several organophosphonates as simulants of nerve agents on a silver-coated microsphere substrate [19]. *Spencer* and coworkers used SERS to measure several chemical agents on electrochemically roughened gold or silver foils [20, 21, 22]. However, in all of these measurements, the sample needed to be dried on the substrates to obtain the

K. Kneipp, M. Moskovits, H. Kneipp (Eds.): Surface-Enhanced Raman Scattering – Physics and Applications, Topics Appl. Phys. **103**, 447–460 (2006)
© Springer-Verlag Berlin Heidelberg 2006

best sensitivity at $0.05\,\text{mg}\cdot l^{-1}$. More recently, *Tessier* et al. obtained SERS of $0.04\,\text{mg}\cdot l^{-1}$ cyanide in a stream flowing over a substrate formed by a templated self-assembly of gold nanoparticles [23]. However, optimum sensitivity required introduction of an acid wash and the measurements were irreversible.

In the past few years, we have also been investigating the ability of SERS to measure chemical agents at $\mu\text{g}\cdot l^{-1}$ in water and with sufficient spectral uniqueness to distinguish the agent and its hydrolysis products [24, 25, 26, 27, 28, 29]. In our work, we have developed silver-doped sol-gels as the SERS-active medium. These sol-gels can be coated on the inside walls of glass vials, such that water samples can be added to perform point analysis, or they can be incorporated into glass capillaries, such that flowing measurements can be performed [30]. Here, SERS measurements using both sampling devices are presented for several classes of chemical agents, specifically, the blood agent hydrogen cyanide, the blister agent sulfur-mustard (designated HD), and two nerve agents, sarin (designated GB) and VX (no common name), as well as their hydrolysis products.

2 Experimental

Sodium cyanide, 2-hydroxyethylethyl sulfide (HEES), 2-chloroethylethyl sulfide (CEES) and methylphosphonic acid (MPA) were purchased from Sigma-Aldrich (St. Louis, MO) and used as-received. Ethyl methylphosphonic acid (EMPA), isopropyl methylphosphonic acid (IMPA), 2-(diisopropylamino) ethanethiol (DIASH), and thiodiglycol (TDG, bis(2-hydroxyethyl)sulfide) were purchased from Cerilliant (Round Rock, TX). Highly distilled sulfur-mustard (HD, bis(2-chloroethyl)sulfide), isopropyl methylphosphonofluoridate (GB), ethyl S-2-diisopropylamino ethyl methylphosphonothioate (VX), and ethyl S-2-diisopropylamino methylphosphonothioate (EA2192) were obtained at the US Army's Edgewood Chemical Biological Center (Aberdeen, MD) and measured on-site. All samples were initially prepared in a chemical hood as 1000 parts per million ($1\,\text{g l}^{-1}$ or $0.1\,\%$ by volume, Environmental Protection Agency definition) in HPLC-grade water (Fischer Scientific, Fair Lawn, NJ) or in some cases methanol or ethanol (Sigma-Aldrich) to minimize hydrolysis.

Once prepared, the samples were transferred into 2 ml glass vials internally coated with a silver-doped sol-gel (*Simple SERS Sample Vials*, Real-Time Analyzers, Middletown, CT) or drawn by syringe or pump into 1 mm diameter glass capillaries filled with the same SERS-active material [31, 32, 33, 34]. In the case of flow measurements, a peristaltic pump (variable-flow minipump, Control Co., Friendswood, TX) was used to flow the various cyanide solutions through a SERS-active capillary at $1\,\text{ml}\cdot\text{min}^{-1}$. The vials or capillaries were placed on aluminum plates machined to hold the vials or capillaries on a standard XY positioning stage (Conix Research, Springfield, OR), such that the focal point of an f/0.7 aspheric lens was positioned just inside the glass wall. The lens focused the beam into the sample and collected the scattered radiation back along the same axis. A dichroic filter (Omega Optical, Brattleborough, VT) was used to reflect the excitation laser to the lens and pass the Raman-scattered radiation collected by the lens. An f/2 achromat was used to collimate the laser beam exiting a 200 μm core diameter source fiber optic, while a second f/2 achromat was used to focus the scattered radiation into a 365 μm fiber optic (Spectran, Avon, CT). A short-pass filter was placed in the excitation beam path to block the silicon Raman scattering generated in the source fiber from reflecting off the sampling optics and reaching the detector. A long-pass filter was placed in the collection beam path to block the sample Rayleigh scattering from reaching the detector. SER spectra were collected using a Fourier transform Raman spectrometer equipped with a 785 nm diode laser and a silicon photoavalanche detector (*IRA-785*, Real-Time Analyzers). All spectra were nominally collected using 100 mW, 8 cm^{-1} resolution, and a 1 min acquisition time, unless otherwise noted. Additional experimental details can be found in [35]. For added safety, all samples were measured in a chemical hood. In the case of actual agents measured at Edgewood, the FT-Raman instrument was placed outside the laboratory and 30 foot fiber optic and electrical cables were used to allow remote SERS measurements and plate manipulation.

3 Results and Discussion

3.1 Cyanide

All of the basic forms of cyanide, HCN, KCN, or NaCN, are extremely soluble in water (completely miscible, $716 \, \text{g} \cdot \text{l}^{-1}$, and $480 \, \text{g} \cdot \text{l}^{-1}$, respectively) [36]. In solution the cyanide ion is formed in equilibrium with the conjugate acid, HCN (Fig. 1a), according to the K_a of 6.15×10^{-10} [37]. This is significant in that only CN$^-$ appears to interact sufficiently with silver to produce a SER spectrum, and no spectral signal is observed below pH 7 (except on electrodes at specific potential conditions [38]). Furthermore, the required detection sensitivity of cyanide in water has been estimated by the United States military using toxicological data at $2 \, \text{mg} \cdot \text{l}^{-1}$ [39,40]. This would result in $1.25 \, \text{mg} \cdot \text{l}^{-1}$ of CN$^-$ and $0.75 \, \text{mg} \cdot \text{l}^{-1}$ of HCN. At lower concentrations even less of the CN$^-$ is available to generate a SER signal.

The SER spectra of cyanide are dominated by an intense, broad peak at 2100 cm^{-1} attributed to the C≡N stretch (Fig. 2). This mode occurs at 2080 cm^{-1} in Raman spectra of solutions, and the frequency shift in SER spectra is attributed to a strong surface interaction, which is supported by the appearance of a low-frequency peak at 135 cm^{-1} due to a Ag–CN stretch (not shown). It is also observed that as the concentration decreases, the CN peak shifts to 2140 cm^{-1}. This shift has been attributed to the formation of

A

$$HCN \xrightleftharpoons{H_2O} CN^- + H_3O^+$$

B

C

D

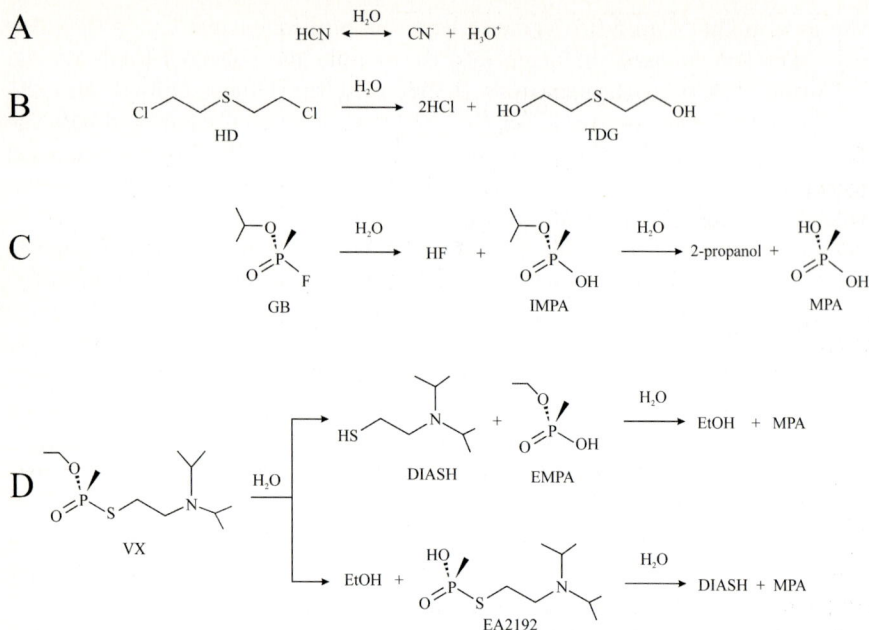

Fig. 1. Hydrolysis reaction pathways for (**a**) HCN, (**b**) HD, (**c**) GB, and (**d**) VX

a tetrahedral $Ag(CN)_3^{2-}$ surface structure [41], as well as to CN adsorbed to two different surface sites [42]. Alternatively, it has also been suggested that at concentrations near to and above a monolayer coverage, the CN species is forced to adsorb end-on due to crowding, and at lower concentrations the molecule can reorient to lie flat. This suggests that the $2100\,cm^{-1}$ and $2140\,cm^{-1}$ peaks correspond to the end-on and flat orientations, respectively. However, a previous concentration study of cyanide on a silver electrode observed the reverse trend, i.e., greater intensity was observed for the $2100\,cm^{-1}$ peak at low concentration [43].

Repeated measurements of cyanide in the SERS-active vials consistently allowed measuring $1\,mg \cdot l^{-1}$ (1 ppm), but rarely below this concentration (Fig. 2a). Nevertheless, this sensitivity is in general sufficient for point sampling of water supplies. In the case of continuous monitoring of water, the capillaries are a more appropriate sampling format, and they also allowed routine measurements at $0.01\,mg \cdot l^{-1}$ and repeatable measurements at $0.001\,mg \cdot l^{-1}$ (1 ppb, Fig. 2b). Employing this format, a 50 ml volume of $0.01\,mg \cdot l^{-1}$ cyanide solution was flowed at $2.5\,ml \cdot min^{-1}$ through a SERS-active capillary, and spectra were recorded every 20 s. As Fig. 3 shows, the cyanide peak was easily discerned as soon as the solution entered the capillary and remained relatively stable over the course of the experiment. It is worth noting, as indicated above, that the SERS peak in Fig. 3 is in fact due to $210\,ng \cdot l^{-1}$!

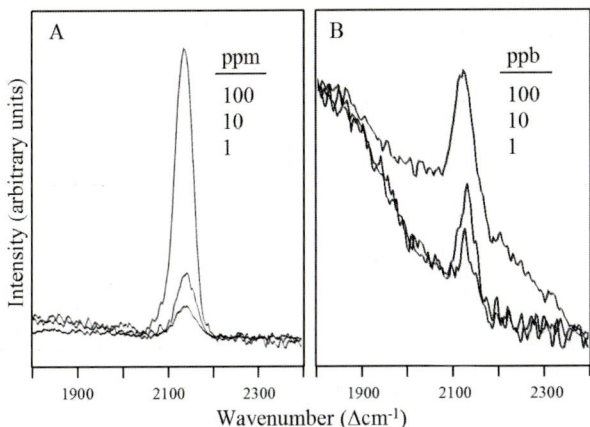

Fig. 2. Surface-enhanced Raman spectra of CN in water in silver-doped sol-gel (**a**) coated glass vials and (**b**) filled glass capillaries. All spectra were recorded using 100 mW of 785 nm in 1 min and at a resolution of $8\,cm^{-1}$

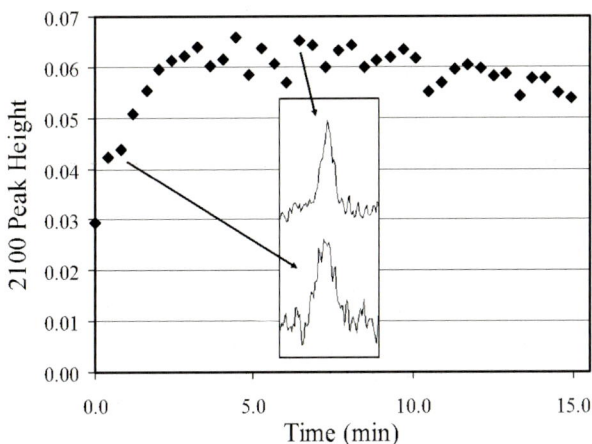

Fig. 3. $2100\,cm^{-1}$ peak height measured during continuous flow of a $0.01\,mg \cdot l^{-1}$ (10 ppb) cyanide in water. Surface-enhanced Raman spectra are shown for 1 min and 6 min after sample introduction. A $2.5\,ml \cdot min^{-1}$ flow rate was used and spectra were recorded every 20 s using 100 mW of 785 nm

3.2 HD and CEES

HD is marginally soluble in water tending to form droplets, and hydrolysis occurs at the droplet surface. This property has made measuring the hydrolysis rate constant difficult, and half-lives anywhere from 2 h to 30 h are reported [44]. Hydrolysis involves the sequential replacement of the chlorine atoms by hydroxyl groups through cyclic sulfonium ion intermediates to form

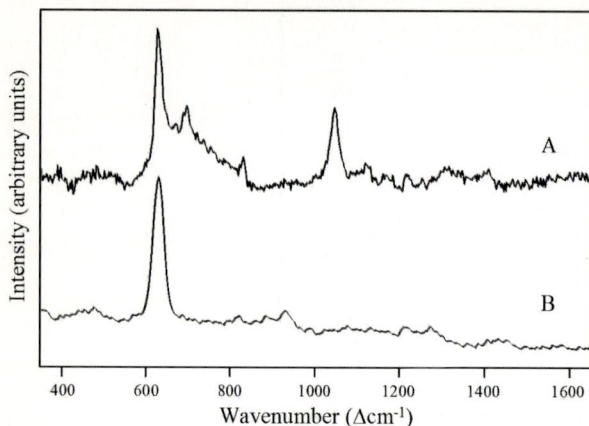

Fig. 4. Surface-enhanced Raman spectra of (**a**) HD in methanol and (**b**) TDG in water. Spectral conditions as in Fig. 2, samples were $1 \, \mathrm{g} \cdot \mathrm{l}^{-1}$

thiodiglycol (Fig. 1b) [45]. If a median hydrolysis rate is assumed, then early detection of poisoned water will require measuring HD, while postattack or downstream monitoring will require measuring TDG. The required detection sensitivity of sulfur-mustard has been set at $0.047 \, \mathrm{mg} \cdot \mathrm{l}^{-1}$ [39, 40], and a similar sensitivity can be assumed for TDG.

The surface-enhanced Raman spectrum of HD is dominated by a peak at $630 \, \mathrm{cm}^{-1}$ with an extended high-frequency shoulder composed of at least two peaks evident at $695 \, \mathrm{cm}^{-1}$ and $830 \, \mathrm{cm}^{-1}$, as well as a moderately intense peak at $1045 \, \mathrm{cm}^{-1}$ (Fig. 4a). The latter peak is assigned to a CC stretching mode, based on the assignment for a peak at $1040 \, \mathrm{cm}^{-1}$ in the Raman spectrum of HD [46]. The assignment of the $630 \, \mathrm{cm}^{-1}$ peak is less straightforward, since the Raman spectrum of HD contains five peaks in this region at 640, $655 \, \mathrm{cm}^{-1}$, $700 \, \mathrm{cm}^{-1}$, $740 \, \mathrm{cm}^{-1}$, and $760 \, \mathrm{cm}^{-1}$ [26, 46]. Theoretical calculations for the Raman spectrum of HD indicate that the first three peaks are due to CCl stretching modes, and the latter two peaks to CS stretching modes [47]. Based on these calculations, and the expected interaction between the chlorine atoms and the silver surface, it is reasonable to assign the $630 \, \mathrm{cm}^{-1}$ SERS peak to a CCl mode [26]. However, recent SERS measurements of diethyl sulfide produced a very simple spectrum with an intense peak at $630 \, \mathrm{cm}^{-1}$ [48, 49], strongly suggesting CS or CSC stretching modes as the appropriate assignment for this peak [50]. The authors of the theoretical treatment concede that the CCl and CS assignments could be reversed [47]. The CS assignment also indicates that HD interacts with the silver surface through the sulfur electron lone pairs. But, interaction between chlorine and silver is still possible and may be responsible for the $695 \, \mathrm{cm}^{-1}$ peak. The $830 \, \mathrm{cm}^{-1}$ peak is left unassigned.

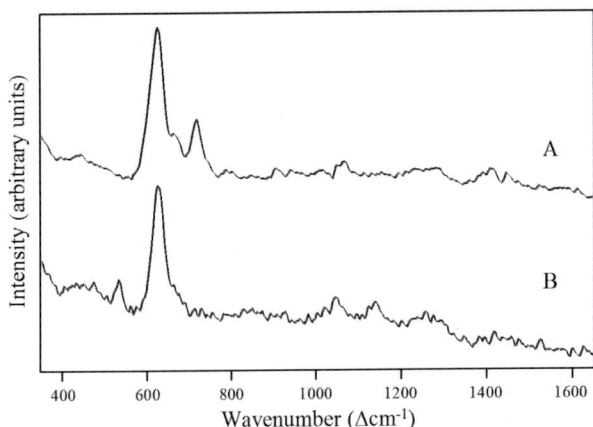

Fig. 5. Surface-enhanced Raman spectra of (a) CEES and (b) HEES. Spectral conditions as in Fig. 2, samples were $1\,g\cdot l^{-1}$ in methanol

The surface-enhanced Raman spectrum of TDG is also dominated by a peak at $630\,cm^{-1}$ with minor peaks at 820, 930, 1210, and $1275\,cm^{-1}$ (Fig. 4b). Again, the $630\,cm^{-1}$ peak is preferably assigned to a CSC stretching mode versus a CCl mode, especially since the chlorines have been replaced by hydroxyl groups. Furthermore, the lack of a $695\,cm^{-1}$ peak in the TDG spectrum supports the assignment of this peak in the HD spectrum to a CCl mode. The 930, 1210 and $1275\,cm^{-1}$ SERS peaks are assigned to a CC stretch with CO contribution, and two CH_2 deformation modes (twist, scissors, or wag) based on the assignments for the corresponding peaks at 940, 1230 and $1290\,cm^{-1}$ in the Raman spectrum of TDG [46, 48]. It is worth noting that irradiation at high laser powers or for extended periods produces peaks at $715\,cm^{-1}$ and $1010\,cm^{-1}$, which are attributed to a degradation product, such as 2-hydroxy ethanethiol [48].

The SERS of CEES, also known as half-mustard, is very similar to HD, dominated by a peak at $630\,cm^{-1}$ that is accordingly assigned to a CS or CSC stretching mode (Fig. 5a). This peak also has a high-frequency shoulder centered at $690\,cm^{-1}$, and a third peak appears at $720\,cm^{-1}$ in this region. Again, these can be assigned to CCl or CS modes. The quality of this spectrum also reveals weak peaks at 1035, 1285, 1410, and $1445\,cm^{-1}$. Peaks at 1035, 1285, 1425, and $1440\,cm^{-1}$ appear in the Raman spectrum of CEES, and the previous peak assignments are used here [46], i.e., the first peak is assigned to a CC stretch, while the remaining peaks are assigned to various CH_2 deformation modes.

Replacing the chlorine atom of CEES by a hydroxyl group in forming HEES produces SER spectral changes analogous to those cited above for HD and TDG. Again, the SER spectrum is dominated by an intense peak at $630\,cm^{-1}$ attributed to a CS or CSC stretching mode, and the other CEES peaks in this region, specifically the $720\,\text{cm}^{-1}$ peak, disappear (Fig. 5b). Peaks with modest intensity at $1050\,\text{cm}^{-1}$ and $1145\,\text{cm}^{-1}$ are assigned to a CC stretching mode and CH_2 deformation, respectively. A new peak at $550\,\text{cm}^{-1}$ is likely due to a skeletal bending mode, such as CSC, SCC, or CCO. Finally, it is worth stating that HD, TDG, CEES, and HEES all produce moderately intense peaks at 2865 and $2925\,\text{cm}^{-1}$ (not shown), that can be assigned to symmetric and asymmetric CH_2 stretching modes.

Only a limited number of measurements of HD were performed to evaluate sensitivity, due to the safety requirements. HD was repeatedly observed at $1\,\text{g}\cdot\text{l}^{-1}$ and usually observed at $0.1\,\text{g}\cdot\text{l}^{-1}$ (100 ppm) in the SERS-active vials [26]. But even at the latter concentration, substantial improvements in sensitivity are required to approach the required $0.05\,\text{mg}\cdot\text{l}^{-1}$ (50 ppb) sensitivity. More extensive experiments were performed on HD's hydrolysis product, TDG since this chemical is safely handled in a regular chemical lab. Flowing TDG through SERS-active capillaries allowed repeatable measurements at $10\,\text{mg}\cdot\text{l}^{-1}$, and routine measurements at $1\,\text{mg}\cdot\text{l}^{-1}$ (1 ppm) [49]. These SERS measurements of TDG suggest that the required HD sensitivity may be achievable using this technique. Similar flowing measurements in capillaries for HD, CEES, and HEES have not been performed.

3.3 GB

Sarin readily dissolves in water and is stable for a day or more. Therefore, detecting poisoned water will largely require measuring sarin, while monitoring the attack will require detecting its sequential hydrolysis products, isopropyl methylphosphonic acid and methyl phosphonic acid (Fig. 1c) [43, 44, 51]. The other hydrolysis products, hydrofluoric acid and 2-propanol, are too common to provide definitive evidence of water poisoning and their measurement would be of limited value. The required detection sensitivity in water has been estimated at $0.0046\,\text{mg}\cdot\text{l}^{-1}$ [39, 40].

SERS measurements of GB have not been made, but its primary hydrolysis products, IMPA and MPA, have been measured using the SERS-active vials. The SERS of IMPA is very similar to its Raman spectrum [28], which in turn is very similar to the Raman spectrum of sarin [16]. The SER spectrum is dominated by a peak at $715\,\text{cm}^{-1}$ (Fig. 6a), which is assigned to a PC or PO plus skeletal stretching mode, as is a weak peak at $770\,\text{cm}^{-1}$. These assignments are also consistent with a theoretical treatment of the Raman spectrum for sarin [52]. Similarly, a modest peak at $510\,\text{cm}^{-1}$ can be assigned to a PC or PO plus skeletal bending mode. Other SERS peaks of modest intensity occur at 875, 1055, 1415, and $1450\,\text{cm}^{-1}$, and based on the spectral analysis of sarin and the Raman spectrum of IMPA with peaks at 880, 1420, and $1455\,\text{cm}^{-1}$, are assigned to a CCC bend, a PO_3 stretch, a CH_3 bend, and a CH_2 rock, respectively.

MPA has been well characterized by infrared and Raman spectroscopy [53, 54], as well as normal coordinate analysis [55], and the literature assignments

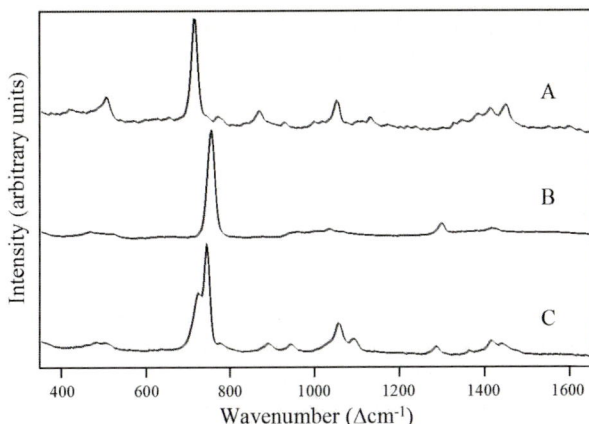

Fig. 6. Surface-enhanced Raman spectra of (**A**) IMPA, (**B**) MPA, and (**C**) EMPA. Spectral conditions as in Fig. 2, samples were $1\,\mathrm{g}\cdot\mathrm{l}^{-1}$ in water

are used here for the SERS of MPA. The SER spectrum is dominated by a peak at $755\,\mathrm{cm}^{-1}$, which is assigned to the PC symmetric stretch (Fig. 6b). In comparison to IMPA, it is clear that removing the isopropyl group shifts this frequency substantially ($40\,\mathrm{cm}^{-1}$), as the mode becomes a purer PC stretch. Additional peaks with comparatively little intensity occur at 470, 520, 960, 1040, 1300, and $1420\,\mathrm{cm}^{-1}$, and are assigned to a PO_3 bending mode, a C-PO_3 bending mode, a PO_3 stretching mode, another PO_3 bending mode, and two CH_3 deformation modes (twisting and rocking).

SERS-active vials allowed repeatable measurements of MPA at $10\,\mathrm{mg}\cdot\mathrm{l}^{-1}$ and routine measurements at $1\,\mathrm{mg}\cdot\mathrm{l}^{-1}$, and repeatable measurements of IMPA at $100\,\mathrm{mg}\cdot\mathrm{l}^{-1}$ and routine measurements at $10\,\mathrm{mg}\cdot\mathrm{l}^{-1}$. Again, however, substantial improvements in sensitivity are required to achieve the minimum requirement of $0.0046\,\mathrm{mg}\cdot\mathrm{l}^{-1}$.

3.4 VX

VX is reasonably soluble, and like sarin, is fairly persistent with a hydrolysis half-life greater than 3 days [56]. Hydrolysis can occur along two pathways (Fig. 1d) [9, 51], either being converted to DIASH and EMPA or EA2192 and ethanol with the former pathway favored four to one. These products also hydrolyze, and EMPA forms MPA and ethanol, while EA2192 forms DIASH and MPA. However, the hydrolysis of EA2192 is exceptionally slow [57], and is considered just as toxic as VX. Consequently, detecting the early stages of poisoning water should focus on measuring VX, while longer-term monitoring should focus on EA2192. The required detection sensitivity in water has been estimated at $0.0025\,\mathrm{mg}\cdot\mathrm{l}^{-1}$ [39, 40]. Here the SER spectra of VX, EA2192 and DIASH are compared, while EMPA is compared to IMPA and MPA.

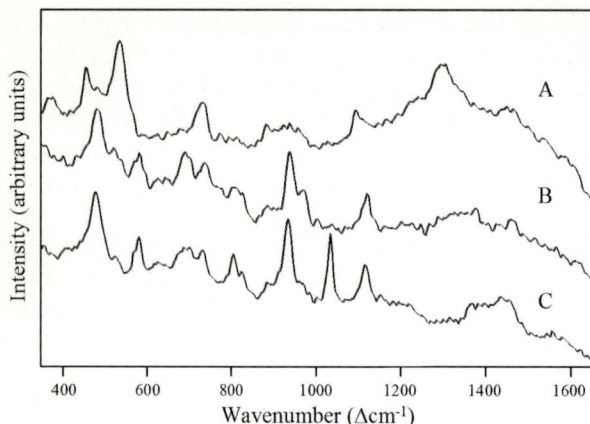

Fig. 7. Surface-enhanced Raman spectra of (**A**) VX, (**B**) EA2192, and (**C**) DIASH. Spectral conditions as in Fig. 2, samples were $1\,g \cdot l^{-1}$ in water

The SER spectrum of VX is similar to its Raman spectrum with corresponding peaks at 375, 460, 540, 730, 1095, 1300, 1440, and $1460\,cm^{-1}$ (Fig. 7a). Since a computer-predicted Raman spectrum contains most of the measured Raman spectral peaks [29, 58], it is used to assign the above SERS peaks, respectively, to an SPO bend, a CH_3–P=O bend, a PO_2CS wag, an OPC stretch, a CC stretch, and three CH_n bends. As previously described for CEES and HD, the $730\,cm^{-1}$ peak could alternatively be assigned to a CS stretch, but the SER spectra of these chemicals suggest otherwise.

The SER spectrum of EA2192 is somewhat different from VX with the PO modes having limited intensity and the NC_3 modes having significant intensity (Fig. 7b). Specifically, the EA2192 spectrum has moderately intense peaks at 480, 585, 940, and $1125\,cm^{-1}$ that can be assigned to an NC_3 breathing mode, an NCC bending mode, another NC_3 stretching mode, and a NCC stretching mode. Two additional peaks with significant intensity at 695 and $735\,cm^{-1}$ are assigned to a CS stretching mode and an OPC stretching mode, respectively. Two peaks of modest intensity at 525 and $970\,cm^{-1}$ are attributed to a PO_2S bending mode and a PO_2 stretching mode.

The SER spectrum of DIASH contains most of the NC_3 modes cited previously for EA2192 (Fig. 7c), specifically peaks appear at 480, 585, 940, and $1120\,cm^{-1}$, and can be assigned as above. Additional peaks at 740, 810, and $1030\,cm^{-1}$, are assigned to CH bending, a combination of SC stretching and NC_3 bending, and SCCN bending modes, based on the Raman spectrum of DIASH [29]. A broad peak centered at $695\,cm^{-1}$ also occurs that has previously been assigned to an SC stretch, but the frequency and intensity of this mode in the HD and CEES spectra above, makes this assignment less certain.

It is worth noting the similarity between the EA2192 and DIASH SER spectra, the principle difference being the addition of the SCCN bending mode at $1030\,cm^{-1}$ for the latter. This may simply be due to the fact that both molecules interact through the sulfur with the metal surface to similar extents resulting in similar spectra. However, it is also possible that the EA2192 spectrum is of DIASH formed either by hydrolysis or photodegradation. Since the sample was measured within one hour of preparation, and the hydrolysis half-life is on the order of weeks [56], the former explanation seems unlikely. Since the peak intensities did not change during these measurements, photodegradation catalyzed by silver also seems unlikely. Further experiments are required to clarify this point.

The SER spectrum of the other hydrolysis product formed from VX, EMPA, is shown in Fig. 6. It is included with MPA and IMPA, the hydrolysis products of GB, for convenient spectral comparison of these structurally similar chemicals. The spectrum is dominated by a peak at $745\,cm^{-1}$ with a substantial low-frequency shoulder at $725\,cm^{-1}$. Both are assigned, similarly to IMPA, to PC or PO plus skeletal stretching modes. In fact, virtually all of the peaks in the SER spectrum correspond to peaks of similar frequency in the SER spectrum of IMPA, and are assigned as follows: the peaks at 480 and $500\,cm^{-1}$ to PC or PO plus skeletal bends; 890, 1415, and $1440\,cm^{-1}$ to CH_n-deformations; 945 and $1060\,cm^{-1}$ to PO_n stretches; and $1095\,cm^{-1}$ to a CO or CC stretch. A peak at $1285\,cm^{-1}$ is assigned to a CH_n-deformation based on the MPA spectral assignment for a peak at $1300\,cm^{-1}$.

In this series of chemicals VX and EA2192 were routinely measured at $100\,mg\cdot l^{-1}$, and on occasion at $10\,mg\cdot l^{-1}$ using the SERS-active vials. Again, however, only a limited number of measurements were attempted. More extensive measurements of EMPA using the SERS-active capillaries allowed repeatable measurements of $10\,mg\cdot l^{-1}$ and routine measurements of $1\,mg\cdot l^{-1}$. No concentration studies of DIASH were undertaken.

4 Conclusions

The ability to obtain surface-enhanced Raman spectra of several chemical agents and their hydrolysis products has been demonstrated using silver-doped sol-gels. Two sampling devices, SERS-active vials and capillaries, provided a simple means to measure water samples containing chemical agents. No sample pretreatment was required and all spectra were obtained in 1 min. It was found that the SER spectra can be used to identify chemical agents by class. Specifically, cyanide contains a unique peak at $2100\,cm^{-1}$, HD and CEES both have a unique peak at $630\,cm^{-1}$, while VX has a unique peak at $540\,cm^{-1}$. In the case of HD and CEES, their hydrolysis products produce very similar spectra, and it may be difficult to determine relative concentrations in an aqueous solution. In the case of the VX hydrolysis products,

EA2192 and DIASH were spectrally similar, as was IMPA and MPA. However, there appear to be sufficient differences when comparing entire spectra, such that chemometric approaches might allow successful compositional analysis of aqueous solutions.

The SERS-active vials and capillaries provided sufficient sensitivity to measure cyanide below the required $2\,mg\cdot l^{-1}$ sensitivity either as a point measurement or as a continuous flowing-stream measurement. Measurements of TDG suggest that the sensitivity requirements for it and HD may be attainable with modest improvements. In contrast, the vials and capillaries did not provide sensitivity sufficient to meet the requirements of VX. In this case substantial improvements in sensitivity are required and are being pursued.

Acknowledgements

The authors are grateful for the support of the US Army (DAAD13-02-C-0015, Joint Service Agent Water Monitor program) and the Environmental Protection Agency (EP-D-05-034). The authors would also like to thank Mr. Chetan Shende for sol-gel chemistry development.

References

[1] S. L. Hoenig: *Handbook of Chemical Warfare and Terrorism* (Greenwood Press, London 2002)
[2] H. Nozaki, N. Aikawa: Sarin poisoning in Tokyo subway. Lancet **345**, 1446 (1995)
[3] R. M. Black, R. J. Clarke, R. W. Read, M. T. Reid: J. Chromat. **662**, 301 (1994)
[4] G. A. Sega, B. A. Tomkins, W. H. Griest: J. Chromat. A **790**, 143 (1997)
[5] S. A. Oehrle, P. C. Bossle: J. Chromat. A **692**, 247 (1995)
[6] J. E. Melanson, B. L.-Y. Wong, C. A. Boulet, C. A. Lucy: J. Chromat. A **920**, 359 (2001)
[7] J. Wang, M. Pumera, G. E. Collins, A. Mulchandani: Anal. Chem. **74**, 6121 (2002)
[8] W. R. Creasy: Am. Soc. Mass. Spectrom. **10**, 440 (1999)
[9] Q. Liu, X. Hu, J. Xie: Anal. Chim. Acta **512**, 93 (2004)
[10] G. A. Eiceman, Z. Caras: *Ion Mobility Spectrometry* (CRC Press, Boca Raton, FL 1994)
[11] N. Krylova, E. Krylov, G. A. Eiceman: J. Phys. Chem. **107**, 3648 (2003)
[12] L. D. Hoffland, R. J. Piffath, J. B. Bouck: Opt. Eng. **24**, 982 (1985)
[13] E. H. J. Braue, M. G. Pannella: Appl. Spectrosc. **44**, 1513 (1990)
[14] C.-H. Tseng, C. K. Mann, T. J. Vickers: Appl. Spectrosc. **47**, 1767 (1993)
[15] S. Kanan, C. Tripp: Langmuir **17**, 2213 (2001)
[16] S. D. Christesen: Appl. Spectrosc. **42**, 318 (1988)
[17] S. Christesen, B. Maciver, L. Procell, D. Sorrick, M. Carrabba, J. Bello: Appl. Spectrosc. **53**, 850 (1999)

[18] D. L. Jeanmaire, R. P. Van Duyne: J. Electroanal. Chem. **84**, 1 (1977)

[19] A. M. Alak, T. Vo-Dinh: Anal. Chem. **59**, 2149 (1987)

[20] K. M. Spencer, J. Sylvia, S. Clauson, J. Janni: Proc. SPIE **4577**, 158 (2001)

[21] S. D. Christesen, M. J. Lochner, M. Ellzy, K. M. Spencer, J. Sylvia, S. Clauson: *23rd Army Sci. Conf.* (ASC, Orlando, FL 2002)

[22] S. D. Christesen, K. M. Spencer, S. Farquharson, F. E. Inscore, K. Gosner, J. Guicheteau: in S. Farquharson (Ed.): *Applications of Surface-Enhanced Raman Spectroscopy* (CRC Press, Boca Raton, FL in preparation)

[23] P. Tessier, S. Christesen, K. Ong, E. Clemente, A. Lenhoff, E. Kaler, O. Velev: Appl. Spectrosc. **56**, 1524 (2002)

[24] Y. Lee, S. Farquharson: Proc. SPIE **4378**, 21 (2001)

[25] S. Farquharson, P. Maksymiuk, K. Ong, S. Christesen: Proc. SPIE **4577**, 166 (2001)

[26] S. Farquharson, A. Gift, P. Maksymiuk, F. Inscore, W. Smith, K. Morrisey, S. Christesen: Proc. SPIE **5269**, 16 (2004)

[27] S. Farquharson, A. Gift, P. Maksymiuk, F. Inscore, W. Smith: Proc. SPIE **5269**, 117 (2004)

[28] F. Inscore, A. Gift, P. Maksymiuk, S. Farquharson: Proc. SPIE **5585**, 46 (2004)

[29] S. Farquharson, A. Gift, P. Maksymiuk, F. Inscore: Appl. Spectrosc. **59**, 654 (2005)

[30] S. Farquharson, P. Maksymiuk: Appl. Spectrosc. **57**, 479 (2003)

[31] S. Farquharson, Y. H. Lee, C. Nelson: (2003), US Patent Number 6,623,977

[32] S. Farquharson, A. Gift, P. Maksymiuk, F. Inscore: Appl. Spectrosc. **58**, 351 (2004)

[33] S. Farquharson, P. Maksymiuk: (2005), US Patent Number 6,943,031

[34] S. Farquharson, P. Maksymiuk: (2005), US Patent Number 6,943,032

[35] F. Inscore, A. Gift, P. Maksymiuk, J. Sperry, S. Farquharson: in S. Farquharson (Ed.): *Applications of Surface-Enhanced Raman Spectroscopy* (CRC press, Boca Raton, FL in preparation)

[36] Material safety data sheets, available at `www.msds.com`

[37] D. R. Lide: *Handbook of Chemistry and Physics* (CRC Press, Boca Raton, FL 1997)

[38] D. Kellogg, J. Pemberton: J. Phys. Chem. **91**, 1120 (1987)

[39] Committee on Toxicology: *Guidelines for Chemical Warfare Agents in Military Field Drinking Water* (Nat. Acad. Press, Washington, DC 1995)

[40] Committee on Toxicology: *Review of Acute Human-Toxicity Estimates for Selected Chemical-Warfare Agents* (Nat. Acad. Press, Washington, DC 1995)

[41] J. Billmann, G. Kovacs, A. Otto: Surf. Sci. **92**, 153 (1980)

[42] C. A. Murray, S. Bodoff: Phys. Rev. B **32**, 671 (1985)

[43] G. Wagner, Y. Yang: Ind. Eng. Chem. Res. **41**, 1925 (2002)

[44] N. B. Munro, S. S. Talmage, G. D. Griffin, L. C. Waters, A. P. Watson, J. F. King, V. Hauschild: Environ. Health Perspect. **107**, 933 (1999)

[45] A. G. Ogsten, E. R. Holiday, J. S. L. Philpot, L. A. Stocken: Trans. Faraday Soc. **44**, 45 (1948)

[46] S. D. Christesen: J. Raman Spectrosc. **22**, 459 (1991)

[47] C. Sosa, R. J. Bartlett, K. KuBulat, W. B. Person: J. Phys. Chem. **93**, 577 (1993)

[48] F. Inscore, S. Farquharson: J. Raman Spectrosc. (submitted)

[49] F. Inscore, S. Farquharson: Proc. SPIE **5993**, 19 (2005)

[50] T. Joo, K. Kim, M. Kim: J. Molec. Struct. **16**, 191 (1987)
[51] W. Creasy, M. Brickhouse, K. Morrissey, J. Stuff, R. Cheicante, J. Ruth, J. Mays, B. Williams, R. O'Connor, H. Durst: Environ. Sci. Technol. **33**, 2157 (1999)
[52] H. Hameka, J. Jensen: *CRDEC Technical Report 326* (DTIC, Aberdeen Proving Ground, MD 1992)
[53] R. Nyquist: J. Molec. Struct. **2**, 123 (1968)
[54] B. J. Van der Veken, M. A. Herman: J. Molec. Struct. **15**, 225 (1973)
[55] B. J. Van der Veken, M. A. Herman: J. Molec. Struct. **15**, 237 (1973)
[56] Y. Yang: Acc. Chem. Res. **32**, 109 (1999)
[57] Y. Yang, J. Baker, J. Ward: Chem. Rev. **92**, 1729 (1992)
[58] H. Hameka, J. Jensen: *ERDEC Technical Report 065* (DTIC, Aberdeen Proving Ground, MD 1993)

Index

Index

Topics in Applied Physics